Birkhäuser

Thomas Sonar

3000 Years of Analysis

Mathematics in History and Culture

Birkhäuser

Thomas Sonar
Institut für Analysis Computational Mathematics
Technische Universität Braunschweig
Braunschweig, Germany

Translated by
Thomas Sonar
Braunschweig, Germany

Morton Patricia
Oxford, UK

Keith William Morton
Oxford, UK

Editor:
Project Group "History of Mathematics" of Hildesheim University, Hildesheim, Germany
H.W. Alten (deceased), K.-J. Förster, K.-H. Schlote, H. Wesemüller-Kock

Original published by Springer-Verlag GmbH Deutschland 2016, 3000 Jahre Analysis

ISBN 978-3-030-58221-0 ISBN 978-3-030-58223-4 (eBook)
https://doi.org/10.1007/978-3-030-58223-4

This book is published under the imprint Birkhäuser, www.birkhauser-science.com by the registered company Springer Nature Switzerland AG
The registered company address is: Gewerbestrasse 11, 6330 Cham, Switzerland

Dedicated to

Eberhard Knobloch

in gratefulness.

'I conceive that these things, king Gelon, will appear incredible to the great majority of people who have not studied mathematics ...'

ARCHIMEDES [Archimedes 2002, p. 360]

About the Author

Thomas Sonar was born 1958 in Sehnde next to Hannover. After studying Mechanical Engineering at the University of Applied Sciences ('Fachhochschule') Hannover he became a laboratory engineer in the Laboratory for Control Theory of the same University for a short time, and founded an engineering office. He then studied mathematics at the University of Hannover (now Leibniz University), after which he worked from 1987 until 1989 at the German Aerospace Establishment DLR (then DFVLR) in Brunswick for the orbital glider project HERMES as a scientific assistant. Next he went to the University of Stuttgart to work as a PhD student under Prof. Dr. Wolfgang Wendland while spending some time studying under Prof. Keith William Morton, at the Oxford Computing Laboratory. His PhD thesis was defended in 1991 and Thomas Sonar went to Göttingen to work as a mathematician ('Hausmathematiker') at the Institute for Theoretical Fluid Mechanics of the DLR; there he developed and coded the first version of the TAU-code for the numerical computation of compressible fluid fields, which is now widely used. In 1995 the postdoctoral lecture qualification for mathematics was obtained from the TU (then TH) Darmstadt on the basis of a habilitation treatise. From 1996 until 1999 Thomas Sonar was full professor of Applied Mathematics at the University of Hamburg and is professor for Technical and Industrial Mathematics at the Technical University of Brunswick since 1999 where he is currently the head of a work group on partial differential equations. In 2003 he declined an offer of a professorship at the Technical University of Kaiserslautern connected with a leading position in the Fraunhofer Institute for Industrial Mathematics ITWM. In the same year Sonar founded the centre of continuing education for mathematics teachers ('Mathelok') at the TU Brunswick which stays active with regular events for pupils also.

Early in his career Thomas Sonar developed an interest in the history of mathematics, publishing in particular on the history of navigation and of logarithms in early modern England, and conducted the widely noticed exhibitions in the 'Gauss year' 2005 and in the 'Euler year' 2007 in Brunswick. Further publications concern Euler's analysis, his mechanics and fluid mechanics, the history of mathematical tables, William Gilbert's magnetic theory, the history of ballistics, the mathematician Richard Dedekind, and the death of Gottfried Wilhelm Leibniz. In 2001 Sonar published a book on Henry Briggs' early mathematical works after intense research in Merton College, Oxford. In 2011 his book *3000 Jahre Analysis* (3000 years of analysis) was published in this series and in December 2014 he edited the correspondence of Richard Dedekind and Heinrich Weber. Altogether Thomas Sonar has published approximately 150 articles and 15 books – partly together with colleagues. He has established a regular lecture on the history of mathematics at the TU Brunswick and has for many years held a lectureship on this topic at the University of Hamburg. Many of his publications also concern the presentation of mathematics and the history of mathematics to a wider public and the improvement of the teaching of mathematics at secondary schools.

Thomas Sonar is a member of the Gesellschaft für Bildung und Wissen e.V. (Society for Education and Knowledge), the Braunschweiger Wissenschaftliche Gesellschaft (Brunswick Scientific Society), a corresponding member of the Academy of Sciences in Hamburg, and an honorary member of the Mathematische Gesellschaft in Hamburg (Mathematical Society in Hamburg).

Preface of the Author

As an author I was very glad that the German edition of this book which was first published in 2011 was very well received. Indeed, a second edition with corrections and additions was published in 2016. After the English translation *The History of the Priority Dispute between Newton and Leibniz* of my book *Die Geschichte des Prioritätsstreits zwischen Leibniz und Newton* was published by Birkhauser in 2018 and my revered 'language editor' Pat Morton did not want to retire but was keen to start on another translation I began translating the second edition of *3000 Jahre Analysis*. Here it is.

The German edition of this book has some precursors in a series of books on the history of mathematics: '6000 Jahre Mathematik', '5000 Jahre Geometrie', '4000 Jahre Algebra', of which only the volume on geometry has been translated into English up to now. It seemed logical to add a volume on the history of analysis to this series and thereby making the history of analysis available to interested non-specialists and a broader audience.

The current volume stands out in the series for the following reason. All books in the series were designed to present scientifically reliable facts in a readable form to convey the delight of mathematics and its historical development. But while a cultural history of mathematics can be presented without much mathematical details, while geometry can be described in the history of its constructions in beautiful drawings, and while the history of algebra, at least until the 19th century, can be developed from quite elementary mathematical reflections, this concept naturally has to fail in the case of analysis. In essence analysis is the science of the infinite; namely the infinitely large as well as the infinitely small. Its roots lie already in the fragments of the Pre-Socratic philosophers and their considerations of the 'continuum', as well as in the burning question of whether space and time are made 'continuously' or made of 'atoms'. Thin threads of the roots of analysis reach even back to the realms of the Pharaohs and the Babylonians from which the Greek received some of their knowledge. But not later than with Archimedes (about 287–212 BC) analysis reached a maturity which asks for the active involvement of my readership. Not by any stretch of imagination can one grasp the meaning of the Archimedean analysis withouth studying some examples thoroughly and to comprehend the mathematics behind them with pencil and paper. Although after Archimedes this knowledge was buried in the dark again it came back to life at latest with the Renaissance where analysis progressed in giant steps; and again this science calls for the attention of the reader! To put it somehow poetically: Analysis turns out to be a demanding beloved and one has to succumb to her in order to gain some understanding.

But have no fear! My remarks are not meant to discourage you; on the contrary: they are meant to increase the excitement concerning the contents of this book. You are required to think from time to time, but then deep and satisfying insights into one of the most important disciplines of mathematics wait as a reward. Without analysis the Technical Revolution and the developments of our highly engineered world relating thereto would have been unthinkable.

There are several books on the history of analysis on the market and the reader deserves a few remarks concerning the position of this book in relation to others. I do not claim to publish the latest and hitherto unknown research results. However, the present book differs significantly from others. First of all historical developments in the settings are given much attention as is usual for books in our German book series. Furthermore I have put weight on the Pre-Socratic philosophers and the Christian middle ages in which the discussion of the nature of the continuum had been decisive. Finally the common clamp encompassing all areas described in this book is the infinite. This clamp allows me not to surrender to the unbelievable breadth of developments in the 20th century; functional analysis, measure theory, theory of integration, and so on, but rather to end in the nonstandard analysis in which we again find infinitely small and large quantities and in which the continuum of the Pre-Socratic philosophers is honoured again. In this sense we come to full circle which connects Zeno of Elea (about 490–about 430 BC), Thomas Bradwardine (about 1290–1349), Isaac Newton (1643–1727), Gottfried Wilhelm Leibniz (1646–1716), Leonhard Euler (1707–1783), Karl Weierstraß(1815–1897), Augustin Louis Cauchy (1789–1857), and finally Abraham Robinson (1918–1974) and Detlef Laugwitz (1932–2000).

It is also due to this encompassing clamp that I have included the development of set theory in the history of analysis which is unusual. In the light of the history of the handling of infinity set theory certainly belongs here.

This book has been made possible by the project group 'History of Mathematics' of the University of Hildesheim, Germany, which I want to thank with gratitude. In particular I have to thank the late Heinz-Wilhelm Alten, my friend Klaus-Jürgen Förster, and Karl-Heinz Schlote for their confidence in me. Heiko Wesemüller-Kock has taken care of the design of this book in his usual, professional manner. One can only sense the enormity of his task if one has drawn pictures and sketches, modified or corrected existing diagrams, and designed hundreds of legends of pictures by oneself. The results of his extensive work can be seen in this book and in all other books in our series. Without the publisher, who encouraged this translation, the book would not have come to life. I have to thank Mrs. Sarah Annette Goob and Mrs. Sabrina Hoecklin of Birkhauser Publishers in particular.

However, I am not a trained historian. My continuing and long-standing interest in history has helped a lot, of course, but reliable books like the 'Der große Ploetz' [Ploetz 2008] or the wonderful little volumes of Reclam Publishers starting their titles with 'Kleine Geschichte ...' or 'Kleine ... Geschichte' [Maurer 2002], [Altgeld 2001], [Dirlmeier et al. 2007], [Haupt et al. 2008] were indispensable. In case of doubts however, only an informed historian is of real help and I am very lucky that my friend and colleague

Gerd Biegel of the 'Institut für Braunschweigische Regionalgeschichte' was at my side although permantly suffering from an overload of work. While we smoked many a cigar and drank innumerable cups of espressos he provided insights into many historical contexts.

Although in the meantime he finished his studies and is currently working on his PhD thesis my LaTeX-wizard Jakob Schönborn, who already cared for the second German edition of this book, the first German edition of *Die Geschichte des Prioritätsstreits zwischen Leibniz und Newton*, and its English version *The History of the Priority Dispute between Newton and Leibniz* stood at my side to also supervise the LaTeXnical side of this book. I can only thank him wholeheartedly for his commitment to this book project!

I am particularly grateful to Prof. Dr. Eberhard Knobloch, not only for his precious time he sacrified while proofreading the German edition but also for numerous constructive criticism and hints concerning correct translations from ancient Greek and Latin. Since he is a true role model not only for me but for a whole generation of scientists this book is dedicated to him.

I am most grateful to the wonderful Patricia (Pat) Morton who offered again to turn my 'Germanic English' into her lovely Oxford English. Without her encouragement to continue our work on book projects I would have dared to even start working on this book.

A book like this costs time; much time! My decision to write this book therefore had serious consequences in particular for my wife Anke. I had to spend a lot of time in my study and in libraries while life went on without me. A lot of money was spend to buy new and second-hand books to enrich my private library on the history of analysis. All this as well as the now meter deep piles of books and manuscripts in our living room, on couches and chairs and on the floor my beloved wife Anke has put up with and she has reacted with humour and only with a few biting remarks. After the two volumes '6000 Jahre Mathematik' by Hans Wußing had been published and were presented to the public during a small ceremony at the town hall of Hildesheim, my wife got to the heart of it: 'My husband has a mistress who is 6000 years old, and he loves her dearly!' For this and since she bears with me and my old mistress I thank her with all my heart.

Thomas Sonar

Preface of the Editors

With great pride we can present here the English translation of the second edition of the German book '3000 Jahre Analysis' for which our author deserves gratitude and appreciation.

Concerning the contents of this books it has to be noted that it is not just a translation of the German book. Again some sections have been reworked and some new material has been carefully added. We have to thank again Heiko Wesemüller-Kock and Anne Gottwald for their ernormous work concerning the pictures and the gathering of publication rights at different license suppliers. Their work ensures that the illustrations appearing in this book share the same high quality as in all other books in our series.

We are also greatful to Birkhauser Publishers and in particular to our partner Sarah Annette Goob who supported us actively. My gratitude also extends to further persons which Thomas Sonar has already mentioned in his preface.

After '5000 Years of Geometry' and 'The History of the Priority Dispute between Newton and Leibniz' the '3000 Years of Analysis' is the third book of our German book series appearing in the English language. We wish this book to also become a real success enjoying a wide distribution. May this book find many readers and may it convey an impression of the beauty and meaning of mathematics in our culture. It may perhaps even arouse their interest in mathematics.

Hildesheim, July 2020, for the editors

Karl-Heinz Schlote *Klaus-Jürgen Förster*

Project group 'History of Mathematics'
at the University of Hildesheim

Advice to the reader

Parentheses contain additional insertions, biographical details, or references to figures.

Squared brackets contain

- omissions and insertions in quotations
- references to the literature within the text
- references to sources in legends of figures

In the figure legends squared brackets mark the author/creator of the particular work. Further specifications appear in common paranthesis.

Figures are numbered following chapters and sections, e.g. Fig. 10.1.4 means the fourth figure in section 10.1 of chapter 10.

The original titles of books and journals appear in italic type, likewise quotations. Further reading or explanations of only shortly described circumstances are marked by references like '(cp. more detailed in...)'.

Literally or textually quoted literature as well as further reading can be found in the bibliography.

Contents

1 Prologue: 3000 Years of Analysis

T. Sonar, *3000 Years of Analysis*, https://doi.org/10.1007/978-3-030-58223-4_1

since 3000 BC	Nomads from the north immigrate to southern Mesopotamia. Sumerian city states emerge and the cuneiform writing on clay tablets. The realms at the Nile unite. Emergence of hieroglyphs
about 2707–2170	Old Kingdom in Egypt. Emergence of pyramids; the step pyramid at Saqqara, the bent pyramid at Dahshur, the great pyramids of Khufu, Chefren and Mykerinos
2170–2020	First interim period in Egypt
about 2235–2094	Realm of Akkad in Mesopotamia founded by Sargon of Akkad
about 2137–1781	Middle Kingdom in Egypt. Mathematical papyri
1850	Presumable time of origin of the Moscow Papyrus
1793–1550	Second interim period in Egypt
1650	Ahmes writes the Rhind Mathematical Papyrus
2000–1595	Ancient Babylonic period in Mesopotamia. Emergence of the first legislative texts of mankind under King Hammurabi (about 1700)
1675	In Mesopotamia a clay tablet is inscribed with the length of the diagonal in a square
about 1550–1070	New Kingdom in Egypt. Temple of Hatshepsut and royal tombs in Thebes. Temple of Amun in Karnak. Sun worship of Akhenaten in Amarna
1279–1213	Ramsses II, Temple of Abu Simbel
1070–525	Third interim period and late period in Egypt
about 1700–609	Assyrian realm. Mathematical cuneiform texts; zikkurates
about 750–620	Neo-Assyrian realm, the first great empire in the history of the world; residences in Nimrud and Nineveh
625–539	Neo-Babylonian realm, heyday of astrology and astronomy
539	Cyrus the Great conquers Babylon
525	Persians conquer Egypt
332	Alexander the Great conquers Egypt

Remark: There are differing chronologies in the literature

Fig. 1.0.2. Egypt and Mesopotamia in the pre-Christian era

1.1 What is 'Analysis'?

Three thousand years of analysis? Did analysis not emerge in the 17th century by Newton and Leibniz?

To answer this question satisfactorily we should look at a definition of 'analysis' first. On the internet the following definition[1] can be found:

> *'Mathematical analysis formally developed in the 17th century ...'*

There you go! According to this definition analysis would be approximately 400 years old, but beware: The definition goes on:

[1] `https://en.wikipedia.org/wiki/Mathematical_analysis`

'... but many of its ideas can be traced back to earlier mathematicians.'

But how far do we have to 'trace back'?

Good old reliable Encyclopaedia Britannica defines 'Analysis (mathematics)' as

> *'a branch of mathematics that deals with continuous change and with certain general types of processes that have emerged from the study of continuous change, such as limits, differentiation, and integration. Since the discovery of the differential and integral calculus by Isaac Newton and Gottfried Wilhelm Leibniz at the end of the 17th century, analysis has grown into an enormous and central field of mathematical research, with applications throughout the sciences and in areas such as finance, economics, and sociology.'*

I do not have any problems whatsoever to follow this definition! Analysis is concerned with the mathematics of continuous changes from which problems of tangents, quadrature problems (i.e. the computation of areas below crooked curves), and eventually the actual differential and integral calculus of Newton and Leibniz developed.

In a narrower sense analysis is but the mathematical branch of infinite processes and of 'infinitely small quantities' and this sense should be the ribbon accompanying us on our journey through history as a kind of Ariadne's thread. However, this is not possible consistently. The notion of 'function' is certainly central to analysis but for a start has nothing to do with infinitely small quantities. Nevertheless a discussion of the concept of functions certainly belongs to the history of analysis.

How come the 3000 years? Well, special numbers like π or $\sqrt{2}$ play a certain role and such numbers (or the approximations thereof) can in fact be found in ancient Egypt and in the cultural region of Mesopotamia.

1.2 Precursors of π

Already in the famous Papyrus Rhind[2] an approximate computation of the area of a circle can be found. Papyrus Rhind was written by a scribe named Ahmes about the year 1650 BC who wrote that he only copied mathematical problems which were at least 200 years older.

In Problem 48 of his papyrus Ahmes depicted a circle which is inscribed in a square. We can infer from the calculations following that the square of edge length of 9 units results in an area of 81 square units, and that the circle with diameter 9 units has an area of 64 square units. In Problem 50 a precise instruction to compute a circle area can be found [Gericke 2003, p. 55]:

[2] Named after the Scotsman Alexander Henry Rhind who bought the papyrus in 1858 in Luxor.

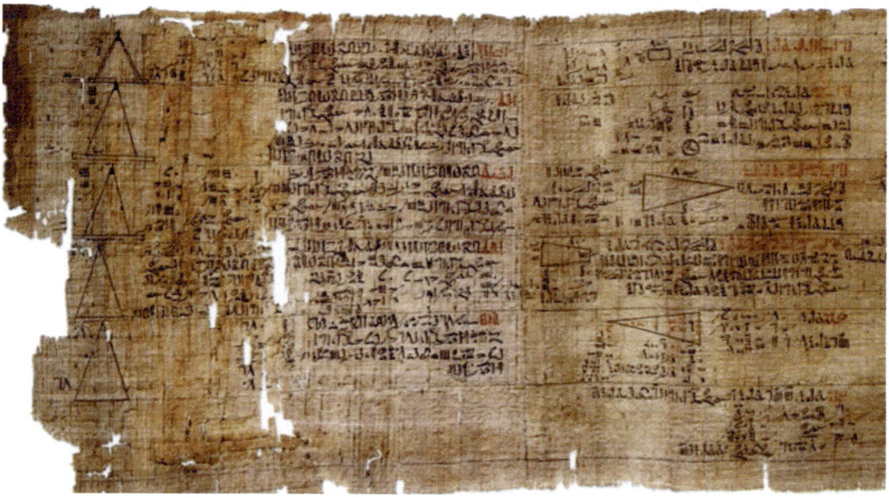

Fig. 1.2.1. The start of the Papyrus Rhind. The Papyrus is 5,5 m long and has a height of 32 cm. It contains problems concerning mathematical themes which nowadays would be called algebra, fractional arithmetic, geometry and trigonometry. It is itself a copy of an original from the 12th dynasty (19th century BC). Scribe Ahmes copied this original about 1650 BC in hieratic writing. (Department of Ancient Egypt and Sudan, British Museum EA 10057, London [Photo: Paul James Cowie])

> *'Example of the computation of a circular field of (diameter) 9. What is the amount of its area? Take 1/9 away from it (the diameter). The remainder is 8. Multiply 8 by 8. It becomes 64.'*
>
> (Beispiel der Berechnung eines runden Feldes vom (Durchmesser) 9. Was ist der Betrag seiner Fläche? Nimm 1/9 von ihm (dem Durchmesser) weg. Der Rest ist 8. Multipliziere 8 mal 8. Es wird 64.)

This calculation rule allows us to conclude that the Egyptians used $\pi_{\text{Egypt}}/4 = (8/9)^2$ as the value for $\pi/4$. Since they did neither know the nature nor the role of π we may ask ourselves how this value was achieved. One possibility would be the use of a grid. Circumscribe a square with edge length d around a circle with diameter d units and divide the square into 9 evenly spaced subsquares as shown in figure 1.2.2 (left). The area of the square would grossly overestimate the area of the circle, hence we divide the four subsquares in the corners of the square into two triangles each and count only one each as contributing to the area as in figure 1.2.2 (right). Therefore 5 subsquares and 4 triangles remain and the area of the circle is approximated by

$$A_{\text{circle}} \approx 5 \cdot \left(\frac{d}{3}\right)^2 + 4 \cdot \frac{1}{2}\left(\frac{d}{3}\right)^2 = \frac{7}{9}d^2.$$

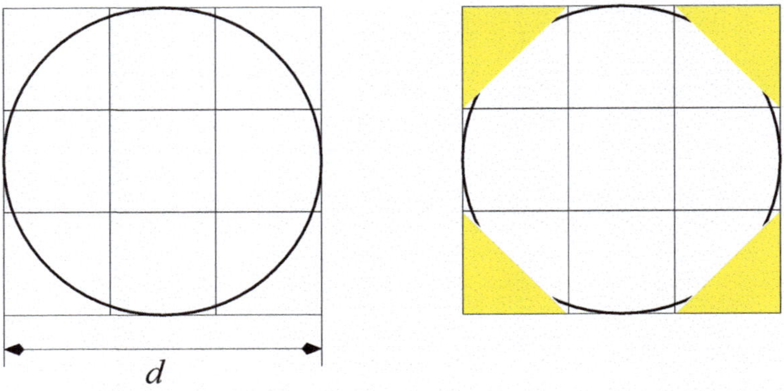

Fig. 1.2.2. Approximation of the area of a circle from the outside

However, Ahmes gives the approximation

$$A_{\text{circle}} \approx \frac{64}{81}d^2 = \left(\frac{8}{9}d\right)^2 .$$

He apparently enlarged the (correct) approximation $\frac{7}{9}d^2 = \frac{63}{81}d^2$ by an area of $\frac{1}{81}d^2$ to finally arrive at square numbers in numerator and denominator! But did he? Somewhat frustrated Otto Neugebauer (1899 – 1990) commented [Neugebauer 1969a, p. 124]:

> *'And it is not understandable how one comes from this term* [$\frac{7}{9}d^2$ for the area of the circle] *to the Egyptian formula. Without new sources it therefore makes little sense to express presumptions concerning this formula since the obvious way obviously does not lead directly to the desired result'*
>
> (Und es ist nicht einzusehen, wie man von diesem Ausdruck zu der ägyptischen Formel hinüberkommen kann. Ohne neues Textmaterial hat es also wenig Sinn, über die Entstehungsgeschichte dieser Formel Vermutungen zu äußern, da der naheliegende Weg offenbar nicht direkt zum Ziel führt.)

Since the true area of a circle is given by $A_{\text{circle}} = \pi r^2 = (\pi/4)d^2$ the ancient Egyptians worked with the approximate value

$$\pi_{\text{Egypt}} = 3.16049$$

which is by no means a bad approximation! At least the relative error is only

$$\frac{\pi_{\text{Egypt}} - \pi}{\pi} \approx 0.00601643,$$

hence about 0.6%!

In the TV production 'The Story of Maths' [Du Sautoy 2008] which is well worth watching, mathematician Marcus du Sautoy (b. 1965) gave another explanation of how the Egyptians might have come up with their formula for the area of a circle. Following his explanation the approximation $\pi_{\text{Egypt}}/4 = (8/9)^2$ stems from an ancient Egyptian board game in which spheres filling hemispherical depressions in a wooden board have to be moved around. Using these spheres a circle can be formed having a diameter corresponding to 9 spheres. Redistributing the spheres so that they form a square then this square happens to have an edge length of 8. If du Sautoy's interpretation is correct then we have here an early attempt to 'square the circle'. This problem, also called 'quadrature of the circle', will occupy us later on.

Fig. 1.2.3. Queen Neferarti (19th dynasty, wife of Ramesses II) playing the game Senet. The rules of the game Senet could be roughly reconstructed. The rules of other games like 'Hounds and Jackals' or the Snake Game are mostly unknown (Wall painting in the burial chamber of Nefertari, West Thebes)

1.3 The π of the Bible

The ancient Egyptian value for π was already much more accurate than the 'biblical' value. In the first Book of Kings, Chapter 7:23 we read

> *'And he made a molten sea, ten cubits from the one brim to the other: it was round all about, and its height was five cubits: and a line of thirty cubits did compass it round about.'*

And in the Second Book Chronicles, Chapter 4:2 we find

> *'Also he made a molten sea of ten cubits from brim to brim, round in compass, and five cubits the height thereof; and a line of thirty cubits did compass it round about.'*

Hence the form of the sea is indeed a circle with diameter $d = 10$ cubits and circumference of $U = 30$ cubits. Since the relation between circumference and diameter of every circle is $U = \pi d$ we arrive at

$$\pi_{\text{Bible}} = \frac{U}{d} = \frac{30}{10} = 3.$$

This was the value which was also used by the Babylonians and Edwards (b. 1937) in [Edwards 1979, p. 4, Ex.5] gave an attempt to explain it which I find appealing. Instead of approximating the area of a circle in the Egyptian manner one could have come up with the idea of not only circumscribing a square to the circle but also to *inscribe* another square as in figure 1.3.1. Then the area of the circle should be approximated by the arithmetic mean of the areas of the squares. The area of the circumscribed square apparently is $A_1 := d^2 = 4r^2$. According to Pythagoras' theorem which was well known in Mesopotamia it follows that the edge length of the inner square is $\sqrt{r^2 + r^2} = \sqrt{2}r$, hence the area of the inscribed square turns out to be $A_2 := 2r^2$. Since the area of the circumscribed square overestimates the area of the circle while the area of the inscribed square underestimates it one can hope that the arithmetic mean might yield a useful approximation to the area of the circle:

$$A_{\text{Circle}} \approx \frac{A_1 + A_2}{2} = 3r^2.$$

And in fact here the biblical value of π appears!

But that seems not to be the end of the story as far as the Babylonians are concerned. According to Beckmann (1924 – 1993) [Beckmann 1971, p. 21 f.] and Neugebauer [Neugebauer 1969b, p. 46 f.] clay tablets were excavated in 1936 some 200 miles east of Babylon at Susa including computations concerning some geometrical figures. One of the tablets was concerned with a regular hexagon inscribed in a circle and stated that the ratio of the perimeter of the hexagon to the circumference of the circumscribed circle would be

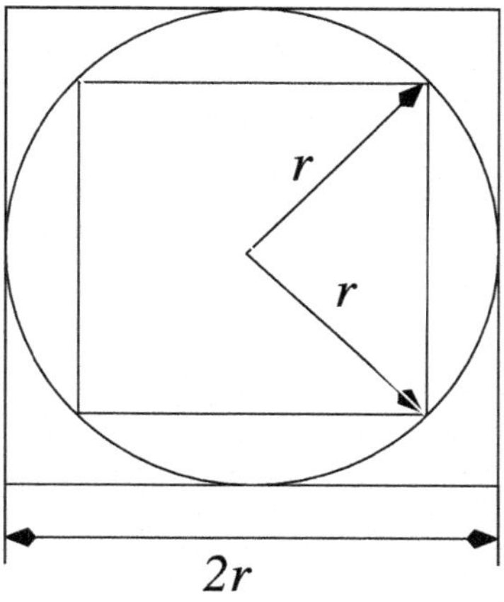

Fig. 1.3.1. Approximating the area of the circle from within

$$\frac{57}{60}+\frac{36}{60^2}.$$

The Babylonians knew that the perimeter of the hexagon is $6r$ if the radius is denoted by r, see figure 1.3.2.

The ratio sought therefore is $6r/C$ if C denotes the circumference of the circle. Since

$$\pi = C/2r$$

we conclude that

$$\frac{3}{\pi}=\frac{57}{60}+\frac{36}{60^2}$$

and hence $\pi = 31/8 = 3.125$.

This shows that also the Babylonians knew better approximations to π than just 3.

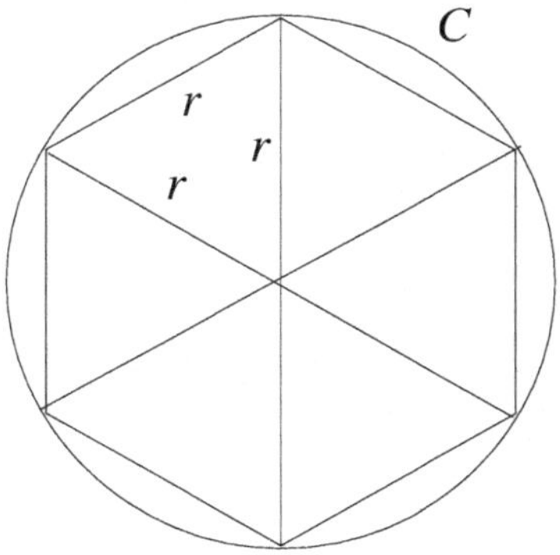

Fig. 1.3.2. Approximating the area of the circle by means of a regular hexagon

1.4 Volume of a Frustum of a Pyramid

In the so called Moscow Mathematical Papyrus located at the Pushkin Museum in Moscow one finds Problem 14 which almost points to one of the basic tasks of analysis. In this problem the volume of a frustum of a pyramid is computed.

Fig. 1.4.1. Computation of the volume of a frustum of a pyramid (Moscow Mathematical Papyrus) in hieratic writing and in hieroglyphs

For the master builders of the pyramids this calculation must have been of particular importance since the pyramids were built in layers. A pyramid is therefore nothing more than the sum of frusta of pyramids with a pyramidal part on top. We do not want to speculate here how the Egyptians arrived at their (correct) formula of the volume of a frustum of a pyramid but refer our readers to the corresponding sections in [Gillings 1982] (see also [Scriba/Schreiber 2000, p. 14 ff.], [Wußing 2008, p. 99 f.]).

Although Problem 14 is only concerned with the computation using concrete numbers the Egyptians must have been aware of the correct formula

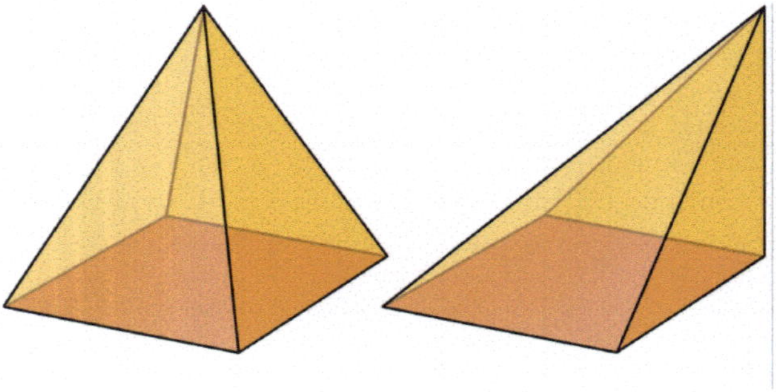

Fig. 1.4.2. A symmetric and a right-angled pyramid with identical base areas and identical heights share the same volume

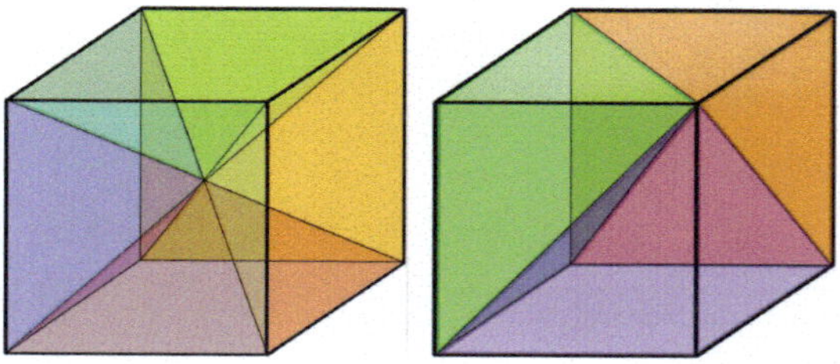

Fig. 1.4.3. Decomposition of a cube into six symmetric pyramids of half the height with the tip points in the centre (left), and in three right-angled pyramids (right)

$$V = \frac{h}{3}\left(a^2 + ab + b^2\right) \tag{1.1}$$

for the volume where a and b are the edge lengths of the two deck areas and h denotes the height of the frustum. Neugebauer [Neugebauer 1969a, p. 126] calls this a 'gem' (Glanzstück) of Egyptian mathematics.

Pointing towards mathematical analysis is the method with which the volume of a pyramid was probably actually computed (in [Gillings 1982] yet another method can be found). For this purpose the Egyptian scribes considered a pyramid in which the top point is located exactly above one of the edge points.

Three of those right-angled (or 'oblique') pyramids with identical height and base edge together form a cube with identical height and base edge; i.e. the volume of each of the pyramids is a third of the volume of the cube. One can alternatively build a cube from 6 congruent symmetric pyramids with half the height of their base edges. Place one of these pyramids top point to top point above another and fill the free space with the remaining four pyramids as

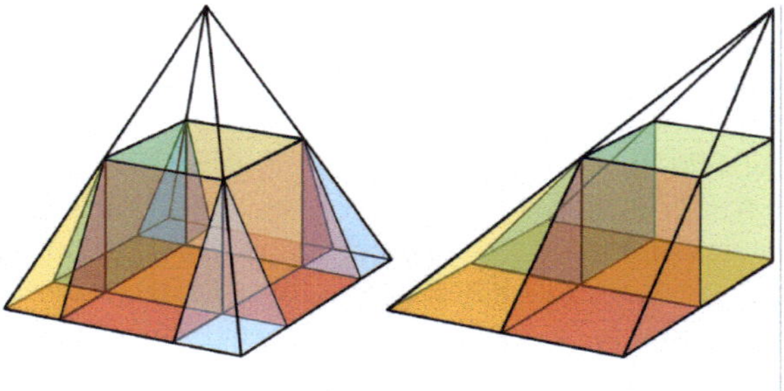

Fig. 1.4.4. The calculation of a frustum of a pyramid can graphically be understood by division into its geometric basic forms: 1 cuboid in the middle, 4 prisms at the sides, and 4 right-angled pyramids at the edges; in case of the right-angled pyramid the same cuboid but only 2 prisms of twice the volume at the sides and 1 right-angled pyramid of fourfold volume at the edge, so that both frusta have the identical volume of $V = \frac{h}{3}(a^2 + ab + b^2)$

Fig. 1.4.5. Step Pyramid of Pharao Djoser in Saqqara (about 2600 BC) [Photo: H.-W. Alten]

shown in figure 1.4.3 (left). Imagining now the right-angled pyramid in figure 1.4.2 cut into very many thin slices parallel to the base area and shifting these slices then a symmetric pyramid of the same volume results where its top point is now above the centre of the square base area. The same is valid in case of the frusta of pyramids in figure 1.4.4, of course. Indeed pyramids have emerged in ancient Egypt from many layers. Already the Mastabas of the kings of the first two dynasties (about 3000 – 2700 BC) show these layers.

King Djoser, second king of the 3rd dynasty, ordered his original three-stage Mastaba to be increased by three further stages where each of the stages consists of many thin layers of stone cuboids. Hence emerged the famous Step Pyramid of Djoser about 2680 BC in Saqqara. Under the rule of King Sneferu of the 4th dynasty the transition from layers of frusta towards the abstract geometrical form of the pyramid took place. That seems to have been a kind of a great gamble since in the first phase of the building up to a height of approximately 49 meters the construction by means of inwardly inclined layers proved unstable. This was the result of a too steep slope angle of approximately 58 degrees as well as the inclination of the layers. In the second phase the base area was enlarged, the slope angle was decreased to 54 degrees, but the techniques of the inwardly inclined layers was kept. As this also turned out to lead to instabilities the slope angle was further decreased to 43 degrees in a third phase and horizontal layers were put on the present frustum of a pyramid. Hence emerged the Bent Pyramid of Sneferu about

Fig. 1.4.6. The Bent Pyramid of Pharao Sneferu at Dahshur
[Photo: H. Wesemüller-Kock]

2615 BC – prototype of the Great Pyramids of Khufu, Khafra, and Menkaura, as well as the other pyramids of the Old Kingdom, all with horizontal layer build-up.

Let us again cite Neugebauer [Neugebauer 1969a, p. 126]:

> *'What is surprising with this formula* [(1.1)] *is twofold: for one thing it is its symmetrical form, on the other hand its mathematical correctness which in particular in the case of this formula, if it should be derived correctly, necessarily requires infinitesimal considerations, i.e. leaves the framework of elementary geometry behind.'*
>
> (Was an dieser Formel überrascht, ist vor allem zweierlei: einerseits die symmetrische Gestalt, andererseits die mathematische Korrektheit, die ja gerade bei dieser Formel, falls sie auch korrekt abgeleitet werden sollte, bekanntlich mit Notwendigkeit Infinitesimalbetrachtungen verlangt, d.h. über den Rahmen der Elementargeometrie hinausführt.)

If the idea of slicing a spatial figure into (principally infinitely many) horizontal layers really was in the heads of the ancient Egyptian scribes then they have anticipated an idea central to analysis, namely the so-called 'principle of Cavalieri'. Millennia after the building of the Egyptian pyramids Bonaventura Cavalieri (1598–1647) discovered the principle of computing

Fig. 1.4.7. Layer structure of the pyramid of Khufu (Giza, Cairo) [Photo: H.-W. Alten]

areas and volumes by 'summing' so-called 'indivisibles'. Later this technique was replaced by the notion of definite integrals where (infinitely many) infinitesimal layers are 'summed'.

1.5 Babylonian Approximation of $\sqrt{2}$

In the *Yale Babylonian Collection* a Babylonian cuneiform clay tablet can be found under the archive number YBC 7289 on which the length of the diagonal of a unit square was approximately computed as an example for using Pythagoras' theorem, cp. [Alten et al. 2005, p. 41, Abb. 1.3.9], see figure 1.5.1. This cuneiform tablet stems from about 1675 BC. Transcribing the cuneiform text as in figure 1.5.1 results in the numbers

$$\begin{aligned} a &:= 30 \\ b &:= 1, 24, 51, 10 \\ c &:= 42, 25, 35, \end{aligned}$$

written in the sexagesimal system, i.e. a number system with base 60. In this system it is not easy to find the right position of each figure since the Babylonians wrote either $1, 2, 3 \equiv 1 \cdot 60^2 + 2 \cdot 60^1 + 3 \cdot 60^0 = 3723$ or $1, 2, 3 \equiv 1 \cdot 60^1 + 2 \cdot 60^0 + 3 \cdot 60^{-1} = 62.0166\ldots$ depending on the context. Cuneiform experts nowadays follow a suggestion of Otto Neugebauer and write 1, 2, 3 in the first case and 1, 2; 3 in the second place; the ';' marking clearly the position of the 'sexagesimal point' and the ',' separating different figures. The separation marks are necessary since there are 59 figures in a sexagesimal system and 587 may mean 5, 8, 7 or 58, 7. If one is used to compute with sexagesimal numbers it is apparent that with the numbers a, b, c above $c = a \cdot b$ is valid. Interpreting the numbers as being the edge length a of the square and the length of the diagonal c Pythagoras' theorem tells us $c^2 = 2a^2$, hence $c = \sqrt{2}a$. The number b thus should be an approximation of $\sqrt{2}$ and indeed we compute

$$1;24,51,10 \equiv 1 \cdot 60^0 + 24 \cdot 60^{-1} + 51 \cdot 60^{-2} + 10 \cdot 60^{-3} = 1.414213$$

so that the square $(1;24,51,10)^2 = 1;59,59,59,38,1,40$ is fairly close to 2, cp. [Aaboe 1998]. Apparently the Babylonians knew very well that the length of the diagonal of a square is the $\sqrt{2}$-fold of the edge length and they had superb approximations of $\sqrt{2}$ at hand. Irrational numbers like $\sqrt{2}$ or π were very important in the development of analysis and it is just this class of

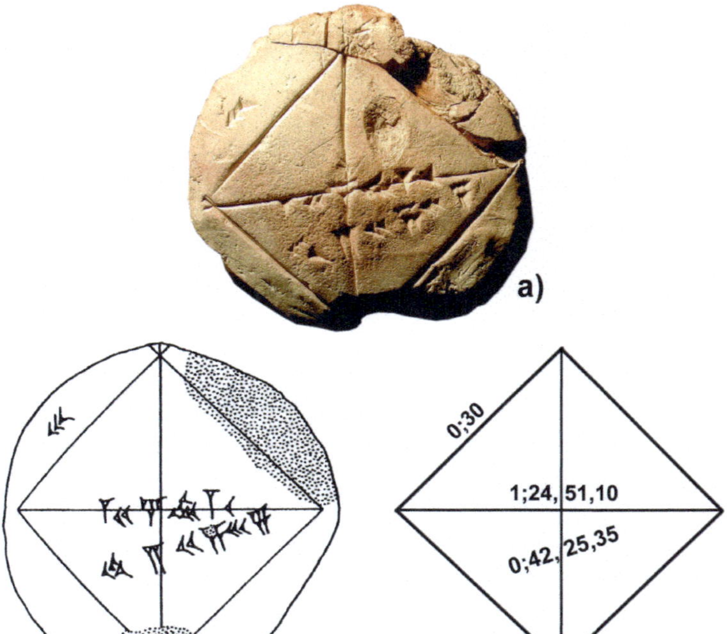

Fig. 1.5.1. Concerning the computation of $\sqrt{2}$: a) Cuneiform table YBC 7289 of the Yale Babylonian Collection, b) Reproduction of the text YBC 7289 after Resnikoff, c) The text in Indian-Arabic numerals in the sexagesimal system [Photo: William A. Casselman]

numbers making an analysis in real numbers possible at all. We have no hint of a deeper understanding of irrational numbers in the Mesopotamian culture area and this must not surprise us. A true understanding of the structure of the real number system will come not before the end of the 19th century.

Have we justified the ‘3000 years’ of the title? Khufu’s pyramid was built about 2600 BC. If we concede that the ancient Egyptians already had analytical methods for the computation of volumes of pyramids at their disposal then analysis would be 5000 years old. If we accept that the actual beginning of analysis (in our modern understanding) can be found – as so many other things – in ancient Greece, then the discovery of irrational numbers by Hippasus of Metapontum about the year 500 BC surely is important for the development of analysis. In that case analysis would be 2500 years old. The true answer to the question of the age of analysis is: we simply don’t know! This is the reason why we have chosen 3000 years as a kind of compromise.

2 The Continuum in Greek-Hellenistic Antiquity

T. Sonar, *3000 Years of Analysis*, https://doi.org/10.1007/978-3-030-58223-4_2

3500 BC	First traces of minoic settlements on Crete. Large influence on the Aegean and on south-western areas of Asia Minor
2nd c. BC	Indo-German tribes of the Acheans and Ionians immigrate to the southern Balkan peninsula and merge with the proto-Greek tribe of the Thracians. The Mycenaean culture develops under Minoian influence
about 1630	Volcanic eruption of Santorini. Beginning of the destruction of the Minoian culture
from 1200	The so-called Sea Peoples devestate the Mediterranean region. The Minoian culture disappears
about 1000	Dorian invasion. The tribe of the Dorians gains predomonance in the Peleponnes. Further merging of all Greek tribes. Foundation of cities like Miletus or Ephesos
1200–750	'Dark age' between the end of the Mycanaean culture and the beginning of the archaic time
750–500	Archaic time. Colonisation of the Meditteranian region. Foundation of the Greek *Poleis* (city-states)
about 550	Sparta founds the Peloponnesian League
about 500–494	The Ionian Revolt leads to conflicts with the Persian Empire under Darius I
497	The Greek defeat the Persian army in the Battle of Plataea
490	Greek victory at Marathon. Athen massively rearmaments
480	Battle of Thermopylae against the Persians under Darius' son Xerxes I
Sept. 480	Decisive Battle of Salamis. The Persian fleet is whitewashed
478/477	Athen founds the Delian League. Development of the Attic democracy on the basis of the reforms of Solon and Cleisthenes
431–404	Peloponnesian War Between the Delian League and Sparta ends with the voctory of the Spartans
395–387	In the Corinthian War Sparta has to stand up against an alliance of the city-states of Athen, Thebes, Corinth, and Argos
371	Eventually the Spartans succumb to Thebes in the Battle of Leuctra. Thebes becomes the Greek centre of power for a short while
about 382–336	Philip II of Macedon. Macedonia achieve predominance in Greece
356–323	Alexander the Great, son of Philip. Victory over the Persian armies, advance to India. Beginning of the Hellenistic age
218–201	2nd Punic War against Carthage
200–197	2nd Macedonian War ends with the defeat of the Macedonians
168	Battle of Pydna. Macedonia is finally beaten by the Romans
146	Greek is completely incorporated into the Roman Empire
30	Ptolemaic Egypt, the last hellenistic enclave, is annexed by Rome

Ancient Greece: Birthplaces and central places of selected mathematicians (red)

2.1 The Greeks Shape Mathematics

Already when Egypt and Mesopotamia were blossoming a further advanced civilisation developed on the island Crete. This civilisation founded the Minoan culture. At the beginning of the 2nd millennium BC Indo-German tribes had immigrated to the southern Balkan peninsula. Under the influence of the Minoan culture the culture of the Mycenaeans developed between about 1600 and 1000 BC.

After the so-called Sea Peoples devastated large parts of the Mediterranean region about 1200 BC the Dorian Invasion set in about 1000 BC. The tribe of the Dorians left their native region in northwest Greece (Dalmatia) and spread over the Peloponnese. Further Greek tribes merged and eventually formed a conglomerate of peoples which we today call 'the Greek'.

A 'dark age' about 750–500 BC in which the whole Mediterranian region was colonised by Greek peoples was followed by the foundation of the Peloponnesian League founded by Sparta in about 550 BC. This initiated the way into the 'classic Greek period' which transformed into the Hellenistic age only with Alexander the Great.

Fig. 2.1.1. Throne Hall in the Palace of Knossos on Crete [Photo: H.-W. Alten]

2.1.1 The Very Beginning: Thales of Miletus and his Pupils

At the time of Thales, Miletus in Asia Minor was one of the large Ionian trading towns but already the year of birth of Thales is uncertain. Following Diogenes Laertius who wrote his *Lives of Eminent Philosophers* [Diogenes Laertius 1931] probably as late as the first half of the third century AD Thales lived between 640 and 562 BC; Gericke [Gericke 2003] mentions 624–548/545 BC as biographical data. Meanwhile the dates of his birth about the year 624 BC and his death about 546 BC seem to be accepted. It is certain that Thales travelled Egypt and that he brought along Egyptian mathematical knowledge to Asia Minor, hence to the cultural area of the Greeks.

We do not have any written documents by Thales but his abilities were praised early on so that he was called the first of the seven wise men of antiquity. If we believe in tradition then Thales must have been a Jack of all trades. It is said that he predicted a solar eclipse in the year 585 BC and thereby terminated a long-standing war between the Medes and the Lydians. Schramm has clarified that no paramount knowledge of astronomy was necessary to predict regular eclipses of the moon [Schramm 1994, p. 572f.] since time grids could be found already in Babylonian cuneiform tables. However, this is not true in case of eclipses of the sun and we have to banish this story to the realm of myths. We follow Neugebauer [Neugebauer 1969b, p. 142]:

> *'There exists no cycle for solar eclipses visible at a given place* [...] *No Babylonian theory for predicting a solar eclipse existed at 600 B.C.* [...]*',*

Fig. 2.1.2. Thales of Miletus and Detail from the Gate of Miletus (Vorderasiatisches Museum, SMB [Photo: H.-W. Alten])

and even sharper [Neugebauer 1975, p. 604]:

> *'Hence there is no justification for considering the story of the "Thales eclipse" as a piece of evidence for Babylonian influence on earliest Greek astronomy. All available sources point to no such contacts until three centuries later.'*

Very recently Otta Wenskus (b. 1955) has taken up the case of the 'Thales eclipse' again in [Wenskus 2016] and showed clearly that we are concerned with a tale and not with a true story.

It is said that Thales was also working as an engineer and that he had a reputation for *proving* theorems. The latter statement is somehow explosive! The mathematics of the Egyptians and the Mesopotamians was characterised by a strict culture of *solving mathematical problems*, i.e. pupils learned to calculate via concrete examples. We do not have any evidence that *general* mathematical theorems were stated (let alone proved) in the archaic cultures. Only concrete problems were solved and even the Pythagorean theorem stating that in any right-angled triangle the square of the hypotenuse c equals the sum of the squares of the two legs a and b was not known in the form of the formula $c^2 = a^2 + b^2$ but only in form of concrete (pythagorean) number triples like $5^2 = 4^2 + 3^2$. That the ancient Greek felt the necessity of a *derivation* by means of *deduction* can simply not be overestimated. Thales is therefore seen by Aristotle (384–322 BC) as the inventor of the natural philosophy way of reasoning [Mansfeld 1999]. This means in particular that Thales demythologised the phenomena of nature and opened them to rational explanations.

Thales marks the beginning of the triumphal way of mathematics as a deductive science and with Thales mathematics became an important part of Greek philosophy. Even the word 'mathematics' is of Greek origin and means 'that what belongs to learning'. Hence mathematics was seen as a fundamental part of classical education.

Fig. 2.1.3. Anaxagoras and Anaximenes on coins

The most famous pupils of Thales were Anaximander (b. 611 BC) and Anaximenes (b. 570 BC). Both expanded the natural philosophy of their master but both do surely not play any role in the prehistory of analysis. Only Anaxagoras (500–428 BC) being a pupil of Anaximenes contributed to the history of analysis and that he did so fiercely that his work occupied mathematicians until well into the 19th century! Anaxagoras stemmed from Klazomenai (Klazomenae), an Ionian town which lay about 40 km west of the modern Turkish city of İzmir. With his cosmological theories he rubbed his contemporaries up the wrong way. He stated that the sun was a blazing hot fiery mass of iron which led to him being prosecuted for godlessness and he was sent to jail. Diogenes Laertius [Diogenes Laertius 1972, p. 143] gives us two different versions concerning the outcome of the trial. Since the great statesman Pericles (c. 495 – 429 BC) was a pupil of Anaxagoras he could spare him from the worst so that Anaxagoras got away with a fine and was banished. In the other version Anaxagoras was additionally accused of treason and sentenced to death. You may choose the version which appeals most to you. As things might have been Anaxagoras seemed to have suffered from boredom in prison and hence came up with the following task:

The quadrature of the circle:
Given a circle with radius r. Construct a coextensive square.

The word 'construct' contains some dynamite here since shortly after Anaxagoras it went without saying that 'construct' would mean a construction with compass and straightedge alone. The straightedge must not show any marks and has to be thought of as an idealised abstract instrument. The compass must not show any reference scale either and is an abstract tool, too. In addition, it is thought of as a collapsible compass; every time you move it away from the paper it will collapse so that it may not serve as a divider.

Since a collapsible compass is cumbersome to use (think of transferring a given length to another part of your drawing paper) Euclidhad already in the 3rd century BC in his famous *Elements* proved a 'circle equivalence theorem' as Proposition 2 in Book I [Euclid 1956, Vol. I, p. 244] which eventually allows the use of an ordinary compass *not* collapsing every time it is lifted from the paper.

The quadrature of the circle has occupied generations of mathematicians and the problem vaporised only in the 19th century when Ferdinand Lindemann (1852 – 1939) proved that the number π is a *transcendental* irrational number, i.e. is not the root of an equation of type

$$a_n x^n + a_{n-1} x^{n-1} + \ldots + a_1 x + a_0 = 0$$

with rational coefficients $a_0, a_1, \ldots, a_n$. This finally proved that the quadrature of the circle was simply impossible.

We shall come back to Anaxagoras since he played a further important role in the history of analysis, but let us mention two remarkable things here:

(1) The problem of the quadrature of the circle finally turns out to be a problem concerning properties of irrational numbers; (2) The problem of the existence of irrational numbers seems to have played an important role in the mathematics of the Greek. It is this point we want to illuminate now.

2.1.2 The Pythagoreans

In his overview of the history of Greek philosophy Luciano de Crescenzo (b. 1928) called Pythagoras (about 570 – about 496 BC) simply a 'superstar' [De Crescenzo 1990, Band 1, p. 61 ff.]. Pythagoras was born about 570 BC on the Ionian island of Samos lying close to the coastal town of Miletus. Following Diogenes Laertius [Diogenes Laertius 1931, S. 323] Pythagoras travelled Egypt and Mesopotamia where he learned mathematics; probably also the famous theorem which now bears his name, the Pythagorean theorem. Iamblichus (c. 250–c. 325) in his *Life of Pythagoras* [Guthrie 1987, p. 57 ff.] wrote that Pythagoras was kidnapped by soldiers of the Persian King Cambyses (probably the older one with this name) to Babylon. There he is said to have finished his studies of arithmetic, music, and 'all other sciences' [Guthrie 1987, p. 61]. When Pythagoras returned to Samos he found the island under the control of the tyrant Polycrates. Although it is said that he acted as teacher of Polycrates' son he eventually went into exile to the city of Kroton (today Crotone in Calabria) in Southern Italy. There he is said to have established a true aristocracy (=rule of the best) and to have been legislating. Besides these political activities he founded a school in Kroton of which he was the headmaster. However, many thought his ideas inflammatory and hence he had to leave Kroton. He found refuge about 510 BC in Metapontum (now Metaponto) where he died about 496 BC. The Dorian temple of Hera at Metapontum was viewed later by the Romans as 'school of Pythagoras'. This 'school' is frequently called a sect or secret society since strange rules had to be obeyed:

- abstain from beanes,
- never break the bread,
- eat not the heart,
- do not primp by torch-light,
- stir up the bed as soon as you are risen; do not leave in it any print of the body,

and many more [Guthrie 1987, p. 159 ff.]. The group must have been very effective: for a long time Pythagoreans were active in leading political positions in large regions of the Greek sphere of influence. We are more concerned with the mathematics of this secret society which was passed from

Fig. 2.1.4. Pythagoras of Samos, medieval wooden figure in the choir stalls of the Ulm Minster [Photo: H. Wesemüller-Kock]

the master down to his pupils. There were at least two kinds of Pythagoreans as we know from Iamblichus: the *mathematikoi* and the *akousmatikoi*. The former were those who actively were concerned with *mathemata* – learning topics – and the latter were the ones who only listened to what they heard (*akousmata*). Today we would say that the former were mathematicians while the latter were admirer of mathematics. The Pythagorean motto was:

'All is number!'

We can be sure that 'number' meant the natural numbers,

$$\mathbb{N} := \{1, 2, 3, \ldots\}.$$

To appreciate the enthusiasm for natural numbers one has to know that Pythagoras left a mathematical theory of harmony.

This theory of harmony probably arose from Pythagoras playing on the monochord which, according to Diogenes Laertius, he also invented. One string was stretched over a scale with twelve equally spaced parts. The choice of this division seems smart since 12 is divisible by 2, 3, 4, and 6 and therefore tones of strongly shortened strings could be compared with each other [van der Waerden 1979, p. 370]. The number 12, its half, two thirds and three quarters of it,

$$12, 9, 8, 6,$$

played a central role in the thinking of the Pythagoreans. They form the proportion

$$12 : 9 = 8 : 6;$$

the number 9 being the arithmetic mean of 12 and 6 ($9 = (12 + 6)/2$), and 8 is the harmonic mean of 12 and 6 ($9 = 2/(1/12 + 1/6)$). Of course such relations were what the numerologicaly inclined Pythagoreans had waited for! If two tones appearing at the same time sounded pleasantly the Pythagoreans called them *symphon*. The octave, the pure fifth and the fourth are such tone differences which are symphon. Now we can assign the relation 2:1 to the octave, 3:2 to the fifth, and 4:3 to the fourth. Starting from these basic ratios the Pythagoreans calculated numerous further ones, cp. [van der Waerden 1979, p. 367 f.].

That a tone can be 'measured' in comparison to another one in form of a ratio was also found by the Pythagoreans in the case of lengths, areas, and volumes. Two quantities a and b were called *commensurable* (jointly measurable) if they both are multiples of a third number c, i.e.

> Two quantities a and b are commensurable $:\Leftrightarrow$ if there exists a quantity c and two numbers m and n so that $a = mc$ *and* $b = nc$ holds.

This may be expressed differently, namely as a proportion:

> Two quantities a and b are commensurable $:\Leftrightarrow$ if $a : b = m : n$ holds for two natural numbers m and n.

We have to note that in the eyes of the Greek, comparable quantities always had to be of the same kind, i.e. true to scale or dimension. It was not allowed to compare an area with a length or a volume with an area, but only lengths with lengths, et cetera.

The Pythagoreans had arrived at positive *rational numbers* via the proportions, i.e. in modern words they arrived at the set $\mathbb{Q}^+$ of the positive fractions[1]. Other numbers did not exist for them – they were not allowed to exist: the natural numbers ('all is number') and the proportions comprised the very nature of reality. They were the epitome of being. If 'all' is number then everything is commensurable, hence measurable on the same scale. Nothing in their world was incommensurable! However, this fundamental conviction of the Pythagoreans was shattered by one of their own members! The Pythagorean Hippasus of Metapontum, dated by van der Waerden to have lived between 520 and 480 BC[2] [van der Waerden 1979, p. 74], is said to have shown the existence of incommensurable quantities and thereby he triggered the first fundamental crisis in the history of mathematics! Very often one reads that his existence proof was carried out at the secret symbol of the

[1] However, there is no evidence that the ancient Greek knew how to compute with fractions; they knew only proportions!

[2] K. v. Fritz dates his lifetime around 450 BC [von Fritz 1971]

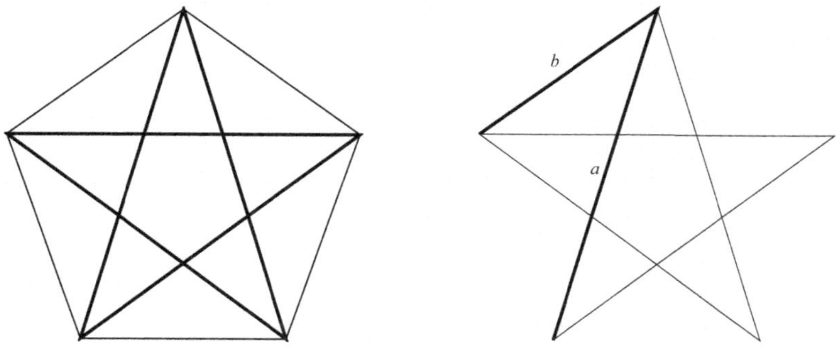

(a) The pentagram inscribed in a pentagon

(b) Concerning infinite reciprocal subtraction at the pentagram

Fig. 2.1.5. Symbol of the Pythagoreans: The pentagram

Pythagoreans: the pentagram. This would have raised Hippasus' sacrilege to the level of monstrosity. Actually, the edge of a pentagram a and the edge b of the circumscribed pentagon in figure 2.1.5(b) are incommensurable, i.e. for all natural numbers m and n we have

$$a : b \neq m : n;$$

hence there is no common measure c so that $a = mc$ and $b = nc$ holds simultaneously. To show the incommensurability of a and b Hippasus probably used the ingenious method of reciprocal subtraction which was known long before his times [Scriba/Schreiber 2000, p. 36 ff.]. The method of reciprocal subtraction is nothing else than the well known Euclidean algorithm in geometric outfit. Let us describe the method in form of an algorithm in pseudocode for simplicity:

- As long as $a \neq b$ do:
  - if $a > b$ set $a := a - b$;
  - otherwise set $b := b - a$;
- print a.

The algorithm terminates after finitely many steps if and only if a and b are commensurable. If a and b are incommensurable the algorithm does *not* terminate! In the first step of the infinite reciprocal subtraction we subtract b from a. Noting that the edge of the pentagon and the edge of the pentagram are parallel as depicted in figure 2.1.6(a) the remaining part of a will be a_1. This remaining part is now subtracted from b and we get b_1 as shown in figure 2.1.6(b). By a parallel shift into the interior of the pentagram we now see that a_1 is just the edge length of a smaller pentagram while b_1 is the edge length of the associated circumscribed pentagon. Hence already in the second step of the infinite reciprocal subtraction we arrive at a situation identical to

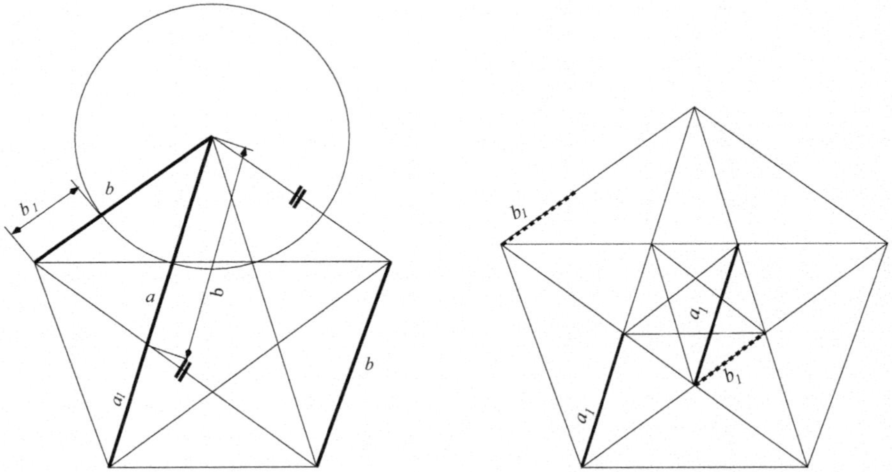

(a) First step of the infinite reciprocal subtraction

(b) Preparations for the second step

Fig. 2.1.6. Infinite reciprocal subtraction to prove incommensurability of edge b and diagonal a of a pentagon

the point of departure! The infinite reciprocal subtraction can therefore not terminate in a finite number of steps and the incommesurability of a and b is proven.

Whether Hippasus in fact chose the sacred symbol of the Pythagoreans or if he did not prove anything at all we do not know. At least the diagonal in the unit square has length $\sqrt{2}$ after Pythagoras' theorem and also in this simpler case the method of infinite reciprocal subtraction would have been the method of choice to prove the incommensurability of 1 and $\sqrt{2}$.

It belongs to the folklore of the history of mathematics that the discovery of incommensurable quantities should have shattered the Pythagoreans to the bones. Iamblichus tells us [Guthrie 1987, p. 116]:

> *'It is accordingly reported that he who first divulged the theory of commensurable and incommensurable quantities to those unworthy to receive it, was by the Pythagoreans so hated that they not only expelled him from their common association, and from living with him, but also for him constructed a [symbolic] tomb, as for one who had migrated from the human into another life. It is also reported that the Divine Power was so indignant with him who divulged the method of inscribing in a sphere the dodecahedron, one of the so-called solid figures, the composition of the* icostagonus. *But according to others this is what happened to him who revealed the doctrine of irrational and incommensurable quantities.'*

A dodecahedron is a solid built from 12 pentagons and here is where the link to Hippasus comes in if we read Iamblichus [Guthrie 1987, p. 79]:

> *'As to Hippasus, however, they acknowledge that he was one of the Pythagoreans, but that he met the doom of the impious in the sea in consequence of having divulged and explained the method of forming a sphere from twelve pentagons;* [...] *'*

And here is the full folklore: Hippasus of Metapontum discovers the existence of incommensurable numbers. This is a sacrilege so hideous that his own Pythagorean friends devise a devious assassination in that they sink the ship carrying Hippasus so that Hippasus drowned.

We have good reasons to mistrust this folklore tradition. Leonid Zhmud argues in [Zhmud 1997, p. 173] that the notion *arrhetos* used by Plato (428/427 – 348/347 BC) in his dialog *Hippias major* for irrational numbers indeed translates as 'secret' or 'unspeakable enigma', but that it just means not more, than 'not expressible in numbers'. Zhmud explains the reports on Hippasus of Metapontum as coming from translating errors of earlier commentators. Substantiated is only a quarrel between Hippasus and the Pythagoreans which was politically motivated. It is also possible that the discovery of irrational numbers constituted no violation of unwritten Pythagorean laws at all! Another modern critic was David Fowler (1937 – 2004) [Fowler 1999, p. 289 ff.] who presented a painstaking research on the history and analysis of the folklore of the foundational crisis. Although Fowler's book [Fowler 1999] can also not give definitive answers it offers some interesting and thought-provoking ideas not being in the focus of historians of mathematics before.

Where did the folklore of the fundamental crisis due to the discovery of irrational numbers come from? At the height of the really shattering crisis concerning the foundations of mathematics at the beginning of the 20th century mathematician Helmut Hasse (1898 – 1979) and logician and philosopher Heinrich Scholz (1884 – 1956) wrote a paper on the foundational crisis of Greek mathematics [Hasse/Scholz 1928]. It is this paper which most likely triggered the story of the mathematical crisis of the Greeks which is still alive today.

Be that as it may the discovery of the incommensurability of certain lengths seems to have brought the Pythagorean program of arithmetising geometry to a halt. It may be that this halt may also have been responsible for the good shape of geometry in Greek mathematics while Arithmetic and Algebra were only treated shabbily. At least the great philosopher Plato declared in his *Theaitetos* 147d–148a and his *Laws* 819d–822d that he felt it a shame for the Greek not to have taken up the problem of incommensurable quantities, see [Popper 2006, p. 329].

2.1.3 The Proportion Theory of Eudoxus in Euclid's Elements

The *Elements* of Euclid [Euclid 1956] comprise a superb summary of Greek mathematics. This famous work divided into 13 books is attributed to Euclid of Alexandria (about 300 BC) although we barely know anything about his existence. Book X contains an axiomatic treatment of commensurable and incommensurable quantities and explains the method of (infinite) reciprocal subtraction. At the end of Book X we find the proof that the diagonal in the unit square and its edge are incommensurable. This proof conducted on a square is the original one if we follow Kurt von Fritz [von Fritz 1971, p. 562].

Even more interesting is Book V, however, containing the proportion theory of Eudoxus of Cnidus. Eudoxus is thought to have been the most ingenious mathematician in Plato's academy to which he held close contact. At the foundation of modern analysis lies what we now call the *Archimedean axiom*:

Axiom: To every ever so small positive real number ε there exists a natural number n so that:

$$0 < \frac{1}{n} < \varepsilon$$

holds.

Fig. 2.1.7. Images of Euclid of the Italian renaissance [Panel painting by Joos von Wassenhove about 1474] and of the early modern era [Fantasy image of an unknown artist]

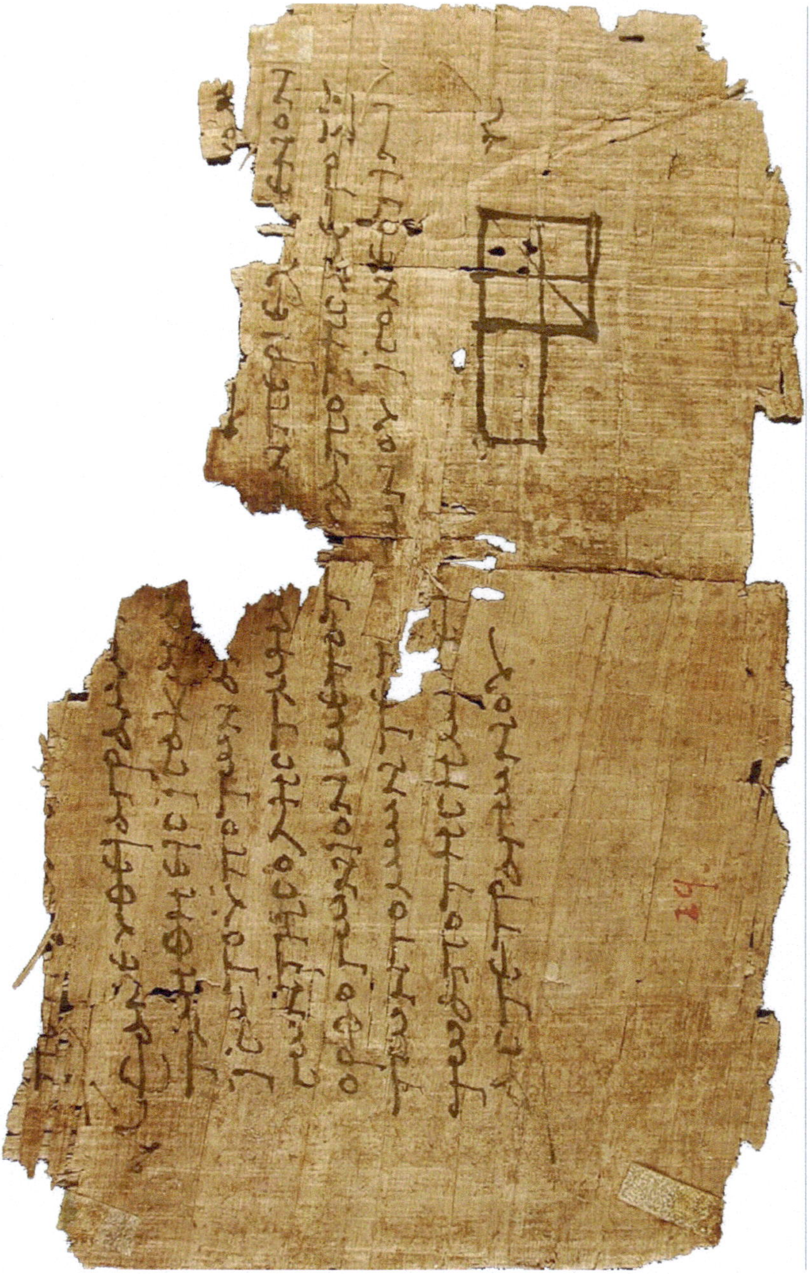

Fig. 2.1.8. Fragment of the Elements of Euclid (Papyrus dating back to 75–125, one of the oldest preserved fragments of the *Elements* of Euclid) [University of Pennsylvania, P.OXY.l 29]

THE ELEMENTS
OF GEOMETRIE
of the moſt aunci-
ent Philoſopher
EVCLIDE
of Megara.

Faithfully (now firſt) tran-
ſlated into the Engliſhe toung, by
H. Billingſley, Citizen of London.
Whereunto are annexed certaine
Scholies, Annotations, and Inuenti-
ons, of the beſt Mathematici-
ens, both of time paſt, and
in this our age.

With a very fruitfull Præface made by M. I. Dee,
ſpecifying the chiefe Mathematicall Sciẽces, what
they are, and wherunto commodious: where, alſo, are
diſcloſed certaine new Secrets Mathematicall
and Mechanicall, vntill theſe our daies, greatly miſſed.

Imprinted at London by Iohn Daye.

Fig. 2.1.9. Famous English edition of the *Elements* of Euclid by Henry Billingsley of 1570 with a preface by John Dee

This axiom, nowadays often shown to be a conclusion of the supremum principle, gives a decisive insight into the structure of the real numbers. For example, following this axiom there can not be a smallest real positive number. For if $\delta > 0$ were this smallest number a natural number n could be found such that $1/n$ would fit in between 0 and δ. Hence $1/n$ would be even smaller than δ. This axiom is the genuine Greek answer to the question of the infinite small: There are no infinitely small numbers! Hence all problems would have been solved but of course the idea of the infinitely small was much too interesting to leave it alone.

We can restate the Archimedean axiom equivalently in the following way: To any two quantities $y > x > 0$ there exists a natural number N so that $N \cdot x > y$ holds. It is this form in which the Archimedean axiom has found its way into Euclid's *Elements*. The axiom ought to have been called *Eudoxus' axiom*. At the beginning of Book V we find Definition 4 [Euclid 1956, Vol. II, p. 114]:

> *'4. Magnitudes are said to* **have a ratio** *to one another which are capable, when multiplied, of exceeding one another.'*

And it goes on:

> *'5. Magnitudes are said to* **be in the same ratio**, *the first to the second and the third to the fourth, when, if any equimultiples whatever be taken of the first and third, and any equimultiples whatever of the second and fourth, the former equimultiples alike exceed, are alike equal to, or alike fall short of, the latter equimultiples respectively taken in corresponding order.'*

Today we would formulate this as follows:

Definition 5 Eudoxus/Euclid: We define $a : b = c : d$ if and only if for all natural numbers m, n it holds:

$$\begin{array}{lll} \text{If } n \cdot a > m \cdot b & \text{then} & n \cdot c > m \cdot d \\ \text{If } n \cdot a = m \cdot b & \text{then} & n \cdot c = m \cdot d \\ \text{If } n \cdot a < m \cdot b & \text{then} & n \cdot c < m \cdot d. \end{array}$$

Now it looks as if Eudoxus had done nothing but to blur the seemingly clear definition of proportions or ratios and had left an incomprehensible definition. Nothing is further from the truth! This Definition is a breakthrough of a special kind. What happens in case of two incommensurable quantities a and b? Then Eudoxus' definition divides the rational numbers m/n into two disjoint set; namely a set U for which the first possibility in Definition 5 above holds. Members of U are those m/n for which

$$n \cdot a > m \cdot b, \quad \text{hence} \quad \frac{m}{n} < \frac{a}{b}$$

holds. Since a and b are assumed to be incommensurable the middle possibility in Definition 5 above is out of the question, but not the last possibility. Here now is a set O defined containing all those m/n for which

$$n \cdot a < m \cdot b, \quad \text{hence} \quad \frac{m}{n} > \frac{a}{b}$$

holds. Hence we have established the dissection

$$\mathbb{Q} = U \cup O$$

of the rational numbers into a lower set U and an upper set O, both sets being disjoint:

$$U \cap O = \emptyset.$$

Each number in U is smaller than any number in O. At the 'interface' between U and O a new number may be defined which obviously has to be an irrational one. We had to wait well until the second half of the 19th century before Eudoxus' theory of proportions could be utilised for the construction of the real numbers. This fundamental step was finally carried out by the mathematician Richard Dedekind (1831–1916) from Brunswick, Germany. The foundation of his 'Dedekind cuts' is the disjoint dissection of the rational numbers by means of Eudoxus' definition of proportionals as described above! We shall come back to Dedekind and his cuts in section 9.7.1.

Another advance in the definition of proportionality is that now quantities of 'different kind' can be compared. Now a and b can be lengths and c and d volumes; their rations are now comparable.

But for which tasks has Eudoxus used the Archimedean/Eudoxus' axiom? This becomes clear also in Book V of Euclid's *Elements* when

$$a : c = b : c \quad \Longrightarrow \quad a = b$$

should be proven (Proposition 9, [Euclid 1956, Vol. II, p. 153 f.]). We follow [Edwards 1979, p. 14]: Assume $a > b$. Following the Archimedean axiom there exists a natural number N so that

$$N \cdot (a - b) > c$$

holds. Furthermore there exists a smallest natural number M with the property $M \cdot c > N \cdot b$, but it also holds $N \cdot b \geq (M - 1) \cdot c$ since M is the *smallest* natural number satisfying $Mc > Nb$. Adding the inequalities $N(a - b) > c$ and $Nb \geq (M - 1)c$ yields

$$Na > Mc,$$

but $Nb < Mc$ according to our assumption! This contradicts the definition of proportionality. Hence our assumption $a > b$ must have been wrong.

Fig. 2.1.10. Euclid (Statue in the Oxford University Museum of Natural History) [Photo: Thomas Sonar]

This kind of proof is called *reductio ad absurdum* since an assumption leads to a contradiction. Now only two possibilities are left: either it is $a < b$ or $a = b$. Following the principle of *tertium non datur* (principle of the excluded third) we only need to falsify $a < b$ which works analogously to the proof above. All in all the proof is of the kind of a *double reductio ad absurdum*. We shall come back to this technique shortly.

2.1.4 The Method of Exhaustion – Integration in the Greek Fashion

With the Archimedean axiom Eudoxus also brought a method for the computation of areas to life: the method of exhaustion. The name *method of exhaustion* was invented by Grégoire de Saint-Vincent (cp. section 7.2), who coined it in 1647 [Jahnke 2003a, p. 18]. The area of a curvilinearly bounded figure is filled by polygons and a sequence of such polygons is studied which exhaust the figure better and better.

The foundation of this method can be found in Book X of Euclid's *Elements* in Proposition 1 [Euclid 1956, Vol. III, p. 14]:

> *'Two unequal magnitudes being set out, if from the greater there be subtracted a magnitude greater than its half, and from that which is left a magnitude greater than its half, and if this process be repeated continually. there will be left some magnitude which will be less than the lesser magnitude set out.'*

Reading carefully we see that this is an obvious reformulation of the Archimedean axiom. In modern terms we start with two positive quantities G_0 and ε and then calculate intermediate quantities

$$G_1 < \frac{1}{2}G_0, \quad G_2 < \frac{1}{2}G_1, \quad G_3 < \frac{1}{2}G_2, \quad \text{etc.}$$

until we reach $G_n < \varepsilon$ for some index n. The proof that such a strategy leads to success follows from a multi-step consideration:

In step 1 the Archimedean axiom warrants the choice of a natural number n such that $(n+1)\varepsilon > G_0$ holds. Since ε is positive it follows $n\varepsilon \geq \varepsilon$ and $n\varepsilon + \varepsilon \geq 2\varepsilon$, hence

$$\varepsilon \leq \frac{(n+1)\varepsilon}{2}.$$

Therefore we have

$$n\varepsilon > G_0 - \varepsilon \geq G_0 - \frac{(n+1)\varepsilon}{2}.$$

Dividing by 2 and combining related terms yields

$$\frac{3}{4}n\varepsilon + \frac{\varepsilon}{4} \geq \frac{G_0}{2}.$$

But now $n\varepsilon \geq \varepsilon$ so that

$$n\varepsilon \geq \frac{3}{4}n\varepsilon + \frac{\varepsilon}{4} \geq \frac{G_0}{2},$$

i.e $n\varepsilon \geq \frac{1}{2}G_0 > G_1$ follows and we have accomplished the first step.

In the second step we start with $n\varepsilon > G_1$. Since $(n-1)\varepsilon \geq \varepsilon$ holds it follows $n\varepsilon \geq 2\varepsilon$, or

$$\varepsilon \leq \frac{n\varepsilon}{2}.$$

Therefore we get

$$(n-1)\varepsilon > G_1 - \varepsilon \geq G_1 - \frac{n\varepsilon}{2}.$$

Division by 2 and combining terms yields

$$\frac{3}{4}(n-1)\varepsilon + \frac{\varepsilon}{4} \geq \frac{G_1}{2},$$

and since $(n-1)\varepsilon \geq \varepsilon$ it follows

$$(n-1)\varepsilon \geq \frac{1}{2}G_1 > G_2,$$

so that the second step is also accomplished. Proceeding in this way eventually leads to

$$\varepsilon > G_n$$

in step n.

We can now apply this version of the Archimedean axiom to the computation of areas by means of exhaustion.

Theorem (Application of the method of exhaustion to the calculation of the area of a circle): *Given a circle C with area $A(C)$ and a number $\varepsilon > 0$. Then there exists an inscribed regular polygon$P$ with the property*

$$A(C) - A(P) < \varepsilon.$$

In other words, increasing the number of edges of the polygons inscribed in the circle leads eventually to a polygon the area of which differs from the area of the circle not more than a given arbitrary small positive ε. The method of exhaustion now allows proving the theorem *without* falling back on a rigorous limit definition. To this end we start with an inscribed square $P_0 = EFGH$ as in figure 2.1.11(a). To reference the Archimedean axiom we write

$$G_0 := A(C) - A(P_0)$$

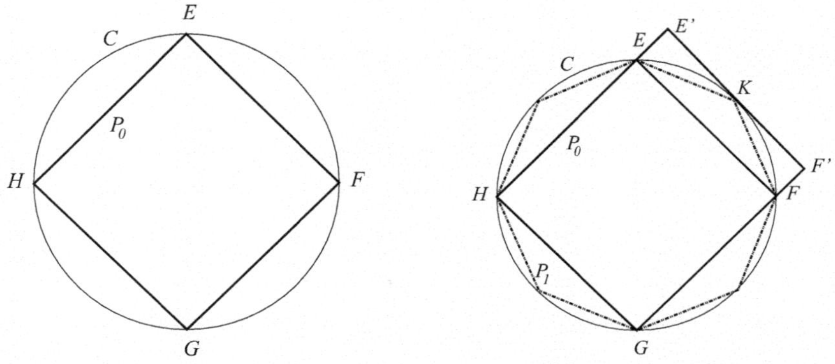

(a) Initial polygon P_0 in the circle C (b) Second step: P_1

Fig. 2.1.11. The method of exhaustion exemplified at a circle

for the difference of areas between circle C and square P_0. Doubling the number of edges results in an octagone P_1 as the next regular polygon. Further doubling the number of edges leads to a sequence $P_0, P_1, P_2, \ldots, P_n$, $\ldots$ where the regular polygon P_n has exactly 2^{n+2} edges. If we now could show that for

$$G_n := A(C) - A(P_n)$$

the inequality

$$G_{n+1} < \frac{1}{2} G_n$$

is valid, then it would follow from the above version of the Archimedean axiom that $G_n < \varepsilon$ holds for sufficiently large n.

We start with $n = 0$. Fig. 2.1.11(b) tells us that

$$\begin{aligned} G_0 - G_1 &= A(P_1) - A(P_0) = 4 \cdot A(\text{triangle } EFK) \\ &= 2 \cdot A(\text{rectangle } EFF'E') > 2 \cdot A(\text{segment } EKF) \\ &= \frac{1}{2} \cdot 4 \cdot A(\text{segment } EKF) = \frac{1}{2} \underbrace{(A(C) - A(P_0))}_{=G_0}; \end{aligned}$$

hence we have shown $G_0 - G_1 > \frac{1}{2} G_0$ and this is nothing but

$$G_1 < \frac{1}{2} G_0.$$

Analogously we get for arbitrary n

$$G_n - G_{n+1} = A(P_{n+1}) - A(P_n) > \frac{1}{2} (A(C) - A(P_n)) = \frac{1}{2} G_n,$$

where $A(C)-A(P_n)$ is the sum of the areas of the 2^{n+1} segments of the circle which are cut off by the edges of P_n.

As a final example of the method of exhaustion let us consider the following theorem which can be found in Euclid's *Elements* as Proposition 2 in Book XII [Euclid 1956, Vol. 3, p. 371]:

Theorem: *If* C_1 *and* C_2 *are circles with radii* r_1 *and* r_2*, respectively, then it holds*

$$\frac{A(C_1)}{A(C_2)} = \frac{r_1^2}{r_2^2}. \tag{2.1}$$

(*'Circles are to one another as the squares on the diameters.'*)

The proof of this theorem is accomplished via the method of *double reductio ad absurdum.* There can only be three possibilities since either

$$\frac{A(C_1)}{A(C_2)} = \frac{r_1^2}{r_2^2}, \quad \text{or} \quad \frac{A(C_1)}{A(C_2)} < \frac{r_1^2}{r_2^2}, \quad \text{or} \quad \frac{A(C_1)}{A(C_2)} > \frac{r_1^2}{r_2^2}.$$

We state *assumption 1*: $\frac{A(C_1)}{A(C_2)} < \frac{r_1^2}{r_2^2}$, hence we assume

$$A(C_2) > \frac{A(C_1)r_2^2}{r_1^2} =: S.$$

Then the number $\varepsilon := A(C_2) - S$ would be positive, i.e. $\varepsilon > 0$. Following the theorem on the exhaustion of the area of a circle above there exists a polygon P inscribed in circle C_2 so that $A(C_2) - A(P) < \varepsilon = A(C_2) - S$ holds, i.e. $A(P) > S$. We now inscribe a polygon P corresponding to P into C_1. We now have (cp. fig. 2.1.12)

$$\frac{A(Q)}{A(P)} = \frac{r_1^2}{r_2^2},$$

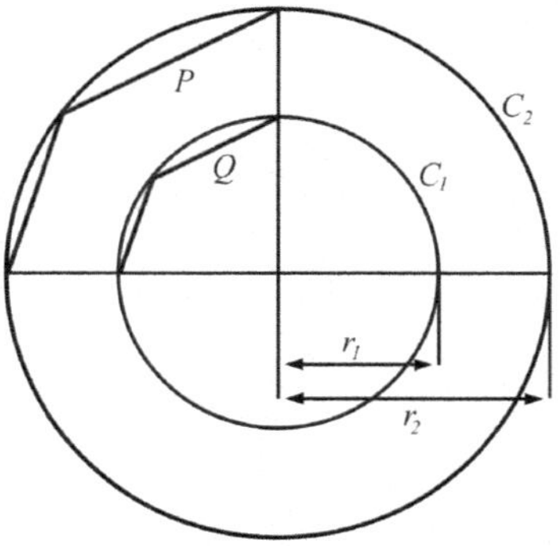

Fig. 2.1.12. Regular polygons in circles

hence

$$\frac{A(Q)}{A(P)} = \frac{r_1^2}{r_2^2} = \frac{A(C_1)}{\frac{A(C_1)r_2^2}{r_1^2}} = \frac{A(C_1)}{S}.$$

But from this it follows

$$\frac{S}{A(P)} = \frac{A(C_1)}{A(Q)} > 1,$$

hence $S > A(P)$, and this contradicts our assumption $A(P) > S$. Thus *assumption 1* must be wrong and this we have proven with the method of *reductio ad absurdum.*

Now we state *assumption 2*: $\frac{A(C_1)}{A(C_2)} > \frac{r_1^2}{r_2^2}$ and show that a contradiction follows as above. We have now completed the *double reductio ad absurdum* and only the case $\frac{A(C_1)}{A(C_2)} = \frac{r_1^2}{r_2^2}$ remains as correct possibility.

2.1.5 The Problem of Horn Angles

The mathematics of the Greeks was ruled by *Archimedean* number systems since the days of Eudoxus, i.e. number systems in which the Archimedean axiom holds: To any two positive quantities $x < y$ a natural number n can always be found so that $n \cdot x > y$ holds. But already Eudoxus knew that also other number systems – so-called non-Archimedean number systems – were conceivable [Becker 1998, p. 104]. Such a system of quantities which was already known to the Greeks were *cornicular angles* or *horn angles* [Thiele 2003, p. 1f.]. These are angles between two circles touching each other or between a circle and its tangent as shown in figure 2.1.13. In Book III of Euclid's *Elements* we find Proposition 16 [Euclid 1956, Vol. II, p. 37] on cornicular angles:

> *'The straight line drawn at right angles to the diameter of a circle from its extremity will fall outside the circle, and into the space between the straight line and the circumference another straight line cannot be interposed; further the angle of the semicircle is greater, and the remaining angle less, than any acute rectilineal angle.'*

Between the circumference and the tangent simply no further straight line can be drawn which remains outside the circle. Cornicular angles comprise a non-Archimedean system of numbers since in the case of any two cornicular angles the Archimedean axiom does obviously not hold. If we define cornicular angles as angles between the tangents then every cornicular angle is simply zero and the Archimedean number system is again restored. As Körle writes [Körle 2009, p. 29]:

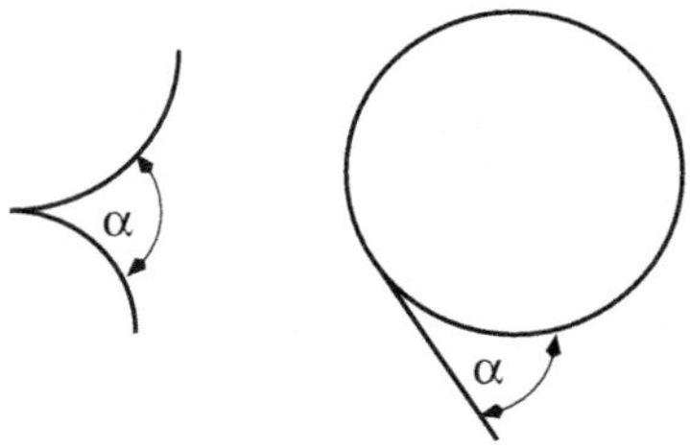

Fig. 2.1.13. Cornicular or horn angle

> *'For us there simply are no cornicular angles. Their problem is of a psychologigal nature. One did not know how to defend oneself against the idea that something had to fill this gap. They could not be argued away but became invalid with the notion of the limit. By any stretch of imagination concerning the interpretation there remains only the quantity zero for cornicular angles. The controversey concerning cornicular angles remained for a long time and even Leibniz was concerned with them.'*
>
> (Für uns gibt es schlichtweg keine Kontingenzwinkel. Ihr Problem ist psychologischer Natur. Man wusste sich nicht gegen die Vorstellung zu wehren, irgendwas müsse jene Öffnung doch ausfüllen. Wegdiskutieren ließen sie sich nicht, gegenstandslos wurden sie mit dem Begriff des Grenzwerts. Bei bestem Willen zur Interpretation bliebe den Kontingenzwinkeln nur die Größe Null. Die Kontroverse um sie hielt lange an, noch Leibniz beschäftigte sich mit ihnen.)

We shall see later that analysis is also possible in non-Archimedean number systems. In such systems different cornicular angles can be of different sizes!

2.1.6 The Three Classical Problems of Antiquity

We have already reported on the quadrature of the circle and its authorship by the imprisoned Anaxagoras. We have to mention two other problems which, together with the quadrature of the circle, have played an important role in the history of analysis. All three problems became know as the 'classical problems of mathematics'; compare [Alten et al. 2005] and [Scriba/Schreiber 2000].

> **The Trisection of the Angle:**
> Given an angle α. Construct an exact trisection of this angle by means of straightedge and compass.

Heath [Heath 1981, p. 235] has suspected that this problem originated at a time when one was able to construct the pentagon by means of straightedge and compass and wanted to construct further polygons. The construction of a regular polygon of 10 edges in fact requires the trisection of an angle. As the quadrature of the circle is an unsolvable problem the trisection of the angle also is unsolvable as modern algebra has revealed. Generation of mathematicians have tried to solve the problem nevertheless.

> **The Doubling of the Cube:**
> Given a cube with volume V. Construct a cube with double volume by means of straightedge and compass.

Legend has it that the Deloians, inhabitants of the Cycladic island of Delos in the Aegean Sea, were haunted by the plague. The oracle was asked for advice and suggested the doubling of the cube-shaped altar in the temple of the Deloians. As this was unsuccessful the Deloians asked the great philosopher Plato who answered that the god did not actually require a new altar but that he had posed this problem to put the Deloians to shame because they were not interested in mathematics at all and despised geometry, cp. [Heath 1981, p. 245f.].

Concerning the Quadrature of the Circle

The attempt to square the circle seems to have been a temptation for the Greek mathematicians. An approach not based on the method of exhaustion was developed by Hippocrates of Chios (middle or second half of the 5th c BC) which has had an impact even on the maths books used in schools in recent times. The method relies on the 'lunes of Hippocrates'. Hippocrates thereby followed the less ambitious task of computing the area of surfaces which are bounded by parts of circles.

If an isosceles triangle is drawn inside a half circle as shown in figure 2.1.14 this triangle is right-angled after the theorem of Thales. Drawing two further half circles about the legs of the triangle then circular shaped areas M appear which remind on lunes. With the notations in figure 2.1.14 we introduce the areas

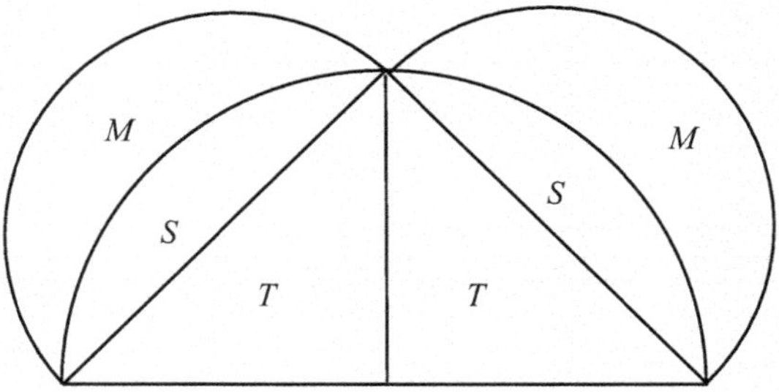

Fig. 2.1.14. Lunes of Hippocrates

$$C_1 := M + S$$
$$C_2 := 2 \cdot (S + T).$$

Now we know from the theorem on page 37 that ratios of areas of circles are like the ratios of the square of the radii. This, of course, holds true also in the case of the areas of half circles. If the radius of the large half circle is r then it follows from Pythagoras' theorem that the radii of the small circles are $\frac{\sqrt{2}}{2}r$. The squares of these radii are r^2 and $\frac{1}{2}r^2$, respectively. Hence for the areas of the half circles C_1 und C_2 it follows:

$$\frac{C_1}{C_2} = \frac{1}{2}.$$

Inserting the definition of C_2 it follows $2(T + S) = 2C_1$, hence $T + S = C_1$. By definition we have $C_1 = M + S$, i.e.

$$T + S = C_1 = M + S$$

and it is shown that $M = T$ holds true. The area of one of the lunes hence is exactly the area of the triangle T.

There is no doubt that Hippocrates' results encouraged not only himself but also others to go on and finally tackle the quadrature of the circle. More complicated lunes can thus be found in the repertoire of the method of lunes, cp. [Baron 1987, p. 32f.], [Scriba/Schreiber 2000, p. 47f.].

Heath [Heath 1981, Vol.1, p. 225f.] cites some ancient authors reporting on some of the mathematical developments which arose from the many futile attempts to solve the three great problems by means of a construction with straightedge and compass. Iamblichus wrote concerning the quadrature of the circle:

> '*Archimedes effected it by means of the spiral-shaped curve, Nicomedes by means of the curve known by the special name* quadratrix [...], *Apollonius by means of a certain curve which he himself calls "sister of the cochloid" but which is the same as Nicomedes's curve, and finally Carpus by means of a certain curve which he simply calls (the curve arising) "from a double motion".*'

Pappus of Alexandria (about 290–about 350) is cited with:

> '*for the squaring of the circle Dinostratus, Nicomedes and certain other and later geometers used a certain curve which took its name from its property; for those geometers called it* quadratrix.'

Proclus (412–485) wrote concerning the trisection of the angle:

Fig. 2.1.15. Sphinx and pillar of Pompeius in Alexandria; the town with the largest library of antiquity and many scholars. Pappus of Alexandria was one of them [Photo: H.-W. Alten]

> *'Nicomedes trisected any rectilineal angle by means of the conchoidal curves, the construction, order and properties of which he handed down, being himself the discoverer of their peculiar character. Others have done the same thing by means of the* quadratrices *of Hippias and Nicomedes ... Others again, starting from the spirals of Archimedes, divided any given rectilineal angle in any given ratio.'*

Proclus then goes on and mentions explicitly the mathematicians who explained the properties of different curves:

> *'thus Apollonius shows in the case of each of the conic curves what is its property, and similarly Nicomedes with the conchoids,* Hippias with the quadrices, *and Perseus with the spiric curves.'*

The three great problems are unsovable in their original formulation, i.e. with a construction by means of straightedge and compass alone. However, they can be solved with the help of the curves mentioned above. Therefore it is worth looking at at least two of those curves in the context of the trisection of the angle.

Concerning the Trisection of the Angle

Hippias of Elis (5th c BC) was born in Western Greece. He is attributed with the discovery of the quadratrix which is defined pointwise by means of a mechanical model as shown in figure 2.1.16(a).

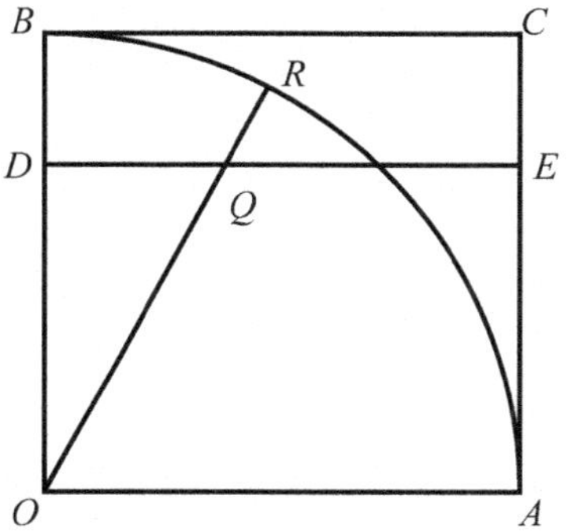

(a) Definition of the quadratrix

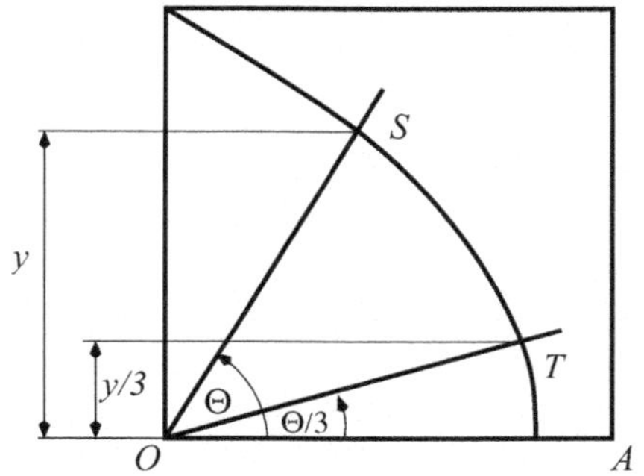

(b) Trisection of an angle by means of the quadratrix

Fig. 2.1.16. The quadratrix – an auxiliary curve to trisect an angle

Let a square $OACB$ with edge length 1 and an inscribed quarter circle BRA be given. We imagine the edge BC moving with constant speed down to OA. Simultaneously the edge OB rotates with constant speed about the point O in radial direction also towards OA. Both edges are assumed to reach OA at exactly the same time. At any time in between BC has reached DE and OB is at the position OR. The point of intersection Q is the defined to be a point of the quadratrix.

Now follow the movement of the point R. Its coordinates are described by

$$x = \cos \Theta$$
$$y = \sin \Theta,$$

where the angle Θ changes from $\Theta = \pi/2$ (=90°) to $\Theta = 0$ as shown in figure 2.1.16(b). Edge BC moves with constant speed in y-direction. Let us denote this speed by v_y and connect the movement of BC with the angle Θ by

$$y = v_y \cdot \Theta.$$

If $\Theta = 0$ then also $y = 0$. If $\Theta = \pi/2$ then $y = 1$. These conditions lead to the equation

$$y = 1 = v_y \cdot \frac{\pi}{2},$$

hence the speed of BC has to be $v_y = 2/\pi$. Hence the relation between angle and y-coordinate is

$$\Theta = \frac{\pi y}{2}.$$

But

$$\frac{y}{x} = \frac{\sin \Theta}{\cos \Theta} = \tan \Theta,$$

so that $\tan(\pi y/2) = y/x$ follows, hence

$$x = y \cdot \cot \frac{\pi y}{2}.$$

This then is the equation of the quadratrix which, of course, was unknown to Hippias. Since the cotangent appears in the equation it is a *transcendent* function. The trisection can then be accomplished as follows. To an angle Θ there corresponds a certain value of y, cp. figure 2.1.16(b), and since Θ and y are connected via $\Theta = \pi y/2$ we only need to intersect a horizontal line at height $y/3$ with the quadratrix (the point of intersection is T) in order to trisect the angle Θ. The use of the quadratrix to square a circle is much more involved, cp. [Heath 1981, Vol.1, p. 227f.].

The second curve mentioned in the citations above is the conchoid of Nicomedes (about 280–about 210 BC), also called 'shell curve' because its outer branches resemble the shape of conch shells. It seems a number of different curves were known as cocleoids and the conchoid is just one of them.

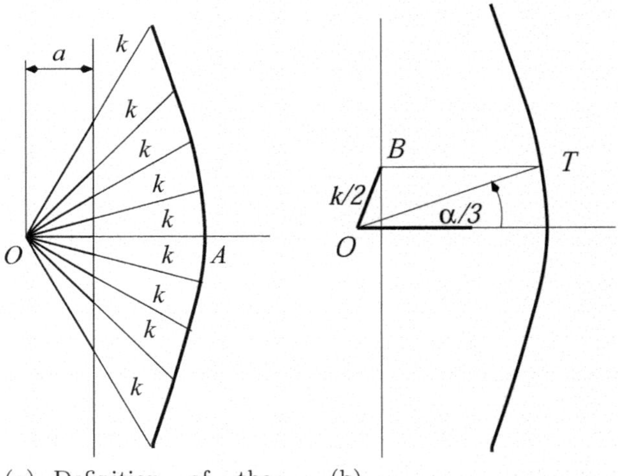

(a) Definition of the conchoid

(b) Trisection of an angle by means of a conchoid

Fig. 2.1.17. The conchoid – a further auxiliary curve to trisect the angle

It is an algebraic curve and can be constructed mechanically. Choose two positive numbers a and k. In a Cartesian coordinate system draw a vertical line at distance a to the origin. Then a point of the conchoid is defined as follows. Draw a line from O to the vertikal line and extend it by a line segment of length k. At the end of this line segment lies a point of the conchoid, cp. figure 2.1.17(a).

One can describe the conchoid of Nicomedes either in Cartesian form

$$y^2 = \frac{x^2(k+a-x)(k-a+x)}{(x-a)^2}$$

or in polar form

$$r = k + \frac{a}{\cos\Theta},$$

where r denotes the length of the line segment from O to a point of the conchoid lying under the angle Θ measured counterclockwise from the x-axis.

The trisection of an arbitrary angle α can now be accomplished as follows. As shown in figure 2.1.17(b) one leg of the angle is put on the horizontal axis and the vertical axis is shifted so that the second leg from O to the point of intersection with the vertical line has length $k/2$. A parallel line to the horizontal axis through this point of intersection B results in the intersection point T on the conchoid. Connecting T with the origin results in the trisection of the angle α.

In case of the conchoid there is also a mechanical construction shown in figure 2.1.18. A pointer with tip point P slides in a groove N of a horizontal rail with a pin C. Perpendicular to this rail and firmly attached to it is a lug with a fixed pin K to support the groove of the pointer. Moving the pointer its tip point will describe a conchoid.

As further methods Archimedes and later Pappus of Alexandria described two insertion methods, known as 'neusis', which were popular in ancient Greek geometry. Details can be found in [Scriba/Schreiber 2000, p. 45f.].

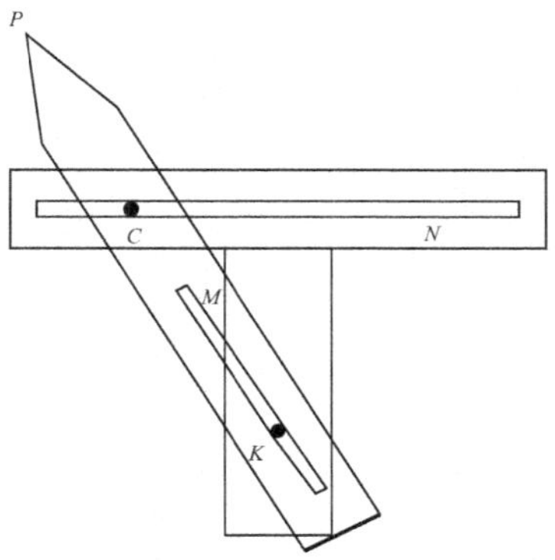

Fig. 2.1.18. A mechanical construction of the conchoid

Concerning the Doubling of the Cube

Let a cube with edge length a be given so that its volume will be $V = a^3$. If a new cube with double the volume has to be constructed than the new edge length x has to satisfy $x^3 = 2 \cdot a^3$, or

$$x = \sqrt[3]{2} \cdot a.$$

Modern algebra as developed only in the 19th century tells us that this number x can *not* be constructed by means of straightedge and compass since $\sqrt[3]{2}$ is not a constructable number. The honour having given the first rigorous proof of the unsolveability of the problems of doubling the cube and trisecting the angle is due to the French mathematician Pierre Laurent Wantzel (1814–1848) who published his proofs in 1837 in Liouvilles's 'Journal de Mathématiques Pures et Appliquées' [Cajori 1918].

We do not want to describe the many ingenious attempts of the Greeks to tackle this problem but Hippocrates of Chios earned the honour of immortality due to a groundbreaking discovery. He succeeded to reduce the problem of doubling the cube to a problem of the determination of two *mean proportionals.* A cube with edge length $2a$ obviously does not satisfy our conditions since it has twice the edge length of the cube we started with but its volume is $8 \cdot a^3$. Nevertheless the doubling of the edge length must have got stuck somehow in Hippocrates's mind. He looked for two numbers in between the lengths a and $2a$ which are called two mean proportionals. Here x is a mean proportional of two numbers a and b if

$$a : x = x : b.$$

Solving this proportionalty for x results in $x^2 = a \cdot b$ or $x = \sqrt{a \cdot b}$, hence the mean proportional is nothing but the geometric mean of a and $2a$ defined by

$$a : x = x : 2a,$$

hence $x = \sqrt{2} \cdot a$. There is no way to arrive at a solution of our problem of doubling the cube, i.e. $x = \sqrt[3]{2} \cdot a$, by means of just *one* mean proportional. It is highly likely that this perception was already of pythagorean origin since Plato writes in his dialogue *Timaios* [Plato 1929, p. 59, 32A-B]:

> *'But it is not possible that two things alone should be conjoined without a third; for their must needs be some intermediary bond to connect the two. And the fairest of bonds is that which most perfectly unites into one both itself and the things which it binds together; and to effect this in the fairest manner is the natural property of proportion. For whenever the middle term of nay three numbers, cubic or square, is such that as the first term is to it, so is it to the last term, – and again, conversely, as the last term is to the middle, so*

is the middle to the first, – then the middle term becomes in turn the first and the last, while the first and last become in turn middle terms, and the necessary consequence will be that all the terms are interchangeable, and being interchangeable they all form a unity. Now if the body of the All had had to come into existence as a plane surface, having no depth, one middle term would have sufficed to bind together both itself and its fellow-terms; but now it is otherwise: for it behoved it to be solid of shape, and what brings solids into unison is never one middle term alone but always two.'

Hippocrates was also concerned with a solid, namely a cube. One mean proportional hence did not suffice. Therefore he sought *two* mean proportionals x and y of a and $2a$ so that

$$a : x = x : y = y : 2a$$

holds. From these proportionalities three equations follow, namely

$$x^2 = a \cdot y, \quad y^2 = 2a \cdot x, \quad x \cdot y = 2a^2.$$

Solving the last equation for y and inserting the result into the first equation yields the equation of the doubling of the cube,

$$x^3 = 2 \cdot a^3 \quad \Rightarrow \quad x = \sqrt[3]{2} \cdot a.$$

It is obvious that Hippocrates did not succeed coming closer to the doubling of the cube. But the insight that the problem of doubling the cube is completely equivalent to finding two mean proportionals of two line segments can only be named a strike of genius!

Of course further Greek mathematicians tried to solve the problem of doubling the cube and achieved impressive advances in the development of their mathematical methods. We have to name Diocles (about 240–about 180 BC) and the cissoid (ivy curve) named after him, with which two mean proportionals can be constructed geometrically. In this context we also have to mention Archytas of Tarentum (428–347 BC) who presented a remarkable three-dimensional construction to determine the edge length x of the sought new cube. He succeeded in intersecting no less than three bodies of revolution the unique point of intersection being the sought x.

In the treatment of the problem of doubling the cube by means of Hippocrates's two mean proportionals Menaechmus (380–320 BC) discovered the conic sections. However, conic sections got their name and were analysed only later by Apollonius.

Details concerning the constructions mentioned above can be found in [Alten et al. 2005], [Scriba/Schreiber 2000] und in [Heath 1981, Vol. I].

Remarks

The insight and creativity of the Greek mathematicians is impressive even from a modern point of view. Although they were able to solve the problems of squaring the circle, doubling the cube and trisecting the angle by means of certain curves like the conchoid, the conic sections and the quadratrix with arbitrary order of accuracy the actual task – exact construction with straightedge and compass alone – proved to be not solvable at all. However, even today, after the development of analysis, algebra and number theory, there are some amateur mathematicians who believe that they have succeeded in solving at least one of the three great problems of antiquity or in proving the rationality of π. Splendid examples can be found in Underwood Dudley's book [Dudley 1987]. Every attempt to put a stop to the game of these pseudo-mathematicians is unfortunately doomed to failure; they simply either do not understand the problem definition, or necessary notions and mathematical knowledge are missing. Often they invent approximate methods leading to astonishingly good results in a finite number of steps but refuse to acknowledge that the exact solution would only result after an infinity of steps (and therefore can not be the solution the Greeks sought for).

2.2 Continuum versus Atoms – Infinitesimals versus Indivisibles

The discovery of incommensurable quantities, i.e. the existence of irrational numbers, may have unsettled the Greek mathematicians and may even have been responsible for the Greeks withdrawing to geometry. The irrational was at the same time the unspeakable, incomprehensible, nonpictorial [Lasswitz 1984, Vol. 1, p. 175]. However, a philosophical quarrel concerning items of *being* (existence) has shattered analysis almost more and this shock can be felt even today. We do not want to dive too deeply into philosophical problems but refer the reader to the literature, e.g. [von Fritz 1971]. However, a few words may certainly be in order for the sake of a better understanding of the mathematical and historical background.

2.2.1 The Eleatics

Due to warlike struggles with the Persians at the Ionian coast took some Greeks to the South-Italian west coast in the year 545 BC. There they founded the settlement Elea which today is Velia. In Elea there evolved a community of philosophers called the Eleatics. The poet and natural philosopher Xenophanes (about 570–about 475 BC) is seen as its founding

Fig. 2.2.1. Parmenides; Zeno of Elea [Photo: Sailko]

father. One of the truelly great Eleatics was Parmenides (about 540/535–about 483/475 BC) who introduced a new thinking into Greek philosophy. While philosophers before Parmenides were keen on understanding the world Parmenides now introduced the claim of absolute certainty of non-empirical theories into philosophical thinking whereby these theories *can not* serve directly to describe the world [Parmenides 2016, p. 4ff.]. The 'existing' ('being') became a central point of Parmenides' philosophy and the being (or the logos, the one, or god [De Crescenzo 1990, Vol. 1, p. 112]) is something unique, a whole, and an immovable. There is no void and no 'becoming'; 'non-being' is inconceivable. Since the being is immobile Parmenides obviously doubted the possibility of movement at all – we only see apparent movement of human beings whereas the actual being is static – and this is splendidly acknowledged by his most famous pupil, Zeno of Elea (about 490–about 430 BC). Plato in his dialogue *Parmenides* [Plato 1939, p. 205] reports in 128d that Zeno wanted to come to his teacher's defence against the accusation of absurd consequences if movement would be rejected. But what is behind all that mathematically?

2.2.2 Atomism and the Theory of the Continuum

Almost all we know about Zeno has come down to us in the writings of the great philosopher Aristotle who shaped the thinking in the Western world for many centuries. Of all philosophers from Thales to some who lived in Socrates's days no written lore is extant. These philosophers are called the *Pre-Socratics*. The only material we have has come to us in form of fragments [Early Greek Philosophy 2016] which were only written down by philosophers

Fig. 2.2.2. Democritus of Abdera; Detail of a banknote (100 Greek drachma 1967)

of later generations. In Aristotle's *Physics* the great philosopher dedicated a whole book – Book VI – to the problem of the continuum [Aristotle 1995, p. 390–407]. It is there where Zeno gets a chance to speak.

Two schools of thought concerning the passage of time and the structure of space were popular with the Greek thinkers: atomism and the theory of the continuum, respectively. The philosophers Leucippus (5th c BC) and his pupil Democritus (about 460–about 370 BC) are regarded as having invented atomism. Following their thoughts everything consists of infinitely small quantities, the 'atoms' (atomon = indivisible). We must not, however, confuse our modern understanding of atoms with what Democritus understood when he talked about atoms. Democritus's 'atoms' in their primal meaning were probably still further divisible but concerning our discussion of their *mathematical* implications we should imagine an atom as being a point lying in a straight line. This point is an atom and following Democritus the whole straight line is made up of infinitely many points. Aristotle and many others overwhelmingly rejected this theory of atoms that in part can be ascribed to Zeno as we shall see. Following the theory of the continuum a straight line is a 'continuum' which is arbitrarily divisible. Even if a continuum is divided arbitrarily often there always remains a continuum which is still further divisible. Never will the process of division results in a point, however! A point can therefore not be an element of a straight line!

In Book V of his *Physics* Aristotle introduces the notions of 'together', 'apart', 'contact', 'succession', continuity', and others [Aristotle 1995, p. 383]

> *'Let us now proceed to say what it is to be together and apart, in contact, between, in succession, contiguous, and continuous, and to show in what circumstances each of these terms is naturally applicable.*

> *Things are said to be together in place when they are in one primary place and to be apart when they are in different places. Things are said to be in contact when their extremities are together.'*

And he goes on [Aristotle 1995, p. 383]:

> *'A thing is in succession when it is after the beginning in position or in form or in some other respect in which it is definitely so regarded, and when further there is nothing of the* same *kind as itself between it and that to which it is in succession, e.g. a line or lines if it is a line, a unit or units if it is a unit, a house if it is a house (there is nothing to prevent something of a* different *kind being between).* [...] *A thing that is in succession and touches is contiguous. The continuous is a subdivision of the contiguous: things are called continuous when the touching limits of each become one and the same and are, as the word implies, contained in each other: continuity is impossible if these extremities are two.'*

This definition is used in Book VI to give a mortal blow to the idea of atomism [Aristotle 1995, p. 390f.]:

> *'Now if the terms 'continuous', 'in contact', and 'in succession' are understood as defined above – things being continuous if their extremities are one, in contact if their extremities are together, and in succession if there is nothing of their own kind intermediate between them – nothing that is continuous can be composed of indivisibles: e.g. a line cannot be composed of points, the line being continuous and the point indivisible.'*

At the beginning of his *Elements* Euclid defines [Euclid 1956, Vol. I, p. 153]:

1. A **point** is that which has no part.
2. A **line** is breadthless length.

Thereby he cleverly avoided any subtle discussions. Using 'breadthless length' as a paraphrase for a line simply excludes any form of critique concerning atoms or the continuum. But if a point has no part, so Aristotle asked, how then can a line be built from points? In which sense should two points on a line then be adjacent?

Questions like these have fascinated thinkers up to our present days. We remind our readers of the mathematician Hermann Weyl (1885–1955) wrote on the continuum already in 1917 [Weyl 1917] (English translation see [Weyl 1994]) and discussed philosophical problems of mathematics still in 1946 when he was in old age. In [Weyl 2009, p. 41] he looks at the continuum from a modern point of view and builds a bridge to modern analysis when he writes:

> *'The individual natural numbers form the subject of number theory, the possible sets (or the infinite sequences) of natural numbers are the subject of the theory of the continuum.'*

He also cites Anaxagoras to characterise the nature of the continuum [Weyl 2009, p. 41]:

> *'Among the small there is no smallest, but always something smaller. For what is cannot cease to be no matter how far it is being subdivided.'*

This citation refers to the so-called *Fragment 3* of Anaxagoras [Schofield 1980, p. 80]:

> *'For of the small there is no least but always a lesser (for what is cannot not be)'*

and is seen in connection with Zeno's paradoxes which we shall discuss in the following.

2.2.3 Indivisibles and Infinitesimals

Among other reasons atomists and the supporters of the continuum collided was because infinity was concerned, cp. [Heuser 2008, p. 59ff.]. Democritus and the atomists stated nothing less than the existence of an actual infinity since a line (or even a line segment only) consists of an actual infinity of points, hence atoms. Aristotle and many others already in the life time of Zeno rejected the existence of an actual infinity and postulated the 'potential infinity' so that the process of division can always be continued. This dispute, how 'ancient' it may seem to us, has not ceased even today! It was only Georg Cantor (1845–1918) who introduced the actual infinity rigorously into mathematics by the launch of set theory. In Cantor's mathematics a line (e.g. the real number line) is indeed called a 'continuum', but Cantor's continuum is defined by means of single points! Aristotle as well as Democritus would shudder! Only in the 1960s the idea of the continuum aroused again like phoenix from the ashes with the invention of nonstandard analysis. We shall report on this development at the end of this book.

Concerning Democritus it is said [Edwards 1979] that he found the volume formula

$$V = \frac{1}{3} A \cdot h$$

for the cone and the tetrahedron, where A denotes the base area and h the height of these solids. However, this was proven only by Eudoxus. Democritus imagined that solids consisted of infinitely many slices of zero thickness. Today we call these slices 'indivisibles'. It is to be noted, however, that

Democritus certainly did not have an idea concerning a theory of indivisibles [Heath 1981, Vol. I, p. 181]. A point, a line, and a surface are indivisibles in one-, two-, and three-dimensional space, respectively, since one of their dimensions is nil. On the other end of the spectrum and quite contrary to the ideas of the atomists the supporters of the continuum believed that solids consisted of continua – hence solids again – which were themselves arbitrarily divisible again. Those slices of finite thickness are today called 'infinitesimals'. The proof of Democritus's volume formula by Eudoxus in Euclid's *Elements* XII.5 relies on a subdivision of the tetrahedron but we have every reason to believe that Democritus arrived at his result when he imagined a pyramid being built up from infinitely many indivisibles (plane cuts parallel to the base area as in figure 2.2.3(a)). Baertel van der Waerden (1903–1996) in his book *Science Awakening* [van der Waerden 1971, p. 138] cites Plutarch (compare also [Heath 1981, Vol. I, p. 179f.]) who attributed the following argument to Democritus:

> *'If a cone is cut by surfaces parallel to the base, then how are the sections, equal or unequal? If they were unequal* [(and, we might add mentally, if the slices are considered as cylinders)][3], *then the cone would have the shape of a staircase; but if they were equal, then all sections will be equal, and the cone will look like a cylinder, made up of equal circles; but this is entirely nonsensical.'*

Hence Democritus seemed to have very obscure ideas of a solid being an accumulation of two-dimensional cuts as Eberhard Knobloch has pointed out in [Knobloch 2000]. Knobloch even called it a 'pseudo-problem' [Knobloch 2000, p. 86]. The intersection of a plane with the cone results in *two* cuts A and B; one belonging to the lower frustum of the cone and the other belonging

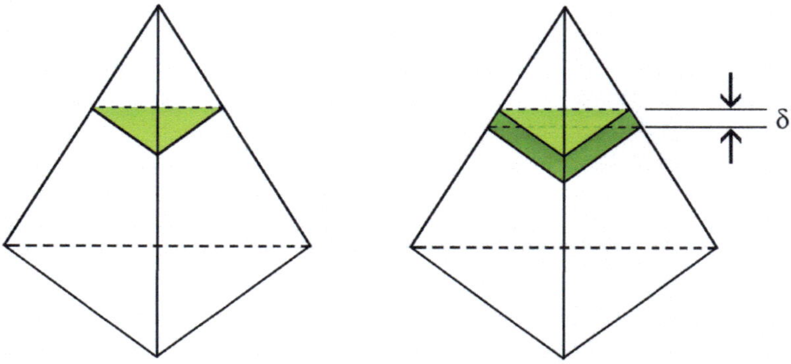

(a) Tetraeder of indivisibles (b) Tetraeder of infinitesimals

Fig. 2.2.3. Indivisible and infinitesimal

[3] Remark by van der Waerden.

Fig. 2.2.4. Sunset above the tetrahedron in Bottrop. It was designed by the architect Wolfgang Christ (Bauhaus University Weimar) and errected in 1995 with a viewing platform [Photo: H. Wesemüller-Kock]

to the upper cone. These two cuts may be identical, $A = B$, without implying the identity of *all* cuts. Only a further slice at another height resulting in two cuts C and D and the use of the law of transitivity would lead from $A = B$ and $C = D$ to $A = C$ and would show that the cone actually is a cylinder. Democritus's arguments are therefore of a non-rigorous, physical type, while the arguments of Archimedes would be of a mathematical rigorous type.

However it is only a small step now to assume that Democritus already had the principle of Cavalieri (Bonaventura Cavalieri (1598–1647)) at his disposal:

Principle of Cavalieri: *Suppose two solids are included between two parallel planes. If every plane parallel to these two planes intersects both regions in cross sections of equal area, then the two regions have equal volumes.*

It is evident that Democritus could also have seen that a prism with triangular base area A can be dissected into three tetrahedra of equal size and that, following the principle of Cavalieri, the volumes of the three pyramids would exactly match the volume of the prism. The transfer of this argument to the cone would also have been fairly easy for an atomist like Democritus [Heath 1981, Vol. I, p. 180]; he certainly would have argued that the cone could be constructed from the tetrahedron by infinite addition of lateral surfaces.

2.2.4 The Paradoxes of Zeno

What is the role played by Zeno? Remember that he wanted to defend his teacher Parmenides and that he wanted to show that all movement is nothing but an illusion. Aristotle reports four paradoxes of Zeno which we will now start to discuss.

The most widely known paradox is that of Achilles and the tortoise but the dichotomy, the flying arrow, and the stadium have become immortal through the reports of Aristotle.

Achilles and the tortoise: The fast runner Achilles is asked to compete against a tortoise. Since Achilles is much faster than the tortoise the latter gets a considerable lead. Now Zeno states: *Achilles will never catch up with the tortoise!* He reasons as follows: when Achilles will be at the starting point of the tortoise the latter will be a short distance in front of him. When Achilles reaches this point the tortoise is again a (very) short distance in front of him, and so on. To illustrate the argument let us assume that the tortoise gets a lead of 10m and that Achilles is 10 times faster than the tortoise. Achilles needs only 1 second to run a distance of 10m. If the competition has started Achilles will be in the starting position of the tortoise after 1 second. During this second the tortoise is 1m ahead of Achilles. Now Achilles has to run 1m to arrive at the position of the tortoise (he needs only 1/10 seconds for this), but then the latter is still 10cm in front of him. When Achilles has covered the 10cm the tortoise will still be 1cm ahead of him, and so on ad infinitum.

The dichotomy: Zeno states: one can not move from a point A to a point B, $A \neq B$. For to get from A to B one has to cover half the distance first. To cover half of the distance one has to cover a quarter of the distance; to cover a quarter one has to cover one eighth of the distance, and so on. Hence an infinity of distances has to be covered to come from A to B and this is not possible in finite time. Therefore movement is impossible.

The flying arrow: Following Zeno an arrow shot from a bow does not fly. Suppose the arrow is flying and freeze time at a certain point during the flight. At this point of time the tip of the arrow is at a fixed point of space and its velocity is zero (because at this point of time the arrow is fixed). Since this point of time can be chosen anywhere during the flight of the arrow the arrow has everywhere nil velocity. Hence the arrow does not fly at all.

The stadium: Imagine being an observer of an ancient chariot race between two chariots manned by 8 persons each, cp. figure 2.2.6. One of the chariots is manned by persons B, the other with persons C, while further 8 persons A are watching from a fixed stand. The B-chariot moves to the right while the C-chariot moves to the left with the same velocity. When the B-chariot has travelled one A-position then B and C-chariot have travelled two positions!

Following Zeno this means, however, that half of the elapsed time is equal to the elapsed time and this contradiction again implies that movement is not possible at all.

Now one can immediately argue against the paradox of the stadium that Zeno obviously was not aware of the notion of relative velocity which would resolve the paradox. The paradox of Achilles and the tortoise is also not a real problem for modern mathematicians since the whole distance travelled (in meters) by Achilles is

$$10 + 1 + 0.1 + 0.01 + \ldots = 10 + \sum_{k=0}^{\infty} \left(\frac{1}{10}\right)^k$$

and this infinite series is a convergent geometric series achieving the value of 10/9. The distance $10 + 10/9 = 11.11111\ldots$ is exactly the distance at which Achilles would overtake the tortoise. But this is not the point here! Firstly we are using here knowledge of the 19th century, and secondly such arguments fail to explain the actual problem. Let us concentrate on Achilles and the tortoise for the sake of illustration: the actual question raised by Zeno is the question concerning the structure of space (in this case the structure of the racecourse) and time. If we take the real numbers as a basis the paradox can be resolved, but who tells us that the *real* space and the *real* time can in fact be modelled by real numbers? In an essay of the year 1992 the late Jochen Höppner [Höppner 1992, p. 59–69] has examined this

Fig. 2.2.5. Stadium in Delphi. The stadium is also an ancient measure of length of 600 feet. Depending on the local measure of a foot this is between 165 and 195m (Olympia 192.28m, Delphi 177.35m) [Photo: J. Mars]

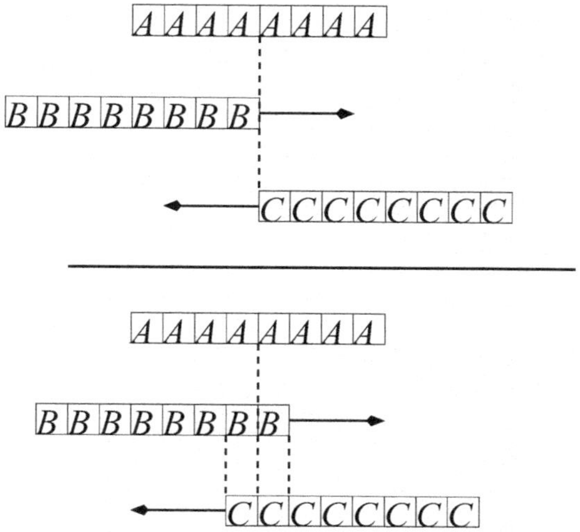

Fig. 2.2.6. Concerning the paradox of the stadium

subtle point in detail and has taken other structures as the basis of Achilles's racecourse than the real numbers. Besides an imagined Minkowski space, a probabilistic racecourse, and a representation with obstacles Höppner also examines the Cantor set as a racecourse. The Cantor set is the set of those numbers between 0 and 1 which have representations in the ternary numeral system (i.e. only with digits $0, 1, 2$) where no digit 1 appears. This set is often called 'Cantor dust'. It can be constructed by recursively removing the middle thirds starting with $[0, 1]$. The Cantor set has length zero but still contains denumerable points, hence it is a set 'as large as' the starting set $[0, 1]$! On this set neither Achilles nor the tortoise can move at all – as Höppner writes: *'they sink irrecoverably in the Cantor dust'* (... versinken sie unrettbar im Cantor-Staub).

If we now look at the paradoxes of Zeno from the point of view of the difference between atomism and the theory of the continuum then we recognise two groups of paradox: *Achilles and the tortoise* and the *dichotomy* are aimed against the assumption of a continuum and show that serious problems occur if a continuum is assumed (i.e. arbitrary divisibility). The *flying arrow* and the *stadion* are aimed against the assumption of atomism and show that this assumption also leads to severe problems. Imagining time being build up from atoms, hence as a collection of *points* of time, then, as Zeno says, we can look at the arrow in one of these points of time and we find it standing still. This observation does not seem to be in accordance with the movement of the arrow! Imagining on the other hand the racecourse in the paradox of Achilles and the tortoise being a continuum the runner would have to pass an infinity of parts of that continuum getting progressively smaller. This, says Zeno, can not be done in finite time.

Fig. 2.2.7. Marble bust of Aristotle (National Museum Rome). Roman copy after the Greek bronze original by Lysippos (330 BC). The alabaster mantle was added in modern times [Photo: Jastrow 2006]

Translating this into the language of analysis we stand here at the cradle of two different views having an effect even today. The great Leibniz (1646–1716) will turn out to be a mathematician of the infinitesimal and he therefore is not troubled by computing with infinitely small quantities. Great Isaac Newton (1643–1727, 1642–1726 old style) became inclined to atomism in his thoughts on physics and even thought about light in an atomistic way.

How did the fundamental ideas of the continuum and of atomism find its way into the Western cultural hemisphere? We owe this to Aristotle, to his translators, and to the overwhelming interest of medieval Christian scholastic philosophers in Aristotle. We will have to report on that later.

Zeno and his paradoxes are discussed controversially even now. The great English mathematician and philosopher Bertrand Russell (1872–1970) got so excited about the ideas of Zeno that he saw Zeno as a precursor of the mathematics of the 19th century, in particular of Karl Weierstraß [Russell 1903, p. 346ff.]. This certainly drives things too far. At the other end of the spectrum stands Baertel van der Waerden who wanted to marginalise the role of Zeno. He argued in [van der Waerden 1940, p. 141ff.] that atomism developed only after Zeno and as a counter-reaction against the Eleatics. Furthermore he claimed that the Pythagoreans never showed a verifiable interest in infinitesimal methods. This opinion also takes it too far in the other direction.

It is certain that the question concerning the structure of a straight line lies at the root of analysis and that all researchers like Newton and Leibniz and their successors were influenced by it. However, the Aristotelian continuum *cannot* be reconciled with our recent set theoretic opinion of the real line being a 'continuum'. We also have to remark that the continuum in Aristotle's writings is always closely linked to the problem of *movement*, cp. [Wieland 1965]. It is this thinking about the nature of movement which will bring the question of the continuum into Christian scholastics.

2.3 Archimedes

As splendid as the mathematics of the ancient Greeks may seem; the star outshining everything else – a universal genius – was Archimedes (about 287–212 BC). His domain was the town of Syracuse on Sicily which then belonged to the Greek realm. Probably Archimedes was even born in Syracuse.

2.3.1 Life, Death, and Anecdotes

We know disconcertingly little about the life of this genius. However, some anecdotes have come to us where we have to be very cautious concerning their truthfulness. When he discovered the law of the lever he is said to have stated: 'Give me a place to stand on, and I will move the Earth' (quoted by Pappus of Alexandria). Even more famous is the story concerning the crown of King Hiero II (about 306–215 BC). Archimedes was situated at the court of this king and probably even his relative. Hiero II is said to have ordered a second crown as an exact copy of the original one and, although both crowns were of the same weight, was wondering whether the goldsmith had betrayed him concerning the mass of gold in the copy. Archimedes was assigned to examine the case. In order to allow for relaxed thinking he went to a bathhouse and laid down in the warm water.

In the tub all of a sudden the idea of the 'Archimedean principle' is said to have come to him: every body displaces exactly as much water as it has volume. He immediately jumped out of the bath and ran home naked, crying 'Eureka! Eureka!' ('I have found [it]!'). The Roman architect and engineer Marcus Vitruvius Pollio (Vitruvius) (1st c BC) describes this event in [Vitruvius 1914, Book IX, p. 253f.]:

> *9. In the case of Archimedes, although he made many wonderful discoveries of diverse kinds, yet of them all, the following, which I shall relate, seems to have been the result of a boundless ingenuity. Hiero, after gaining the royal power in Syracuse, resolved, as a*

consequence of his successful exploits, to place in a certain temple a golden crown which he had vowed to the immortal gods. He contracted for its making at a fixed price, and weighed out a precise amount of gold to the contractor. At the appointed time the latter delivered to the king's satisfaction an exquisitely finished piece of handiwork, and it appeared that in weight the crown corresponded precisely to what the gold had weighed.

10. But afterwards a charge was made that gold had been abstracted and an equivalent weight of silver had been added in the manufacture of the crown. Hiero, thinking it an outrage that he had been tricked, and yet not knowing how to detect the theft, requested Archimedes to consider the matter. The latter, while the case was still on his mind, happened to go to the bath, and on getting into a tub observed that the more his body sank into it the more water ran out over the tub. As this pointed out the way to explain the case in question, without a moment's delay, and transported with joy, he jumped out of the tub and rushed home naked, crying with a loud voice that he had found what he was seeking; for as he ran he shouted repeatedly in Greek, 'Ευρηκα, ευρηκα'.

11. Taking this as the beginning of his discovery, it is said that he made two masses of the same weight as the crown, one of gold and the other of silver. After making them, he filled a large vessel with water to the very brim, and dropped the mass of silver into it. As much water ran out as was equal in bulk to that of the silver sunk in the vessel. Then, taking out the mass, he poured back the lost quantity of water, using a pint measure, until it was level with the brim as it had been before. Thus he found the weight of silver corresponding to a definite quantity of water.

12. After this experiment, he likewise dropped the mass of gold into the full vessel and, on taking it out and measuring as before, found that not so much water was lost, but a smaller quantity: namely, as much less as a mass of gold lacks in bulk compared to a mass of silver of the same weight. Finally, filling the vessel again and dropping the crown itself into the same quantity of water, he found that more water ran over for the crown than for the mass of gold of the same weight. Hence, reasoning from the fact that more water was lost in the case of the crown than in that of the mass, he detected the mixing of silver with the gold, and made the theft of the contractor perfectly clear.

And also further Archimedean inventions are concerned with water. Even today the Archimedean screw shown in the left part of figure 2.3.2 is used to

Fig. 2.3.1. Archimedes [Oil painting by Domenico Fetti, 1620] (Gemäldegalerie Alter Meister, Staatliche Kunstsammlungen Dresden)

pump water from a lower reservoire to a higher level, for example on rice fields in Asia. One can study these screws even on modern playgrounds where they appear in form of tubes bent into screwshaped form as shown in the right part of figure 2.3.2.

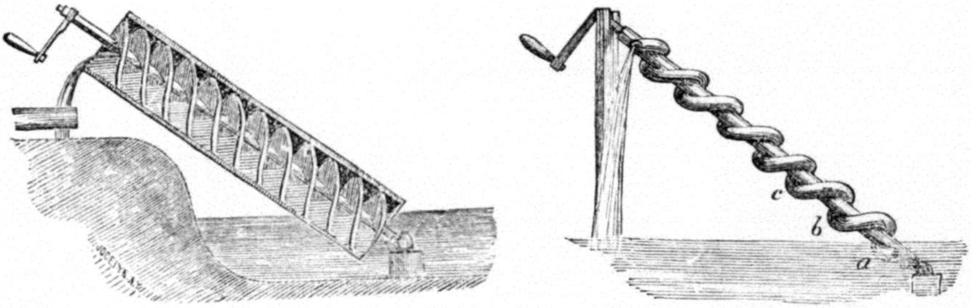

Fig. 2.3.2. Archimedian screw [*Chambers's Encyclopedia Vol. I.* Philadelphia: J.B. Lippincott & Co. 1871, S. 374]

The most famous of all Archimedean anecdotes is woven around his death. Our knowledge comes firstly from Plutarch (about 45–about 125) and secondly from Titus Livius (Livy) (about 59 BC–about AD 17). Plutarch in [Plutarch 2004, p. 437–523] describes the life of the Roman consul and general Marcus Claudius Marcellus, called Marcellus (about 268–208 BC) who besieged Syracuse with his troops from land and sea in 214 BC. The siege happened during the Second Punic War (218–201 BC) and was directed agains the Carthaginians. Although King Hiero II was a supporter of the Romans until his death his successor turned to the side of the Carthaginians what made Syracuse a target for Roman attacks. Plutarch reports about an artillery on eight galleys bound together with which Marcellus attacked. But [Plutarch 2004, p. 471]:

> *'... all this proved to be of no account in the eyes of Archimedes and in comparison with the engines of Archimedes. To these he had by no means devoted himself as work worthy of his serious effort, but most of them were mere accessories of a geometry practised for amusement, since in the bygone days Hiero the king had eagerly desired and at last persuaded him to turn this art somewhat from abstract notions to material things, and by applying his philosophy somehow to the needs which make themselves felt, to render it more evident to the common mind.'*

Marcellus and his troops now had to experience the war machines of Archimedes at first hand and also here the genius of Archimedes shone. Besides the law of the lever Archimedes employed pulleys and therewith built machines never seen before. These machines made it difficult for the Romans to conquer Syracuse. A pulley allowed a claw tied to a long rope to be brought under the bow of a ship; the ship was lifted up and then dropped back so that it broke. Archimedes is said to even have constructed parabolic mirrors with which Roman ships could be set to fire. Livy reports [Livy 1940, Book XXIV, p. 283ff.]:

Fig. 2.3.3. Archimedes's contribution to the defence of Syracuse, collage of modern depictions (among others of the Renaissance). The lack of authentic drawings caused artists to produce fantasy images

'And an undertaking begun with so vigorous an assault would have met with success if one man had not been at Syracuse at that time. It was Archimedes, an unrivalled observer of the heavens and the stars, more remarkable, however, as inventor and contriver of artillery and engines of war, by which the least pains he frustrated whatever the enemy undertook with vast efforts. The walls, carried along uneven hills, mainly high positions and difficult to approach, but some of them low and accessible from level ground, were equipped by him with every kind of artillery, as seemed suited to each place. The wall of Achradina, which, as has been said already, is washed by the sea, was attacked by Marcellus with sixty five-bankers. From most of the ships archers and slingers, also light-armed troops, whose weapon is difficult for the inexpert to return, allowed hardly anyone to stand on the wall without being wounded; and these men kept their ships at a distance from the wall, since range is needed for missile weapons. Other five-bankers, paired together, with the inner oars removed, so that side was brought close to side, were propelled by the outer banks of oars like a single ship, and carried towers of several stories and in addition engines for battering walls. To meet this naval equipment Archimedes disposed artillery of different sizes on the walls. Against ships at a distance he kept discharging stones of great weight; nearer vessels he would attack with lighter and all the more numerous missile weapons. Finally, that his own men might discharge their bolts at the enemy without exposures to wounds, he opened the wall from bottom to top with numerous loopholes about a cubit wide, and through these some, without being seen, shot at the enemy with arrows, others from small scorpions. As for the ships which came closer, in order to be inside the range of his artillery, against these an iron grapnel, fastened to a stout chain, would be thrown on to the bow by means of a swing-beam projecting over the wall. When this sprung backward to the ground owing to the shifting of a heavy leaden weight, it would set the ship on its stern, bow in the air. Then, suddenly released, it would dash the ship, falling, as it were, from the wall, into the sea, to the great alarm of the sailors, and with the result that, even if she fell upright, she would take considerable water. Thus the assault from the sea was baffled, and all hope shifted to a plan to attack from the land with all their forces. But that side also had been provided with the same complete equipment of artillery, at the expense and the pains of Hiero during many years, by the unrivalled art of Archimedes.'

Finally Syracuse fell after a siege of two years. Now Marcellus wanted to talk to Archimedes whom he had learned to admire. The soldier ordered to fetch Archimedes found him absorbed in thoughts over a drawing. Archimedes refused to go with the soldier until he had finished a certain proof and was thereupon slayed with the sword by the angry soldier. But Plutarch even

Fig. 2.3.4. Copperplate on the title page of the Latin edition of the *Thesaurus opticus* by Alhazen (Ibn Al-Haytham). Archimedes sets fire to Roman ships by mean of parabolic mirrors

gave two further versions of this murder. In the second version a soldier is said to have killed him immediately while in the third version Archimedes wanted to follow the soldier to Marcellus but wanted to take along with him some of his mechanical models to present them to Marcellus. The soldier panicked because he never saw such models before and assumed that they were weapons which Archimedes could use against him; therefore he killed Archimedes. When Marcellus heard about the death of Archimedes he was grief-stricken and turned his back to the murderer. Livy writes [Livy 1940, Book XXV, p. 461]:

'While many shameful examples of anger and many of greed were being given, the tradition is that Archimedes, in all the uproar which the alarm of a captured city could produce in the midst of plundering soldiers dashing about, was intent upon the figures which he had traced in the dust and was slain by a soldier, not knowing who he was; that Marcellus was grieved at this, and his burial duly provided for; and that his name and memory were an honour and a protection to his relatives, search even being made for them.'

Many artists took up the death of Archimedes as a motif. Figure 2.3.5 shows a mosaic from the Städtische Galerie Liebieghaus in Frankfurt am Main. It was long assumed that it dates back to antiquity but today experts assume it being either a forgery or a copy from the 18th century.

It belongs to the treasure trove of anecdotes that the last sentence spoken by Archimedes before the sword of the soldier pierced him was 'Noli turbare circulos meos' (Don't disturb my circles). However, neither Plutarch nor Livy noted such a sentence. Only Valerius Maximus, a Latin writer of the first century AD, let Archimedes say: 'Noli obsecro istum disturbar' (Please do not disturb this), cp. [Stein 1999, p. 3]. In the 12th century this turns into 'Lad, stay away from my drawing' (Bursche, bleib' von meiner Zeichnung weg). Hence we have to dismiss this sentence to the realm of fantasy.

Fig. 2.3.5. Death of Archimedes (Mosaic Städtische Galerie Frankfurt a. M.)

Without exaggeration it can be said that Archimedes certainly was the greatest engineer and physicist of antiquity; but he is a giant when it comes to mathematics and analysis in particular. However, mankind had pure luck that writings of Archimedes have come upon us at all!

2.3.2 The Fate of Archimedes's Writings

In an unprecedented bloodshed in April 1204 the city of Constantinople fell and went down. Christian crusaders who actually wanted to 'release' Jerusalem misappropriated the most radiant European town; they defiled the Hagia Sophia, plundered, pillaged, raped, and – they destroyed and displaced books which had been collected in Constantinople for centuries. Among them there were also three books by Archimedes; the so-called codices A, B, and C.

Codices A and B found their way to Sicily but after the battle of Benevento in 1266 they were sold to the pope [Dijksterhuis 1987, p. 37]. Codex B was mentioned for the last time in 1311. After that codex A disappeared; in 1491 it was in the possession of the Italian humanist Giorgo Valla. After his death is was bought by the Prince of Capri, Alberto Pio, then went into possession of his nephew Cardinal Rodolfo Pio in 1550 who died in 1564. After that year the trace of codex A vanished. The Renaissance masters draw their knowledge of the Archimedean writings from codices A and B. Only codex C remained lost. Already with the writings contained in codices A and B Archimedes could be identified as a great mathematician and physicist, but it is codex C that catapulted Archimedes into the heaven of immortals and gave him a place of honour at the side of Newton and Leibniz. The history of codex C is a crime story – no, a thriller – which Arthur Conan Doyle could not have thought up better. The story is described in [Netz/Noel 2007] and we want to follow it in their main features. In the summer holiday of the year 1906 the Danish philologist Johan Ludvig Heiberg (1854–1928) travelled to Constantinople to examine a strange manuscript in the Metochion (ecclesiastical embassy church). Before that he got the information about a palimpsest from a catalogue of 1899 which immediately enthralled him. Palimpsests are parchments – tanned goatskin – which are reused after they had been already written on. Since parchment was an expensive raw material it takes no wonder that authors and writers fell back to already inscribed older parchments. They scraped the old inscription, cut the parchment into a new format, and inscribed it again with their writings. The author of the aforementioned catalogue, a certain Papadopoulos-Kerameus, did not enjoy a permanent position but was paid according to the number of pages he wrote for the catalogue. Therefore he delivered quite extensive descriptions. He not only described the new text on the palimpsest but also the imperfectly scraped parts of the original parchment which he could still read. Philologist

Fig. 2.3.6. Manuscript from the Archimedes palimpsest [Auction catalogue of Christie's, New York 1998]

Heiberg immediately recognised that the original text on the parchment could only have been a writing of Archimedes. Attempts to get the palimpsest to Copenhagen by the help of diplomatic channels failed so that the scholar had to undertake the task of travelling with it himself. In Constantinople Heiberg could meet his greatest hopes: he had located the lost codex C! The New York Times headlined on the 16th July 1907: 'Big Literary Find in Constantinople – Savant Discovers Books by Archimedes, Copied about 900 A. D.'. The whole newspaper article is reproduced in [Stein 1999, p. 28]. Already from older translations of writings of Archimedes his genius could be seen but how he came up with his mathematical theorems he left in the dark. The palimpsest now contained a letter by Archimedes to his friend Eratosthenes of Cyrene. This letter became falsely[4] known as the *The Method of Archimedes Treating of Mechanical Theorems* in which the master explained how he derived his theorems – by means of an ingenious method of indivisibles which we will have to discuss in some detail.

Heiberg deciphered the palimpsest as good as it was possible by naked eye and magnifying glass. He published a translation of the *Method* in a scientific journal and compiled a completely new edition of the works of Archimedes between 1910 and 1915 – based on the codices A and B which are extinct today and codex C which he just had found again. This new edition by Heiberg became the basis of the English translation by Sir Thomas Heath (1861–1940) [Heath 2002] which made the works of Archimedes internationally known.

The palimpsest contained seven more or less complete works:

1. *On the equilibrium of planes or the centres of gravity of planes,*
2. *On floating bodies,*
3. *The method,*
4. *On spirals,*
5. *On the sphere and cylinder,*
6. *Measurement of a circle,*
7. *Stomachion* (A fragment on a tangram-like game).

Three further books have been preserved from other sources, transcriptions, and extracts of codices A and B, respectively:

1. *Quadrature of the parabola,*
2. *The sand-reckoner,*
3. *On conoids and speroids.*

[4] As Eberhard Knobloch, himself being a renowned philologist, told me there is no Greek word in the title meaning 'method' but 'access' instead, [Knobloch 2010]. However, it is too late to change the title – the work is known as *The method* all over the world.

Fig. 2.3.7. Eratosthenes of Cyrene

Except of the *Stomachion* which is of no interest to us these works comprise all of the works of Archimedes concerning mathematics and physics we know today. They can be found in [Heath 2002].

The history of codex C is not finished here, however. In 1938 the manuscripts and books of the Metochion were removed to Athens under the eyes of the Turks; however, codex C was not among the lot. Research described in [Netz/Noel 2007] revealed that the palimpsest found its way into a private collection of a French collector. After his death in 1956 his daughter who inherited the palimpset became interested in it in the 1960s. Around 1970 this daughter seemed to have realised the importance of the manuscript in her possession because she was looking at some of the pages being cleared from fungal infestation. She unsuccessfully tried to sell the palimpset for quite some time until it appeared at an auction at Christie's in New York in the year 1998. The estimated price was given as 800 000 US Dollars. The Greek ministry of education and cultural affairs was one of the bidders but there was also a middleman of an unknown bidder. Eventually the Greek ministry had to drop out and the palimpset was sold for the unbelievable sum of 2 200 000 US Dollars to the great unknown.

However, the palimpset had suffered a lot since the days of Heiberg. There was fungal infestations all over and damages from humidity so serious that even the parts which could be read by Heiberg by the naked eye were devastatingly damaged.

The unknown buyer remains unknown even to this day (Bill Gates has credibly declared that it is not him), but fortunately he offered the palimpset for use in science. It is now on loan in the Walters Art Museum in Baltimore where it is conserved and examined by means of the latest methods in image processing. I can only strongly recommend the web page `http://www.archimedespalimpsest.org` which was built accompanying the works on the palimpset. The discovery of the palimpset and its first publication by Heiberg as well as the recent results of the research group in Baltimore concerning the resurfaced palimpset have clearly shown Archimedes's important role in the history of analysis. Now is the time for us to present some of his works.

2.3.3 The Method: Access with Regard to Mechanical Theorems

Following Heiberg *The Method* of Archimedes was called *The Method of Archimedes Treating of Mechanical Problems* but we have already remarked that the word 'method' does not actually appear in the title. Instead, one should rename this work of Archimedes *The Access* instead of calling it *The Method.* Hence the correct title would be [Knobloch 2010]

> *Access with regard to mechanical theorems*
>
> (Zugang hinsichtlich der mechanischen Sätze).

The *Access* in the edition of Heath begins with the words [Heath 2002, p. 12, Appendix after p. 326]:

> *'Archimedes to Eratosthenes greeting.*
>
> *I sent you on a former occasion some of the theorems discovered by me, merely writing out the enunciations and inviting you to discover the proofs, which at the moment I did not give.'*

Then he starts off telling Eratosthenes that he is going to deliver the proofs in the following. At the beginning he repeats some theorems on the centre of gravity for which he had given proofs already in his work *On the equilibrium of planes or the centres of gravity of planes* [Heath 2002, p. 189ff.]. In this work particularly we find the law of the lever which is derived in a purely axiomatic manner by Archimedes. The whole theory of the lever rests on only three axioms [Heath 2002, p. 189]:

1. Equal weights at equal distances are in equilibrium, and equal weights at unequal distances are not in equilibrium but incline towards the weight which is at the greater distance.
2. If, when weights at certain distances are in equilibrium, something be added to one of the weights, they are not in equilibrium but incline towards that weight to which the addition was made.
3. Similarly, if anything be taken away from one of the weights, they are not in equilibrium but incline towards the weight from which nothing was taken.

Eventually Archimedes proves the law of the lever in Proposition 6 and 7 [Heath 2002, p. 192]. In our words:

A large weight G being a distance D away from the pivot of the lever is in equilibrium with a smaller weight g at distance d from the pivot, if

$$D : d = g : G \tag{2.2}$$

holds.

With the help of this mechanical method Archimedes now goes on and weighs indivisibles!

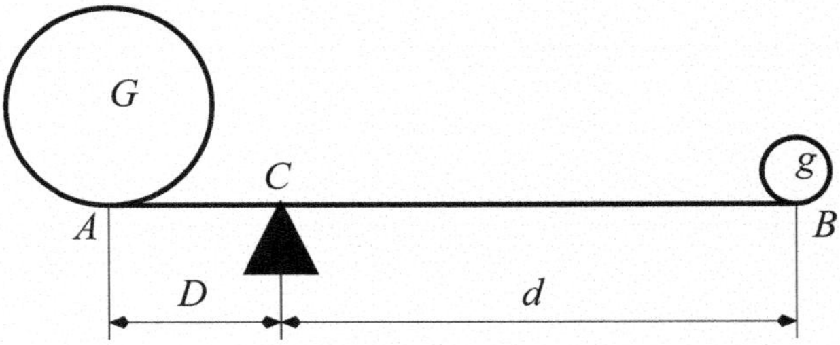

Fig. 2.3.8. The law of the lever

Weighing the Area Under a Parabola

To illustrate Archimedes's 'Access' we study his computation of the area of a parabolic segment as shown in figure 2.3.9(a). Archimedes even considered a parabolic segment lying arbitrarily in the plane but the simpler case suffices us. In the parabolic segment we place a triangle ABC where BD is the axis of symmetry.

Drawing the tangent to the parabola at point C and erecting the perpendicular in A we denote the point of intersection of tangent and perpendicular

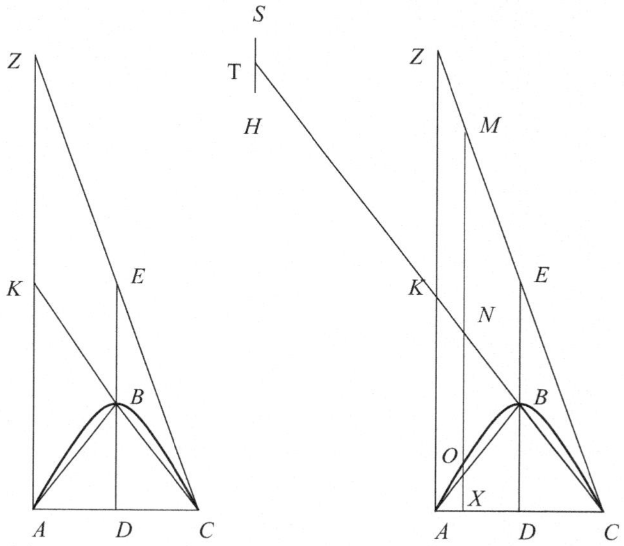

(a) Auxiliary construction (b) The process of weighing

Fig. 2.3.9. Weighing a parabolic segment

by Z. Extending the segment BC gives the point K while the extension of DB results in point E as shown in figure 2.3.9(a). Archimedes proved and employed the following facts we will take for granted:

1. K lies exactly in the middle of AZ.
2. B lies exactly in the middle of DE.
3. The area of the triangle AKC is just half of the area of triangle AZC.
4. B lies exactly in the middle of KC.
5. The area of the triangle ABC is just half of the area of triangle AKC.

From this fact it follows that the area of the triangle ACZ is exactly four times as large as the area of triangle ABC. Now Archimedes used a particular property of the parabola. If we draw a line parallel to to BD, say MX in figure 2.3.9(b), then

$$\frac{MX}{OX} = \frac{AC}{AX} \tag{2.3}$$

always holds. We do not prove this property but refer the reader to the explanations in [Stein 1999].

We now extend the segment CK to the point T which is defined by the fact that the segments KT and CK are of equal length. This is our lever or balance beam and the point K is the pivot point. Then we shift the segment OX into the point T so that SH runs through T. Note that SH and OX share the same length. *We have thus shifted the indivisible OX which is a part of the parabolic segment to the other side of the lever.* Now it follows from the intercept theorem

$$\frac{AC}{AX} = \frac{KC}{KN}$$

and together with (2.3) this becomes

$$\frac{MX}{OX} = \frac{KC}{KN}.$$

Since $TK = KC$ we also have

$$\frac{MX}{OX} = \frac{TK}{KN}.$$

But HS is nothing more than the segment OX shifted to T, hence it also has to hold

$$\frac{MX}{SH} = \frac{TK}{KN}. \tag{2.4}$$

And this is the actual outrageous! Archimedes treats the *line segment MX* and SH as *weights* which are fixed at distances TK and KN, respectively, from the pivot. Their respective 'weights' are proportional to their lengths. Equation (2.4) obviously is nothing else than the law of the lever! The two *segments* MX and SH are obviously in balance on our lever. Since we have not stated any particular condition concerning the point X this balance holds for all segments MX and $SH = OX$, regardless of where the point X is chosen between A and C. Since $SH = OX$ is an indivisible of the parabola and MX an indivisible of the triangle ACZ the area of the parabola has to be in balance with the area of the triangle if they are located at their respective distances from the pivot point. If we imagine the whole triangle shifted, the centre of gravity of the parabolic segment is located in the point T. But where is the centre of gravity of the triangle ACZ? it is located on the median KC in a distance of two thirds of K and Archimedes knew that, of course. But this means that the lever arm of the triangle is only one third as long as the lever arm TK of the parabola. Since triangle and parabola are in balance the area of the triangle has to be thrice as large as the area of the parabolic segment since the triangle 'weighs' thrice as much. Hence Archimedes arrived at the result:

> The area of the parabolic segment is one third of the area of the triangle ACZ.

It is now not difficult to see that the triangle ABC inscribed in the parabola is only one quarter as large as the triangle ACZ. This gives:

> The area of the parabolic segment is four thirds the area of the triangle ABC.

The Volume of a Paraboloid of Rotation

The method of 'weighing' indivisibles naturally works in the case of solids. We consider the simple parabola $y = x^2$ on a segment $[-a, a]$ of the x-axis and ask for the volume which arise when this parabola rotates around the y-axis. This solid obviously is a paraboloid of rotation of height a. As can be seen in figure 2.3.10 we place our paraboloid on the right side of a lever with pivot point A. The segments AH and AD are assumed to be of equal length. We imagine the paraboloid being enclosed in a circular cylinder with volume

$$\text{Vol(cylinder)} = \text{base area} \times \text{height}.$$

The height of the cylinder is AD, its base area $\pi \cdot BD^2$, and its centre of gravity is point K located exactly in the middle of AD. Our y-axis now points to the right since we have rotated the paraboloid by 90°. The segments from A up to the paraboloid are hence our y-values. Therefore

$$\frac{BD^2}{OS^2} = \frac{AD}{AS},$$

since AD is just the y-value of the parabola if the x-value is $BD = MS$. Then also

$$\frac{MS^2}{OS^2} = \frac{AD}{AS}$$

has to hold and hence

$$AS \cdot MS^2 = AD \cdot OS^2.$$

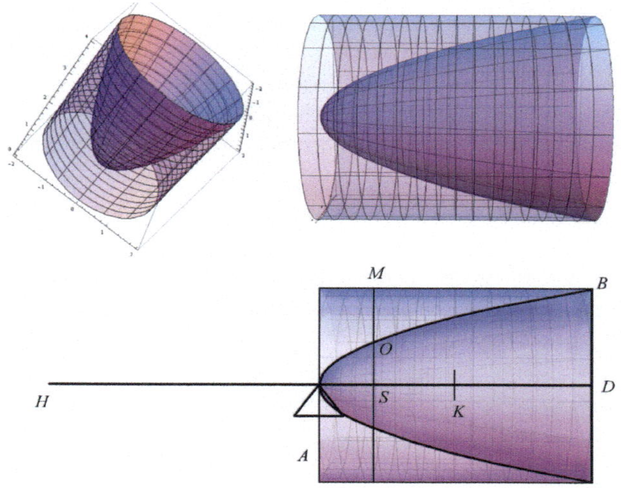

Fig. 2.3.10. The paraboloid in the cylinder on the lever

Since $AD = AH$ due to our premises we can also write

$$AS \cdot MS^2 = AH \cdot OS^2.$$

But we are working with solids and not with areas and our indivisibles are not lines but discs. Hence actually we have

$$AS \cdot (\pi \cdot MS^2) = AH \cdot (\pi \cdot OS^2),$$

i.e. the cross sections of the cylinder at S balance the cross sections of the paraboloid at H. If we assume (like Archimedes did) that a solid consists of indivisibles then we arrive at the equation

$$AH \cdot \text{Vol(paraboloid)} = AK \cdot \text{Vol(cylinder)}.$$

Now it is $AK = \frac{1}{2}AD$ and $AH = AD$ so that

$$AD \cdot \text{Vol(paraboloid)} = \frac{1}{2}AD \cdot \text{Vol(cylinder)}$$

follows and hence:

> The volume of the paraboloid of revolution is exactly half of the volume of the including cylinder.

2.3.4 The Quadrature of the Parabola by means of Exhaustion

We may justifiably assume that Archimedes did not accept the method of weighing indivisibles himself, cp. [Cuomo 2001]. The method simply was too outrageous. Therefore only classical proofs based on the double reductio ad absurdum are contained in the works of Archimedes. In his work *Quadrature of the parabola* he gave one further proof of the area of a parabolic segment which is fundamentally different from the one he gave in the *Method (Access)*. He employed a proof by exhaustion in that he filled the parabolic segment with triangles.

The first triangle is constructed as shown in figure 2.3.11. The parabolic segment is bounded by the chord AC. Let B be the point on the parabola at which the slope of the tangent equals the slope of the chord. Then ABC comprises the first triangle of the exhaustion. We construct further triangles following the same pattern. In the next step the triangle BCP appears together with its 'sister triangle' over the segment AB. The point P is defined to be the point on the parabola at which the slope of the tangent equals the slope of the segment BC.

Joining the point B with the midpoint D of the segment AC and drawing a parallel line to BD through P defines the points M and Y. A parallel line to AC through P defines the point N as shown in the right part of figure 2.3.11.

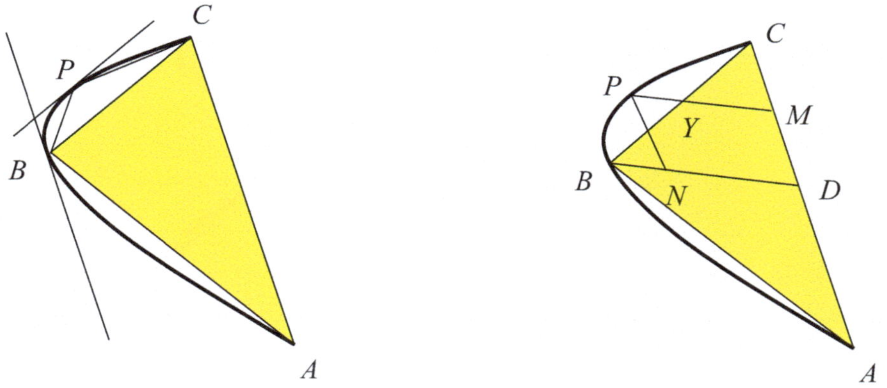

Fig. 2.3.11. Quadrature of a parabolic segment

M divides the segment CD halfway, hence $CD = 2PN$. Furthermore it holds

$$\frac{BD}{BN} = \frac{CD^2}{PN^2},$$

since the length of BD is just the function value $y = x^2$ at the location x which is the distance CD apart from the origin if we use a coordinate system. Inserting $CD = 2PN$ yields

$$\frac{BD}{BN} = \frac{4 \cdot PN^2}{PN^2} = 4,$$

hence $BD = 4 \cdot BN$. But it also holds $PM = 3 \cdot BN$ and $YM = \frac{1}{2}BD$ and this implies $YM = \frac{1}{2} \cdot 4 \cdot BN = 2 \cdot BN$ by insertion. Hence we have reached our first intermediate result:

$$YM = 2 \cdot PY.$$

Our aim is now to prove the relation

$$\text{Area}(BPC) = \frac{1}{4} \cdot \text{Area}(BDC)$$

between the triangles BPC and BDC. To do so we introduce the segment BM as shown in figure 2.3.12.

The point M is the midpoint of the segment CD and this yields

$$\text{Area}(BCD) = 2 \cdot \text{Area}(BMC).$$

Since $YM = 2 \cdot PY$ it follows $\text{Area}(BMC) = 2 \cdot \text{Area}(BPC)$, which immediately yields

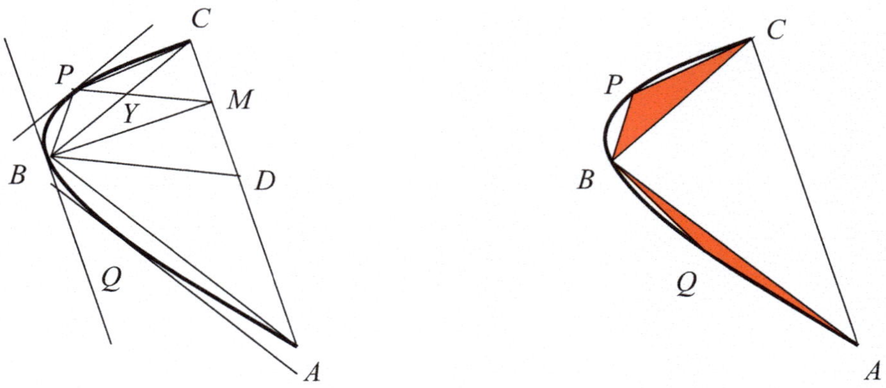

Fig. 2.3.12. Quadrature of a parabolic segment

$$\text{Area}(BCD) = 4 \cdot \text{Area}(BPC).$$

Repeating the same inspections for the other half of the parabolic segment it follows analogously:

$$\text{Area}(BAD) = 4 \cdot \text{Area}(BQA).$$

We now start to fill up (i.e. exhaust) the area with triangles. We know already from our earlier investigations that

$$\text{Area}(BPC) + \text{Area}(BQA) = \frac{1}{4} \cdot \text{Area}(ABC).$$

The areas of the two triangles inserted in the first step are shown in the right part of figure 2.3.12. In the second step four triangles have to be inserted. Their total area is $\frac{1}{4} \cdot (\text{Area}(BPC) + \text{Area}(BQA))$ or, in other words,

$$\frac{1}{4^2} \cdot \text{Area}(ABC).$$

In the third step eight triangles have to be inserted the total area of which is 1/4 of the total area of the four triangles inserted in the previous step. Their total area hence is

$$\frac{1}{4^3} \cdot \text{Area}(ABC).$$

In this way we proceed further so that after n steps we get a total area of

$$\left(1 + \frac{1}{4} + \frac{1}{4^2} + \frac{1}{4^3} + \ldots + \frac{1}{4^n}\right) \cdot \text{Area}(ABC).$$

The finite sum (and its limit for $n \to \infty$) is easy to compute by every contemporary student of mathematics and Archimedes also had no problems whatever to compute it. First of all he proved (cp. [Stein 1999, p. 45f.])

$$1 + \frac{1}{4} + \frac{1}{4^2} + \ldots + \frac{1}{4^n} + \frac{1}{3} \cdot \frac{1}{4^n} = \frac{4}{3}. \tag{2.5}$$

Since

$$\frac{1}{4^n} + \frac{1}{3} \cdot \frac{1}{4^n} = \frac{1}{4^n}\left(1 + \frac{1}{3}\right) = \frac{1}{4^n} \cdot \frac{4}{3} = \frac{1}{3} \cdot \frac{1}{4^{n-1}}$$

the sum (2.5) reduces to

$$1 + \frac{1}{4} + \frac{1}{4^2} + \ldots + \frac{1}{4^{n-1}} + \frac{1}{3} \cdot \frac{1}{4^{n-1}}$$

and now we can again reduce because it holds

$$\frac{1}{4^{n-1}} + \frac{1}{3} \cdot \frac{1}{4^{n-1}} = \frac{1}{4^{n-1}}\left(1 + \frac{1}{3}\right) = \frac{1}{4^{n-1}} \cdot \frac{4}{3}.$$

If we apply this reduction further we end up with

$$1 + \frac{1}{3} \cdot \frac{1}{4^0} = 1 + \frac{1}{3} = \frac{4}{3},$$

which proves the sum (2.5).

But Archimedes has not yet finished! He has proven:

$$\text{Area (parabolic segment)} \geq \frac{4}{3} \cdot \text{Area}(ABC),$$

but it might have happened that in the process of inserting triangles gaps remained the total area of which does not vanish. Now Archimedes considered the first step as shown in figure 2.3.13 and then further steps of exhaustion. The area of the triangle XYZ is exactly half of the area of the parallelogram drawn. Hence the area of XYZ is more than half of the area of the parabolic segment. The shortfall in area resulting from the next step (these are again smaller parabolic sections) is therefore smaller than half of the shortfall in the

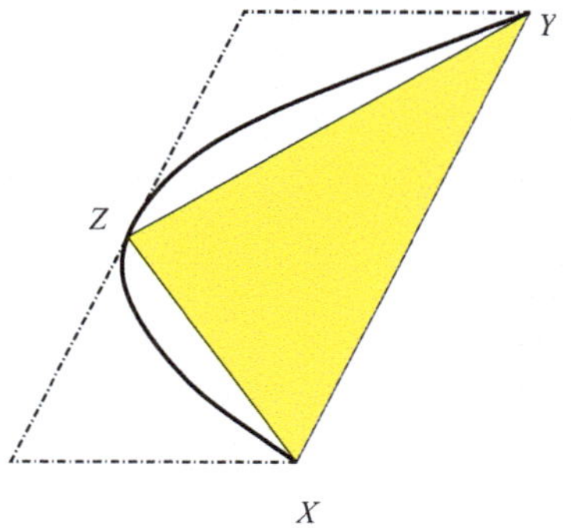

Fig. 2.3.13. Concerning the calculation of the shortfall in area

recent step. Hence the shortfall in area is decreasing with every exhaustion step and it finally follows:

$$\text{Area of the parabolic segment} = \frac{4}{3} \cdot \text{Area of the triangle } ABC.$$

2.3.5 On Spirals

In his work *On Spirals* Archimedes considered the spiral which now carries his name. In polar form it is given as

$$r = \frac{a}{2\pi}\varphi, \quad \varphi \geq 0.$$

The parameter $a > 0$ may be freely chosen and φ is the angle of the position vector which has length r depending on φ. For every angle φ we hence find the corresponding point of the spiral at the tip of the position vector. Archimedes computed the area under the spiral and we want to describe his solution of this quadrature problem in some detail.

Archimedes considers the spiral after one full turn, i.e. $r = \frac{a}{2\pi}\varphi, \varphi \in [0, 2\pi]$, as well as the circumscribing circle C as shown in the left part of figure 2.3.14. Circle C obviously has radius a since the length of the position vector after one full turn of the spiral ($\varphi = 2\pi$) is $r = a$. We denote the area of the circle with radius a by A and the area of any other circle with radius b by B. We know from (2.1) that the ratio of the areas of the circles equals the ratio of the squares of their radii, i.e.:

$$\frac{A}{B} = \frac{a^2}{b^2}.$$

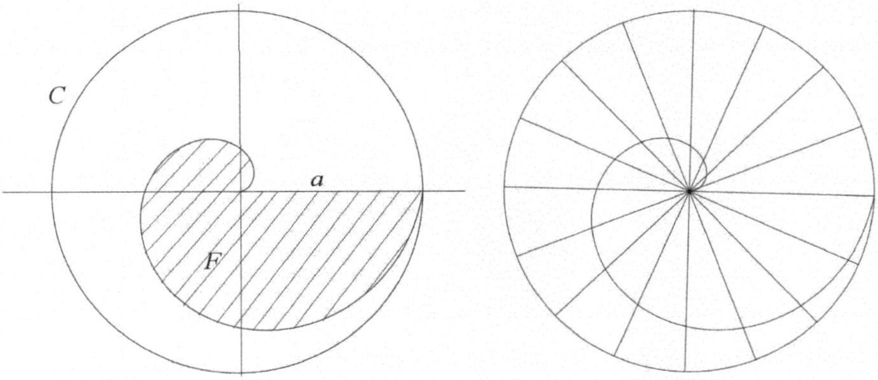

Fig. 2.3.14. Computing the area under the spiral. Left: the area of the spiral, right: the partition of the circle into n parts

We now partition circle C into n sectors and demonstrate this in the right part of figure 2.3.14 for the case of $n = 16$.

Every sector has an area of

$$A_{\text{sector}} = \frac{A}{n}.$$

In every sector the situation shown in figure 2.3.15(a) occurs. By A_k we denote the area of a circle with radius $\frac{k \cdot a}{n}$ which circumscribes the spiral segment in the k-th sector.

The angle φ_k as shown in figure 2.3.15(a) obviously is

$$\varphi_k = k \cdot \frac{2\pi}{n},$$

hence it follows for the radius l of the circumscribing circle

$$l = r(\varphi_k) = \frac{a}{2\pi} \cdot k \cdot \frac{2\pi}{n} = \frac{k \cdot a}{n}$$

and hence for its area

$$A_k = \pi \cdot l^2 = \pi \cdot \left(\frac{k \cdot a}{n}\right)^2.$$

This yields

$$\frac{A_k}{A} = \frac{\left(\frac{k \cdot a}{n}\right)^2}{a^2} = \frac{k^2}{n^2},$$

hence

$$A_k = \frac{k^2}{n^2} A,$$

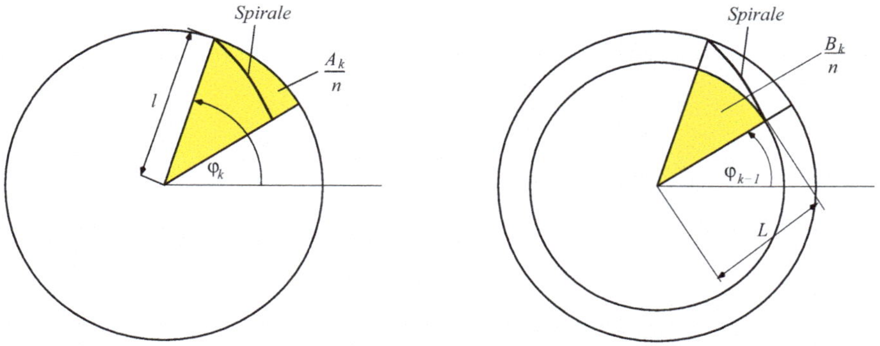

(a) Overestimating the segment area

(b) Underestimating the segment area

Fig. 2.3.15. Estimating the spiral segments in the k-th sector

or

$$\frac{1}{n}A_k = \frac{k^2}{n^3}A.$$

The sum of the areas of all sectors for all $k = 1, 2, \ldots, n$ then yields the value

$$\frac{1^2 + 2^2 + \ldots + n^2}{n^3} \cdot A. \tag{2.6}$$

At this point Archimedes is in need of the following result.

Lemma: *The inequalities*

$$1^2 + 2^2 + \ldots + (n-1)^2 < \frac{n^3}{3} < 1^2 + 2^2 + \ldots + (n-1)^2 + n^2,$$

hold.

The lemma is rigorously proven by Archimedes, of course. Applying this lemma to (2.6) we get

$$\frac{1^2 + 2^2 + \ldots + n^2}{n^3} \cdot A > \frac{n^3}{3n^3}A = \frac{1}{3}A.$$

We now consider the sector with regard to figure 2.3.15(b). Let us denote the area of the circle meeting the spiral in the k-th sector at the right edge (hence towards sector $k-1$) by B_k. The angle φ_{k-1} is obviously

$$\varphi_{k-1} = (k-1)\frac{2\pi}{n}$$

and hence it follows for the radius of the circle:

$$L = \frac{a}{2\pi}(k-1)\frac{2\pi}{n} = \frac{(k-1)\cdot a}{n}.$$

Therefore we arrive at

$$B_k = \pi \cdot L^2 = \pi \cdot \left(\frac{(k-1)\cdot a}{n}\right)^2.$$

Thereby it follows

$$\frac{B_k}{A} = \frac{\left(\frac{(k-1)\cdot a}{n}\right)^2}{a^2} = \frac{(k-1)^2}{n^2}$$

or

$$\frac{1}{n}B_k = \frac{(k-1)^2}{n^3} \cdot A.$$

The sum of all such sector areas from $k = 1$ to $k = n$ is therefore

$$\frac{1^2 + 2^2 + \ldots + (n-1)^2}{n^3} \cdot A. \tag{2.7}$$

We now apply the Lemma and get

$$\frac{1^2 + 2^2 + \ldots + (n-1)^2}{n^3} \cdot A < \frac{n^3}{3n^3} \cdot A = \frac{1}{3}A.$$

Evidently the area (2.6) is larger than the sought spiral area F and as well as evident the area (2.7) is smaller than F. Hence it holds

$$\frac{1^2 + 2^2 + \ldots + (n-1)^2}{n^3} \cdot A < F < \frac{1^2 + 2^2 + \ldots + n^2}{n^3} \cdot A. \tag{2.8}$$

We also have shown with our computation that

$$\frac{1^2 + 2^2 + \ldots + (n-1)^2}{n^3} \cdot A < \frac{1}{3} \cdot A < \frac{1^2 + 2^2 + \ldots + n^2}{n^3} \cdot A \tag{2.9}$$

holds. We may now proceed in two different ways: firstly the difference between over- and underestimated area is

$$\frac{1^2 + 2^2 + \ldots + n^2}{n^3} \cdot A - \frac{1^2 + 2^2 + \ldots + (n-1)^2}{n^3} \cdot A = \frac{1}{n} \cdot A$$

and the larger n the smaller will be the difference. In the limit (a notion unknown to Archimedes) $n \to \infty$ the difference will vanish. Hence this limit is just $\frac{1}{3}A$ as follows from (2.9) and hence it follows from (2.8) that

$$F = \frac{1}{3} \cdot A.$$

Secondly we could fall back on the formula for the sum of squares which can be found in every formulary. It holds

$$1^2 + 2^2 + \ldots + n^2 = \frac{2n^3 + 3n^2 + n}{6}$$

and

$$1^2 + 2^2 + \ldots + (n-1)^2 = \frac{2n^3 - 3n^2 + n}{6}.$$

Hence (2.9) can be written as

$$\frac{2n^3 - 3n^2 + n}{6n^3} \cdot A < \frac{1}{3} \cdot A < \frac{2n^3 + 3n^2 + n}{6n^3} \cdot A,$$

and after reducing the fractions by n^3 we get

$$\left(\frac{1}{3} - \frac{1}{2n} + \frac{1}{6n^2}\right) \cdot A < \frac{1}{3} \cdot A < \left(\frac{1}{3} + \frac{1}{2n} + \frac{1}{6n^2}\right) \cdot A.$$

If now n grows beyond limits the left and the right side converge towards $\frac{1}{3}A$ and we arrive again at the result that the spiral area F equals exactly one third of the area of the circle A.

Archimedes had to travel on a more laborious path to achieve this last step and employed his method of the double *reductio ad absurdum*.

2.3.6 Archimedes traps π

We have used the number π already a few times before in a naive manner; for example in the formula πr^2 giving the area of a circle of radius r. Archimedes has trapped π by means of compression and exhaustion with regular polygons applied to a circle in his *Measurement of a circle*. He could not be aware of the nature of this number being transcendent and irrational. *Measurement of a circle* starts with the proposition that the area of a circle with radius r corresponds to a right-angled triangle where one of the legs has length r and the length of the other is just the circumference U of the circle. Archimedes proves this again with a double *reductio ad absurdum* but in this case we can easily see how he may have come up with the idea underlying the proposition. To this end we partition a circle in different sectors as shown in figure 2.3.16.

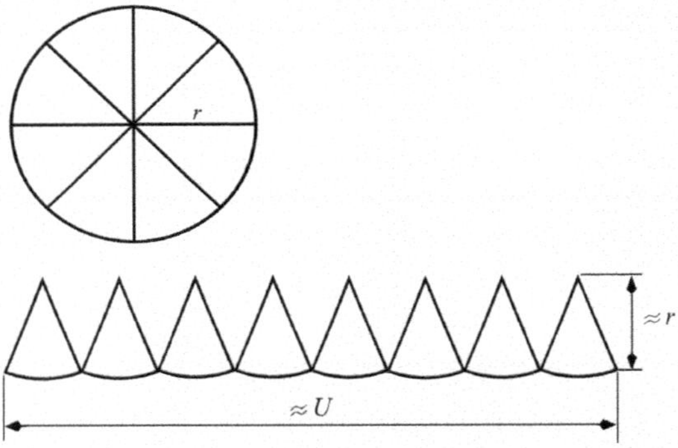

Fig. 2.3.16. Possible idea concerning the computation of the area of a circle

Lining up these sectors as shown in the figure we nearly get half of a rectangle if the number of segments is high enough. The half rectangle has an edge length of U and height r and this corresponds to a right-angled triangle with area

$$A = \frac{1}{2}U \cdot r.$$

To get an estimate of the number π Archimedes eventually worked with regular polygons with 96 corners and arrived at the estimates

$$3\frac{1}{7} > \frac{14688}{4673\frac{1}{2}} > \pi > \frac{6336}{2017\frac{1}{4}} > 3\frac{10}{71}.$$

It took several centuries until these estimates could be improved for the first time!

In the light of Archimedes's achievements this is a good place to take a look at the bloodcurdling nonsense written by modern authors concerning

π. The physician Edwin J. Goodman from Solitude, Posey County, in the State of Indiana petitioned a law concerning a 'new mathematical truth' on 18th January 1897 to the Indian House of Representatives. He offered his 'mathematical truth' to be used free of charge by the State of Indiana while all other users would have to pay licence fees. As Petr Beckmann in [Beckmann 1971, p. 174ff.] deduced from this petition it follows from Goodman's descriptions the 'true' value of

$$\pi = \frac{16}{\sqrt{3}} \approx 9.2376.$$

Thereby Goodman has given the largest known overestimation of the value of π although π does not occur explicitly in the petition! Since the offer seemed to be too tempting for the State of Indiana the petition passed all impediments until the last reading in parliament when accidentally the mathematician Clarence Abiathar Waldo (1852–1925) was present and stopped the nonsense. However, the petition was *not* officially rejected and there is some risk that it may easily be activated again.

Further examples of spirited men believing they had proven the 'true value' of π are the US-Americans John A. Parker and Carl Theodore Heisel. Parker publishes 1874 in New York a book with the title *The Quadrature of the Circle. Containing Demonstrations of the Errors of Geometers in Finding Approximations in Use.* In his attempt to square the circle he presented the value

$$\pi = \frac{20612}{6561} \approx 3.14159$$

as the exact value and he even 'proved' it. Carl Theodore Heisel lived in Cleveland, Ohio. In his book *Behold! The Grand Problem – The Circle Squared Beyond Refutation – No Longer Unsolved*, published in 1931, he presented to the world his greatest discovery, namely π being

$$\frac{256}{81} \approx 3.16049.$$

This value is exactly the one which Ahmes noted in ancient Egypt about 1700 BC!

Even as late as 1960 Heinz-Wilhelm Alten, editor of this book series, had by request of the Office of the Federal President (!) to inspect a paper by an amateur mathematician proposing he had found the 'true number π' (needless to say, a rational number). The amateur was a Jewish emigrant who claimed his discovery suppressed by the Nazis on racist grounds. Therefore his discovery had not found its way into German school books. He claimed that he not only had suffered a financial loss but also a health damage so that he asked the Federal President for 10000 German mark[5] to start a treatment in Israel.

[5] Note that the English pound was to the German mark as 1 to 11.8 back in 1960.

2.4 The Contributions of the Romans

We can be very brief here. The inhabitants of the Roman Empire which kept a tight hold of the Mediterranean region, built aqueducts and conquered large parts of the then-known world were not interested in mathematics at all. However, they created the Roman numerals with which the Central Europeans calculated until the 16th century. They are used even today as page numbers and can be seen on old watches and tombstones. But we know of no Roman contributions to analysis except one:

As Plutarch tells us in his *Life of Marcellus* [Plutarch 2004, p. 481] Archimedes had instructed his friends to put a sculpture on his grave showing a sphere and a cylinder, and, in form of an inscription, the formula of the ratio of their respective volumes. Apparently his wish was fulfilled. When the great Marcus Tullius Cicero (106–43 BC) was quaestor[6] in Sicily in the year 75 BC he searched for the tomb, found it in bad shape, and ordered to uncover and restore it.

Cicero himself reported himself in [Cicero 1877, p. 186f.]:

Fig. 2.4.1. Cicero discovers the tomb of Archimedes ([Painting by Benjamin West of the year 1797] Yale University Art Gallery, New Haven)

[6] A public official in ancient Rome; the lowest rank in the career finally becoming a senator.

> *'I will present you with an humble and obscure mathematician of the same city, called Archimedes, who lived many years after; whose tomb, overgrown with shrubs and briers, I in my quæstorship discovered, when the Syracusians knew nothing of it, and even denied that there was any such thing remaining; for I remembered some verses, which I had been informed were engraved on his monument, and these set forth that on the top of the tomb 187 there was placed a sphere with a cylinder. When I had carefully examined all the monuments (for there are a great many tombs at the gate Achradinæ), I observed a small column standing out a little above the briers, with the figure of a sphere and a cylinder upon it; whereupon I immediately said to the Syracusi ans – for there were some of their principal men with me there – that I imagined that was what I was inquiring for. Several men, being sent in with scythes, cleared the way, and made an opening for us. When we could get at it, and were come near to the front of the pedestal, I found the inscription, though the latter parts of all the verses were effaced almost half away. Thus one of the noblest cities of Greece, and one which at one time likewise had been very celebrated for learning, had known nothing of the monument of its greatest genius, if it had not been discovered to them by a native of Arpinum.'*

As Cicero wrote in the fifth book of [Cicero 1931, p. 395] he had also visited the house and place of death of Pythagoras in Metapontum:

> *'But I, Piso, agree with you; it is a common experience that places do strongly stimulate the imagination and vivify our ideas of famous men. You remember how I once came with you to Metapontum, and would not go to the house where we were to stay until I had seen the very place where Pythagoras breathed his last and the seat he sat in.'*

To be honest, this was not the only contribution of Rome to analysis. Roman surveyors – Agrimensoris, as they were called – had to survey areas of curvilinear shapes in landscapes. They accomplished their task by means of 'linearising' the boundary. The curvilinear boundary was replaced piecewise by finitely many secants so that some parts of the area were inside, and some others outside of the linearised boundary. In this way one arrived at computable parts of the area. The idea of linearisation lies at the foundation of the integral calculus, [Cantor 1875, p. 96], [Hinrichs 1992]. However, there were all the preliminary works by the Greeks and this was examined and published recently by Lelgemann in [Lelgemann 2010].

Approaches to Analysis in the Greek Antiquity

about 600 BC	THALES founds mathematics as deductive science and proves theorems
about 579	PYTHAGORAS is born. He becomes the founder of a sect following the motto 'All is number'
about 520	The Eleatics PARMENIDES and his pupil ZENO start discussions on the continuum and on atomism
about 500	HIPPOCRATES OF CHIOS succeeds to compute the areas of lunes. HIPPIAS OF ELIS finds the quadratrix to trisect any angle
about 470	ANAXAGORAS formulates the problem of squaring the circle. Further classical problems are the doubling of the cube and the trisection of the angle. ZENO formulates his paradoxes
about 450	The Pythagorean HIPPASOS discovers incommensurable (irrational) numbers
about 400	EUDOXUS is the greatest mathematician of his time. He founds the doctrine of proportions. PLATO founds the first university (academy) in Athens. DEMOCRITUS becomes the most well known atomist
about 360	ARISTOTLE writes his book *Physics*
about 300	EUCLID writes *The Elements*
about 250	NICOMEDES invents the conchoide to trisect any angle. ARCHIMEDES, greatest engineer and physicist of antiquity, constructs technical devices (water screw, war machines), discovers the law of the lever and the buoyancy, employed the law of the lever to compute areas and volumes, invents the method of exhaustion and infinitesimal methods to compute areas (quadrature of the parabola and approximations of π, area under a spiral), publishes his results in codices A, B, C
212	ARCHIMEDES is killed by a Roman Legionnaire

3 How Knowledge Migrates – From Orient to Occident

T. Sonar, *3000 Years of Analysis*, https://doi.org/10.1007/978-3-030-58223-4_3

455	Rome is being plundered by the Vandals
about 570	Mohammed is born in Mecca
610	Mohammad has an apparition of archangel Gabriel. Mohammad becomes a prophet
622	Mohammed moves from Mecca to Medina and founds a new religion, Islam
632–661	Reign of the four rightly guided caliphs
634–644	The expansion movement of Islam in the Mediterranean starts under the second caliph Omar I, father-in-law of Mohammed. Omar vollzieht sich die Expansionsbewegung des Islam im Mittelmeerraum. Omar is said to be the creator of the Islamic realm.
705–715	Second large expansion movement of Islam up into Spain under Walid I
732	The conquest of the realm of the Franconians fails. The Arabic troops loose against the troops of Charles Martel in the battle of Tours and Poitiers
800	Imperial coronation of Charlemagne
969–1171	The Fatimieds rule Egypt
1055	The Turkish Seljuks conquer Bagdad
1071	Jerusalem falls to the Seljuks
1076	Damascus falls to the Seljuks
1086	The Almoravids secure the Islamic control in Spain
1095	Pope Urban II proclaims the first crusade at the Councli of Clermont. Six further crusades will follow until the end of the 13th century
1147	The orthodox-islamic Almohads replace the Almoravids in Spain
1299	Year of the foundation of the Ottoman Empire
1453	The Ottomans conquer Constantinople

3.1 The Decline of Mathematics and the Rescue by the Arabs

The Roman Empire finally collapsed with the destruction of the West-Roman realm in the fifth century AD. The shifts and distortions due to the migration of the peoples and the attacks of the Huns shook the Empire and let it tremble in its very foundations. In the battle of Adrianople in 378 Emperor Valens was beaten by the Goths and Rome rapidly lost control. In the middle of the 5th century large parts of Gaul and Spain fell to the invading Vandals, Franconians, and Goths; in 435 the African provinces fell to the Vandals. The West-Goths plundered Rome in 410; in 455 the Vandals followed and plundered again. The Roman Emperor Romulus Augustulus was dismissed in 476 by the Teuton Odoacer and the successor of the latter already behaved like a West-Roman Emperor. However, Odoacer and Theoderic the Great

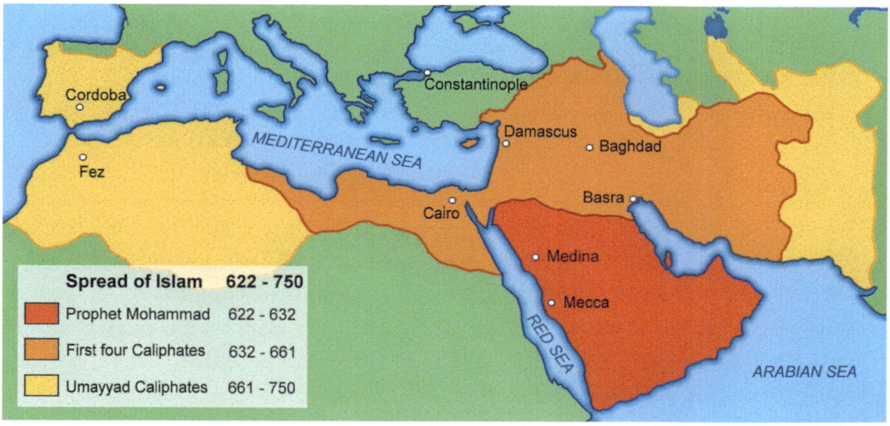

Fig. 3.0.2. The expansion of Islam was rapid in the first 100 years. Only from about 750 until into the 9th century culture and natural sciences had its heyday

were eager to get recognised by the East-Roman Emperor in Byzantium (Constantinople). The Roman Boethius (between 457 and 480– between 524 and 526) worked at Theoderics court. He came from a distinguished family. He lives in our memory because he fell in disgrace, was put in prison, and while imprisoned wrote one of the most famous books of philosophy: *Consolation of Philosophy* [Boethius 1999]. While in prison he had an apparition of

Fig. 3.1.1. Boethius teaches his pupils (University of Glasgow, Library, Special Collections)

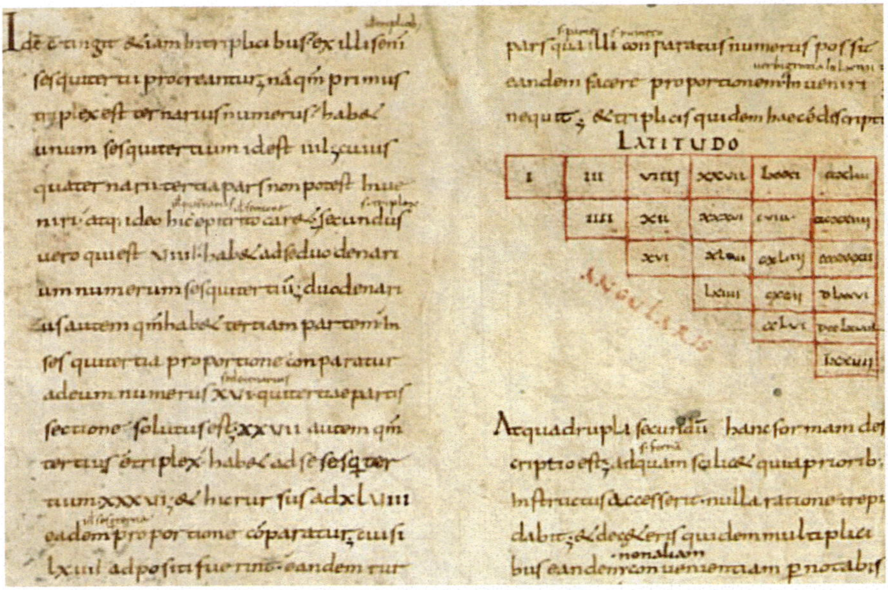

Fig. 3.1.2. Boethius: *De institutione arithmetica*, Manuscript of the 10th c., page 4 left (St. Laurentius Digital Library, Lund University)

philosophy in the allegoric form of a lady and Boethius started a dialogue. It is this dialogue which is written down in *Consolation of Philosophy*.

Boethius also translated a part of the works of Aristotle and commented on them. So he became an important messenger of Greek philosophy in medieval scholasticism since knowledge of Greek language was barely existent in the latinised Western Europe. Once he had planned to translate all of the (then-known) works of Plato and Aristotle into Latin but his imprisonment destroyed this plan. Boethius is also known as the author of some books concerning mathematics [Cajori 2000, p. 67f.].

These books clearly show the decline of mathematical sciences at the end of the Roman Empire. Boethius's *De Institutione Arithmetica* is just a partial translation of a book by the Greek mathematician Nicomachus (about 60–about 120 AD) where the most beautiful results of Nicomachus are missing. In a book on geometry by Boethius we find a translation of a few parts of Euclid's *Elements*.

But Boethius has also had other far reaching impact on the middle ages. He is the one who coined the notion of *quadrivium* meaning the four sciences arithmetic, astronomy, geometry, and music. He himself spoke about the quadrivium as the 'fourfold way to wisdom' [Gilson 1989, p. 97]. Besides the quadrivium stands the *trivium:* grammar, rhetoric, and dialectics.

1	3	9	27	81	243
	4	12	36	108	324
		16	48	144	432
			64	192	576
				256	768
					1024

Fig. 3.1.3. Table from the manuscript *De institutione arithmetica* by Boethius showing Indian-Arabic digits instead of Roman numerals. In the table every number is to the number below it like 3:4. Two of the numbers in the lower right (768 and 576) appear again in the 4:3 TV screen format of the PAL system (before broadband and HD) where the resolution is 768×576

The eastern part of the empire was in contrast far more successful than Rome and here the idea of a Roman Empire survived much longer. Firstly the Byzantine Empire was better protected against the intrusion of foreign peoples simply due to its geographical situation. Secondly in contrast to Western Rome, economy and administration were intact and the East Roman army still stuck together. In the reign of the last Latin-speaking East Roman Emperor Justinian I (about 482–565) the Byzantines even reconquered large parts of Italy, Southern Spain, and Northern Africa. However, East Rome was then involved in devastating wars with the Persian peoples from the east who conquered large parts of the realm in the 7th century.

In the year 529 Emperor Justinian ordered the closing of the 'pagan schools' in Athens. This order included also Plato's academy which had lasted for nine centuries. Any thoughts about mathematics stopped; however, many mathematical manuscripts were kept in Byzantium, Alexandria, and many other places of the Greek-Hellenistic world.

In the Arabic world Mohammad is born in the year 570 in Mecca. In 610 he has an apparition of the archangel Gabriel giving him his life task. Mohammad becomes the prophet and founds Islam as a new religion. In only a few decades Syria, Mesopotamia, Persia, Egypt, Turkestan (the modern countries Turkmenistan, Uzbekistan, Afghanistan as well as Tadzhikistan and Kashgar in the Chinese province of Sinkiang), and the Punjab (today in Pakistan) were conquered under the banner of the prophet, hence large parts of the former realm of Alexander and of the Greek-Hellinistic world. In an unprecedented triumph North Africa fell under Arabic rulership up to the Atlantic Ocean as well as large parts of the Iberian Peninsula. The victory march could be stopped only on French soil by Charles Martel's army in the battle of Tours and Poitiers in the year 732. Basically, the Arabic expansion aimed at completely embracing the Byzantine Empire and to conquer Constantinople.

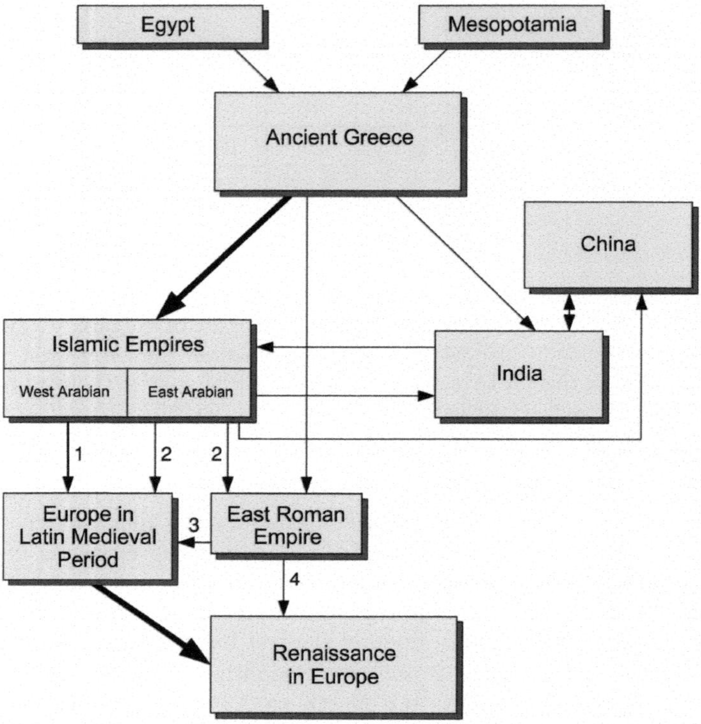

Fig. 3.1.4. How knowledge migrated – main streams of handing down mathematical knoweldge. (Legend: 1: Propagation from Western Arabic countries via Spain and Sicily to Europe in 11th–13th c; 2: Encounter with the East Arabic world during the crusades; 3: Handing down of mathematical works of the Greek antiquity and islamic scholars of the middle ages via the Byzantine Empire to Europe; 4: Direct influence of Byzantine sources on the European Renaissance. After that propagation of European mathematics starting in the 16th c; Development of a global and consistent mathematics concerning terminology and symbolism in the 19th c.)

In the world empire of the Arabs many different cultures and peoples were united and the Arabic language was their unifying bond. The scientific writings of these peoples were widely scattered. The Arabs started to collect them – in particular the extensive lot of manuscripts written by Greek, Persian, and Indian mathematicians – during the 8th and 9th century in the east of their huge realm and nurtured their translation into Arabic. The Abbasid caliphs had moved the capital of their empire from Damascus to Bagdad which was founded by the caliph al-Mansur. Under his successors Bagdad became a centre for the arts and sciences which started to flourish in all parts of the Arabian realm.

Caliph al-Ma'mun (Caliph between 813 and 833), son of Harun al-Rashid, built the 'Bayt al-Hikma' (House of Wisdom) in which the writings of Greek, Syrian, Indian, and Persian scholars were collected in the library, translated into Arabic, and copied. In the west of the realm the ruler of the Ummayads, Abd al-Rahman (756–788), had founded an independent kingdom in Spain which later became the caliphate of Córdoba. Beside Bagdad Córdoba became the second important cultural centre of the Islamic world. Among others the philosopher Ibn Rushd (1126–1198) (Averroes, see section 3.2.3) worked here. In the nearby summer residence Medina Azahara of the caliphs a large library containing 400 000 volumes developed. Córdoba became of paramount importance concerning the transmission of the mathematical knowledge of the Greeks and the Islamic cultural area into Western Europe.

Western European scholars got in contact with the Islamic culture in the regions which were reconquered on the Iberian Peninsula during the Reconquista. Arabic translations of old manuscripts containing the results and knowledge of scientists of antiquity and of the Orient were now available; for example the Indian-Arabic digits in the decimal notation of numbers which came in the 'Siddanthas' from India to the court of the Caliphs of Bagdad, the Indian trigonometric functions sinus and cosinus and the astronomical calculations and methods of approximation for the solution of algebraic equations developed by Persian scholars [Alten et al. 2005, p. 116ff.]. Above all the European scientists got to know the set of concepts, the methods of proof, and the plethora of results by Greek mathematicians. The works of Euclid, Archimedes, Apollonius, Diophantus, Hero, and many others were now available in Arabic translations.

Schools of translations and interpretations were founded at many places in re-Christianised parts of Spain in which works in Arabic language were translated mostly into Latin, the language of scholars in the middle ages. Often Christian translators worked closely with Jewish scholars so that many Hebrew translations resulted as well. The school of translation in Toledo, a Christian town again since 1085, became most well-known and famous. Gerard of Cremona (about 1114–1187) translated about 80 works in Toledo. In Segovia Robert of Chester (around 1150) translated the famous work by Muḥammad ibn Mūsā al-Khwārizmī (about 780–about 850) concerning algebra whose name can be found today in the modern word of 'algorithm' which is nothing but a malapropism of the name al-Khwārizmī.

Hence Spain, besides the re-Christianised Sicily, became the most important bridgehead for the transmission of Greek-Hellistic and Oriental mathematics to Europe. It was a loading station for the migration of knowledge from the Orient to the Occident.

In [Alten et al. 2005] the independent achievements of Arabic scientists in the field of algebra were thoroughly described; their achievements in geometry were examined in [Scriba/Schreiber 2000]. But a look at their contributions to analysis will be worthwhile.

3.2 The Contributions of the Arabs Concerning Analysis

Already in the year 773 there were contacts between the governmental seat in Bagdad and India [Alten et al. 2005, p. 161f.]. In this way our modern digits 1, 2, 3, 4, 5, 6, 7, 8, 9 and the digit 0 found entrance into Arabic mathematics. The first work in which this system of numbers was exploited was a work of al-Khwārizmī concerning arithmetic [Juschkewitsch 1964, p. 187] which is extant only in a Latin translation.

In analysis the Arabic scientists distinguished themselves by a continuation of Archimedean ideas concerning quadrature and cubature, but analysis was not only enriched by these explicit works. Arabic scientists were also concerned with the philosophy of Aristotle and his physics and thereby had a far reaching impact on the scholastic philosophy of the European middle ages.

3.2.1 Avicenna (Ibn Sīnā): Polymath in the Orient

Abū ʻAli al-Ḥusayn ibn ʻAbd Allāh ibn al-Ḥasan ibn ʻAli ibn Sīnā or simply Ibn Sīnā was born about 980 in Afshan in Persia and died 1037 in Hamadān in the region of what today is Iran. He became one of the most famous scholars of his times and worked as mathematician, astronomer, alchemist, and physician. In the West he is known under his latinised name Avicenna.

The historian of science George Sarton (1884–1956) became so excited about Avicenna's achievements that he valued him as the most famous scientist of Islam and perhaps the most famous scientist of all times.

It is said that Avicenna knew the Quran by heart at the age of ten and that he had already read numerous works of literature. In the following years he studied the works of Euclid and the *Almagest* of Claudius Ptolemy. He is also said to have masterly used the Indian numerals in calculations. At the age of seventeen his attention turned to medicine and hence found the field to make his name immortal. Avicenna collected medicinal herbs, examined their effects, and documented his investigations. The entire medieval medicine in Europe later relied mainly on Avicenna's medical writings. The interest in his writings is still high even today since his recipes are nowadays investigated by means of modern methods to reveal the secrets of 'ancient' natural medicine. In the 12th century Gerard of Cremona translated *The Canon of Medicine* which remained a relevant textbook in the West until the 17th century.

It is also of interest to us that he commented on the works of Aristotle in a genuine style. His critique of some points in Aristotle's writings will later serve as the basis of a new reception of Aristotle in Western Europe. In particular he wrote on logic and also developed an own logical system which became known as 'Avicenna's logic'. Hence he became a thinker independent of Aristotle who remained critical in his views. Together with Averroes he

Fig. 3.2.1. Ibn Sīnā (Avicenna) was a great polymath. He was well versed in many fields and is especially well known in Europe due to his knowledge in medicine. Among other things he described the construction of a highly accurate measuring device for astronomical purposes [Photo: H. Wesemüller-Kock]

became the most important intermediary of Aristotle's nature studies and philosophy [Strohmaier 2006], [Gutas 1988].

3.2.2 Alhazen (Ibn al-Haytham): Physicist and Mathematician

Abū 'Alī al-Ḥasan ibn al-Ḥasan ibn al-Haytham, known with his Latinised name Alhazen in the Western hemisphere, was born about 965 in Basra and died 1039 or 1040 in Cairo. He is known as the father of optics and as a great geometer but he also worked to surpass some of the results of Archimedes. While Archimedes computed the volume of a solid of revolution by rotating a parabolic segment about its axis of symmetry Alhazen computed the volume in the case that a parabola rotates about some other axis, for example about axis AB in figure 3.2.3. Much later, in the year 1615, Johannes Kepler will denote the solid as 'parabolic spindle' in his work *Nova Stereometria Doliorum Vinariorum* (New stereometry of wine barrels).

A parabola is defined as the set of all points which share the same distance to a fixed point (the focus F) and a given line (the directrix l). In figure 3.2.3 the point P on the parabola therefore satisfies

Fig. 3.2.2. Banknote with the portrait of Abū ʻAlī al-Ḥasan ibn al-Ḥasan ibn al-Haytham (about 965–1039/40) (Iraq 1982)

$$\text{length of } FP = \text{length of } PQ.$$

This has naturally to hold for the focus point S as well, i.e.

$$\text{length of } FS = \text{length of } SR =: f.$$

Introducing the usual (x, y)-coordinate system and denoting the length of PQ by x (x and y are hence the Cartesian coordinates of point P) the length of the segment PQ apparently is $y + f$. It follows from Pythagoras's theorem

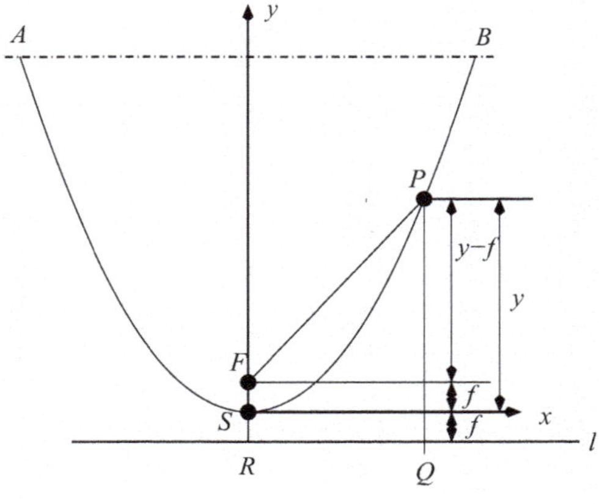

Fig. 3.2.3. Figure showing the emergence of a parabolic spindle by means of rotation of a parabolic segment

that

$$(y-f)^2 + x^2 = (\text{length of } FP)^2,$$

and since length of FP = length of $PQ = y + f$ it follows

$$(y-f)^2 + x^2 = (y+f)^2.$$

Solving this equation for y we arrive at the equation of the parabola in the form

$$y = x^2 \frac{1}{4f}. \tag{3.1}$$

Now this relation does not only hold in case of the parabola being described with respect to a coordinate system but holds generally. Fixing two points P and Q on the parabola, drawing a tangent in O and then a parallel through P as shown in figure 3.2.4, then it holds

$$OX = (XP)^2 \frac{1}{4OF},$$

in full analogy to (3.1). We have written OX instead of 'length of OX' for simplicity. Alhazen was also concerned with the solid of revolution ocurring if the parabola is rotated about the axis OX. If $O = S$ we are back at the problem which was treated by Archimedes. But also rotation about PX may be considered [Baron 1987, p. 68] and in this case the choice of $O = S$ leads to rotation about the axis AB in figure 3.2.3.

Let us consider the case of the parabolic spindle, hence rotation about the axis AB. Baron has shown in 3.2.3 that all other cases can be reduced to this one. Alhazen inscribes rectangular strips in the parabola as shown in the left part of figure 3.2.5. During rotation about AB these strips become cylindrical discs. The sum of all discs underestimates the volume of the parabolic spindle however thin the rectangular strips may be chosen. Then he circumscribes the parabola by rectangular strips as shown in the right part of figure 3.2.5. The sum of these circular discs overestimates the actual volume. Let us consider the case of the inscribed discs somewhat closer where we only need to look at

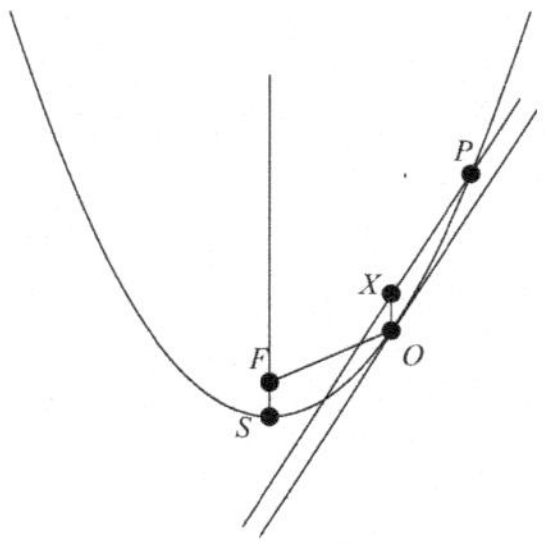

Fig. 3.2.4. A further axis of revolution for the parabola as used by Alhazen

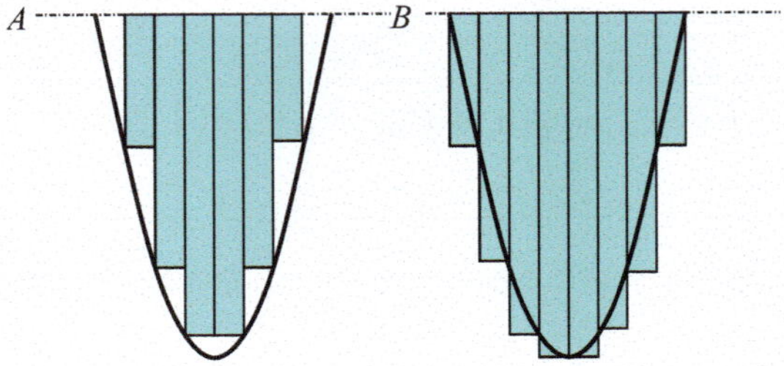

Fig. 3.2.5. Computation of volume if a parabolic segment is rotated about AB

one half of the parabola as shown in figure 3.2.6. In the presentation chosen in this figure the equation of the parabola apparently is

$$x = ky^2,$$

where we have defined $k := 1/4f$. If we traverse the parabola in y-direction until $y = b$ then the value of x at this point is $x = a = kb^2$. On the y-axis we subdivide the segments from 0 until b into n equal parts of width

$$h := \frac{b}{n} = y_i - y_{i-1}, \quad i = 1, \ldots, n.$$

The crosshatched rectangle therefore has an area

$$A_i := (a - x_i) \cdot h$$

and if this rectangle rotates about the axis at $x = a$ we get a cylindrical disc of volume

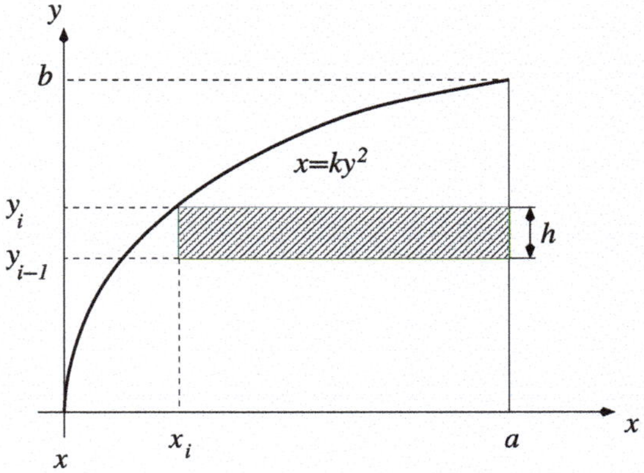

Fig. 3.2.6. Figure concerning the calculation of the volume of the parabolic spindle following Alhazen

$$V_i := \text{base area} \cdot \text{height} = \pi(a - x_i)^2 \cdot h.$$

Now we have $a = kb^2$ and $x_i = ky_i^2 = ki^2h^2$, hence

$$V_i = \pi h(kb^2 - ki^2h^2)^2$$

and since $b = nh$ it follows

$$V_i = \pi h(kn^2h^2 - ki^2h^2)^2 = \pi h^5k^2(n^2 - i^2).$$

Therefore the solid consisting of all n cylindrical discs has the volume

$$\begin{aligned} V_{\text{inner}} &= \sum_{i=1}^{n} \pi h^5k^2(n^2 - i^2)^2 = \pi h^5k^2 \sum_{i=1}^{n}(n^4 - 2n^2i^2 + i^4) \\ &= \pi h^5k^2 \left(n^5 - 2n^2 \sum_{i=1}^{n} i^2 + \sum_{i=1}^{n} i^4 \right). \end{aligned} \tag{3.2}$$

Analogously we see that the volume of the outer cylindrical discs is

$$V_{\text{outer}} = \sum_{i=0}^{n-1} \pi h^5k^2(n^2 - i^2)^2.$$

In order to carry out the calculations Alhazen needed the sum of the first n fourth powers. In case of an oblique axis of rotation also the sum of the first n cubic numbers is required. By contrast Archimedes required the sums

$$\sum_{i=1}^{n} i, \quad \sum_{i=1}^{n} i^2.$$

To get the values of the sums

$$\sum_{i=1}^{n} i^3, \quad \sum_{i=1}^{n} i^4$$

Alhazen came up with an ingenious geometric construction. He represented the sum $\sum_{i=1}^{n} i^k$ geometrically as a juxtaposition of rectangles; see the yellow rectangles in figure 3.2.7. The height of every single rectangle is 1 and there are apparently n such rectangles. Alhazen now completed these rectangular strips to result in a large rectangle as can be seen in figure 3.2.7. The large rectangle has an area of

$$F = (n + 1) \cdot \sum_{i=1}^{n} i^k.$$

This area is comprised of the sum of the areas of the horizontal rectangles

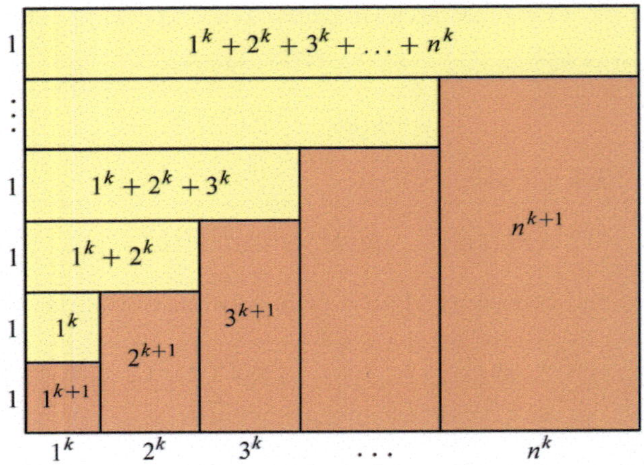

Fig. 3.2.7. Concerning the derivation of the summation formulae

$$F_1 = (1^k)+(1^k+2^k)+(1^k+2^k+3^k)+\ldots+(1^k+2^k+3^k+\ldots+n^k) = \sum_{\ell=1}^{n}\left(\sum_{i=1}^{\ell} i^k\right)$$

and of the sum of the areas of the vertical rectangles

$$F_2 = \sum_{i=1}^{n} i^{k+1}.$$

Since $F = F_1 + F_2$ Alhazen arrived at

$$(n+1)\cdot\sum_{i=1}^{n} i^k = \sum_{\ell=1}^{n}\left(\sum_{i=1}^{\ell} i^k\right) + \sum_{i=1}^{n} i^{k+1}. \tag{3.3}$$

If we consider $k = 1$ then it follows

$$(n+1)\cdot\sum_{i=1}^{n} i = \sum_{\ell=1}^{n}\left(\sum_{i=1}^{\ell} i\right) + \sum_{i=1}^{n} i^2. \tag{3.4}$$

Archimedes already knew the value of the sum

$$\sum_{i=1}^{n} i = 1+2+3+\ldots+n = \frac{n(n+1)}{2} = \frac{1}{2}\left(n^2+n\right).$$

Replacing this value of the sum for $\sum_{i=1}^{n} i$ and for $\sum_{i=1}^{\ell} i$ then from (3.4) the equation

$$(n+1)\cdot\left(\frac{1}{2}\left(n^2+n\right)\right) = \sum_{\ell=1}^{n}\frac{1}{2}\left(\ell^2+\ell\right) + \sum_{i=1}^{n} i^2$$

follows, and if we multiply the left side out and resolve the first sum on the right side it follows

$$\frac{1}{2}n^3 + n^2 + \frac{1}{2}n = \frac{1}{2}\sum_{\ell=1}^{n} \ell + \frac{3}{2}\sum_{i=1}^{n} i^2.$$

Now we have on the right side again a sum of the form $\sum_{\ell=1}^{n} \ell$ which we can replace by its value $\frac{1}{2}n^2 + \frac{1}{2}n$. Tyding this up eventually yields

$$\sum_{i=1}^{n} i^2 = 1^2 + 2^2 + 3^2 + \ldots + n^2 = \frac{1}{3}n^3 + \frac{1}{2}n^2 + \frac{1}{6}n. \tag{3.5}$$

Hence the formula of the sum of the first n square number is derived.

Now we can go on in this manner. Using $k = 2$ in (3.3) and employing (3.5) it follows

$$\sum_{i=1}^{n} i^3 = 1^3 + 2^3 + 3^3 + \ldots + n^3 = \frac{1}{4}n^4 + \frac{1}{2}n^3 + \frac{1}{4}n^2$$

und in case $k = 3$ with the help of the last equation we arrive at

$$\sum_{i=1}^{n} i^4 = 1^4 + 2^4 + 3^4 + \ldots + n^4 = \frac{1}{5}n^5 + \frac{1}{2}n^4 + \frac{1}{3}n^3 - \frac{1}{30}n.$$

Denoting the volume of the cylinder completely covering the parabolic segment by $V_{\text{cylinder}} = a \cdot \pi b^2$ we can show by means of the above sums that

$$V_{\text{inner}} < \frac{8}{15}V_{\text{cylinder}} < V_{\text{outer}}$$

holds [Edwards 1979, p. 85]. Now Alhazen shows in an Archimedean manner, i.e. using the double *reductio ad absurdum*, that the volume of this solid of revolution is just 8/15 of the volume of the circumscribed cylinder.

3.2.3 Averroes (Ibn Rushd): Islamic Aristotelian

The Spanish-Arabic philosopher, mystic, and physician Averroes or Ibn Rushd was born in Córdoba in the year 1126 where much later a monument was erected to honour him (fig. 3.2.8). He died 1198 in Marrakesh. Through his commentary of many works of Aristotle he became one of the central figures of medieval scholastism concerning the reception of Aristotle. In fact he became so famous that he was called 'the commentator'. His main concern besides medicine was logic in which he saw the only way to human happiness. Averroes also saw Aristolian logic as the only possibility to gain knowledge. He thereby exaggerated the figure of Aristotle excessively and his admiration of the philosopher let him often overstep the mark. Thus he criticised Avicenna although he knew his writings only superficially. However, no none commented on the writings of Aristotle in such a complete manner and with such a critical mind. He invited his fellow believers to think about their faith, to penetrate it by means of Aristotelian logic, and to use their own senses. Thereby he moved away from the traditional perception of the Quran and was banished to exile to North Africa where he eventually died. Even today the philosophy of Averroes is rejected by orthodox Muslims.

As in the case of Avicenna we can not prove any direct occupation with analysis. Nevertheless he had an important role in scholastic philosophy since his commentaries of the works of Aristotle were intensely studied and even led to an own school of thought in Aristotelianism, the so-called Averroism, which led to tensions within the Christian philosophy of the middle ages. Averroism was fought against by Thomas Aquinas and was finally prohibited.

Fig. 3.2.8. Averroes (Ibn Rushd), Statue in Córdoba

Fig. 3.2.9. Commentary by Averroes concerning *De anima* by Aristotle (13th c.)

Contributions of Islamic Scholars to Analysis

980	Persian mathematician, astronomer, alchemist, and physician AVICENNA (Ibn Sīnā) is born in Afshana. His medical writings have an impact well into the Western culture
about 935	Birth of ALHAZEN (Ibn al-Haytham) in Basra
about 1000	ALHAZEN writes a book on optics which makes him the 'father' of this science. In his volume computations he supersedes ARCHIMEDES
1126	Spanish-Arabic philosopher, mystic, and physician AVERROES (Ibn Rushd) is born in Córdoba. His comprehensive commentary of the works of Aristotle earn him the name of honour 'the commentator' in Christian scholastics.

4 Continuum and Atomism in Scholasticism

Fig. 4.1.0 [Collage Wesemüller-Kock]

T. Sonar, *3000 Years of Analysis*, https://doi.org/10.1007/978-3-030-58223-4_4

732	The conquest of the Franconian realm by Muslim troops fails. The Arabic troops are defeated by the army of Charles Martel in the Battle of Tours und Poitiers
800	Charlemagne crowned as Emperor
735–804	Alcuin of York. He founds the educational system in the Franconian realm.
	Cloister schools emerge
871	Alfred the Great becomes King of Wessex in England. Cultural heyday starts in England
919–1024	Rule of the Ottonians in German
955	Otto I defeats Hungarian invaders in the Battle of Lechfeld. Otto becomes Roman Emperor in 962
994	Otto III takes the rule as German king. In 996 he becomes Roman Emperor
999–1003	The monk Gerbert of Aurillac becomes Pope Sylvester II by influence of Otto III
1024–1125	Rule of the Salians in Germany
1066	England is conquered by the Norman William the Conquerer
1095	Pope Urban II initiates the first crusade at the Council of Clermont. Six further crusades will follow until the end of the 13th century
12th–14th c	Heyday of Christian scholasticism
1138–1254	Rule of the Staufer in Germany
1159	Begin of the schism
1189	Richard I 'Lionheart' becomes King of England
1220	Friedrich II crowned Roman Emperor
1228–1229	Friedrich II on crusade in the Holy Land
1337	The Hundred Years' War between England and France starts under the English King Edward III
1348–1350	First occurance of the plague in England
1410	Double election to find the German king. The choice between Sigismund and Jobst of Moravia ends with the majority of one vote in favour of Jobst
1414–1418	Council of Constance
1415	The English King Henry V restarts the war with France
1415	In Constance Jan Hus is sentenced and put to death
1417	The election of Pope Martin V ends the schism
	Normandy is conquered by the English
1419	Beginning of the Hussite Wars
1430	Jean d'Arc is captured at Compiègne. She is burned at the stake in Rouen in 1431
1431–1449	Council of Basel
1433	Sigismund elected Roman Emperor. Conciliation with the moderate Hussites
1438	Albert II of Germany becomes German king
1440	Frederick III crowned German king
1453	Constantinople falls to the Turks
1483	Martin Luther born in Eisleben

4.1 The Restart in Europe

The Arabic realm quickly developed into a world empire of unprecedented size. To the East it extended to India and in the West large parts of Spain could be brought under the control of the Caliph. The conquest of Spain was achieved during the reign of Caliph Al-Walid ibn Abd al-Malik or Al-Walid I (668–715) from 705 to 715. In April 711 the leader of the Berber, Ṭāriq ibn Ziyād, landed in the area of Gibraltar where there now is the city of Tarifa, named after Ṭāriq. What followed was a very fast spread of the Arabs in which the Visigoths were routed. Eventually the Pyrenees were crossed over and the realm of the Franconians was threatened. Under this pressure the Franconian Mayor of the Palace Charles Martel (about 686–741) turned his troops against the Islamic Arabs and routed their army in the Battle of Tours and Poitiers in 732. The Arabian expansionism fully came to a stop after the Battles of Avignon and at the River Berre in 737.

Large parts of Spain remained under Arabic influence nevertheless until 1492. The Emirate of Córdoba existed from 750 until 929 and was succeeded by the Caliphate of Córdoba (929–1031) which then disintegrated into numerous

Fig. 4.1.1. Islamic realm on the Iberian Peninsula at the beginning of the 10th century (Map, processed by H. Wesemüller-Kock)

Fig. 4.1.2. Prayer hall in the Mezquita of Córdoba. Entrance to the Mezquita of Córdoba [Photo: H.-W. Alten]

smaller states. The city of Córdoba developed into an active centre of cultural and scientific life and the time between 750 and 1100 was characterised by magnificent achievements in architecture which are still admired today. The famous Mezquita of Córdoba, built as a mosque and then transformed into a cathedral, and the Alhambra in Granada are stone witnesses of these achievements. Despite initial uprisings of the Christian population and although the Christian realms in the north attacked Al-Anadalus from 1050 on, the time of the Arabic reign in Spain was characterised by tolerance as judged with the standards of the times. However, at certain times and under some rulers the treatment of Christians and Jews became awkward, mostly fuelled by Christian fundamentalists calling for Martyrdom.

Fig. 4.1.3. Grave of Charles Martel in St Denis [Photo: J. Patrick Fischer]; Charlemagne (Painting 1512/13 by Albrecht Dürer, Germanisches Nationalmuseum Nuremberg)

While mathematical writings came to Spain with the Arabs the Anglo-Saxon Benedictine monk Bede the Venerable (672/673–735) worked in England and Ireland at the beginning of the Arabic conquest of Spain. He became one of the greatest scholars of his time but we have to be careful with superlatives like that! In his day Bede might have counted as a scholar outside the Arabic cultural area – he was rather insignificant, however, if he is compared to Arabic scholars. During his travels through England and Ireland Bede was irritated and annoyed that different cloisters computed different dates for Easter. The annual commemoration of the resurrection of Jesus Christ was calendrenically determined during the Council of Nicaea in the year 325 after the Jewish Passover and is celebrated on the first Sunday following the first full moon in spring (the Sunday following the full moon that follows the northern spring equinox; the *paschal full moon*). Since we now follow the Gregorian calendar this means that Easter Sunday falls on the 22nd March at the earliest and the 25th April at the latest. One of the main tasks of Christian monks in the cloisters was the calculation of Easter Sunday; a process called *computus*. To avoid different calculations of Easter Bede developed a quite modern kind of chronology in his writings *De temporibus* and *De temporum ratione*, thereby detecting an error of the calendar which was corrected only in the 16th century when the Gregorian calendar was introduced. With *De natura rerum* Bede also wrote on natural sciences.

Fig. 4.1.4. Bede the Venerable; Alcuin in the palace school of Charlemagne (Woodcut from 'Deutsche Geschichte' of 1862)

He has therefore also earned a place in the history of mathematics as the first 'computist' since the feeling of the importance of mathematics which came to the cloisters through him in fact can be seen as the origin of mathematics in the Christian occident.

In the year 800 Charlemagne (747/748–814), grandson of Charles Martel, was crowned Roman Emperor in Rome. Charlemagne felt the value of education and its importance for his realm although he never learned to write and read in his life. The Emperor ruled from the back of his horse – there was no fixed court, but the royal household moved from palatinate to palatinate to solve the local problems. To build up an efficient administration of the growing realm a well educated bureaucracy was needed and Charlemagne put all matters educational in the hands of the English monk Alcuin of York (735–804). Alcuin attended the cathedral school of York where he later became headmaster. In the year 782 he took over the palace school of Charlemagne in Aix-la-Chapelle (Aachen) and thereby introduced the remaining parts of Latin education, secured in England during the migration of the peoples, to the Franconian realm. An important part of this education was *computus*.

Also Rabanus Maurus (about 780–856), a pupil of Alcuin's born in Mainz and later abbot of the cloister Fulda and Archbishop of Mainz, cared for the education of the clergy. He justified the importance of mathematics because

Fig. 4.1.5. Rabanus Maurus in a manuscript from Fulda about 830/40 (Österreichische Nationalbibliothek Vienna). Rabanus Maurus (Alcuin beside him) hands over his work *De laudibus sanctae crucis* to St Martin of Tours. According to other sources this is Otgar, Archbishop of Mainz.

God had ordered everything according to measure, number, and weight[1]. After the death of Charlemagne the Franconian realm disintegrated quickly due to the Franconian inheritance law and the educational campaign, named today as 'Carolingian Renaissance', came to a halt. In this time Rabanus acted as collector and intermediary of the knowledge of his day.

In the realm cloister schools were founded in which the education of the clergy, even to prepare for secular professions, took place. Crucial was the memorising of the psalms by heart, then came the learning of scripture to be able to read and write Latin. The grammar was based on the logic handed down by Boethius. One truelly believed that one could distinguish the truth

[1] This refers to the Book of Wisdom (Wisdom of Solomon), Chapter 11, Verse 20. In the King James Bible: 'but thou hast ordered all things in measure and number and weight.'

Fig. 4.1.6. Gerbert of Aurillac: left as Pope Sylvester II, right as Monument in Aurillac [Photo: H.-W. Alten]

from the false with it. Besides genuine *computus* we also have testimony documenting the further occupation with elementary mathematics in form of task and exercise sheets [Gericke 2003, p. 78]. Seven years at school were seen as sufficient for worldly oriented pupils; clerics had to be educated some years more in philosophy and theology. A few schools even became famous due to the teachers working there. One of those schools was located in Reims where the monk Gerbert of Aurillac (about 950–1003) taught. Gerbert studied at Islamic schools in Sevilla and Córdoba and hence could absorb some of the achievements of the Arabic cultural area. In his time he was seen as one of the leading scholars although we have to be careful in the light of the knowledge of Arabic scholars. From 997 on Gerbert acted as personal advisor to Emperor Otto III. He became bishop of Ravenna in 998 and finally the first French pope in 999 as Sylvester II.

In Reims Gerbert created an atmosphere of great curiosity concerning what philosophical, scientific, and cultural achievements might be lying dormant in the writings of the Arabs. In particular, interest arose in the writings of Aristotle which were already known in parts through the writings of Boethius. Gerbert is said to have been a great collector of scientific books [Brown 2010, p. 30f.]. A book on geometry stemming from 980/982 is attributed to him [Gericke 2003, p. 74] in which he described the geometry of Boethius but at the same time corrected all of Boethius's errors. Hence we may well assume that Gerbert in fact understood mathematics. In the first chapter of

Fig. 4.1.7. Cathedral Notre Dame de Chartres (completed in October 1260) [Photo: H. Wesemüller-Kock]

this book he stated that the benefit of mathematics lies in the 'sharpening of the mind'. It feels almost comforting that such a deeply humanistic perception of the importance and the benefit of mathematics – which in our 'modern' times is seen only in the generation of money by 'applied' mathematics – was expressed more than 1000 years ago.

Gerbert's influence reached far. His pupil Fulbert (about 950–1028/1029) brought the school of Chartres to the front of education. Fulbert's pupil in turn was Franco of Liège (1015/1020–approx. 1083) wrote computistic treatises and another book on the geometry based on Boethius is attributed to him. It is of particular interest to us in that he has left a treatise concerning the quadrature of the circle. The antique ghost of circle squaring here made its first appearance in the mathematics of the Christian Occident and will not leave it for several hundred years! The treatise was written before 1050 and is interesting in many ways since the squaring of the circle was not an actual problem at all – after all one believed that 22/7 was the true value of π. Apparently Franco obtained the fascination of this problem from a remark of Aristotle in the *Categories* [Barnes 1995, Vol. I, p. 3–24]. Aristotle there writes in chapter 7:

> *'For if there is not a knowable there is not knowledge – there will no longer be anything for knowledge to be of – but if there is not knowledge there is nothing to prevent there being a knowable. Take, for example, the squaring of the circle, supposing it to be knowable;*

knowledge of it does not yet exist but the knowable itself exists. Again, if animal is destroyed there is no knowledge, but there may be many 'knowables.'

That something 'knowable' like the squaring of the circle existed but apparently no knowledge about it must have incited Franco. After having written in his treatise about some attempts of 'others' to assign to a circle a coextensive square he starts with his own reflections: If a circle has diameter $d = 14$ then one computes its area by squaring the radius $r = 7$ and multiplying by $22/7$ (that is Franco's π). This results in

$$A_{\text{circle}} = 7^2 \cdot \frac{22}{7} = 7 \cdot 22 = 154.$$

But 154 is no square of an integer and hence the edge of the sought square can not be a whole number. Now Franco proves that this edge can not be a fraction either, since

$$\left(12\frac{4}{12}\right)^2 = \left(12\frac{1}{3}\right)^2 = 152\frac{1}{9} < 154$$

and

$$\left(12\frac{5}{12}\right)^2 = 154 + \left(\frac{5}{12}\right)^2 > 154.$$

However, Franco surpasses this – admittedly insufficient – argument. He *proves* that the number $\sqrt{22/7}$ is irrational, i.e. can not be written as a fraction [Gericke 2003, p. 76]. If the circle were squarable then we could write 154 as $154 = p^2/q^2$. But then the change of the measuring unit of d could not change this. But also if we replace d by $q \cdot d$ the number 154 can not be written as the square of a fraction.

Despite this negative result Franco turned to constructively determine a coextensive rectangle. He subdivides the circumference of the circle into 44 parts and therefore the circle in 44 segments which he then rearranges to give a rectangle. Without any comment he transforms the crooked parts of the pieces of the circumference into straight line segments; thereby changing the sectors into isosceles triangles.

As compared to Archimedes Franco appears to be a tiny light but we can detect a new awakening here! An independent mind was concerned here with a difficult problem of antiquity, and that in fact just *before* the achievements of the Greek mathematicians were spread in the Christian Occident.

Until the end of the 11th century scientific teaching took place exclusively in church institutions like cathedral schools. First institutions similar to universities in 11th century Italy were not yet true universities as we understand them today but rather single faculties of law or medicine.

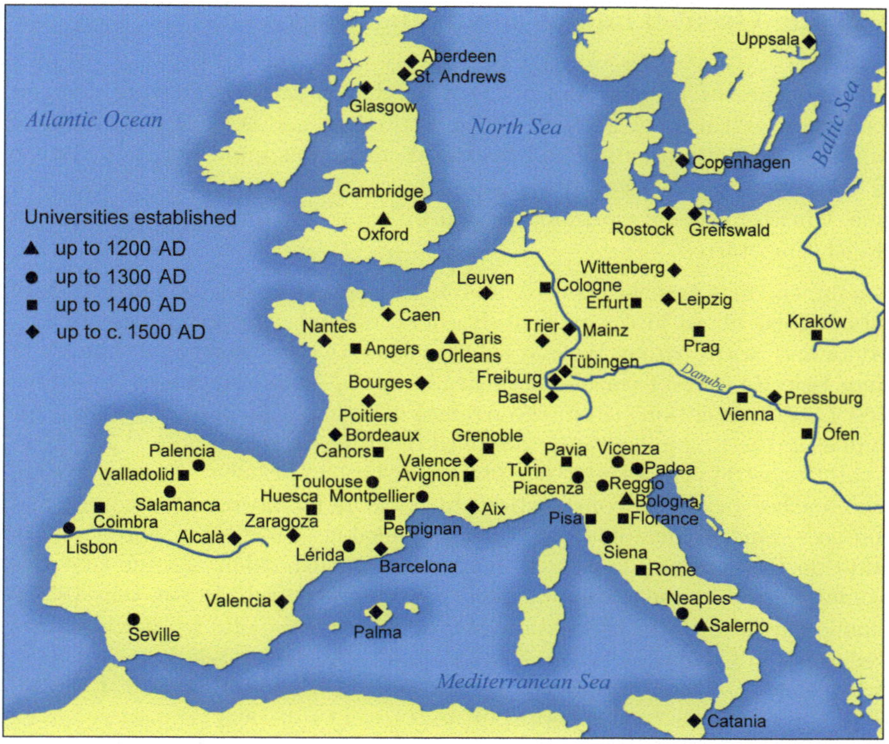

Fig. 4.1.8. European university towns in the Middle Ages [Photo: H. Wesemüller-Kock]

The faculty of Bologna, founded in 1088, counts as one of these law schools; the school at Salerno can count as faculty of medicine and was founded even a year earlier than the law school at Bologna. Holy Roman Emperor Frederick I, called Frederick Barbarossa, issued only in 1155 the *authentica habita* allowing strolling masters and their students to gather in corporations which still remained juridically tied to the Curia. The corporations became detached from the Curia only in the 13th century and first universities started to form in the sense of *universitas magistrorum et scholarium*, the fellowship of teachers and pupils. In a first wave of foundation the Sorbonne in Paris came into existence after 1200 (but there were earlier institutions), the Universities of Oxford (1167) and Cambridge (1209), Salamanca (1218), and Padua (1222), cp. [Wußing 2008, p. 283]. In the German speaking world universities could only be established in a second wave of foundation in the 14th century; the first of them was the Charles University in Prague 1348.

4.2 The Great Time of the Translators

The influence of the Doctors of the Church generated a growing demand for Greek writings; in particular one wanted to get hold of the complete works of Aristotle to study and discuss them. It was known that the Arabs in Spain possessed Greek texts mainly in Arabic translations. Since, as a rule, Christians could work unhindered in Islamic Spain unique activities of translating started.

Among the first translators of Arabic texts was Adelard (Athelard) of Bath (about 1080–about 1152); an English scholar who travelled extensively in Asia Minor and North Africa. We may safely assume that he thereby put himself to some risk. He aimed to master the learning of the Arabic language. One of the first Latin translations (from the Arabic) of Euclid's *Elements* flew from his quill as did astronomical tables compiled by al-Khwārizmī. Figure 4.2.1 shows a painting within the letter *P* from this translation of Euclid. We see a woman who teaches geometric constructions to students by means of set-square and compass. Since female teachers were unthinkable in medieval times (note in addition that the pupils seem to be monks!) we can safely assume that this woman is the personification of 'Lady Geometry' herself. In the year 1857 a manuscript of a Latin translation of a book on arithmetic by al-Khwārizmī was found in Cambridge which is attributed to Adelard [Cajori 2000, p. 118]. A translation of the *Sphaerica* of Theodosius of Bithynia [Heath 2004, p. 394] dealing with circles on a sphere (but no traces of spherical geometry) can be traced back to Adelard.

During the life of Adelard a decisive event took place which, despite of its catastrophic effects, led to further encounters with the Arabic cultural area: in the year 1095 Pope Urban II (about 1042–1099) proclaimed the first crusade (the name 'crusade' appeared not before the 13th century) which began in 1096 and led to the capturing of Jerusalem in 1099. Under false pretences, in particular by inventing stories of the Christian population suffering a lot from the Arabic rulers in Jerusalem and by promising salvation to everyone partaking in the crusade, great masses could be mobilised setting out to the sacred land. They left scorched earth behind. It led to anti-Jewish pogroms and lootings even before the crusaders set out. Much was and still is written on the crusades – following accepted reckoning there were seven crusades until the year 1272; the Children's Crusade included, see [Runciman 1951-4]. Here Barbarians attacked a highly developed and advanced civilisation where the main aim was its annihilation. Only when the crusader states were founded, no definitive military condition could be achieved and the Christian armies had to suffer heavy losses, some of the more educated crusaders got interested in the culture they ought to annihilate. Prominent in this respect was Holy Roman Emperor Frederick II (1194–1250). Frederick was an exceptional man of his time since he was inquisitive and highly educated. People called him *stupor mundi* – amazement of the world. He succeeded in

Fig. 4.2.1. Detail of Adelard's Latin translation of Euclid (The British Library)

the handover of Jerusalem to the Christians just by tedious negotiations since he knew the Arabic mentality very well and respected it. In the eyes of his Christian contemporaries negotiations with Muslims were high treason and he had to face serious difficulties. He was excommunicated twice by the pope although he stayed a quintessential Christian ruler and no Islamic scientist was employed at his court. Nevertheless he was an open and freeminded person who accepted the Arabic culture and promoted the sciences in his realm. Translations of Arabic works were also encouraged by him; he ensured a translation of the *Almagest* of Claudius Ptolemy [Cajori 2000, p. 119]. But let us return to the translators.

Fig. 4.2.2. Capturing of Jerusalem in the first crusade 1099 (Representation about 1300, Bibliothéque Nationale, Paris)

John of Seville (Johannes Hispaniensis or Hispalensis) (12th c) was a converted Jewish scholar who did translation work in Toledo which had developed into an active centre of translators in the 12th century under Archbishop Raymond de Sauvetât. Raymond promoted the translation of Arabic writings and it is possible, but not proven, that he founded the famous school of translators in Toledo. John became famous through translations of works by Aristotle but he also compiled a work entitled *liber algorismi* from Arabic sources in which the division of two fractions is proved in the form

$$\frac{a}{b} \div \frac{c}{d} = \frac{ad}{bd} \div \frac{bc}{bd} = \frac{ad}{bc}.$$

It is interesting that John used the term 'algorismi' which was a translation error of the name al-Khwārizmī. Furthermore he mentioned Hindu mathematics so that it is clear that he worked with Arabic sources.

Somewhat later than John of Seville Gerard of Cremona had his active days. Allegedly his love for the *Almagest* of Claudius Ptolemy draw him to Toledo.

Fig. 4.2.3. Frederick II, left: talking to Al-Kamil (al-Malik al-Kamil Naser ad-Din Abu al-Ma'ali Muhammad), right: as ornithologist with a falcon (from his book *De arte venandi cum avibus*)

In the *Almagest* the geocentric conception of the world is constructed and analysed which has accompanied mankind up until the time of Copernicus. In Toledo Gerard was overwhelmed by the wealth of Arabic manuscripts. He stayed in Toledo, translated some 70 important works, and finally died there. Among his translations is not only the *Almagest* but also 17 further texts in mathematics and optics; for example al-Khwārizmī's *Al-jabr* and Euclid's *Elements*. He is also the translator of *Physics* and *Meteorology* of Aristotle and of 12 astronomical works. He also compiled an influential translation of Avicenna's *Canon of Medicine*.

The translators of the 13th century had to face an inherent problem: non-translatable technical terms of Arabic origin. Therefore they applied a special artifice in that they created certain 'transfer words' which entered the Latin language as artificial words. One such term is 'sinus' which was created by Gerard. It is the transfer of the Arabic word for bag, bay, or bosom.

al-Khwārizmī's book *Al-jabr* was translated again by Robert of Chester as *Liber algebrae et almucabalae*.

In the year of 1260 Campanus of Novara (about 1220–1296), later known as Johannes Campanus, finally achieved the translation of the *Elements* of Euclid which became the basis of all Euclid editions following. Campanus also travelled Arabia and Spain. He was a well known astrologer, astronomer, and mathematician and was in service of Pope Urban IV. He also became the physician of Pope Boniface VIII.

Fig. 4.2.4. Hall of the emissaries in the Alcàzar of Seville – one of the most beautiful examples of the so-called Mudèjar art [Photo: H.-W. Alten]

Fig. 4.2.5. Horseshoe arcades in Santa Maria la Blanca, the first synagogue in Toledo built under the Almohads in the 12th century as a five-aisled pillar basicilica. The wide cultural range of Islamic, Jewish and Christian life in Toledo affected the transfer of knowledge into the Latin West positively. Translators belonging to either culture have worked here side by side [Photo: H.-W. Alten]

Fig. 4.2.6. Scriptorium in a church in Lille

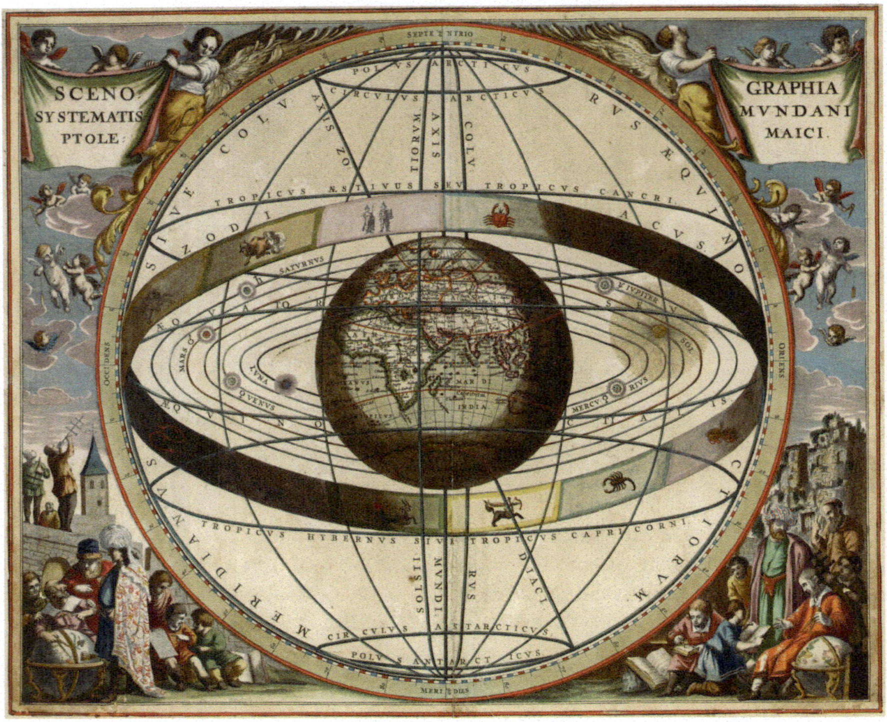

Fig. 4.2.7. World view of Ptolemy from a translation of the Almagest (1661)

The Flemish black friar and later Bishop of Corinth, William of Moerbeke (1215–1286) (Willem van Moerbeke) achieved the complete translation of all works of Aristotle since he doubted the reliability of earlier translations. For instance Gerard of Cremona had re-translated Syrian texts of Aristotle while William could fall back on Greek original texts. William also translated writings of Archimedes about 1270 from codices A and B which were located at this time in the library of the Vatican as Greek manuscripts [Netz/Noel 2007]. We have reported on the fate of codex C in some detail in the section on Archimedes.

Hence at the end of the 13th century substantial works of the Greek antiquity were available in the Latin West in the lingua franca of scholars of the Middle Ages: Latin. Now all prerequisites to further develop mathematics were there. In the form of discussions on the continuum analysis also gained momentum; we are on our way!

4.3 The Continuum in Scholasticism

With the presence of translations of works of Aristotle there was a growing desire of all educated circles in Europe to somehow incorporate Aristotelian philosophy into Christian theology. The Benedictine friar Anselm of Canterbury (about 1033–1109), born in Aosta in the Italian Alps, ranks as the father of early scholasticism because he required to find compelling logical justifications for theological statements. These requirements led to an ontological proof of the existence of God which has become famous. It was published about 1080 in his work *Proslogion* [Anselm 1996] in the form of a prayer. In a somewhat reduced form the argumentation can be described as follows:

> Premise 1: God is the most perfect being ('that-than-which-a-greater-*cannot*-be-thought').
> Premise 2: Existence belongs to perfection.
> Conclusion: God exists.

The ontological proof of God's existence already attracted the attention of critics in the lifetime of Anselm and today belongs to the most disputed problems in the history of philosophy.

Peter Abelard (1079–1142) (Petrus Abaelardus) taught logic and theology at the Cathedral school of Paris; the tragic love story with Héloïse has made both of them immortal.

As a Christian theologian he was convinced that the truth was in the bible; God's word. There were annoying differences in interpretation, however, which led to disputes. Here now Aristotle enters. The Greeks had reduced

Fig. 4.3.1. Anselm of Canterbury; Window in Canterbury Cathedral [Photo: H.-W. Alten]

Fig. 4.3.2. Abelard and Héloïse (from a manuscript of the 14th century, Musée Condé Chantilly)

complicated issues to simpler ones by their mathematical methods to find truth until they arrived at irrefutable axioms. Since this strategy apparently was very successful – and the works of Aristotle document this success – it should be possible to apply these methods to theology (and the natural sciences) as well. Abelard apparently linked the *ratio* in form of Aristotelean logic with faith; it emerges the medieval scholasticism which was prepared already by Boethius. Abelard's work *Sic et non* (that and no, in the sense of: this way or not this way) emerged about 1121/22, laid the foundation of this new direction of Christian philosophy, and fuelled the study of Aristotle. Quickly two centres of scholasticism developed: The University of Oxford and the Sorbonne in Paris.

4.3.1 Robert Grosseteste

According to Alistair C. Crombie [Crombie 1995, II p. 27] Robert Grosseteste (about 1175–1253), born in poor circumstances, was the actual founder of the Oxford scientific tradition and even of all of English intellectual culture.

We know little about the education and the early life of Grosseteste. As a cleric he was given a sinecure in Abbotsley in the diocese of Lincoln in 1225. In Oxford he can be traced as lector of the Franciscans in 1229/30. The Franciscan friars had founded a convent in 1224 and Grosseteste stayed until

Fig. 4.3.3. Robert Grosseteste, Bishop of Lincoln

1235. Besides his work as a teacher he made some progress in the hierarchy of the church in England. He became superintendent of Leicester and canon of Lincoln Cathedral. A serious illness in 1232 forced him to abandon some of his offices but in 1235 he was elected Bishop of Lincoln.

In sciences Grosseteste became immortal as the founder of experimental science [Crombie 1953] and as predecessor of Roger Bacon. He wrote on astronomy, optics, the mathematical method in natural sciences, and on the rainbow. It is highly likely that he was the first scholastic to understand the Aristotelian conceptions of natural sciences and his way to gain knowledge by heart and who could apply it. He also commented on Aristotle's *Physics* and the *Prior analytics*.

As far as we are concerned Grosseteste is interesting on the one hand as the founder of the Oxford scholastic school and on the other hand in his opposition concerning the Aristotelian continuum [Lewis 2005]. He considered that the continuum consisted of an infinity of points and hence he can be counted as one of the very few atomists among the Aristotelians. Of particular interest in scholasticism and controversialy debated was 'the infinite'. If we follow the discussion of the continuum of Aristotle an actual infinity is impossible; it is only the infinity *in potentia* which is conceivable. For a Christian thinker there thus appeared a 'God problem': God is almighty and omniscient, hence God certainly has the ability to create an actual infinity. This conflict will pervade Christian scholasticism. In the eyes of Grosseteste even several stages existed in the infinite. He wrote [Gericke 2003, Part II, p. 140]:

> *'There are infinities of different sizes. For the set of whole numbers is infinite and larger than the also infinite set of the even numbers.'*
>
> (Es gibt verschieden große Unendlich. Denn die Menge der ganzen Zahlen ist unendlich und größer als die ebenfalls unendliche Menge der geraden Zahlen.)

This statement apparently was underpinned by a principle in Euclid's *Elements*:

> *'The whole is greater than the part.'*

But this seemingly obvious principle is only valid in the case of finite sets and *not* in the case of infinite ones. However, Grosseteste apparently never doubted the principle.

4.3.2 Roger Bacon

Whether the Franciscan friar Roger Bacon (1214–1292 or 1294) was indeed a pupil of Grosseteste is by no means certain, but generally assumed. Roger Bacon has entered the history of philosophy as 'Doctor mirabilis' (wonderful teacher).

Bacon studied at Oxford and taught there on the writings of Aristotle. Between 1237 and 1245 he travelled to the scientific centre of his time, the University of Paris, and gave much attended lectures. After his return to Oxford he studied mathematics, alchemy, and optics and committed himself to experimental research probably under the influence of Grosseteste. After ten years of research he entered the Franciscan Order but quickly came under the suspicion of distributing dangerous doctrines. He was even grounded in 1278. His attacks on the scholastics were apparently too spicy and his tendency to mysticism too scary. He was released from arrest only in 1292 and died either in that year or in 1294.

Bacon knew and studied writings of Arabian scientists who saw Aristotle as the measure of all things. He revered the comments of Avicenna. The more Bacon was concerned with Aristotelian doctrines the more did he reject scholastic methods and their hair-splitting arguments. He formulated arguments which he saw as obstacles to gain knowledge. Among them was the respect concerning authorities, habit, and the dependence on the opinion of the masses. He required to reform theologian studies in the direction away from detailed scholastic discussions and towards an unaltered study of the bible in its original language. Additionally he required to study *all* sciences during an education at a university.

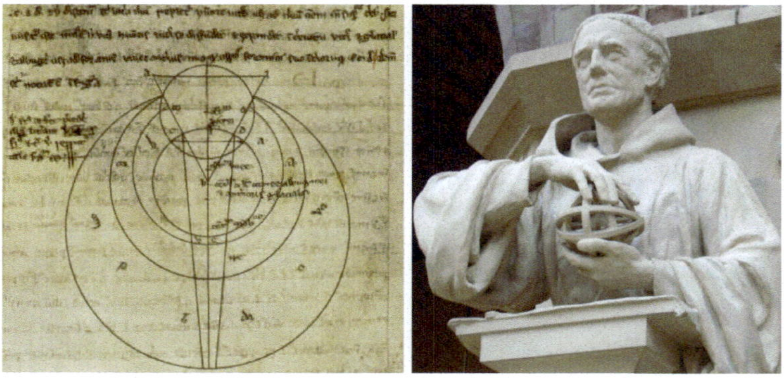

Fig. 4.3.4. Detail of a page of a book on optics from *Opus Maius* by Roger Bacon, published 1267. Ausschnitt aus einer Buchseite zur Optik aus dem 1267 erschienenen Werk *Opus Maius* von Roger Bacon. Dieses Werk enthält auch Kapitel zur Mathematik. Statue von Bacon im Oxford University Museum of Natural History [Photo: Michael Reeve]

Bacon ranks as the inventor of spectacles (he knew Alhazens book on optics) and of gunpowder. It is also said that he was an early critic of the Julian calendar and that he foresaw the microscope and the telescope.

In discussing the infinite Bacon is of the clear opinion that an actual infinity can not exist. His argument is concerned with a straight line extending to infinity in both directions (denoted by D and C in figure 4.3.5) [Gericke 2003, Part II, p. 140]. Points A and B lie on the line. Since, so says Bacon, 'infinity=infinity', we have to have $BD = BAC$. However, $BAC > AC$ since following Euclid the whole is greater than the part and BAC eventually is longer than AC. On the other hand we also have $AC = ABD$, again following the principle 'infinity=infinity'. Hence we arrive at the following assertions:

$$\begin{aligned} BD &= BAC \\ BAC &> AC \\ AC &= ABD. \end{aligned}$$

But it follows from this that $BD > ABD$ holds which is impossible since the part would be greater than the whole.

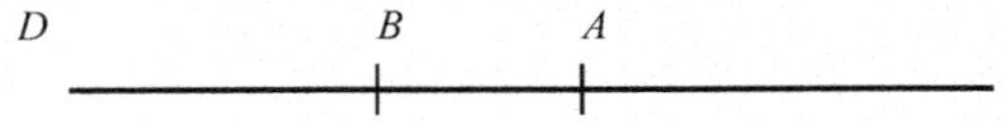

Fig. 4.3.5. Illustrating Roger Bacon's argument against the infinite

Bacon has confronted us with a further imposition of the infinite which he learned from the writings of the Arabian philosopher Abū Ḥāmid Muḥammad ibn Muḥammad al-Ghazālī (about 1058–1111), shortened as Al-Ghazali and known in the West as Algazel. In his work *Opus maius* Bacon repeats the argument in a somewhat modified form [Gericke 2003, Teil II, S. 145]. Imagine a square with a diagonal as shown in figure 4.3.6. Every point on the right edge of the square can be assigned a corresponding point on the diagonal in a one-to-one manner. Although the diagonal in a square is longer than the edge both segments have to consistent of the same number of atoms.

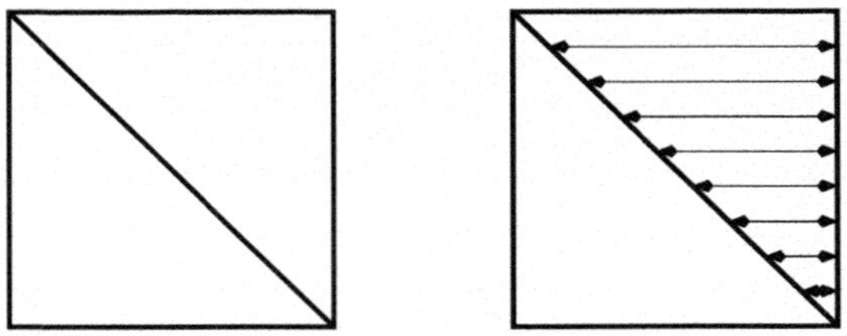

Fig. 4.3.6. Illustrating Roger Bacon's argument against atomism

4.3.3 Albertus Magnus

The great Doctor of the Church Albertus Magnus (about 1200–1280) (Saint Albert the Great) became a pioneering figure concerning Aristotelianism in Christian philosophy.

Born in Lauingen on the left bank of the Danube he studied the liberal arts (grammar, rhethorics, dialectics (logic), arithmetic, music, and astronomy) and probably medicine before he entered the Dominican Order in 1223. He became Master of Theology at the Sorbonne in Paris in 1245; the first master ever in the Dominican Order. During his period of study he intensively worked on Aristotelian and Jewish writings. In 1248 he returned to the Dominican's headquarters in Cologne and led the Cologne Cathedral school to great fame. This school later became the nucleus of Cologne University (founded in 1388).

Fig. 4.3.7. Albertus Magnus (Fresco of 1352 in Treviso)

Albertus has to be understood as a true polymath. He was philosopher, legal expert, natural scientist, and, of course, theologian. His scholarship earned him the surname 'magnus' – the Great. Where the scholastics of the previous generations had problems arguing with Aristotle – after all he was a pagan from the point of view of Christians – Albertus placed Aristotle in the very centre of scholastic philosophy and integrated him into the Christian doctrine. In doing this Albertus heralded the heyday of scholasticism which we nowadays call High Scholasticism.

Continuum and Indivisibles

Naturally Albertus also discussed the continuum. As was Aristotle so was Albertus convinced that the continuum is arbitrarily divisible. However, below a certain threshold Albertus believed that unlimited division would lose its capacity [Neidhart 2007, p. 571f.]. Detecting here an anticipation of the quantum of action appearing in quantum mechanics as Neidhart does in [Neidhart 2007] seems to be wishful thinking of the 21st century in my eyes.

Albertus Magnus can serve as an example of how the doctrine of motion crept into the minds of the medieval thinkers. Aristotle had postulated that both space and time are continua, i.e. both are divisible arbitrarily often and every part will again be a continuum. However, the continuity of time, says Aristotle, at last can be justified with motion [Aristotle 1995, Book VI 235a 11, p. 396]:

> *'Motion is also susceptible of another kind of division, that according to time. For since all motion is in time and all time is divisible, and in less time the motion is less, it follows that every motion must be divisible according to time.'*

Since space and time determine motion motion is itself doubly divisible and therefore doubly continuous. For motion can be affected by dividing time, but also by dividing space [Breidert 1979, p. 23]. Admittedly *the mobile* itself plays a role in the considerations of Aristotle – a fact ignored by Albertus Magnus in his commentary concerning Aristotle's *Physics* [Breidert 1979, p. 24]. A particular problem in the discussion concerning time is the very moment of the presence. The past is a continuum and – to be sure – so is the future. But the *point* of time we call 'now' does not quite fit into a continuum. Aristotle wrote about this problem in [Aristotle 1995, Book IV 217b 29 - 218 a 29, p. 369f.]. To avoid the 'now' as an indivisible Aristotle falls back on our intuition concerning 'movement', namely that time is something limited by different 'nows'. Now Albertus takes a radical step away from Aristotle's conception of the continuum: The *point* is now seen as *producing* a continuum if it starts moving. As Breidert remarks in [Breidert 1979, p. 28]:

> *'In this sense the point is the* primum continuans, *the* principium *and the* causa *of the line.'*
>
> (In diesem Sinne ist der Punkt das *primum continuans*, das *principium* und die *causa* der Linie.)

It is with Albertus that the connection between *movement* and the continuum finally arrived in scholasticism. Thomas Bradwardine and his fellow companions at Merton College Oxford will use this connection to achieve tremendous results.

4.3.4 Thomas Bradwardine

The famous Anglo-Saxon poet and writer Geoffrey Chaucer (about 1343–1400) has given a worthy tribute to the philosopher, mathematician, and theologian Thomas Bradwardine in his book *The Canterbury Tales.* In *The Nun's Priest's Prologue, Tale* [*and Epilogue*] Chaucer let the priest say [Chaucer 2005, S. 613]:

> *'But I ne kan bulte it to the bren*
> *As kan the hooly doctour Augustin,*
> *Or Boece, or the bisshop Bradwardin'*
>
> (But I can't sift it to the bran with pen,
> As can the holy Doctor Augustine,
> Or Boethius, or Bishop Bradwardine,)[2]

and thus puts him on one level with Boethius and even Augustine (354–430)!

Bradwardine was born in the County of Sussex and studied at Balliol College Oxford where he became a fellow in 1321. He became a Doctor of Divinity, earned an excellent reputation as a mathematician, logician, and theologian, and switched over to Merton College Oxford. He was entrusted with the office of the chancellor of Oxford University and with the office of a 'Professor of Divinity'. But his career was still not at an end. He became Dean of St Paul in London and the confessor of King Edward III (1312–1377). Edward's father, Edward II (1284–1327?), had donated the first colleges in Oxford and Cambridge but experienced an ill-fated reign. He was forced out of his office and was either murdered in 1327 or could flee England so that his son could be crowned at age 14. This son became an extraordinary successful English king who stayed in office for 50 years. As were most monarchs of his time he also was mostly interested in campaigns of conquest and his claim to the French crown caused the Hundred Years' War raging from 1337 until 1453 on French soil. Bradwardine accompanied his king at the Battle of Crécy on the 26th August 1346 in which approximately 12 000 English whitewashed a French army of 16 000 soldiers. Bradwardine read the victory mass when the battle was over.

[2] Corresponding modern translation taken from [Chaucer 2003].

Fig. 4.3.8. Merton College, University of Oxford [Photo: Gottwald]

The 'Babylonian Captivity' lasted from 1305 until 1377: Popes resided no longer in Rome but in Avignon due to massive interference by the French king. This time seamlessly led to the Great Schism from 1378 until the Council of Constance in 1417. There are now popes in Avignon as well as antipopes in Rome. Bradwardine is elected Archbishop of Canterbury in 1349 and this is the highest office in the English Curia. To be acknowledged he has to go to Avignon – a long, onerous, and dangerous journey in the 14th century. However, Bradwardine succeeds and travels back to England in the very same year. Shortly after having set foot on English soil he is struck down and dies from plague in Rochester. How much Bradwardine was revered can be seen from the fact that his infectious corpse was not hastily buried in Rochester but was taken to Canterbury to bury the Archbishop in dignity. In fact, Bradwardine was so famous in his time that he is remembered as *doctor profundus*.

Life in the 14th Century: The Black Death

The bubonic plague or the 'Black Death' raged Europe between 1347 and 1352 and caused the life of approximately 20 million humans; one third of the overall population [Cantor 2002]. Just for comparison: The Spanish Influenza raging Europe at the end of the WW I caused 2 million victims. The

Fig. 4.3.9. Entourage of a plague victim who is removed from a town. The beak mask (mostly worn by physicians) was believed to ban the danger of infection (sequence from the movie: *Vom Zählstein zum Computer – Mittelalter*) [Recording: H. Wesemüller-Kock]

social, economical, and political consequences turned out to be radical. Whole regions became deserted; labourers were desperately needed and the surviving peasants and workmen requested higher wages. This led to an uprising in London which was brutally downed by the king.

One particularity of the bubonic plague in the 14th century is the enormous speed with which it spread. Modern research is divided about the reason for this speed which distinguishes the plague of the 14th century from other outbreaks of the pestilence before and after. One group of researchers has conjectured another pathogen spreading much faster than the plague. They suspect an epidemic like anthrax because large parts of the population used to live very close to their cattle and a pathogen could have quickly spread from cloven hoofed animals to humans [Cantor 2002]. The hope that recent excavations of mass graves of the 14th century in London and around Hereford Cathedral would serve to clarify the situation once and for all was not fulfilled (Sloane [Sloane 2011, p. 172] gives an overview on different interpretations). Other researchers propose that the slowly spreading bubonic plague turned into a fast spreading pneumonic plague [Kelly 2005, p. 295ff.]. Pneumonic plague is not vector-borne like bubonic plague but can spread from person to person.

Whatever it was at the end: Bradwardine died a miserable death quite common for his time.

Concerning Infinity

In his work *De causa Dei contra Pelagium* of 1344 Bradwardine employed arguments against the infinity of time which we can not accept today. The British monk Pelagius (about 360–418) had denied the original sin and emphasised the free will of men [Hofmann 1951, p. 303]. This opinion had already been discarded by Saint Augustine (354–430) but returned to Christian scholasticism through Avicenna and Averroes. Bradwardine's arguments against Pelagius are based throughout on mathematical procedures. One point of discussion concerns the question of the duration of the world. Did the world always exist, i.e. for an infinite time, or has God created the world at a fixed point in time? Bradwardine argues as follows: If the world had been always in existence then until this day there must have been an infinity of human bodies and an infinity of souls which God has assigned to the bodies. God could have assigned the first soul to the first body, the second soul to the second body, and so forth. It would also be possible to assign the first soul to the first body, the tenth soul to the second body, the hundredth soul to the third body, and so on. In this last case every body would also possess a soul but God would have too many souls! This seems ridiculous and absurd in the eyes of Bradwardine [Gericke 2003, Part II, p. 141].

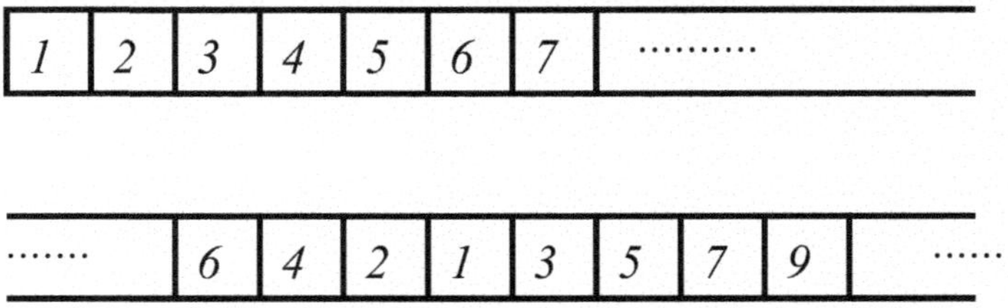

Fig. 4.3.10. Bradwardine's infinity of cubes

Continuous quantities are dissected by Bradwardine into discrete parts (!). He then argues: Imagining an infinity of cubes they could be arranged as in the upper part of figure 4.3.10. Just as well one could place cube 2 left of cube 1, cube 3 to the right, then cube 4 again to the left, cube 5 to the right, and so on. As a result the strip of cubes starting at 1 to the right, 1,3,5,7,9, ... would be exactly as large as the original strip of cubes 1, 2, 3, 4, 5, ... The part would hence be as large as the whole.

Bradwardine's Continuum

Bradwardine's work *De continuo* [Stamm 1937] (extracts also in [Clagett 1961]) contains a detailed analysis of different discussions on the nature of the continuum of his day and of older Greek schools. Some definitions and postulates are followed by 150 *conclusiones* in which we can find almost all of what scholasticism had to say about the continuum. Bradwardine refers to *Henricus modernus* [Gericke 2003, Part II, p. 146] when discussing the question whether a continuum consists of finitely many atoms or yet of an infinity of points lying side by side either gapless or with certain gaps in between. *Henricus modernus* is Henry of Harclay (about 1270–1317), chancellor of the University of Oxford from 1312 to 1317, who pleaded for gapless juxtaposed points comprising the continuum and for an actual infinity. Harclay was concerned with the hotly debated question in scholasticism of whether the existence of the world had a beginning, hence was created by God at a certain point in time, or if it existed 'always', hence for an infinity of time [Maier 1964, p. 41ff.]. Bradwardine refuted all such ideas and accepted only the Aristotelian opinion in the end.

Latitudes of Form: The Merton Rule as First Law of Motion

Between 1328 and 1350 the difference between kinematics and dynamics was clearly carved out at Oxford [Baron 1987, p. 81], [Clagett 1961, Chap. 4]. In Merton College a group of scholars met which we nowadays call 'Oxford calculators'. Besides Bradwardine we know of William Heytesbury (about 1313–1372), Richard Swineshead (about 1340–1354), and John Dumbleton (died about 1349). Swineshead was so famous that he was known as 'the calculator'; a title of honour if we think of Aristotle being 'the philosopher', Averroes 'the commentator' and Saint Paul 'the apostle'. Starting from Aristotelian philosophy one worked on the 'latitudes of forms'. The theory of latitudes of form is the medieval doctrine of the quantification of qualities [Sylla 1973]. A quality (speed, heat, et cetera) is assigned its distribution of intensity, hence its quantity. From these considerations sprang the first rigorous definition of the notion of *uniform* motion. This is a motion where equal distances are traversed in equal times. Today we say that uniform motion means constant velocity. *Uniform acceleration* was defined as motion in which the velocity increases are constant in equal time intervals. To describe this simple motion today we use modern analysis which was not available in the time of Bradwardine, of course. The Oxford scholars were confined with the notions of their time and that they did. A result of their reflections is the famous *Merton rule*:

Theorem: *If a body is uniformly accelerated from velocity v_0 to v_1 within a certain time interval and traverses a distance s, then the same distance is traversed by a body moving with a constant velocity of*

$$v = \frac{v_0 + v_1}{2}.$$

The velocity v obviously is the momentary velocity of the accelerated body in the midpoint of the traversed route. It is interesting that the Merton scholars have indeed *proven* their rule; very laborious and wordy, to be sure, but with the aid of ingenious considerations [Clagett 1961, Chapter 5].

The Merton rule spread exceptionally fast through Europe and was received in particular in France by Nicole Oresme.

Bradwardine's work *Tractatus de proportionibus velocitatum* of 1328 contains revolutionary thoughts on mechanics besides most interesting considerations concerning calculations with proportions. Euclid in his *Elements* had developed fractional arithmetic as a method to handle integer proportions in books VII and IX. Proportions of this kind allow geometrical visualisations. Following Euclid two proportions $a : b$ and $c : d$ can be composed to give $ac : bd$; hence in modern notation:

$$\frac{a}{b} \cdot \frac{c}{d} = \frac{ac}{bd}.$$

Generating the square $a^2 : b^2$ from a proportion $a : b$ is called the 'double' proportion by Euclid since power calculation and its notations was not yet known. Bradwardine felt assured to call the proportion $\sqrt{a} : \sqrt{b}$ the 'half' proportion of $a : b$ [Hofmann 1951]. Thereby Bradwardine also affected Nicole Oresme since the latter will in fact correctly calculate with 'fractional powers'. However, Oresme's notes got lost in the turmoil of the Hundred Years' War and had to be laboriously re-invented in the 16th century [Hofmann 1951, p. 298].

Even more revolutionary in Bradwardine's *Tractatus de proportionibus velocitatum* are his thoughts concerning the law of motion. He wittily examines the motion of a lever and starts to doubt the Aristotelian law of motion. Following Aristotle the velocity v of a body is given by

$$v = \frac{K}{R}$$

where K is the 'moving force' and R the 'inhibiting resistance' [Hein 2010, p. 164ff.]. Thus there is motion even if $K \leq R$ but Bradwardine is in doubt since empirically there can not be any movement if the resistance outweighs the moving force. In a premonition of the notion of functions Bradwardine wants to find the velocity v depending on an appropriate relation involving K/R. To this end he postulates proportions $a : b$ admissible only when $a > b$ holds. Apparently Bradwardine got this idea when he 'composed' $a : b$ und $b : c$, thus at forming the product $\frac{a}{b} \cdot \frac{b}{c} = \frac{a}{c}$. Only Nicole Oresme will 'prove' this requirement about 1350 in his work *De proportionibus proportionum* since if $a : b < 1$ and $b : c < 1$ hold then 'the whole' $a : c$ would be smaller than the 'parts' $a : b$ and $b : c$.

Aristotle had said that if the proportion $K : R$ is doubled than the velocity is doubled. Bradwardine interprets this 'doubling' as being $(K : R)^2$ and generalises this to $(K : R)^v$. In other (todays) words the velocity hence is proportional to

$$\log \frac{K}{R}.$$

This interpretation of the Aristotelian law of motion was eagerly received; first by the Merton scholars, then also by scientists on the Continent as Nicole Oresme and Albert of Saxony (about 1320–1390).

4.3.5 Nicole Oresme

Nicole Oresme (before 1330–1382) (Nicolas Oresme, Nicholas Oreme or Nicolas d'Oresme) studied theology at the College of Navarre in Paris and later became its head.

He got in contact with the family of the King of France who assigned him the translation of the writings of Aristotle into French between 1370 and 1377.

In 1377 he was elected Bishop of Lisieux and died there five years later.

Oresme counts as one of the most important natural scientists and philosophers of the 14th century. He prepared analytical geometry before Descartes, is said to have specified structural theories of organic molecules long before the 19th century, and having analysed or postulated the free fall and the rotation of the earth before Galilei. Needless to say, all of these assessments are wrong as Marshall Clagett (1916–2005) has made clear in [Clagett 1968]. Nevertheless there are some true facts in these exagerations; Oresme was in fact an exceptionally deep thinker in many areas. He eagerly received the works of the Merton scholars and expanded and developed them and he was intensely concerned with infinite series.

Summation of Infinite Serie

In Oxford Richard Swineshead had solved the following problem [Edwards 1979, p. 91]:

> *'If a point moves throughout the first half of a certain time interval with a constant velocity, throughout the next quarter of the interval at double the initial velocity, throughout the following eighth at triple the initial velocity, and so on ad infinitum; then the average velocity during the whole time interval will be double the initial velocity.'*

Fig. 4.3.11. Nicole Oresme. Miniature from the Traité de l'espère (Bibliothéque Nationale Paris, fonds français 565, fol. 15)

Setting the whole time interval to 1 and defining the initial velocity also as 1 Swineshead had proposed

$$\frac{1}{2} + \frac{2}{4} + \frac{3}{8} + \ldots + \frac{n}{2^n} + \ldots = 2 \tag{4.1}$$

and he also gave a long and uncomfortably verbose proof. Oresme 'saw' an extraordinary elegant proof by means of a purely geometric consideration. He juxtaposed rectangles of height n and width $\frac{1}{2^n}$, $n = 1, 2, 3\ldots$, defining the areas $A_1, A_2, A_3\ldots$ as in figure 4.3.12.

How could Oresme know that the length of the horizontal base line

$$\ell := \frac{1}{2} + \frac{1}{4} + \frac{1}{8} + \ldots + \frac{1}{2^n} + \ldots$$

is finite? The infinite series

$$\ell = \frac{1}{2} + \left(\frac{1}{2}\right)^2 + \left(\frac{1}{2}\right)^3 + \ldots$$

is a geometric series and its value was known long before Oresme. Today we would multiply the sum of the first n terms

$$s_n := \frac{1}{2} + \left(\frac{1}{2}\right)^2 + \left(\frac{1}{2}\right)^3 + \ldots + \left(\frac{1}{2}\right)^n$$

by 1/2:

$$\frac{1}{2} \cdot s_n = \left(\frac{1}{2}\right)^2 + \left(\frac{1}{2}\right)^3 + \ldots + \left(\frac{1}{2}\right)^n + \left(\frac{1}{2}\right)^{n+1}$$

and then subtract from the sum s_n,

$$s_n - \frac{1}{2} \cdot s_n = \frac{1}{2} \cdot s_n = \frac{1}{2} - \left(\frac{1}{2}\right)^{n+1}.$$

This way we have proved

$$s_n = 1 - \left(\frac{1}{2}\right)^n.$$

If now n is increased beyond all bounds then $(1/2)^n$ becomes arbitrarily small. In the limit of $n \to \infty$ we therefore get

$$\ell = \frac{1}{2} + \left(\frac{1}{2}\right)^2 + \left(\frac{1}{2}\right)^3 + \ldots = 1. \tag{4.2}$$

The area of one rectangle A_n apparently is $F(A_n) = \frac{n}{2^n}$ and hence the whole

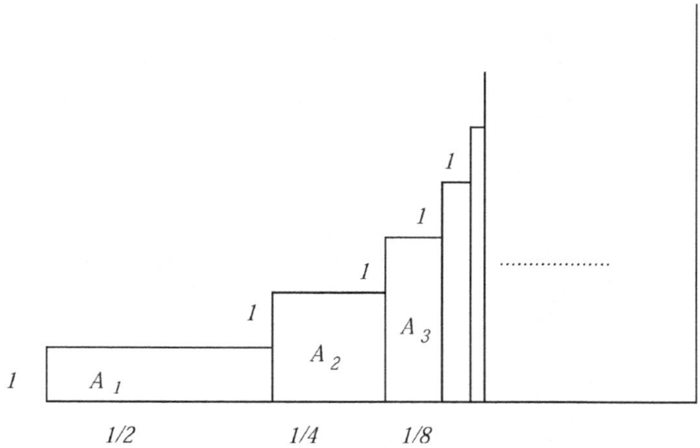

Fig. 4.3.12. Nicole Oresme's proof of Swineshead's sum, Part 1

area is

$$\frac{1}{2} + \frac{2}{4} + \frac{3}{8} + \ldots + \frac{n}{2^n} + \ldots .$$

But this is exactly Swineshead's sum (4.1) which we want to compute. Oresme now simply divides the areas shown in figure 4.3.12 in a different way as shown in figure 4.3.13. The overall area has apparently not changed. Hence

$$\frac{1}{2} + \frac{2}{4} + \frac{3}{8} + \ldots + \frac{n}{2^n} + \ldots = B_0 + B_1 + B_2 + \ldots$$

has to hold. Now $B_0 = 1$, $B_1 = 1/2$, $B_2 = 1/4$ and in general $B_n = (1/2)^n$, hence

$$B_0 + B_1 + B_2 + B_3 + \ldots = 1 + \underbrace{\frac{1}{2} + \left(\frac{1}{2}\right)^2 + \left(\frac{1}{2}\right)^3 + \ldots}_{=1} = 2,$$

since the underbraced infinite series obviously is our geometric series (4.2). Thus Swineshead's formula (4.1) has been rigorously proven.

Oresme also was the first to prove the divergence of the harmonic series

$$1 + \frac{1}{2} + \frac{1}{3} + \frac{1}{4} + \frac{1}{5} + \frac{1}{6} + \ldots$$

namely exactly in the way we still use today. He replaced the series by one with a smaller sum which is already known to be divergent. Thus the harmonic series can not have a finite value.

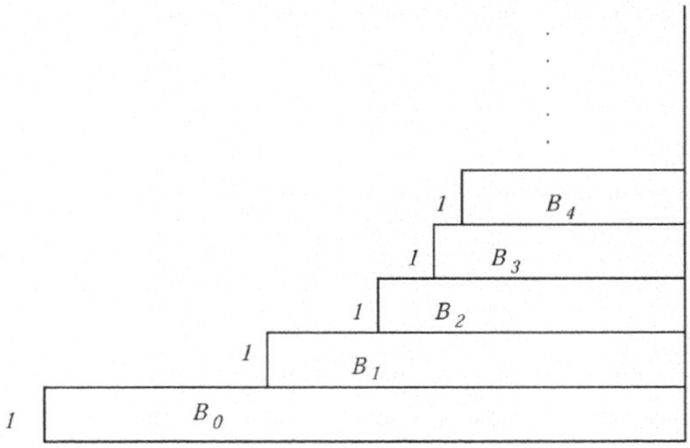

Fig. 4.3.13. Nicole Oresme's proof of Swineshead's sum, Part 2

To arrive at his reference series Oresme recognised:

$$\frac{1}{3}+\frac{1}{4}>\frac{1}{4}+\frac{1}{4}=\frac{1}{2},$$
$$\frac{1}{5}+\frac{1}{6}+\frac{1}{7}+\frac{1}{8}>\frac{1}{8}+\frac{1}{8}+\frac{1}{8}+\frac{1}{8}=\frac{1}{2},$$
$$\frac{1}{9}+\frac{1}{10}+\frac{1}{11}+\frac{1}{12}+\frac{1}{13}+\frac{1}{14}+\frac{1}{15}+\frac{1}{16}>\frac{8}{16}=\frac{1}{2},$$

and so on. Hence he always finds terms which together are larger than 1/2. Thus it holds

$$1+\frac{1}{2}+\frac{1}{3}+\frac{1}{4}+\frac{1}{5}+\frac{1}{6}+\ldots>\frac{1}{2}+\frac{1}{2}+\frac{1}{2}+\frac{1}{2}+\ldots$$

and the geometric series can not have a finite sum.

Latitudes of Form and the Merton Rule

We have to come back once more to the Aristotelian theory of the latitude of forms which Oresme has captured in graphical form. The quality of a body has an extension (*extensio*) and an intensity (*intensio*). Looking at the velocity of a moving body and the Merton rule Oresme represents – in a sense in an anticipation of a Cartesian coordinate system – the extension (=time) as abscissa, and the intensity (= velocity) perpendicular as ordinate axis. Then he classifies the quantity as *uniformis*, *uniformiter difformis*, or *difformiter difformis*. We can think of the uniform motion of the Merton scholars as *uniformis* and accordingly the uniformly accelerated motion as *uniformiter difformis*. The kind of motion which can be subsumed under the third category can not be analysed without the calculus of Newton und Leibniz.

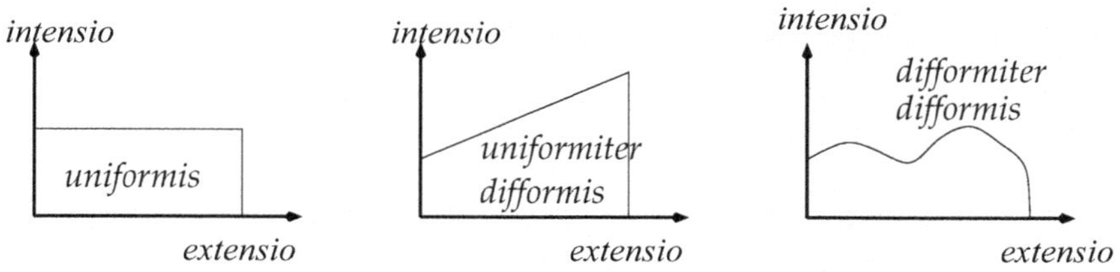

Fig. 4.3.14. Oresme's graphical presentations

A graphical presentation alone would not have been sufficient to secure Oresme a place under the successors of the Merton scholars. He recognises that the Merton rule follows easily in the framework of his diagrams! In his *Tractatus de configuratione intensionum* he writes [Becker 1964, p. 132f.]:

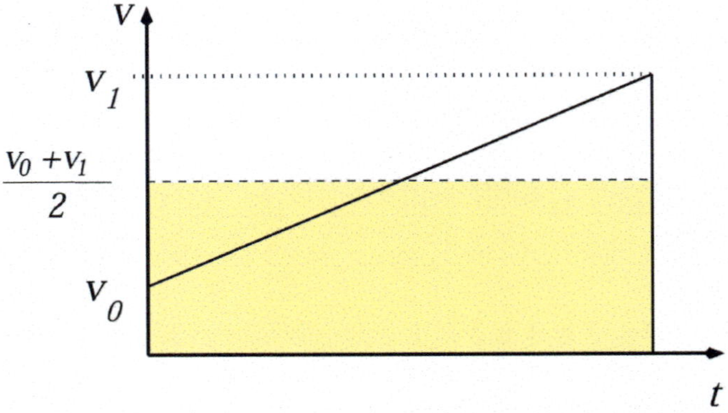

Fig. 4.3.15. The Merton rule in Oresme's diagram

> *'Every uniformly non-uniform quality has the same quantity as if it would belong uniformly to the same object at the degree of the midpoint.'*
>
> (Jede gleichförmigerweise ungleichförmige Qualität hat dieselbe Quantität, als wenn sie gleichförmig demselben Objekt zukommen würde mit dem Grade des mittleren Punktes.)

Here the 'quantity' is nothing else than the area under the curve in Oresme's diagram. Or, in modern language, the integral of the velocity over time. Since velocity is the first derivative of the distance with respect to time the area under the velocity-time curve hence represents the distance travelled. Thus Oresme has given the Merton rule the form of the diagram as shown in figure 4.3.15. In a certain time interval a uniformly accelerated body traverses the same distance as a body moving with uniform speed $v = (v_0 + v_1)/2$.

The Doctrine of Proportions

In the work *De proportionibus proportionum* Oresme is concerned with proportions of the classical form $a : b$ and specifies some rules to calculate with them. His statements come rather close to the rules for calculating with rational exponents and on this field Oresme is directly influenced by Bradwardine. However he remained firmly with the classical notation of proportions. He writes that a proportion $A = a_1 : a_2$ can be assigned to another proposition $B = b_1 : b_2$ so that $B = A \cdot A \cdot A$ if a geometric mean is to be inserted:

$$b_1 : a_1 = a_1 : a_2 = a_2 : b_2.$$

Then apparently it follows that $b_1 = \frac{a_1^2}{a_2}$ and $b_2 = \frac{a_2^2}{a_1}$ and therefore $\frac{b_1}{b_2} = \left(\frac{a_1}{a_2}\right)^3$. Oresme then asks for a representation of A and in modern notation this is $A = B^{\frac{1}{3}}$, of course. However, Oresme could not rely on this modern notation. He expresses this fact rather via proportions of proportions [Gericke 2003, p. 154].

Oresme has also conjectured that not all irrational numbers may be expressible by roots but he can not prove it. He even suspected that there are more irrational than rational numbers.

4.4 Scholastic Dissenters

We have seen that the Aristotelian continuum certainly was much discussed in Christian scholasticism but also widely accepted. However, we find a number of great thinkers who deviated from this opinion. Albertus Magnus already had the idea of a *real* continuum divisible only up to a certain degree. His pupil Thomas Aquinas further developed this idea. The black friar Thomas Aquinas became one of the most influential philosophers of the Christian Middle Ages and one of the most influential philosophers and theologians of all time. Born close to the town of Aquino on Castle Roccasecca as seventh son of a duke he became a member of the Dominican Order against the will of his family. In fact the family even kept him imprisoned for some time. But all attempts to dissuade him from his purpose finally failed. In 1248 he became a pupil of Albertus Magnus in Cologne. This pupil-master relationship lasted until 1252; then we we find Thomas working as a teacher in Paris, Rome, Viterbo, and Orvieto. He quickly pursued a career in his order and went to Naples in 1269 to found a Dominican school. According to one of his male secretaries he dictated three or four different texts at the same time. This could explain the unbelievably large number of his writings but can also serve to illuminate the enormous size of his intellectual gifts. On 7th March 1274 we find Thomas travelling to the Second Council of Lyon. He rests in Fossanova Abbey in the Italian municipality Priverno where he dies under unclear circumstances. In Dante's *Divine Comedy* we find the hint in *Purgatorio XX 69* that the King of Sicily, Charles I of Anjou, is said to have poisoned him [Dante 1920]:

> Charles came to Italy; and there, to make
> amends, a victim made of Conradin; and then,
> to make amends, drove Thomas back to Heaven.

Following other sources it is said that he was offered a poisoned confectionery by the king's personal physician; still other sources do not mention any kind of poisoning. Thomas was canonised in 1323; since 1567 he is revered as a Doctor of the Church.

Neidhart rates him among the atomists in [Neidhart 2007, p. 571]; though he advocated a particular form of atomism which is denoted as 'atomism of form'. This kind of atomism was also advocated by Thomas's pupil Giles of

Fig. 4.4.1. Aristotle, Thomas Aquinas, and Plato in the painting *Triumph of St Thomas Aquinas* by Benezzo Gozzoli (Louvre, Paris)

Rome (Aegidius Romanus) (1247–1316). Matter is indefinitely divisible in an abstract sense but a real body will deny this divisibility if a certain threshold is reached because the substance of matter can only be divided down to a certain minimal quantum. In principle Thomas walked in the footsteps of his teacher Albertus who can therefore also be ranked among the atomists of form [Neidhart 2007, p. 571f.].

There were also 'genuine' atomists, of course, who can be filed in two different groups. One group assumed the continuum consisting of finitely many points; the other favoured the conception of the continuum as consisting of infinitely many points [Maier 1949]. In this second group there was still another distinction. Either the points were directly adjacent to each other (*puncta immediate coniuncta*) or they were separated by other points lying in between (*puncta ad invicem mediata*). These two conception were also described by Bradwardine in *De continuo*. Apparently both conceptions seem strange to us but one should keep in mind that the authors concerned did not discuss a mathematically abstract continuum but one which existed in reality. The 'finite atomists' walked in the footsteps of Democritus and assumed extended particles yet having the property of indivisibility.

4.5 Nicholas of Cusa

In the small village of Kues (Cusa in Latin; today Bernkastel-Kues) at the Middle Moselle in a house close to the river Niklas Chryppfs or Krebs (1401–1464), called Nicholas, was born in 1401 [Flasch 2004], [Flasch 2005], [Flasch 2008]. Nicholas will become the greatest philosopher of the 15th century and as a mathematician he will exert a groundbreaking effect over centuries to come. Eberhard Knobloch has identified Nicholas as a starting point of a chain of conceptions of infinity [Knobloch 2004] reaching out to Galileo Galilei (1564–1642) and culminating in Gottfried Wilhelm Leibniz (1646–1716); but more on this later. Nicholas's father was a wealthy merchant and ship owner and he allowed his son to study at the Faculty of Arts at the University of Heidelberg. At age 15 Nicholas started his studies in 1416.

Fig. 4.4.2. Thomas Aquinas (Painting by Carlo Crivelli, 1476); Nicholas of Cusa (Painting in the hospital of Kues)

In philosophy the quarrel between the nominalists and the realists was an actual theme – we are halfway into the famous problem of universals in which the question is discussed whether there really exist general notions like 'humankind', 'number', and so on. *Realism* does *not* mean that the opinions of the philosophers of this kind of philosophical thinking were particularly 'realistic'. They were Platonists who were convinced of the reality of the 'ideas' in Plato's heaven of ideas. The *nominalists* in contrast were convinced that these notions were rather human constructions (and therefore only names) [Stegmüller 1978]. Among the medieval philosophers we have already discussed Anselm of Canterbury who was a realist – even a 'strong' realist. More moderate realists were Albertus Magnus, Thomas Aquinas, Avicenna and Averroes. One of the founders of the strong nominalism was Roscelin of Compiègne (about 1050–about 1124) from France, one of the teachers of Abelard. In the eyes of Roscelin only those objects exist which can be perceived by our sensory organs. Anyway, in Heidelberg one followed the Parisian fashion at the time and fostered a moderate nominalism.

One year after he started his studies Nicholas already left Heidelberg; as Flasch has conjectured in [Flasch 2004, p. 11] he could not find a home in the intellectual climate of Heidelberg. He went to the University of Padua and began to study law which he finished in 1423 when becoming a Doctor of Canon Law. He was just 22 years old. But why did he feel so much better in Padua than he did in Heidelberg? The final hour of scholasticism had come; the age of the Renaissance began in Italy! Scholars had stopped there to delve delightfully into the writings of Aristotle but began to look for new shores in a free mind and Padua was the centre in this climate of intellectual departure. Sciences were no longer viewed as being 'complete', no longer seen as the field of clever remarks commenting on the works of the ancient philosophers, but as something which had to be thought of again and to discover anew. To be sure, the Greek antiquity was not yet set aside and even attracted an increased interest in architecture, historiography, and also mathematics. One wanted to go back to the sources; not only as far as the idea of man was concerned but also in the natural sciences.

Nicholas met physicians, philosophers, artists, and astronomers at Padua and he made important friends, for example his fellow student Giuliano of the influential family of Cesarini who became a Cardinal already at the age of 28 and to whom Nicholas dedicated two of his books. Nicholas also made friends with Paolo dal Toscanelli (1397–1482) who studied medicine and mathematics. If we believe Flasch [Flasch 2004, p. 15] then Paolo was a better mathematician than Nicholas because he was more precise and critical. He had worked on the construction of the large Dome of Florence Cathedral; ten years after the death of Nicholas he had given advice to attempt a westward sea expedition to get to India in a letter to the Portuguese Martins. This letter became very important to a mariner from Genoa in Spanish service because he copied it with his own hands. This mariner was Christopher Columbus (about

Fig. 4.5.1. (left) Map by Paolo dal Pozzo Toscanelli (right) showing the assumed westward sea route to India (America was unknown then)

1451–1506). It was this intellectual climate in which Nicholas of Cusa could flourish. He experienced an attitude which we nowadays call 'humanism'.

After his doctorate he left Padua and went into the service of the Elector of Trier where he worked as a lawyer and did further studies at the University of Cologne. There he began collecting and searching for rare manuscripts in 1425; an occupation in which Nicholas excelled. Some of the manuscripts he had found served the great humanists later as a foundation for their thinking. He even visited Paris in 1428 to search for old manuscripts. In the time period between 1432 and 1437 Nicholas attended the Council of Basel which was led by his friend Giuliano. Nicholas became an important thought leader of the Council which attempted to restrict the power of the Pope and to bring Christendom closer together again. However, this goal could not be achieved. The Turks threatened Christendom in the East, the Emperor of Byzantium requested military help, and the Pope achieved the surrender of the Eastern Church under Western theology [Flasch 2004, p. 27]. Disappointed by the meagre results of the Council Nicholas and Giuliano changed sides in 1437 and entered the Pope's camp.

From that moment on Nicholas's rapidly advanced in his career. His diplomatic talents were in demand; he travelled a lot on papal missions, worked as a mediator in several cases of conflict, and authored scientific works. Pope Nicholas V, a personal friend of our Nicholas's, made him a Cardinal "in petto" in 1448. The official appointment as Cardinal followed in 1450 when Nicholas became the Bishop of Brixen in South Tyrol. He has again to travel in service of the Holy See and to mediate. In the year 1453 Constantinople fell to the Turks. Nicholas came in conflict with Duke (later Archduke) Sigismund of Austria but in 1458 Pope Pius II ordered Nicholas to come to Rome. An attempt to return to Brixen failed in 1460. On the 11th August 1464 Nicholas died in the town of Todi in Italy while preparing a Crusade against the Turks.

4.5.1 The Mathematical Works

Nicholas of Cusa is *not* a mathematician but he uses mathematics to gain philosophical and theological knowledge. In the eleventh chapter of the first book of his work *De docta ignorantia* (On learned ignorance) he writes [Hopkins 1985, p. 18f.] (translated from [Nikolaus 2002a, p. 43]):

> *'But all perceptible things are in a state of continual instability because of the material possibility abounding in them. In our considering of objects, we see that those which are more abstract than perceptible things, viz., mathematicals, (not that they are altogether free of material associations, without which they cannot be imagined, and not that they are at all subject to the possibility of changing) are very fixed and are very certain to us. Therefore, in mathematicals the wise wisely sought illustrations of things that were to be searched out by the intellect. And none of the ancients who are esteemed as great approached difficult matters by any other likeness than mathematics. Thus, Boethius, the most learned of the Romans, affirmed that anyone who altogether lacked skill in mathematics could not attain a knowledge of divine matters.'*

And somewhat further on [Hopkins 1985, p. 19] (translated from [Nikolaus 2002a, p. 45]):

> *'Proceeding on this pathway of the ancients, I concur with them and say that since the pathway for approaching divine matters is opened to us only through symbols, we can make quite suitable use of mathematical signs because of their incorruptible certainty.'*

All of Nicholas's mathematical works aim at two targets: The quadrature of the circle and the rectification of the circumference of the circle (which is the calculation of the circle's circumference). Hofmann [Nikolaus 1952, p. X] is certain that Nicholas knew Bradwardine's *Geometria speculativa* although neither Bradwardine nor Euclid are even mentioned in *Docta ignorantia*. It is clear to Nicholas that the existence of a square coextensive to a circle does by no means follow from the existence of circumscribed and inscribed squares of the circle. He also knows of Aristotle's opinion that circular arcs can not be in proportion with straight lines because the proportion (ratio) would be irrational and therefore a quadrature of the circle impossible. Apparently he is unaware of the quadrature of the lunes.

Knowledge in the sense of Nicholas of Cusa is threefold: *sensibilis, rationalis, intellectualis.* A mathematician is not concerned with entities of the real world because they all are imperfect. He is rather concerned with ideal, purely intellectual entities which fall in the realm of rational knowledge. But if rational means fail, for example in examining the infinitely small or the infinitely large, then *visio intellectualis* is needed which supersedes *visio rationalis.* Imagining a circle with infinite radius the arc becomes equal to a straight line. Increasing the number of edges of a polygon infinitely then finally a circle will emerge, and so forth. The method used by Nicholas of Cusa hence is a two-step method: First carry out a 'limit' to infinity, then advance to theological interrogations. It is in this sense that 'God' is the name of the infinite circle. In the eyes of Nicholas all contradictions cease to exist. The infinitely large is the absolute maximum, the infinitely small is the maximum of smallness. Abstracting from the notions of 'largeness' and 'smallness' Nicholas arrives at the conception that maximum and minimum coincide in the infinite. On the one hand Nicholas builds on the Aristotelian doctrine of quantities and on the other hand on Euclid's doctrine of extent [Knobloch 2004, p. 491]. Quantity is everything divisible – indivisibles therefore can not be quantities. Mathematics is the science of finite quantities and every quantum allows for a more or less. A quantum is arbitrarily divisible but a smallest part can never be reached. In the same manner a quantum can be arbitrarily enlarged without ever reaching the largest. The infinitely large as well as the infinitely small (=indivisibles) hence are 'non-quanta' in the eyes of Nicholas. As the indivisibles also the infinite has no parts so that Euclid's doctrine

> The whole is larger than the part

is *not* valid at infinity (large or small!) in the opinion of Nicholas! Indivisibles therefore are not mathematical objects since an advance to infinity cannot actually occur. Knobloch summarises this fact in modern terminology as [Knobloch 2004, p. 492]: 'The limit of a convergent sequence which is not constant is not one of its elements', or, even stronger: 'A transfinite number[3] is not just a special case of a real number'. Nicholas thereby arrived at the conclusion that an exact quadrature of the circle is not possible. However, approximately it is possible up to an arbitrary accuracy.

We will find these thoughts on infinity again in Galileo Galilei who resumed Nicholas's idea of the dichotomy between 'quanta' and 'non quanta'.

---

[3] We will come back to transfinite numbers only in the later sections when discussing the mathematical achievements of Georg Cantor. For now it suffices to translate 'transfinite' as 'beyond all numbers'.

Contributions to Analyis in the European Middle Ages

betw. 475/480–betw. 524/526	BOETHIUS writes commentaries of the works of Aristotle and left own mathematical writings
about 700	BEDE VENERABILIS demands that in every cloister in England the date of Easter should be computed correctly
about 800	ALCUIN OF YORK organises the education in the Franconian realm. He founds cloister schools and introduces *computus* as teaching subject
about 1000	The monk GERBERT OF AURILLAC, later Pope SILVESTER II, collects Arabian knowledge and manuscripts. He raises the curiosity concerning sciences in the West
about 1100–1300	Heyday of the translators, mainly in Spain. Since the conquest of Toledo in 1085 by the Christians antique knowledge is available on a large scale
about 1175–1253	ROBERT GROSSETESTE founds the scientific tradition of Oxford
about 1200–1280	ALBERTUS MAGNUS discusses the nature of the continuum and the doctrine of motion of Aristotle
1214–1292 or 1294	ROGER BACON discusses the paradoxes of the infinite and the structure of the continuum as his teacher Robert Grosseteste before him at Oxford
about 1290–1349	THOMAS BRADWARDINE shows an acute mind when discussing properties of the continuum. Together with other scholars of Merton College he works on the 'Merton rule' in which kinematical ideas are expressed for the first time
before 1330–1382	NICOLE ORESME receives the new ideas from England eagerly and develops them further. Brilliant treatment of infinite series
1438–1440	*De docta ignorantia* is written by NICHOLAS OF CUSA
1445–1459	NICHOLAS OF CUSA writes his mathematical works

5 Indivisibles and Infinitesimals in the Renaissance

T. Sonar, *3000 Years of Analysis*, https://doi.org/10.1007/978-3-030-58223-4_5

1434–1498	Florence under the rule of the House of Medici
about 1445	Invention of book printing with moveable types by Johannes Gutenberg
1452–1519	Leonardo da Vinci
1453	Turks conquer Constantinople; End of the Byzantine Empire
1471–1528	Albrecht Dürer
1475–1520	Michelangelo
1492	Rediscovery of America by Columbus
1509–1547	Henry VIII King of England
1517	Start of the German Reformation. Martin Luther posts the Ninety-five Theses against the sale of indulgences at the entrance door of the Castele Church (Schloßkirche) in Wittenberg
1519–1556	Charles V, Holy Roman Emperor
1521	Diet of Worms. Luther suffers the Imperial ban
1524–1525	Peasant's war in the German Lands
1530	Augsburg Confession at the Diet of Augsburg
1534	The Spanish Ignatius of Loyola founds the 'Society of Jesus' (Jesuits)
1546–1547	Schmalkaldic War
1555	Augsburg Settlement
1545–1563	Council of Trent; Reform of the Catholic Church
1548–1603	Elisabeth I Queen of England
1556–1598	Philipp II King of Spain
1556–1612	Reign of Emperors Ferdinand I, Maximilian II, Rudolf II
1560	The first Academy in early modern Europe is founded in Naples
1582	Calendar reform by Pope Gregory XIII: The Julian calendar is (initially in Catholic countries only) replaced by the Gregorian calendar
1588	Loss of the Spanish Armada
1590	The dome of St Peter's Basilica in Rome is completed
1571	On 7th October the Christian Holy League under Don Juan d'Austria defeats the Turkish fleet at the Battle of Lepanto
1579	Union of Utrecht; Alliance of the Northern Netherlands against Spain
1593–1609	War with the Turks
1610	Galileo Galilei publishes sensational astronomical discoveries made by means of a telescope
1612–1619	Matthias emperor of the Holy Roman Empire
1618	Defenestration of Prague; The Thirty Years' War begins
1619–1637	Ferdinand II emperor of the Holy Roman Empire
from 1615	Wallenstein's meteoric rise as warlord on the Catholic side
from 1630	The Swedes under Gustav II Adolf intervene in the Thirty Years' War
1634	Wallenstein assassinated
1648	End of the Thirty Years' War

5.1 Renaissance: Rebirth of Antiquity

The time period from the beginning of the 15th to the end of the 16th century is called the epoch of the Renaissance. A more precise temporal localisation is difficult. The period from about 1420 to 1500 is called the early Renaissance. Very significant is the perspective in painting and drawing and the admiration of the visual arts of the Greeks which fuelled the new curiosity concerning antiquity initiated in Italy. Between about 1500 to 1530 the High Renaissance can be located in which Leonardo da Vinci (1452–1519) created his most famous paintings, Michelangelo Buonarroti (1475–1564) created the frescoes in the Sistine Chapel, and Albrecht Dürer (1471–1521) attracted attention north of the Alps with his copper engravings. The period starting with the death of Raphael (1483–1520) until about 1600 is known as the Late Renaissance which became known as 'manierism' in the arts; a word coined by Jacob Burckhardt (1818–1897).

In the 15th century a fundamental change in European societies emerged due to the influence of the Renaissance. People tried to free themselves from the bonds of an other-directed life dictated by the church and of hierarchically organised structures. They turned to an existence which was more firmly characterised by the individual. Strong impulses came from

Fig. 5.1.1. The School of Athens ([Fresco by Raphael 1510/1511] in the Vatican in the Stanza della Segnatura). Many well known scholars of Greek antiquity are shown on this fresco. Plato and Aristotle can be found in the centre of the painting in the background; Aristotle holds his Ethics in his hand. In the lower left Pythagoras reads a book. With Averroes, also depicted in this painting, scholars having transferred knowledge from the world of antiquity are appreciated

Fig. 5.1.2. The great humanist Erasmus of Rotterdam (right) [Painting by Hans Holbein the Younger 1523] prepared the Reformation by means of his writings but distanced himself when Martin Luther (left) [Painting by Lucas Cranach the Older 1529] carried out the separation from the Roman-Catholic Church

the scholars of the Byzantine Empire who fled from Constantinople to the towns of Northern Italy after Constantinople fell to the Turks in 1453. The change was also supported by the emerging humanism which put man back into the centre of the world according to the admired philosophy of antiquity. Humanism also propagated the unfolding of man's powers in the ethical-philosophical field. Humanism became a comprehensive educational movement well up until our time in which the French philosopher and writer Jean-Paul Sartre (1905–1980) founded the 'existentialist humanism'. In the Renaissance (Desiderius) Erasmus of Rotterdam (1465 or 1469–1536) became the leading figure of humanism. Especially in Germany humanism had a strong political component directed against the pope and the Roman-Catholic Church and finally leading to Martin Luther's (1483–1546) Protestant Reformation. However, most of the humanists could not follow further due to the radicalism of a new founded Church.

In the year 1492 the Genoese mariner Christopher Columbus (about 1451–1506) in Spanish service discovers America which he thought was India. New horizons literally open up. About 1450 Johannes Gensfleisch (1400–1468), called Gutenberg, invents book printing with moveable types. This is a true breakthrough as far as the reproduction of books and other writings of all sort is concerned. There were, however, already a lot of techniques of printing books around but only Gutenberg puts them all together, refines and expands them, so that printing books now becomes feasible. Without this type of book printing neither the Reformation nor the fast spread of scientific knowledge in the Renaissance would have been possible.

Nicolaus Copernicus (1473–1543), canon at Frombork Cathedral and astronomer, publishes in *De revolutionibus orbium caelestium* (On the Revolutions of the Celestial Spheres) the world model with the sun being located in the centre and the planets revolving around it. The book is published in the year of the death of its author and supersedes the geocentric world model of Claudius Ptolemy (about 100–about 170) which was accepted since about 150 AD. Although many authors like to speak about a 'scientific revolution' when it comes to the publication of *De revolutionibus* – the invention of this phrase going back to Thomas S. Kuhn [Kuhn 1957], [Kuhn 1962] – the Copernican system was by no means simpler than the one given by Ptolemy. Ptolemy had had to introduce epicycles to bring the orbits of the planets in accordance with the observations. Now Ptolemy needs twice the number of epicycles in his heliocentric model [Neugebauer 1969b, p. 204]! It is only Johannes Kepler (1571–1630) who will free the theory from circular orbits and will recognise ellipses as being the true orbits.

Not surprisingly the church had great problems with a heliocentric system since contradictions with some verses of the bible were inevitable. At last man was at the centre of God's creation and therefore had to be located in the centre of the cosmos. In the Book of Joshua of the Old Testament in the King James Bible we read in Chapter 10, Verses 12–13:

> (12) Then spake Joshua to the LORD in the day when the LORD delivered up the Amorites before the children of Israel, and he said

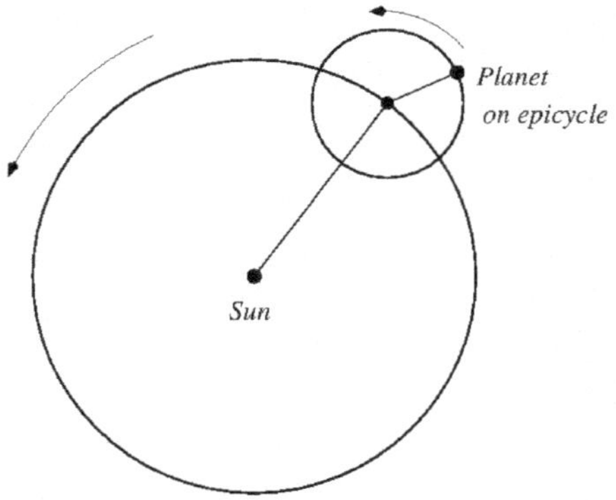

Fig. 5.1.3. Ptolemaic system with epicycles. Besides other auxiliary constructions Ptolemy introduced epicycles going back probably to Apollonius of Perga (about 262–about 190 BC). In observation some planets, e.g. Mars, showed retrograde motion. To realise such retrograde movement in the framework of circular orbits placed the planet on a smaller circle on which he rotates while the smaller circle – the epicycle – itself rotates on a larger circle around earth. The larger circle is called the *deferent*. It was Copernicus (1473–1543) who placed the sun in the centre of the cosmos

> in the sight of Israel, Sun, stand thou still upon Gibeon; and thou, Moon, in the valley of Ajalon.
>
> (13) And the sun stood still, and the moon stayed, until the people had avenged themselves upon their enemies. *Is* not this written in the book of Jasher? So the sun stood still in the midst of heaven, and hastened not to go down about a whole day.

If God could order the sun to complete standstill how than could the sun *not* orbit around the earth? However, publications concerning the Copernican system were possible if the authors would not forget to clearly call it a *model*! Galileo Galilei (1564–1642) got in great trouble only when he declared the Copernican cosmos to be the true, actual system.

In mathematics the view also turned back to the ancients. In analysis the computation of barycentres of certain figures and bodies by Archimedes were taken up, newly reflected, and superseded.

5.2 The Calculators of Barycentres

In the second half of the 16th century a certain culture emerged concerning the calculation of barycentres. This emergence rested on the admiration of the work *On the equilibrium of planes or the centres of gravity of planes* written by Archimedes, see chapter 2. Interest focused on the calculation of barycentres of three-dimensional bodies since little was handed down by Archimedes in this area. Although it may seem that we are discussing a secondary battlefield now the calculation of barycentres have indeed played an important role in the development of the differential and integral calculus. One can recognise the importance of this field for example when reading the work *Analysis tetragonistica ex centrobarycis* by Gottfried Wilhelm Leibniz which was written in October 1675 [Leibniz 2008, p. 263–269], English translation in [Child 2005, p. 65ff.].

Francesco Maurolico (1494–1575) was a Benedictine abbot from Sicily stemming from a Greek family which had left Constantinople for Sicily when the capital of the Byzantine Empire was conquered by the Turks. Maurolico is ranked as a polymath of the 16th century who translated numerous works of ancient writers; among others Euclid and Archimedes. He certainly was a great geometer in that he introduced new methods of surveying; he even made a name as an historian, an astronomer, and, of course, as a mathematician. Only a few of his works were published during his lifetime; most of them are in the form of handwritten manuscripts as is the work *De momentis aequalibus*. A complete edition of his works has recently started in Pisa. In *De momentis aequalibus* Maurolico starts with a theorem on the barycentre of several weights mounted on a rod [Baron 1987, p. 91f.].

Fig. 5.2.1. Left: Detail of a painting of the 16th or 17th c. [probably by Hendrick van Balen]. It could bear the title 'every thing has its measure', but learning and the transmission of knowledge is also made a subject. In Europe different measures of length (e.g. foot, ell), weights, and solid measures were employed. As far as the applications are concerned measurement is clearly in the focus. Right: Francesco Maurolico [Engraving: M. Bovis after Polidoro da Caravaggio]

If in figure 5.2.2 it holds

$$AE = CG, EF = GH, BF = DH$$

and

$$I : O = K : P = M : Q = N : R,$$

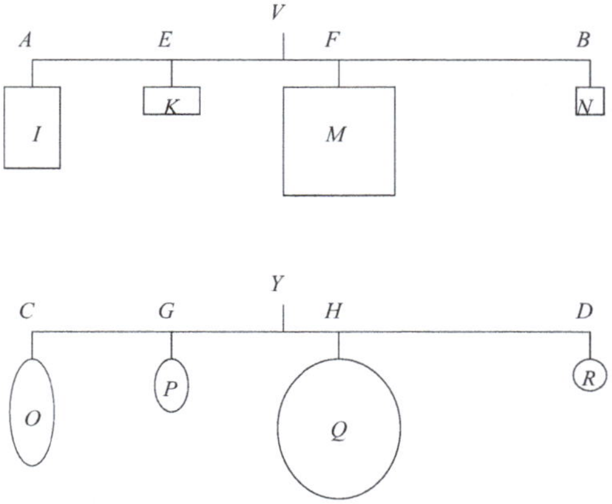

Fig. 5.2.2. Different weights at a rod

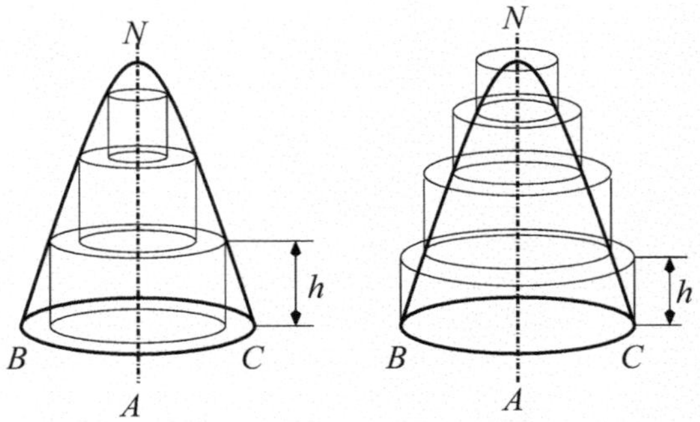

Fig. 5.2.3. Inscribed and circumscribed cylindrical pieces at a paraboloid of rotation

then also $VB = YD$ holds. This theorem serves to prepare a subtle process of weighing which Maurolico explains in the case of a paraboloid of rotation. To this end a section of a paraboloid of rotation is looked at where the longitudinal axis AN is divided in n equal parts of the height h as shown in figure 5.2.3. Therefore the height AN is just nh. Then Maurolico imagines cylindrical pieces of height h being inscribed in and circumscribed around the paraboloid as in figure 5.2.3. Let us first discuss the inscribed cylinders. Let the volumes of the inscribed cylinders be denoted by $v_1, v_2, \dots v_{n-1}$ and the corresponding radii by $r_1, r_2, \dots, r_{n-1}$. Since all cylindrical pieces have the same height h their volume is proportional to r^2 and it holds

$$\frac{v_1}{r_1^2} = \frac{v_2}{r_2^2} = \frac{v_3}{r_3^2} = \dots = \frac{v_{n-1}}{r_{n-1}^2}.$$

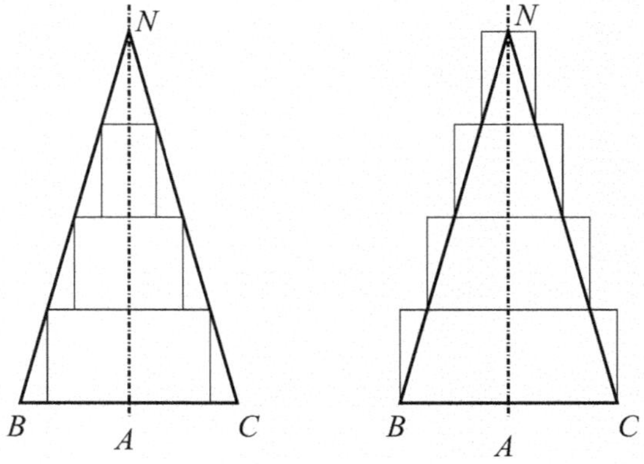

Fig. 5.2.4. Comparison areas at the triangle

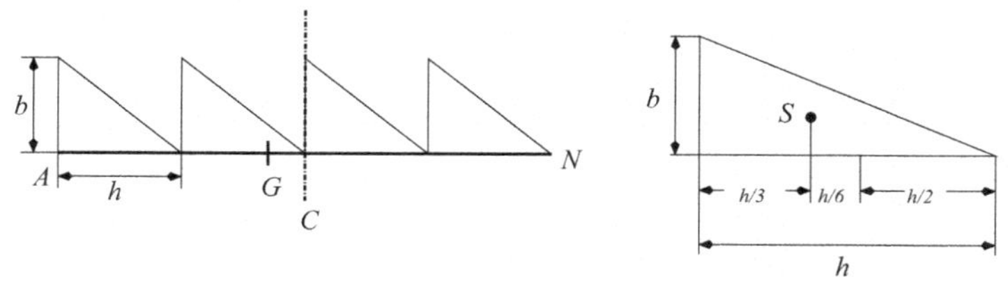

Fig. 5.2.5. Triangles on the lever arm

Now the basic body is a paraboloid, hence also

$$r_1^2 \sim h, \quad r_2^2 \sim 2h, \quad r_3^2 \sim 3h, \quad \ldots$$

holds, so that eventually

$$\frac{v_1}{1} = \frac{v_2}{2} = \frac{v_3}{3} = \ldots$$

follows. And here comes an ingenious idea of Maurolico. He looks for comparison areas in the cylindrical pieces inside his paraboloid and finds them in a cone and a triangle, respectively, fitting the same sizes as found at the paraboloid. Namely, in case of the rectangles inside the triangle shown in figure 5.2.4 their areas relate as

$$1 : 2 : 3 : 4 : \ldots.$$

Hence on a lever arm one can replace the cylindrical pieces by the corresponding rectangles.

Up to now we have only discussed the case of the inscribed cylinders; however, in the case of the circumscribed ones all what was just said applies analogously.

Now Maurolico has to determine the actual position of the barycentre and places n right-angled triangles on a lever arm of length nh; this being just the length of AN.

Since the barycentre S of a single triangle is located at one third of its height as depicted in figure 5.2.5 the barycentre G of the n triangles on the lever arm has to be in a distance of $h/6$ from the midpoint C, i.e.

$$GN = \frac{nh}{2} \pm \frac{h}{6},$$

the sign depending on the arrangement of the triangles on AN. The same result holds even if the triangles are arbitrarily shifted vertically, for example as in figure 5.2.6. Therewith Maurolico can now view the triangular area in figure 5.2.4 as being the area of the rectangles from which triangles are being subtracted and added, respectively. The area of one shaded triangle is $h \cdot b/2$

where b is obviously given by $b = r/n$ if r denotes the length of the segment $AC = AB$. Hence the total area of all the small triangles is

$$2 \cdot n \cdot \frac{h \cdot b}{2} = nhb = hr.$$

Denoting the total area of the large triangle by F we have

$$F = nhr.$$

Thus the area of the rectangles in the inscribed case result in

$$nhr - hr = F - \frac{F}{n} = F\left(1 - \frac{1}{n}\right) \tag{5.1}$$

while in the circumscribed case we get

$$nhr + hr = F + \frac{F}{n} = F\left(1 + \frac{1}{n}\right).$$

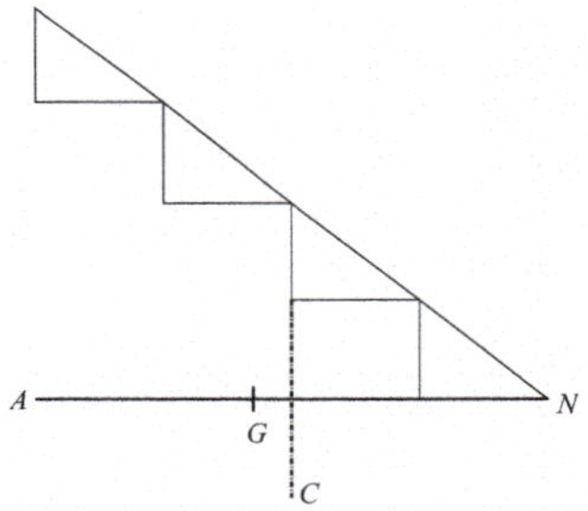

Fig. 5.2.6. Shifted triangles on the lever arm

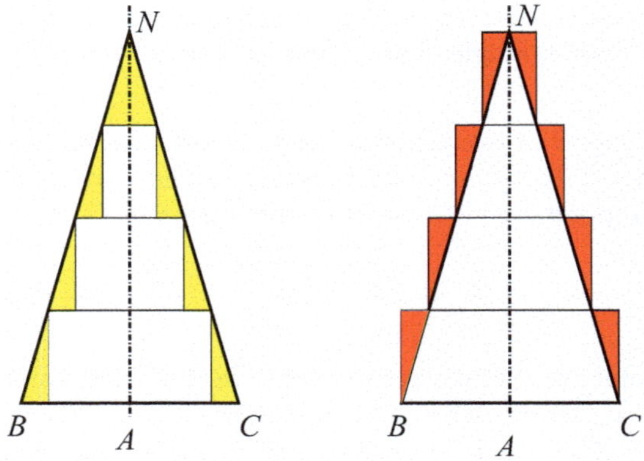

Fig. 5.2.7. Triangular area as sum or difference

Computing the moments of the lever arm with respect to the point N we get

$$M := \underbrace{F \cdot \frac{2}{3} nh}_{\text{moment of the large triangle}} \mp \underbrace{\frac{F}{n} \cdot \left(\frac{nh}{2} \pm \frac{h}{6}\right)}_{\text{moment of the small triangles}(=\frac{F}{n}\cdot GN)}$$
$$= F \cdot \left(\frac{n \mp 1}{n}\right) \cdot G_{\square} N,$$

where $G_{\square}$ now denotes the barycentre of the rectangles. It hence follows

$$h\left(\frac{2}{3}n \mp \frac{1}{2} - \frac{1}{6n}\right) = \frac{n \mp 1}{n} G_{\square} N$$

or

$$G_{\square} N = \frac{2}{3} nh \pm \frac{h}{6}.$$

We can now clearly see two things: The location of the barycentre of the inscribed rectangles differs from that of the circumscribed rectangles by $2 \cdot h/6 = h/3$ and in case of $n \to \infty$, $h \to 0$ and nh being constant, it follows for the barycentre G of the paraboloid that

$$G_{\mathrm{P}} N = \tfrac{2}{3} AN, \quad \text{hence} \quad AG_{\mathrm{P}} : G_{\mathrm{P}} N = 1 : 2.$$

Fig. 5.2.8. Federico Commandino [Artist of the Scuola del Barocco, 16th c.] and the title page of his translation of the works of Pappus, 1589

Maurolico has carried out this step not by taking limits, of course, but by means of a classical *reductio ad absurdum*. Besides Maurolico we have to mention the Italian mathematicians Federico Commandino (1509–1575) and Luca Valerio (1552–1618) who were actively engaged in the computation of barycentres of bodies.

Commandino studied medicine in Padua and Ferrara. He made a name for himself by translating ancient works of Euclid, Aristarchus of Samos, and Pappus of Alexandria. He edited the works of Archimedes and wrote *Liber de centro gravitatis solidorum* in 1565 in which the 'Theorem of Commandini' concerning the geometry of the tetrahedron can be found. Luca Valerio was a Jesuit who corresponded with Galileo Galilei. He studied philosophy and theology and was a student of Christopher Clavius. When the case of Galilei escalated and the Catholic Church officially declared the doctrine of Copernicus being heresy he terminated the correspondence with Galilei and withdrew from the Academia dei Lincei. The works of Commandini and Valerio are described in some detail by Margaret Baron in [Baron 1987, p. 94ff.].

The most interesting calculator of barycentres in these days certainly is the ingenious Fleming Simon Stevin (1548–1620). Stevin was born in Bruges

Fig. 5.2.9. Simon Stevin and his land yacht which he developed for Prince Maurice of Orange, 1649

but little is known concerning his life. He travelled Poland, Denmark, and further countries in Northern Europe. When he came back from his journeys he became the tutor of Maurice of Nassau (later: of Orange) who made him the first director of the 'waterstaat' (the government authority of public works, in particular water management). Stevin was not only an ingenious mathematician but also an excellent engineer and physicist. He probably was the first to use decimal numbers in his computations and also taught their use. Numerous inventions are to his credit and in mechanics he proved to be an ingenious discoverer, for example in the case of the hydrostatic paradox, a theory of tides, and in rejecting the existence of a *perpetuum mobile*. Here we only have to appreciate his achievements in analysis which can be found in his writings on mechanics. His significance lies in his endeavour to simplify the Archimedean arguments concerning the reductio ad absurdum. He exploited the following argument: If the difference of two quantities can be made smaller than any given quantity, then both quantities are equal. We fall back on the triangle with the inscribed rectangles used by Maurolico as shown in figure 5.2.10. Stevin attempts to prove that the barycentre of the triangle is located at the median AD. If the area of the triangle ABC is denoted by F and the sum of the areas of the n inscribed rectangles by I_n, then we saw already in the calculations of Maurolico in (5.1), that

$$F - I_n = \frac{F}{n}$$

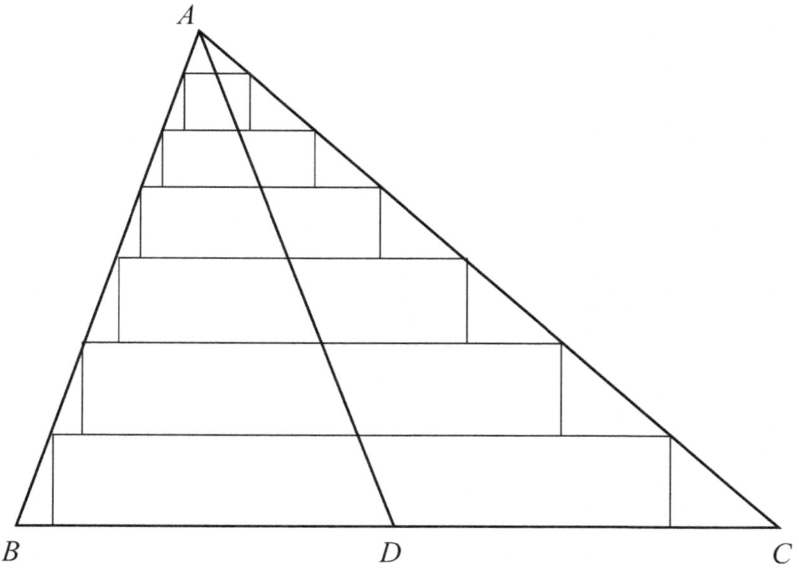

Fig. 5.2.10. A triangle with inscribed rectangles

holds. Following Euclid (*Elements* X.1) n can be chosen to make F/n arbitrarily small, hence

$$\lim_{n\to\infty} \frac{F}{n} = 0.$$

If F_1 and F_2 are the areas of the triangular parts ABD and ADC, respectively, then it holds

$$F - I_n = \frac{F}{n}$$
$$F_1 - \frac{I_n}{2} < \frac{F}{n}$$
$$F_2 - \frac{I_n}{2} < \frac{F}{n}$$

and

$$|F_1 - F_2| < \frac{F}{n}.$$

Hence prescribing a small area $\varepsilon > 0$ a number n can always be found so that

$$|F_1 - F_2| < \varepsilon.$$

Now Stevin does not use a *reductio ad absurdum* but instead the logical rule of 'contraposition'

$$(p \to q) \Rightarrow (\neg q \to \neg p)$$

in the following form [Baron 1987, p. 99]:

1. *If two quantities differ they differ by a finite quantity.*
2. *These quantities* [i.e. F_1 and F_2] *differ by less than a finite quantity.*
3. *These quantities do not differ.*

Stevin therewith has made a decisive step beyond the Archimedean reductio ad absurdum: By means of syllogistic logic he has shown that he does not need the Archimedean detour in order to proceed to the limit. His method was quite accepted in his time but then the influence of Cavalieri became so large that a rigorous transition to the limit was no longer thought of as being feasible.

5.3 Johannes Kepler

On the 27th December 1571 Johannes Kepler is born in the small town of Weil der Stadt in Württemberg.

He will become one the greatest astronomer of all times – his 'Kepler's laws' today are the basis of space flight – and he will become famous as a natural scientist especially in the field of optics [Caspar 1993]. It may be less well known that he also left an excellent contribution to analysis, namely on the occasion of his second wedding!

When Johannes was born Weil der Stadt was Lutheran and among the approximately 1000 inhabitants of the small free imperial town his parents belonged to the wealthier residents. However, Johannes was probably baptised in the Catholic church since there was not yet a Lutheran church nor a Lutheran minister [Lemcke 1995]. The father Heinrich was the fourth child of the mayor of Weil der Stadt who also ran a public house. The mother Katharina Guldenmann came from the family of the mayor of Eltingen who

Fig. 5.3.1. Johannes Kepler ([unknown painter, Copy of a lost original of 1610], Benedictine Monastery of Krems) and his birthplace in Weil der Stadt. Today it is the domicile of the Kepler museum [Photo: Markus Hagenlocher]

Fig. 5.3.2. The Great Comet of 1577 seen over Prague ([Woodcut by Jiri Daschitzsky, 1577] Georgium Jacobum von Datschitz, Zentralbibliothek Zürich)

also ran a bar. Apparently Johannes's mother had a difficult time in the house of her parents-in-law and the whole marriage cannot be called a happy one. Father Heinrich seems to have been a right good-for-nothing. As a mercenary he went to war in the Spanish Netherlands and fought on the Catholic side. The mother followed him, probably to get him back home into the responsibilities of a family man, and left little Johannes with her parents-in-law were he fell ill with smallpox. The child survived despite his fragile constitution but cicatrisations remained on his corneas and throughout his life he suffered from defective vision and myopia. In 1575 we find both parents back in Weil der Stadt where the father now buys a house in the market square. After only one year he again seeks his fortune as a mercenary. The school education of little Johannes starts in 1577 with reading and writing classes in the neighbouring town of Leonberg. In 1578 he went to the grammar school which was also located in Leonberg. The grammar school (or Latin school) was either reserved for children of wealthier families or for very gifted children.

Kepler later wrote that his mother once had led him up a hill in 1577 to show him a comet. The comet could be clearly seen during the day and was observed by many astronomers of the time. There is no doubt that this comet has strongly fuelled Johannes's interest in phenomena of the heavens.

But what were comets in these days? While astronomers were still puzzled whether comets were phenomena within the atmosphere or of extraterestrial origin. In the eyes of the common people they were harbingers of bad news. They were seen as being 'telegrams of God' with which drastic penalties were announced to sanction man for his misconducts.

Many schoolchildren in those days had to help their parents in their work at home. Hence the attendance at school was often infrequent. This is also true in the case of Johannes Kepler. After his father returned home in 1577 he again sold the family house in Weil der Stadt in 1579 as the family moved to Ellmendingen where the father took over the inn 'Zur Sonne'. Not later than this date Johannes had to start helping his family working on the fields. He could attend grammar school only during the winter but he stayed a good student. In the time in Ellmendingen Johannes's father showed him an eclipse of the moon – a further experience which might have fuelled the interest of the little lad in heavenly phenomena. In the year of 1583 the father went bankrupt, abandoned his inn, and moved with his family to Leonberg. It is the same year in which Johannes passes the 'Landexamen', an exam necessary to start a clerical career. On 16th October he is accepted at the monastery school of Adelberg. Life at those schools was hard and relied on a merciless system of spying and oppression. Kepler reacts to the untenable conditions with psychosomatic disorders – he developed high fever and headaches. Now also his theological doubts started which will accompany him during his whole life. The confrontations between Lutherans and Calvinists were at their climax. While Lutherans believed that Jesus Christ was bodily present at the Lord's Supper Calvinists were confident that he is only present 'in spirit'. Kepler being a Lutheran eventually arrived at the conviction that he can only associate himself with the beliefs of the Calvinists. This, of course, was heresy! He could not talk about his convictions to any of his friends or family and had to shut this conflict inside himself. The young lad was concerned with mathematics and poetry in Adelberg in order to flee from the conflicts with his classmates. After two years and having passed the final examination he went to the secondary monastery school in Maulbronn where he had to suffer even more from the reprisals of his classmates. At one time he developed a fever so high that his life was at stake. Centuries later Hermann Hesse will report in his novel *Unterm Rad* (Beneath the Wheel) on the still hideous conditions in Maulbronn which he himself had to suffer between 1891 and 1892 and he finally ran away!

After two years in Maulbronn Kepler passed the Baccalaureat examination in Tübingen with flying colours and returned to Maulbronn for two years as a 'veteran'; a word used for the older pupils. In 1589 he entered the Tübinger Stift from where he should study at the Faculty of Arts of the University of Tübingen. Thereafter he would be able to start studying theology. In this year his mother did something outrageous: she threw out her husband who thereupon disappeared for ever.

Kepler, among other lectures, also attended lectures on mathematics and astronomy presented by the already well known astronomer Michael Mästlin (1550–1631), professor of mathematics at the University of Tübingen. But Kepler also got to know Greek philosophy and he studied the writings of Nicolas of Cusa.

Already at this time Kepler ranked as a smart astrologer who issued horoscopes for his fellow students. Also groundwork concerning the first Science Fiction novel in history – the *Somnium* (The dream) (commenced in 1610) – resulted, and Kepler will work on it until the end of his life. The novel will only be published posthumously. In this novel Kepler wrote about a fictitious journey to the moon and many astronomical details were described. On 11th August 1591 Kepler passed the master degree examination as second of 15 candidates. He applied for a stipend which was granted and took up studies of theology at the university. When Kepler studied in his third year the faculty in Tübingen was asked for a teacher of mathematics. The maths teacher at the Stiftsschule in Graz had died and a successor was asked for. Kepler was ordered to go but he did not want to. Why did Tübingen choose a student in his third year? Probably because of his known mathematical abilities, but it seems more likely that Kepler's heretical theological convictions were leaked and the faculty wanted to get rid of their heretic student.

After twenty days of travel Kepler arrived in Graz on 11th April 1594. A teacher of mathematics was of not much worth back then and had little (if any) prestige. Additionally the protestants in Graz were in a precarious situation. In the course of the Counter-Reformation Jesuits had come to Graz; they had founded a college (the Jesuit equivalent of a cloister) in 1572, a high school in 1573, and a university in 1585. The Stiftsschule at which Kepler now had to teach was the counterpart of the Jesuit university. The situation between Protestants and Catholics had tightened due to the stubbornness of both sides, but mainly of the Protestant side. The Catholic Archduke Karl had begun to put pressure on the Protestants who answered with smear campaigns. Under this impression Kepler reacted in the familiar way – he developed high fever and was able to start his teachings only at 24th May. His teaching was so unskilful and probably so clumsy that in the second year all of his pupils had disappeared. Nevertheless the school inspectors were apparently very satisfied. They stated that [Caspar 1993, p. 56], [Oeser 1971, p. 31]

> *'because the study of mathematics is not everyone's meat'*
>
> (Mathematicum studium nit Jedermans thuen ist)

and stated in the second year that he had [Caspar 1993, p. 57], [Oeser 1971, p. 31]

> *'distinguished himself first as a speaker* [perorando]*, then as a teacher* [docendo] *and finally also as a debater* [disputando]*, in such a manner*

> *that we cannot judge otherwise than that he was, considering his youth, a well trained and modest master and professor, and one who fitted well into this worthy district school.'*
>
> (anfangs perorando, hernach docendo und dann auch disputando dermassen erwisen, was wir anders nit indiciren khönen, dann dz Er bey seiner Jugent ein gelehrter und in moribus ein bescheidener und dieser Einer Ersamen Landschafft Schuell allhie ein wolanstehund Magister und profeßor wie nit ohn ist.)

Besides his occupation as teacher in Graz he was also the mathematician of the Styrian district, the 'Ehrsame Landschaft der Steiermark', and had to produce a yearly calendar containing 'prognostica' concerning the weather and general horoscopes. Kepler's yearly income as a maths teacher was 150 guilders; for the calendar he earned additional 20 guilders. For the sake of comparison: His predecessor in office, Georg Stadius, had a salary of 200 guilders and got additionally 32 guilders for a calender!

Here now the actual career of Kepler as an astrologer started! Already his very first prognostica for 1595 were very successful. He prognosticated a Turkish invasion, strong cold, and peasant revolts and all became reality. Like Shakespeare's Hamlet also Kepler strongly believed that there were more things in heaven and earth than just the visible ones. However, concerning the power of astrology to *accurately* predict fate he was much restrained! Some of his remarks can be consulted to testify his restraint, for example [Gerlach/List 1987, p. 29][1]:

> *Now, this astrology is a foolish daughter (as I have written in my book de Stella, p. 59 [KGW I:211, line 6f.]), but dear Lord, what would happen to her mother, the highly reasonable astronomy, if she did not have this foolish daughter, since the world is even more foolish ...*
>
> (Es ist wol diese Astrologie ein närrisches Töchterlin (hab ich geschrieben in meinem Buch *de Stella*, S. 59 [KGW I:211, Zeile 6f.]). Aber lieber Gott, wo wolt ihr Mutter die hochvernünfftige Astronomie bleiben, wann sie diese närrische Tochter nit hette? Ist doch die Welt noch viel närrischer ...)

or [Kepler 2005, p. 197]:

> *And are otherwise the mathematician's salaries so strange and so low that the mother had to suffer hunger if the daughter would not earn anything.*
>
> (Und seynd sonsten der Mathematiker Honorar so seltzam und so

[1] From *Tertius interveniens* Nummer 7, [Kepler 2005, p. 197].

gering, daß die Mutter gewißlich Hunger leyden müste, wann die Tochter nichts erwürbe.)

Kepler became more involved with astronomy; particularly with the question of whether there are exactly six planets and no more or less (Uranus, Neptune and Pluto were still not found). Kepler was convinced that God had designed the cosmos following a mathematical plan. He tried different ratios for the distances of planetary orbits but his efforts eventually failed. Then suddenly he was struck with an idea: There are exactly five Platonic solids, tetrahedron, hexahedron, octahedron, dodecahedron and octahedron. Could not God have created exactly six planets *because* there are exactly five Platonic solids? Kepler writes [Kepler 2005a, p. 17]:

> *The orbit of the earth is the measure for all others. To this circumscribe a dodecahedron; the encompassing circle is* [the orbit of] *Mars. To Mars circumscribe a tetrahedron; the encompassing circle is Jupiter. To the circle of Jupiter circumscribe an hexahedron; the encompassing circle is Saturn. Now place an icosahedron in the circle of the earth; the inscribed circle is Venus. In the circle of Venus place an octahedron; the inscribed circle is Mercury. There you have the reason for the number of planets.*
>
> (Der Erdkreis ist das Maß für alle anderen. Ihm umschreibe ein Dodekaeder; der diesen umspannenden Kreis ist der Mars. Dem Mars umschreibe ein Tetraeder; der dieses umspannende Kreis ist der Jupiter. Dem Jupiterkreis umschreibe einen Würfel; der diesen einbeschriebene Kreis ist der Saturn. Nun lege in den Erdkreis ein Ikosaeder; der diesem einbeschriebene Kreis ist die Venus. In den Venuskreis lege ein Oktaeder, der diesem einbeschriebene Kreis ist der Merkur. Da hast Du den Grund für die Anzahl von Planeten.)

We today look at this seemingly naive ‘result’ with some concern and reservations, but not only Kepler stood in the Middle Ages with one leg and in the modern era with the other. Others will take up this model of the cosmos with enthusiasm. Kepler goes on [Kepler 2005a, p. 17]:

> *To describe what pleasure I drew from my discovery will be never possible for me.*
>
> (Den Genuß, den ich aus meiner Entdeckung geschöpft habe, mit Worten zu beschreiben, wird mir nie möglich sein.)

Kepler enthusiastically wrote to his former teacher Mästlin in Tübingen and wrote down his results in the book *Mysterium cosmographicum* (The secret of the world), the manuscript of which he personally brought to Tübingen in 1596 to get it printed. The book appeared in print in 1597.

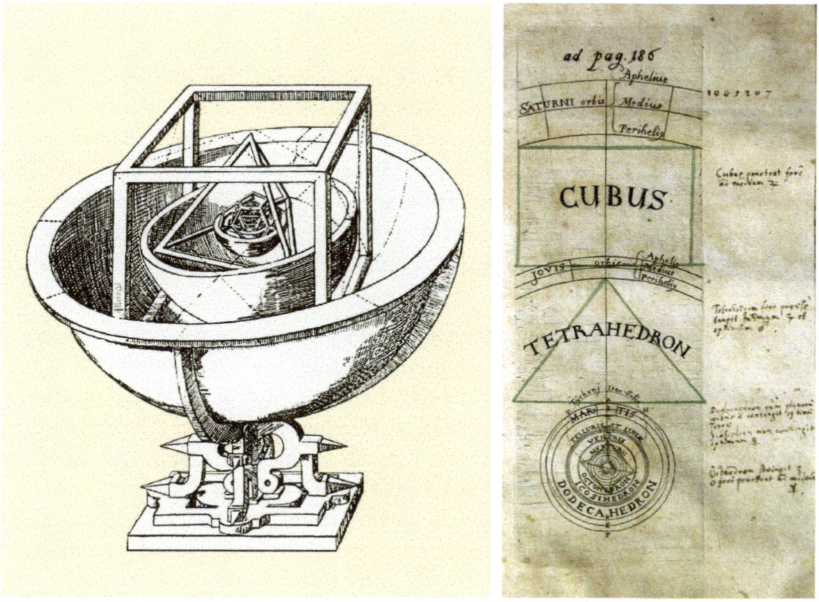

Fig. 5.3.3. Kepler's world model from *Mysterium Cosmographicum*, 1596. Representation of this model in *Harmonices mundi*, 1619

It was in this same year that Kepler married the 23 year-old twice widowed Barbara Müller on whom he had kept an eye in Graz. The marriage will not be a happy one. Of four mutual children two died early; in addition Barbara apparently never had not much sympathy for the scientific ambitions of her Johannes. She died in 1611. The Counter Reformation in Graz reached a climax in 1598: All Protestants employed at schools, churches, or within the Stift were expelled; only Kepler was allowed to return. He desperately addressed Mästlin to beg his former teacher to look for a probable position in Tübingen – Mästlin did not answer. Between 1598 and 1600 he was writing a new manuscript, the *Harmonices mundi* (The harmony of the world). Meanwhile he had received many positive reactions concerning his *Mysterium cosmographicum*; among others by Galileo Galilei and the most famous astronomer of his time, Tycho Brahe (1546–1601) [Thoren 1990].

Tycho was the offspring of a Danish nobleman and studied philosophy, rhetoric and law in Copenhagen, Leipzig, Rostock, Basel and Wittenberg. Aged 20 he got involved in a dispute with a fellow student which resulted in a duel with swords in which Tycho lost his nose. It is said that the dispute leading to that duel started off as a discussion concerning a mathematical formula. Fortunately he survived this gruesome physical injury but had to wear a prosthesis, an artificial nose made probably from copper sheet and fixed by means of vaseline.

The Danish king paid for no less than two observatories for Tycho, Uraniborg and Stjerenborg on the Island of Hven, in which Tycho became the greatest observing astronomer of his day. There are not yet telescopes in existence – Tycho observed the heavens with his naked eye and by means of a large mural quadrant like the one shown in figure 5.3.4. After the death of the king his successor shortened the money for Tycho considerably so that the latter followed an invitation of his friend Heinrich Rantzau (1526–1598) [Oestmann 2004] and left Denmark for Hamburg. However, in 1599 he went to Prague to become the Imperial court mathematician. The Emperor had promised to build a new observatory.

Since Kepler did not own any astronomical instruments for observation (and what could he have done with them regarding his visual defects) so that he was dependent on the observations of others. Tycho, having praised the *Mysterium cosmographicum*, now invited Kepler to Prague. At the beginning of 1601 we find Kepler there. However, both men soon started to argue and parted. Tycho had looked for an assistant to help him with the computational treatment of his data. Kepler on the other hand wanted to get a hand on the observational data of the great astronomer and now had to experience that the latter was not willing to let him participate. Additionally all of the data were still in raw form, hence completely untreated. Only weeks later did Kepler write a letter of apology to Tycho from Graz which Tycho accepted. Together with his family Kepler arrived again in Prague on 16th October 1600. He suffers from malaria and cough. Again he addressed his former teacher Mästlin and begs fervently for a position in Tübingen – again Mästlin did not answer and will not answer for the following four years, although Kepler continued to write.

The death of his father-in-law drew him back to Graz. From there he wrote to Emperor Rudolf II asking for a position. The Emperor had ordered Tycho and Kepler to compute new planetary tables and Tycho suggested calling them 'Rudolphine Tables' (Tabulae Rudolphinae). The Emperor had bought a house at the Loreto Square in Prague where Kepler lived. But suddenly Tycho Brahe died on 24th October 1601 after having suffered from strong renal colics. In recent times there were investigations of the beard of Tycho after the opening of his tomb in the Church of Our Lady before Týn. These investigations by modern analytical methods show a massive poisoning by mercury and in a concoction [Gilder 2005] which is hard to bear. Kepler was even called Tycho's murderer.

Kepler quickly got his hands on Tycho's data material to save it from the struggle of the heirs; however, quarrels could not be avoided. The Emperor had Kepler assigned the position of the new court mathematician and ordered him to continue Tycho's work. Kepler immediately started to work on the orbit of Mars. He is also seriously concerned with optics – certainly also with regard to his own visual problem. In his writing *Ad Vitellionem paralipomena quibus astronomiae pars optica traditur* (Remarks concerning Witelo in which

Fig. 5.3.4. Tycho Brahe with the mural quadrant from *Astronomiae instauratae mechanica*, Wandsbek 1598, later colouring

the optical part of astronomy is communicated) of 1604 Kepler also examined the human eye and discovered that the image on the retina was upside down. He was also able to explain myopia and the functioning of spectacles. Witelo

Fig. 5.3.5. Rudolf II ([Painting about 1590 by Joseph Heintz the Elder] Kunsthistorisches Museum Wien) and in a [painting by G. Arcimboldo, 1590]

(or Vitello) was a contemporary of the 13th century who about 1270 wrote a book on perspective which came in circulation again in Kepler's time. In a letter of May 1603 Kepler wrote [Lemcke 1995, p. 61]:

> *I have taken up the thorny optics;* [...] *Good God, how dark the thing is!*
>
> (Ich habe die dornenvolle Optik vorgenommen; [...] Guter Gott, wie dunkel ist die Sache!)

Due to his work concerning optics Kepler later became one of the scientific godfathers of the telescopes.

Meanwhile work on the orbit of Mars continued. He was familiar with the book *De magnete* of the English physician William Gilbert (1544–1603) which had been published in 1600. Gilbert thought that magnetism was the force keeping the planets on their orbits and Kepler was fascinated by the idea. He asked himself whether the sun actually could be at the centre of *circular* planetary orbits since Tycho's data did not match circles. Since all life comes from eggs (this was at least Kepler's conviction) God might have created the cosmos like an hen's egg: The orbits of the planets may be egg-shaped and the sun's position is where the yolk is located! Kepler needed a full year of calculations, until at Easter 1605 he dismissed the egg-shaped orbits and had to accept elliptical orbits instead where the sun is in one of the two focal points. Thus the first of Kepler's laws saw the light of day. That the radius vector, i.e. the straight line segment connecting the sun with a planet, covers equal areas in equal periods of time was also already clear to him. This is Kepler's second law, indeed. At the beginning of the year

Fig. 5.3.6. Frontispiece of the Rudolphine Tabels 1627
[Engraving: Georg Keller following a design by Johannes Kepler]

1606 a new book was finished, *Astronomia nova* (New Astronomy) [Kepler 2005b]. Unfortunately, quarrels with Tycho's heirs and financial bottlenecks prevented publication until 1609. We can study the almost unbelievable mixture of intuition, knowledge and compensation distinguishing Kepler's astronomical works when we look at his derivation of the second law. We thereby follow [Edwards 1979, p. 100f.]. Kepler considered the orbit of a planet P around the sun S. The radius vector r is shown in figure 5.3.7.

The points P_1 and P_2 are the perihelion and aphelion (words actually coined by Kepler), respectively, of the orbit. Both points are called the apsides. Astronomers in Kepler's time knew from observations that in the apsides the velocity of a planet satisfies

$$v_1 = \frac{k}{r_1}, \quad v_2 = \frac{k}{r_2}$$

where k is some positive factor. Kepler thereupon makes the assumption (which he called 'Theorem') that *everywhere* on the orbit

$$v = \frac{k}{r}$$

holds. This assumption is simply wrong! Kepler would have arrived at the correct formula had he drawn a tangent at point P of the elliptic orbit, a parallel to it passing through S, and computed the distance R between both straight lines as shown in figure 5.3.7. Then, in fact,

$$v = \frac{k}{R}$$

holds.

But let us continue with Kepler's computations. Since the integral calculus was not yet in existence Kepler had to divide the planet's orbit into many tiny discrete steps of angle difference $\Delta\theta$ as shown in figure 5.3.8. In every discrete step an element of arc length Δs is traversed.

As is well known velocity=distance/time; hence in the discrete case

$$v_i = \frac{\Delta s}{t_i},$$

and hence $t_i = \frac{\Delta s}{v_i}$. If all of the orbit is divided into n discrete steps it apparently holds

$$t = \sum_{i=1}^{n} t_i = \sum_{i=1}^{n} \frac{\Delta s}{v_i}.$$

Now, following Kepler's wrong assumption $v_i = k/r_i$, we get

$$t = \frac{1}{k}\sum_{i=1}^{n} r_i \cdot \Delta s. \tag{5.2}$$

Kepler wrote [Koestler 1989, p. 331f.]:

> *'Since I was aware that there exists an infinite number of points on the orbit and accordingly an infinite number of distances [from the sun] the idea occurred to me that the sum of these distances is contained in the* area *of the orbit. For I remembered that in the same manner Archimedes too divided the area of a circle into an infinite number of triangles.'*

Apparently he was referring to the division of the circles into triangles as Archimedes had accomplished, cp. 2.3.16. Every little discrete part of the orbit between the two encompassing radius vectors is – assuming smallness of Δs – approximately a right-angled triangle as shown in figure 5.3.8. Such a triangle has area $\frac{1}{2}r_i \cdot \Delta s$ and the whole covered area of the orbit thus is

$$A = \frac{1}{2}\sum_{i=1}^{n} r_i \cdot \Delta s. \tag{5.3}$$

Together with (5.2) we get

$$t = \frac{1}{k}\sum_{i=1}^{n} r_i \cdot \Delta s = \frac{1}{k} \cdot 2 \cdot A =: \frac{A}{h}$$

where $h := k/2$. And here Kepler's second law has emerged:

$$A = h \cdot t.$$

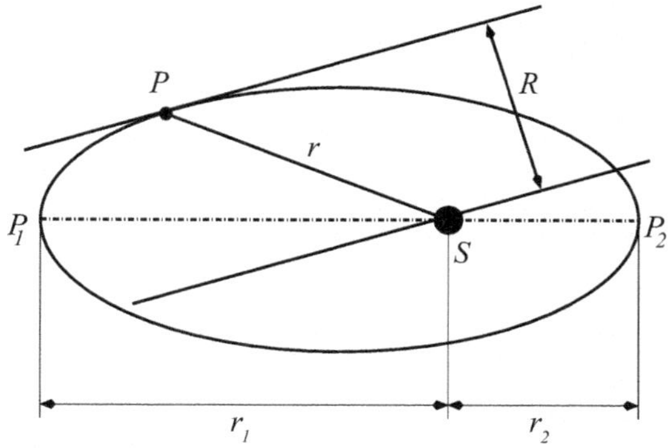

Fig. 5.3.7. Sketch concerning the correct computation of the velocity of a planet orbiting the sun

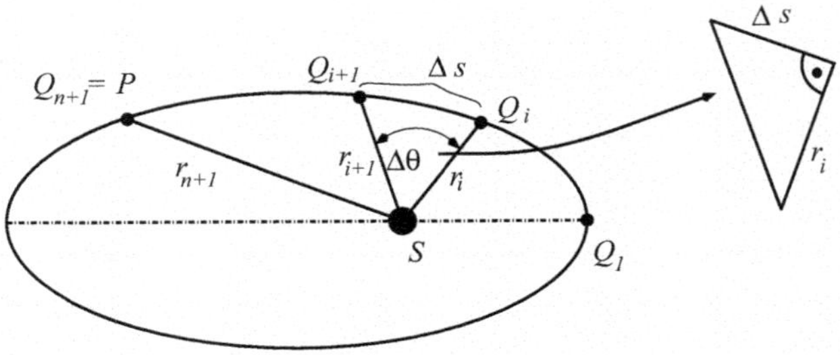

Fig. 5.3.8. Discretisation of a planetary orbit using the Archimedean idea of dividing a circle into triangles

The area covered by the radius vector is proportional to the time the radius vector had to move. In other words: The radius vector covers equal areas in equal periods of time. But, alas, also (5.3) is wrong in general! From Kepler's equation (5.3) it follows for the area in the limit $n \to \infty$:

$$A = \frac{1}{2} \int r\, ds,$$

but

$$A = \frac{1}{2} \int r^2\, d\Theta$$

would be correct! Fortunately arc length s and angle Θ depend on each other as

$$s = r \cdot \Theta$$

so that approximately we have $\Delta s \approx r \cdot \Delta\Theta$. It is breathtaking how everything finally falls in place in Kepler's computations!

In 1610 Galileo Galilei wrote the treatise *Sidereus nuncius* (Siderial messenger) in which he reported his observations of four new planets found with the help of his new telescope. But as Kepler had 'proven' in the *Mysterium cosmographicum* that there *can only* be six planets he supposed that Galileo had only found four moons of Jupiter, and that turned out to be true!

Also in 1610 Kepler was so far advanced in his optical studies that he published his book *Dioptrice*. To Elector Ernest, who lent his telescope to Kepler he explained the name 'diopter' [Lemcke 1995, p. 82]:

> *Since Euclid has created catoptics as part of optics dealing with reflecting rays* [of light] *in which he considered the name of the main tools of this kind, the mirrors and their wonderful and pleasing variety. In that way following this process the name dioptric emerged for my little book because it is concerned mainly with the refracted rays in dense, transparent media; in the natural media of the human eye as in the different artificial glasses.*

(Da nun Euklid einen Teil der Optik, die Katoptrik, geschaffen hat, welche von den zurückgeworfenen Strahlen handelt, indem er den Namen von den Hauptwerkzeugen dieser Art, den Spiegeln und ihrer wunderbaren und erfreulichen Vielgestaltigkeit hernahm, so entstand nach diesem Vorgange für mein Büchlein der Name Dioptrik, weil es hauptsächlich von den in dichten, durchsichtigen Medien gebrochenen Strahlen handelt, sowohl in den natürlichen Medien des menschlichen Auges, als den künstlichen verschiedener Gläser.)

In the preface Kepler gives us an impression of the importance the telescope had for him as an astronomical instrument of observation. We have to remember that Kepler himself could never have been an observational astronomer due to his defective vision [Caspar 1993, p. 201], German in [Kepler 2003]:

'O you much knowing tube, more precious than any sceptre. He who holds you in his right hand, is he not appointed king or master over the work of God!'

(O du vielwissendes Rohr, kostbarer als jegliches Szepter! Wer dich in seiner Rechten hält, ist der nicht zum König, nicht zum Herrn über die Werke Gottes gesetzt!)

Meanwhile political problems had emerged in Prague. Matthias, the brother of the Emperor, tried to restrict Rudolf's reign by different alliances. In turn, Rudolf II allied with the Catholic Archduke Leopold of Passau against Matthias. The Passau soldiers invaded Bohemia; Rudolf lost even more of his reputation in the eyes of the population and finally had to abdicate as the king of Bohemia. Abominably, the alien troops had brought smallpox to the city and Kepler's children fell ill. Friedrich, darling of his parents, died on 19th February 1611. Kepler's wife Barbara lost all her courage to face life and died shortly after in the year 1611. Already before this stroke of fate Kepler had begun negotiations with the Landstände (political representatives) of Linz and had been offered a position. When after the death of his wife also Emperor Rudolf II died in 1612 nothing kept him in Prague and he went to Linz. In Linz he should be concerned further on with the Rudolphine Tables, teach mathematics, philosophy and history to the children of the aristocracy, and to create a map of the region. This became possible only because new Emperor Matthias confirmed his position as Court mathematician on 18th March with a monthly salary of 25 guilders and had allowed him to move to Linz. For work for the Stände of Linz he should receive a yearly salary of 400 guilders. It took only a few weeks until Kepler found himself in the midst of religious quarrels again. Somewhat naive he had spoken about his doubts on the bodily presence of Jesus at the celebration of the Last Supper to the Protestant minister Daniel Hitzler who thereupon excluded Kepler from communion. When Kepler invoked the consistory in Stuttgart he received another damper: The consistory demanded unconditional obedience to the good shepherd Hitzler. Kepler's reputation of being a heretic made it also hard to find the new wife he was looking for. Eleven women were shortlisted. Kepler finally chose – probably remembering his experiences in his first marriage – a totally penniless woman, but with good education and without destitute family members annexed. She promised to be proficient, parsimonious, and to be a good mother to Kepler's children. She was woman number 5 on the list, Susanna Reuttinger. In Eferding close to Linz, the birthplace of the bride, the wedding took place on 30th October 1613. Kepler wrote [Oeser 1971, p. 80f.]:

> *'When I celebrated my remarriage in November of last year, 1613, at a time when, on the Danube banks near Linz, the wine barrels brought from Lower Austria had been piled up and could be bought at a decent price, it was the duty of the new husband and caring family man to get the necessary beverages for his house. On the fourth day, when some of the barrels were encased, the vendor came with the measuring rod with which he measured all the barrels with respect to their content, irrespective of their shape and without any further calculations. I* [...] *doubted the correctness of the method, for one very low barrel with slightly wider bottoms and therefore much smaller content could have the same visor length. It did not seem inappropriate to me as a newly wed to explore a new principle of mathematical work, namely to do research on the accuracy of these convenient and generally important principles and to bring to light the laws that are in existence.'*

> (Als ich im November des letzten Jahres, 1613, meine Wiedervermählung feierte, zu einer Zeit, da an den Donauufern bei Linz die aus Niederösterreich herbeigeführten Weinfässer nach einer reichlichen Lese aufgestapelt und zu einem annehmbaren Preis zu kaufen waren, da war es die Pflicht des neuen Gatten und sorglichen Familienvaters, für sein Haus den nötigen Trunk zu besorgen. Als einige Fässer eingekellert waren, kam am 4. Tag der Verkäufer mit der Meßrute, mit der er alle Fässer, ohne Rücksicht auf ihre Form, ohne jede weitere Überlegung oder Rechnung ihrem Inhalte nach bestimmte. [...] Ich [...] bezweifelte die Richtigkeit der Methode, denn ein sehr viel niedrigeres Faß mit etwas breiteren Böden und daher sehr viel kleinerem Inhalt könnte dieselbe Visierlänge besitzen. Es schien mir als Neuvermähltem nicht unzweckmäßig, ein neues Prinzip mathematischer Arbeiten, nämlich die Genauigkeit dieser bequemen und allgemein wichtigen Grundsätze zu erforschen und die etwa vorhandenen Gesetze ans Licht zu bringen.)

The measuring rod was simply inserted into the bunghole until it reached the bottom of the barrel. After removing the rod the length of the wet part was measured. The (well founded!) criticism of this method is the natal hour of Kepler's works in analysis.

In 1614 Kepler was again working on the Rudolphine Tables. That year he published a pamphlet in which he claimed to have computed Jesus's date of birth exactly. One year later a terrible event overshadowed Kepler's life: In Leonberg his mother was accused of being a witch and she was immediately arrested. Apparently the social interaction with this woman must have been difficult; Kepler himself referred to 'addiction to gossip, inquisitiveness, violent temper, malignancy, addiction to complaints' (Schwatzsucht, Neugier, Jähzorn, Bösartigkeit, Klagsucht [Lemcke 1995, p. 97]) when describing his mother. A denunciation of a neighbour sufficed to bring her into the mills of a witch trial which almost always ended with the execution of the accused. Kepler intervened with all of his power and his titles; he travelled to Württemberg and could at least prevent his mother from being tortured. She was acquitted in 1621 but died a year later; probably suffering from the aftermath of the conditions of her imprisonment.

We do not want to follow the sad story of the many births and deaths of Kepler's children – that was nothing exceptional in these days, although the death of a child always was (and still is, of course) a heavy stroke for father and mother. Kepler worked on *Epitome astronomiae Copernicanae* concerning Copernican astronomy, on his *Harmonices mundi*, and on the duties his position brought with it: Among other duties it was the yearly star catalogue – the ephemerides. The year 1617 saw him in Prague to fulfil court service. There in Prague he saw the first table of logarithms by the Scotsman John Napier (see page 288) for the first time and immediately recognised them being an invaluable means for his works on the star catalogues. However, he shared the imperial table with another inventor of logarithms, Jost Bürgi.

The Swiss clockmaker and instrument builder Jost Bürgi is a tragic figure in the history of logarithms in more than one respect.

We know very little about his life before he was employed as a court clockmaker by the Landgrave of Hesse-Kassel, William IV, in 1579, cp. [Clark 2015, p. 1–12]. He could not have had enjoyed a higher education since he could not read or write Latin, but he must have caught the attention of the Landgrave while working on the second clock in Strasbourg Cathedral so that William called him to his court in Kassel. He became naturalized in Kassel in 1591 and bought a house there. Two marriages remained childless. At the beginning of 1592 Emperor Rudolf II ordered a 'Bürgi globe' from his uncle, the Landgrave of Hesse-Kassel. The globe was a planetary globe which is lost today and which was presented to the Emperor by the instrument maker himself. Following the wish of the Emperor and with the approval of Landgrave Maurice – William IV had meanwhile died – Bürgi went into imperial service at the end of 1604 and obtained an own workshop with an assistant where he also did some work for Johannes Kepler. It is possible that Kepler was introduced to Napier's logarithms by Bürgi during his stay

Fig. 5.3.9. Jost Bürgi and his astronomic ornamental clock. On the inside of the lid first and second disparity of the motion of the moon after Copernicus are displayed as are the equation of time and the exact lunar phase (Astronomisch-Physikalisches Kabinett Kassel) [Photo: H. Wesemüller-Kock]

in Prague, but Bürgi apparently had not told Kepler anything about his own logarithms. Later Kepler talked about Bürgi as of a 'procrastinator' and a 'guardian of his secrets' who had 'abandoned his child at birth and did not rear it for publicity' [Goldstine 1977, p. 22f.]. Franz Hammer in his contribution *Die logarithmischen Schriften Keplers* (The logarithmic writings of Kepler) in volume 9 of Kepler's collected works gives the following German translation (cited after [Tropfke 1980, p. 309]; my English translation) from the Rudolfine Tables :

> *The procrastinator and mystery-monger has degenerated the newly born child instead of raising it for the general good.*
>
> (Allerdings hat der Zauderer und Geheimtuer das neugeborene Kind verkommen lassen, statt es zum allgemeinen Nutzen groß zu ziehen.)

Ernst Kühn, ex member of the board of the Kepler Society Weil der Stadt, translated even more drastically [Kühn 2009]:

> *There has the poor wretch – as a procrastinator and guardian of his secrets – ruined the foetus at birth and not raised it for the public good!*

(Da hat doch der Tropf – als ein Zauderer und Wächter seiner Geheimnisse – den Fötus bei der Geburt eingehen lassen, und hat ihn nicht zu öffentlichem Nutzen aufgezogen!)

Unfortunately and to make matters worth, Bürgi published barely nothing; probably because he was not in command of the language of the scholars in his days – Latin. He deserves credit for having created logarithms even before Napier; however, his work was outpaced by Napier's publications. Bürgi's table of logarithms [Clark 2015] appeared in print only in 1620 in Prague and was not received in scholarly circles.

Today Bürgi is ranked highly as an important astronomer and mathematician. As recent as 2013 the German historian of mathematics Menso Folkerts rediscovered a manuscript by Bürgi which was thought to be lost in the University Library of Wroclaw in Poland. This manuscript is *Fundamentum Astronomiæ* and it contained an ingenious algorithm, called 'artificium' (skilfull/artful method) build on finite differences, to efficiently compute sine tables of all subdividing angles. The algorithm has recently been described in modern mathematical terms by Waldvogel [Waldvogel 2016] which rather serves to look at Bürgi's ingenuity in awe. There may even be some connections to the computation of logarithms by Henry Briggs (1561–1631) which we will describe in chapter 6.

Bürgi returned to Kassel in 1631 where he died in January 1632 and was buried.

Bürgi published his table of logarithms in Prague in 1620 under the title *Aritmetische und Geometrische Progress Tabulen, sambt gründlichem unterricht, wie solche nützlich in allerley Rechnungen zugebrauchen und verstanden werden sol* (Arithmetic and geometric progress tables including a thorough instruction of how they should be understood and used handily in all kinds of calculations), see [Clark 2015]. A two-coloured print was employed throughout; logarithms were printed in red and the corresponding numbers in black. Actually it is *not* a table of logarithms but a table of anti-logarithms instead [Goldstine 1977, p. 20], i.e. to a given logarithm one finds the number. In case of Bürgi's logarithms the normalisation

$$\log 1 = 0$$

already holds and these logarithms, in contrast to Napier's, are already monotomically increasing [Lutsdorf/Walter 1992, p. 13].

Although Bürgi's logarithms are of great mathematical interest they did not play a role in the history of analysis. Incidentally they share this fate with Kepler's logarithms which he published in his *Chilias logarithmorum* in 1624. Kepler had been somewhat unsatisfied with Napier's derivation of the logarithms and created his own logarithms (cp. [Tropfke 1980, p. 307ff.]) being similar to Napier's but not identical.

In the horoscope for 1618/19 Kepler had warned of quite a few comets 'which mean nothing good' (die nichts guts bedeutten) [Lemcke 1995, p. 11]. Indeed three comets appeared as sinister harbingers of things to come. Kepler wrote that [Oeser 1971, p. 71]:

> *in the next spring not alone the weather but mostly the course of the planets will make the hearts of fresh roosters blossom and give them a martial courage.* [...] *Then truly in May it will* [...] *not pass without great troubles.*
>
> (auff künfftigen Früling nit allein das Wetter, sondern auch viel mehr der Lauff der Planeten manchem sonst frischem Hanen das Hertz blühen und einen kriegerischen Muth machen werde.[...] Dann warlich im Mayen wirdt es [...] ohne große Schwürigkeit nicht abgehen.)

On 23rd May 1618 the Defenestration of Prague took place – the Thirty Years' War had been initiated. On 15th May, just eight days before the outbreak of the Thirty Years' War, Kepler had found the third law of planetary motion: 'The square of the orbital period of a planet is proportional to the cube of the semi-major axis of its orbit'. Work on the *Harmonices mundi* could be finished and the book was eventually published in 1619 [Kepler 2006]. Work on the Rudolphine Tables will be dragging on well into the year 1624. Then Kepler travelled to Vienna asking Holy Roman Emperor Ferdinand II for money necessary for giving these tables to the printer. But money was no longer available in the cash offices of the Emperor and (probably to get rid of the astronomer) Ferdinand provided Kepler with written orders for some cities to pay money to the astronomer. Among others were the cities of Memmingen, Kempten, Nuremberg and Augsburg. After some very frustrating experiences in these cities – Nuremberg had just been 'eased' of 100 000 Guilders by Wallenstein – Kepler bought the paper necessary for the printing out of his own pocket. When Albrecht Wenzel Eusebius von Waldstein (1583–1634), known as 'Wallenstein', the great military leader of the imperial party, got message of Kepler being in Vienna, he asked for a horoscope. One such horoscope Kepler had already provided in 1608 for Wallenstein; now Wallenstein asked for a new one. However, Wallenstein was not satisfied with this new horoscope and therefore did not pay Kepler. He wanted to have more details which Kepler was not willing to give.

In Linz the situation of the Protestants became tight. After the reformation patent of 20th October 1625 all Protestants had to leave the country at least until Easter. Although Kepler was explicitly exempted his library was sealed. He wrote to the Jesuit Paul Guldin (1577–1643) who stood up for him so that Kepler was allowed to use his library at least during the nights. To make matters worse the printing plant was burnt during a siege of Linz in 1626. To still publish the Rudolphine Tables Kepler moved with his household to Ulm, leaving his family with friends in Regensburg. When finally the Rudolphine Tables were printed he travelled to the Frankfurt Autumn Fair and, after a

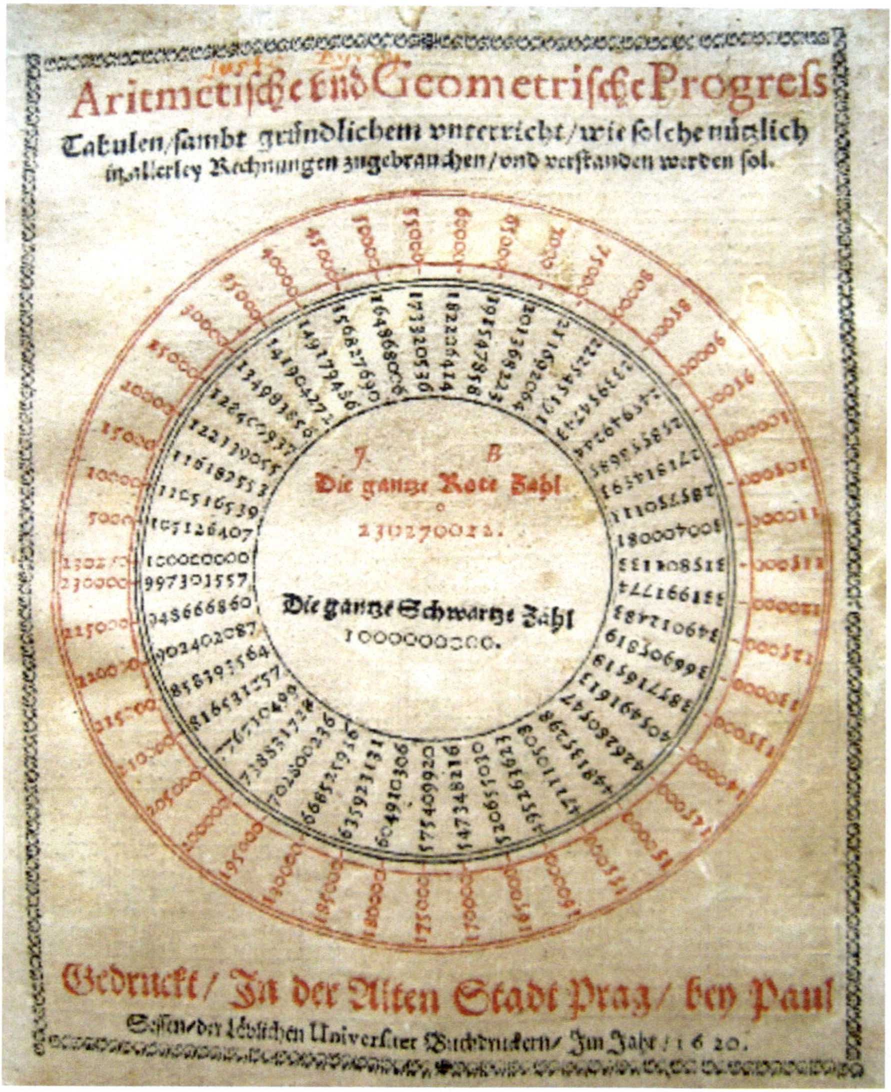

Fig. 5.3.10. Title page of the Progress Tabulen 1620 (Rara collection University Library Graz)

short stay with his family in Regensburg, to Prague in order to present the work to the Emperor. In Prague Kepler met Wallenstein who is looking for a successor to his personal astrologer at that time. He promised to establish a personal printing plant for Kepler at his estate in Sagan in Silesia and pay him well; hence Kepler accepted and went to Sagan. Kepler was now able

Fig. 5.3.11. Tilly and Wallenstein. Both were military leaders of the Catholic League in the Thirty Years' War. They fought, among others, against Gustav II Adolf of Sweden

to take care of the equipment of his printing plant himself and parts of his *Somnium* were actually already printed in Sagan. On 13th September 1630 Wallenstein was dismissed by the Emperor. He had become more mighty than the Emperor and there were forceful enviers on the Catholic side. Kepler was alarmed and travelled to Regensburg via Leipzig on 8th October 1630 where he met with his family. He arrived in Regensburg at the 2nd November 1630 and immediately got a fever. He suffered occasionally from delirium and died on 15th November 1630 around midday. As a Protestant he must not be buried within the city walls of Regensburg and was laid to rest at the city wall. The burial dictum he had chosen himself was [Caspar 1993, p. 359]:

> *I used to measure the heavens, now I shall measure the shadows of the earth.*
> *Although my soul was from heaven, the shadow of my body lies here.*
>
> (Mensus eram coelos, nunc terrae metior umbeas.
> Mens coelestis erat, corporis umbra jacet.)

A short time later Regensburg was besieged by Swedish troops and finally stormed in 1633. During this chaos of war the grave of the great mathematician and astronomer was destroyed. Its exact location is unknown today.

5.3.1 New Stereometry of Wine Barrels

After having judged the measurements of the contents of wine barrels by measuring rods as being faulty at his remarriage in 1613 Kepler wrote his little book *Nova Stereometria doliorum vinariorum* (New stereometry of wine barrels) which appeared in print in Linz in the year 1615. I am not aware of an English translation but a German one appeared in 1908 in Leipzig [Kepler 1908]. Since the printer would not take the risk Kepler paid the printing costs out of his own pocket. The book turned out to be a shelf warmer so that Kepler published a revised German edition *Außzug auß der uralten Messekunst Archimedis* (Extract of the age-old art of the measurement of Archimedes) already in 1616.

In computing the area of a circle he reasoned that the circumference has as many parts as it has points [Struik 1969, p. 194].

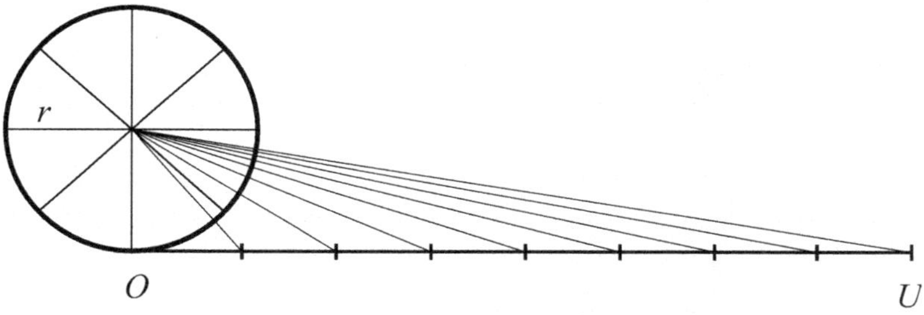

Fig. 5.3.12. Computation of the area of a circle

Unwinding the circumference of a circle with radius r as a line segment OU results in partition points between O and U, the number of which depending on the refinement of the partition of the circumference. If we now join each of the partition points with the circle's centre results in the 'pieces of pie' as shown in figure 5.3.12- Taken together they form a right-angled triangle. Hence the area of the circle is

$$A = \frac{1}{2} U \cdot r.$$

In case of a sphere with radius r he employed small cones having their apexes in the midpoint of the sphere. Their base areas consist of circular pieces of the surface of the sphere, see figure 5.3.13. The volume of such a cone is

$$\frac{1}{3} \cdot \text{ part of the sphere's surface } \cdot r,$$

hence the overall volume of the sphere turns out to be

$$V = \frac{1}{3} \cdot \text{ surface of the sphere } \cdot r.$$

These considerations were not particularly revolutionary but could be found in one or the other form in the works of Kepler's predecessors. But now Kepler turned to solids of revolution where he allowed for different locations of the axis of rotation.

If he imagined a circle rotating around an external axis he got a 'ring' – we would call it 'torus' today; see left part of figure 5.3.14. If the axis of rotation lies within the circle and if the part with larger area rotates (middle part of figure 5.3.14) then Kepler called the resulting solid of revolution an 'apple'. If the part with smaller area rotates instead then he got a 'lemon' as shown in the right part of figure 5.3.14. In order to determine the volume of these solids Kepler developed methods which can be seen as inspiration of other geometric integration methods which occurred only after his death [Baron 1987, p. 112].

Kepler determined the volume of the 'ring' by means of dissecting it into infinitely many parts. As he wrote in [Kepler 1908, p. 19] (see figure 5.3.15):

> *If one actually dissects the ring [...] by intersections from the centre A into infinitely many and very thin slices then one part of the slice will be all the more smaller towards the midpoint A as this part of the slice, e.g. E, lies closer to the midpoint A than F or a normal drawn with respect to the intersection plane ED, and the more wider in the outer point D. It followed that if D and E are considered simultaneously the thickness at these two parts will together be twice as large as in the middle of the slices.*

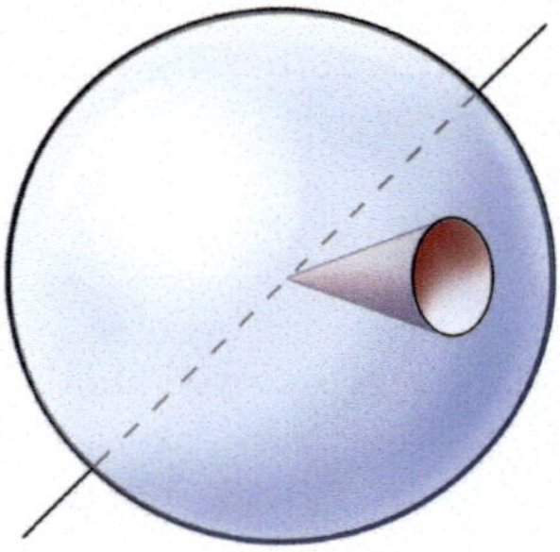

Fig. 5.3.13. Regarding the computation of the volume of a sphere according to Kepler

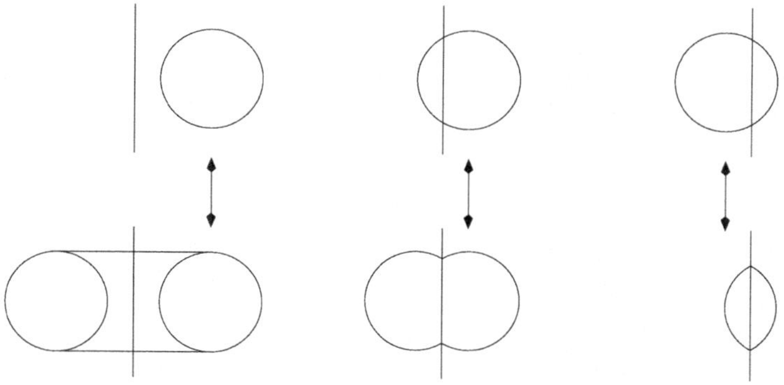

Fig. 5.3.14. Different solids of revolution as sections: ring, apple, and lemon

> (Wenn man nämlich den Ring [...] durch Schnitte aus dem Zentrum A in unendlich viele und sehr dünne Scheiben zerschneidet, so wird eine Stelle der Scheibe gegen den Mittelpunkt A hin um so viel schmäler sein, als diese Stelle, z. B. E, dem Mittelpunkt A näher liegt als F oder eine durch F in die Schnittebene zu ED gezogene Normale, und um so viel breiter in dem äußeren Punkte D. Danach wird, wenn man D u. E zugleich betrachtet, die Dicke an diesen beiden Stellen zusammen doppelt so groß sein wie in der Mitte der Scheiben.)

Denoting the thickness at E by t_1 and by t_2 at D, then the thickness of the slice in the middle is

$$t = \frac{t_1 + t_2}{2}.$$

Every infinitesimal slice hence has the volume $A_{\text{circle}} \cdot t$ where A_{circle} is the area of the circle. Thus every slice can be replaced by a circular cylinder of height t. Since this height t is measured on the centre line of the ring the overall volume of the ring is

$$A_{\text{circle}} \cdot \text{ length of the centre line.}$$

This reasoning is valid only in the case of the ring and Kepler was well aware of this. However, the idea to compare the volume of a solid of revolution with a certain circular cylinder turned out to be fruitful. Take a closer look at the 'apple' in figure 5.3.16. Imagining the apple as being stretched out as a cylinder then the base area is no longer a full circle but a part of a circle cut off by the segment MN, as can be seen in the left part of figure 5.3.17. The height PS of the cylinder is the circular path which is traversed by the point P during a full rotation around MN.

The volume of the apple hence is equivalent to the cylindrical section $MSPN$. Kepler imagined the cylindrical section as being built from indivisible rectangles $IJLK$ as shown in the right part of figure 5.3.17. These rectangles

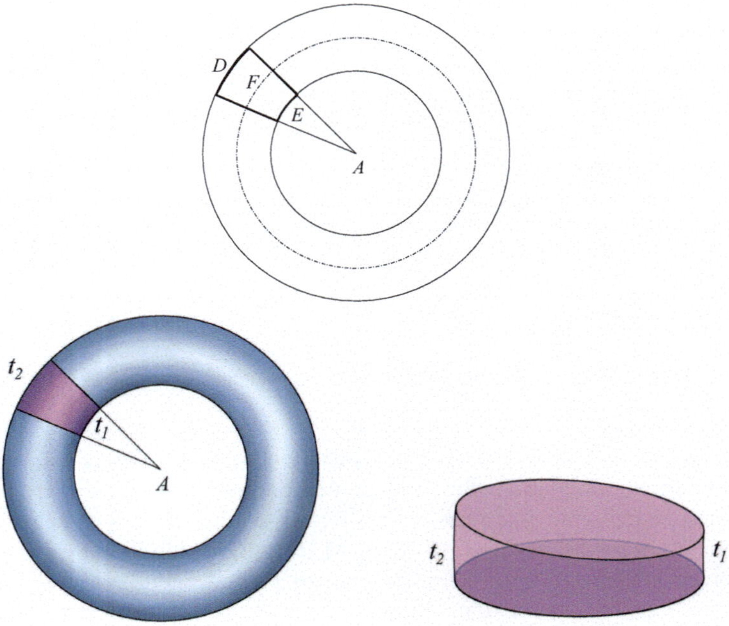

Fig. 5.3.15. 'Infinitesimal' slices of a ring

are formed by segments IK parallel to MN and heights $IJ = KL$ up to the intersection area MSN. Being well aware that it is *not* a sum in the usual sense we nevertheless write:

$$\sum IJLK = \text{volume of the cylindrical segment bounded by } MJSLN, \qquad (5.4)$$

where the symbol of the sum has to be read as 'entirety of all indivisibles'. For the 'lemon' and other solids of revolution Kepler also found formulae of the volume as entirety of all indivisibles in that way, see [Baron 1987, p. 114].

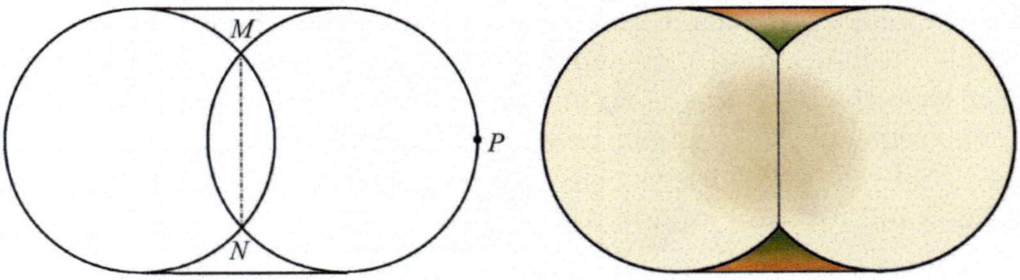

Fig. 5.3.16. Cut through Kepler's apple

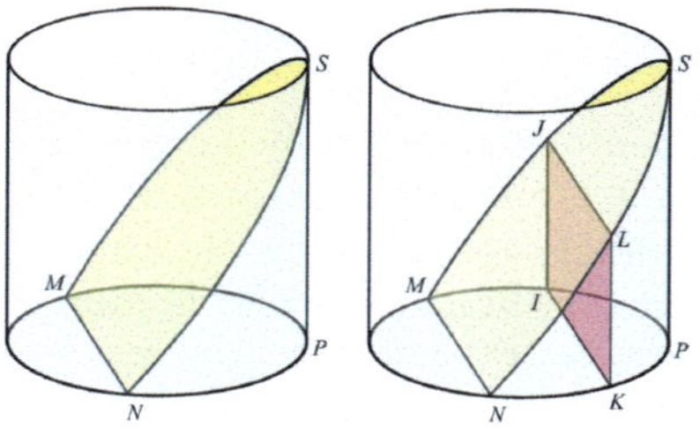

Fig. 5.3.17. (left) The apple as a cylinder; (right) The apple as entirety of indivisibles

Let us now come back to the actual wine barrels. Their hulls can be almost exactly[2] described by parabolas rotating around an axis as shown in figure 5.3.18. Computing the volume of the solid of revolution by means of Guldin's rule, see page 208, then the area under the parabola is needed. Kepler found the formula

$$A = \frac{b-a}{6}(u + 4v + u),$$

or, if more general values $f(a)$ and $f(b)$ are admitted,

$$A = \frac{b-a}{6}\left(f(a) + 4f\left(\frac{a+b}{2}\right) + f(b)\right).$$

This formula for the computation of the area under a parabola is today known as 'Kepler's barrel rule' (Kepler'sche Faßregel) in Germany (and as 'Simpson's rule' in English speaking countries). In case of parabolas it gives the exact area but is used more generally in numerical mathematics as a numerical quadrature rule for the approximate computation of integrals.

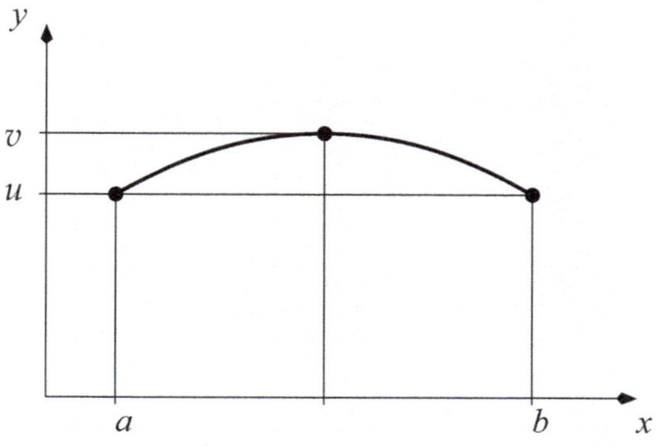

Fig. 5.3.18. Figure concerning Kepler's barrel rule

[2] This results from the art of forming wooden planks by the coopers. The planks were pressed in form by two rings at the ends and one (or two) in the middle.

5.4 Galileo Galilei

Galileo Galilei was born into an impoverished family of patricians in the small Italian city of Pisa in 1564. His father came from Florence and was a cloth merchant but also very interested in music. He wondered how the tone pitch might depend on the tension in a string and he taught his son independant thinking and the importance of the exact sensory perception [Hemleben 2006, p. 16].

The father went back to his home city Florence quite early and let his family follow in autumn 1574. Galileo's mother is said to have been rather cantankerous who outlived her husband by 30 years; six or seven children sprang from this marriage. Only four of them (two girls and two boys) survived infancy. Soon after having moved to Florence Galileo came to a monastery for education where he felt so much at home that, so it is said, he wanted to become a monk himself. All his life – even in the desperate quarrels with the Curia concerning the Copernican system at the end of his life – he stayed a devout Catholic. However, the father had other plans regarding his son. On the 5th September 1580 Galileo enrolled at the University of Pisa to study medicine. Studying medicine in these days is by no way comparable to what it is today – one had to study Aristotle and the writings of Galen (about 129–about 216). We know from Galileo's notes that he knew the Copernican (1473–1543) model of the world as explained in the book *De Revolutionibus Orbium Caelestium* (On the Revolutions of the Celestial Spheres) [Copernicus

Fig. 5.4.1. Galileo Galilei ([Painting by Justus Susterman, 1636] National Maritime Museum, Greenwich, London)

Fig. 5.4.2. Galilei's thermometer [left Photo: Fenners 2006; right Photo: Grin 2005]

1992], in which the geocentric world model of Claudius Ptolemy (about 100–about 170) was replaced by a heliocentric model. However, the 19-year-old Galileo still refused this world model; only later will he become a protagonist of the Copernican system and he will suffer for this. During mass in the Cathedral of Pisa he observed the chandelier swinging. By means of his own heartbeats he measured its period of oscillation and astutely concluded that this period stays constant over time. Galileo had discovered that time can be measured by the pendulum!

After four years Galileo gave up his studies and returned to Florence to learn mathematics from Ostilio Ricci (1540–1603). He worked as a private tutor, gave lectures, and made quite an impression through investigations concerning mechanics and hydrostatics, so much that he was offered a position as university lecturer of mathematics at the University of Pisa. The salary is miserable but leaves enough time for Galileo to develop (and sell) instruments. Among other things he worked on an air thermometer (thermoscope) but apparently *not* on the fluid thermomenter bearing his name to this day and decorating many living rooms (although not many may know how to read off temperature). The now famous 'Galilean thermometer' was developed later by a group of scientists including one of Galilei's pupils [Loyson 2012].

A closed glass cylinder contains a fluid (alcohol) in which glass spheres with temperature marks, also being partly filled by some fluid, experience buoyancy. Changes in temperature lead to changes in density of the outer liquid and the masses of the spheres are adjusted so that the lowest free floating sphere shows the actual temperature. If some spheres are at the bottoms and the others at the top and no sphere free floating in between then the temperature is the arithmetic mean of the temperatures shown by the lowest sphere at the top and the highest sphere at the bottom. The production of such a thermometer is very complicated and has to be performed with high precision. Galileo also invented a water level; he investigated the laws of free falling bodies and concluded that the period of oscillation of a pendulum does only depend on the pendulum length. All of his life he tried to construct a pendulum clock in vain but only the Dutch scientist Christiaan Huygens (1629–1695) will have success. Galileo conducted experiments with spheres rolling on an inclined plane with different inclination angles and in this way derived his laws of falling bodies. Whether Galileo actually conducted experiments on the leaning tower of Pisa is uncertain and rather not likely [Drake 2003, p. 19ff.]. He summarised his results in the treatise *De motu antiquiora* which is explicitly directed against the physics of Aristotle and therefore was not going down very well with his expert colleagues. His contract was not renewed in 1592 and since his father had died the year before Galileo had to look for some income.

Already at this time Galileo had not only critics but also patrons. They provided him with a professorship for mathematics in Padua which he occupied for the next 18 years to come. Padua was a part of the Republic of Venice where one had liberal attitudes. Besides his professorships Galileo gave private lessons for money and two future cardinals were among his pupils. He is furthermore still engaged in building instruments. He designed and built a proportional compass and let a mechanic build an improved 'compasso' which was a predecessor of the slide rule. From a letter to Kepler we know that Galileo meanwhile had become an ardent adherer of the Copernican world model. The observation of the supernova of 1604 led him to further publicly attacking the established Aristotelean doctrine of the structure of the cosmos.

In the year 1609 Galileo came to know the invention of the telescope by the Dutchman Hans (Jan) Lipperhey (about 1570–1619) and he was enthused. He bought glass lenses and rebuilt the telescope. In only a few years he succeeded in perfecting it resulting in a magnification factor of more than thirty. A telescope would surely be of great importance in warfare and so Galileo presented his telescope to the Venician government. He even gave the sole right of use to the government in exchange of a substantial increase of his salary. However, you can't sell rights of use for an instrument which you have not invented, so Galileo was penalised by a drastic shortening of his salary in the following year when the government found out that telescopes could be bought on the market places. Anyway, Galileo is yet the founder of an astronomy with the telescope if we leave out the Englishman Thomas Harriot (1560–1621) who observed the moon in all secrecy with his own telescope already in 1609; some months *before* Galileo, and he observed the

Fig. 5.4.3. The leaning tower of Pisa beside the Cathedral. It is said that Galileo had conducted his free fall experiments from this tower but that is only an often repeated legend [Photo: K. Anne Gottwald]

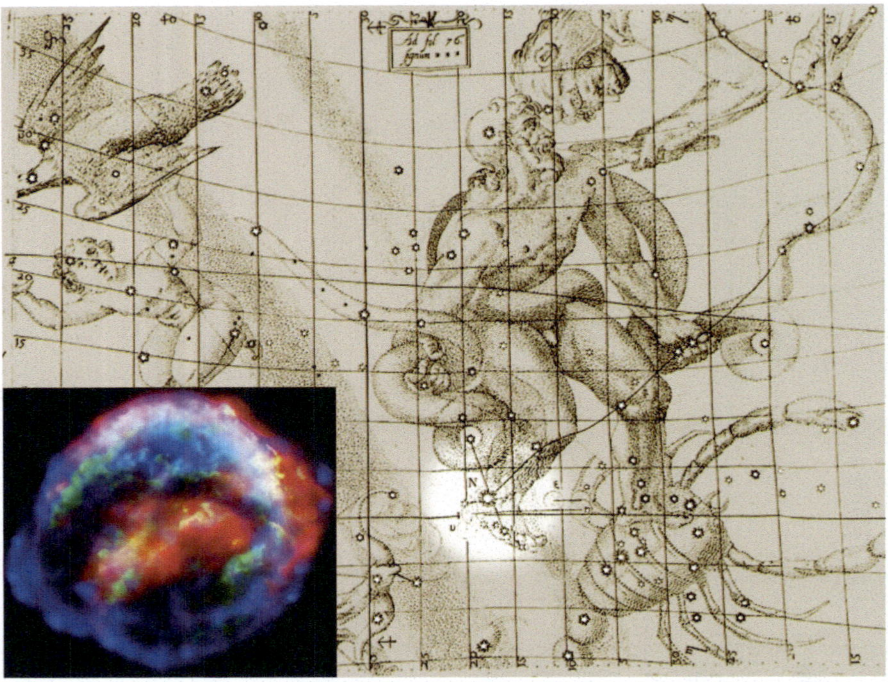

Fig. 5.4.4. The supernova (N) of 1604 in a drawing by Johannes Kepler from his book *De Stella Nova in Pede Serpentarii.* The combination of the recordings of three telescopes (lower left Chandra X-ray, Hubble and Spitzer space telescopes) shows the cloud of gas (brighter in the drawing) 400 years later [Photo: NASA 2000-2004, NASA/ESA/JHU/R.Sankrit&W.Blair]

moons of Jupiter after Galileo [Shirley 1981, p. 283-308]. Galileo discovered the four largest moons of Jupiter and thought them new planets, bringing him in disagreement with Kepler. His report on this and other discoveries was published in 1610 under the title *Sidereus nuncius* (Siderial messenger) [Galilei 2015] which was sold out within a few days.

One of his former students had meanwhile become the Grand Duke of Tuscany: Cosimo II de' Medici. In 1610 this Grand Duke appointed Galileo court mathematician, court philosopher, and first professor of mathematics in Pisa, where Galileo had no teaching responsibilities whatsoever. In the 47th year of his life all his dreams had seemingly come true and he moved to Florence. On this occasion he seemed to have separated from his partner and housekeeper Marina Gamba with whom he had three children. One of his admirers was Cardinal Barberini, later to become Pope Urban VIII, who helped to settle his illegitimate daughter in cloisters while his son came to his father to Florence in 1613.

This is not the right place to discuss Galileo's wonderful findings in physics and astronomy which changed the world and made the work of Isaac Newton possible. Instead we refer to the biography [Drake 2003]. There also can be found a detailed report on the trial of the Curia against Galileo (but do not miss [de Santillana 1955]) which led to a house arrest of the old man and to his revocation of the Copernican doctrine. The first problems started in 1616 when a cleric published a book in which he tried to prove that the Copernican doctrine did *not* contradict the Holy Bible. The inquisition initiated a lawsuit against the author Paolo Antonio Foscarini and his book was banned. Until 1822 (!) within the sphere of influence of the inquisition Copernicus's *De revolutionibus* could only be published in editions in which it was made crystal clear that nothing but a mathematical model was discussed! Galileo's attitude towards Copernicus was clear and publicly known. After the trial of 1616 Galileo had got the warning and he held back himself.

To honour his supporter Cardinal Barberini and on the occasion of the latter's election as pope Galileo dedicated to him the work *Saggiatore* which contained a polemic against one Jesuit who had fantasised on the appearance of comets. In this writing Galileo laid the foundation stone for a mathematisation of natural sciences and he expelled alchemy and astrology once and forever from the realms of science. The work was anonamously denounced because of 'atomism' and only a friendly report and the help of some of his friends within the Roman clergy prevented larger trouble for Galileo.

When in 1624 Galileo was received six times by Pope Urban VIII in Rome who encouraged him to write on the Copernican system as a mathematical model(!) Galileo apparently saw the situation somewhat overoptimistic. In 1630 *Dialogo di Galileo Galilei sopra i due Massimi Sistemi del Mondo Tolemaico e Copernicano* (Dialogue Concerning the Two Chief World Systems – Ptolemaic & Copernican) [Galilei 1967] was published.

Galileo had deliberately chosen his mother language Italian instead of Latin; apparently he wanted the book to be read! Stillman Drake even wrote in the preface [Galilei 1982, p. XX] to a German edition that Galileo had given up trying to convince the professionals. Additionally, there was not even a trace of the world system of Tycho Brahe in this book which was cherished by the Jesuits. For several days three persons, Salviati, Sagredo, and Simplicio, talk to each other and discuss problems. It was the clear aim of the book to unambiguously show the superiority of the Copernican world system. The characters Salviati and Sagredo were based on existing people who were close friends of Galileo. Filippo Salviati, born in 1582 in Florence, acted as Galileo's own spokesman [Galilei 2014, p. xxxiii]. The Venetian Giovanni Francesco Sagredo, born in 1571, studied with Galileo at Padua and was his close friend since 1600 [Galilei 2014, p. xxxiii]. The third person called Simplicio (= the simple-minded) carried all characteristics of a conservative, submissive to authority, 'book savant' of the time and has it's real role-model in Simplicius of Cilicia (about 490–about 560) who commented on the writings of Aristotle

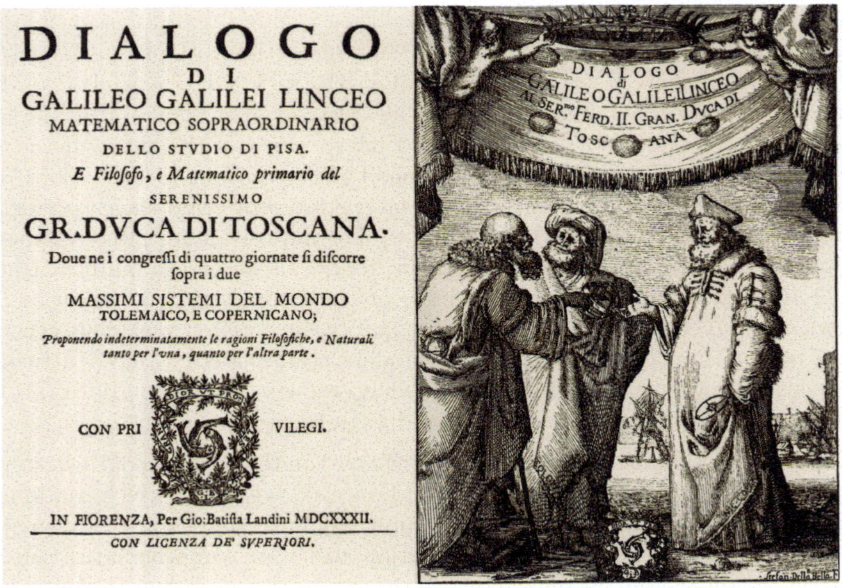
DIALOGO
DI
GALILEO GALILEI LINCEO
MATEMATICO SOPRAORDINARIO
DELLO STVDIO DI PISA.
E Filosofo, e Matematico primario del
SERENISSIMO
GR.DVCA DI TOSCANA.
Doue ne i congressi di quattro giornate si discorre sopra i due
MASSIMI SISTEMI DEL MONDO
TOLEMAICO, E COPERNICANO;
Proponendo indeterminatamente le ragioni Filosofiche, e Naturali tanto per l'vna, quanto per l'altra parte.
CON PRI VILEGI.
IN FIORENZA, Per Gio:Batista Landini MDCXXXII.
CON LICENZA DE' SVPERIORI.

Fig. 5.4.5. Two titles of *Dialogo* by Galilei, 1632 and 1635

in late antiquity. Nevertheless Galileo could not help but put some of the favourite ideas of Pope Urban VIII in the mouth of Simplicio. This eventually led Urban, Galileo's long-standing supporter, to withdraw his confidence and to stand against him. But for the time being Galileo got the imprimatur and the book could be published in Florence.

Already in 1632 the inquisitor of Florence was ordered to prevent the distribution of the book. Galileo was ordered to appear before the pope but shirked by means of medical certificates for a long time. Finally the pope threatened to have him presented in chains. In 1633 the lawsuit against him took place – he had to revoke and was held under house arrest which was not repealed until his death. His health had deteriorated for quite some time; in 1638 he went blind probably due to diabetes, probably due to his observations of the sun. Shortly before he went completely blind he discovered the libration of the moon, i.e. its apparent tumbling motion. In 1633 Galileo started working on his main physical-mathematical work: *Discorsi e Dimostrazioni Matematiche intorno a due nuove scienze attenti alla mecanica e i movimenti locali* (Discourses & Mathematical Demonstrations Concerning Two New Sciences Pertaining to Mechanics & Local Motions) [Galilei 2014] in which the same personelle as in the *Dialogo* performed discussion. A publication with imprimatur by the clergy was unthinkable and hence the book became known north of the Alps only in a Latin translation by Matthias Bernegger (1582–1640). The Italian text appeared in Leiden in 1638. The book founded modern physics on a mathematical basis which Newton later exploited in

Fig. 5.4.6. Tomb of Galilei in Santa Croce, Florence [Photo: Jebulon 2011]

his *Principia*. Galileo treated problems of the theory of elasticity and of kinematics; among other gems a correct description of projectile motion.

In 1641 the highly gifted Evangelista Torricelli became Galileo's private secretary. Galileo died on 8th January 1642. The Vatican prevented a magnificent funeral. His grave can be found within the church Santa Croce in Florence.

5.4.1 Galileo's Treatment of the Infinite

We follow Galileo's contributions concerning infinitely large and small quantities in his *Discorsi e Dimostrazioni Matematiche intorno a due nuove scienze* of 1638.

Aristotle's Wheel

From page 28 on in [Galilei 2014] Galileo is concerned with an apparent paradox which is known as 'Aristotle's wheel', see [Knobloch 1999]. It is simply a wheel with n edges as shown in figure 5.4.7(a) for the case $n = 6$.

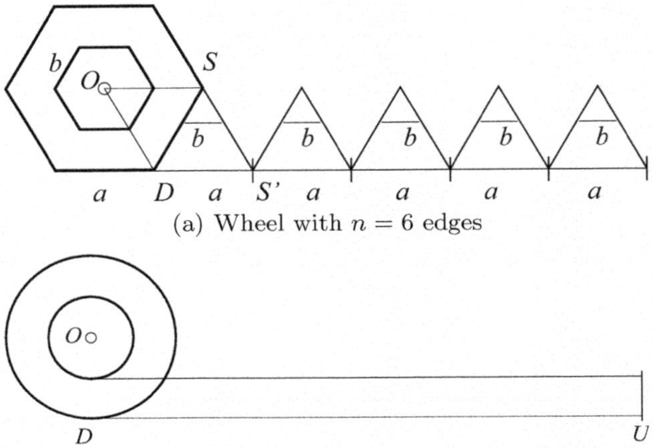

(a) Wheel with $n = 6$ edges

(b) Wheel for $n \to \infty$

Fig. 5.4.7. Wheel of Aristotles

On an axis O imagine two hexagonal wheels fixed to each other where the large wheel has edge length a and the small one edge length b. Unrolling the wheel a straight line segment of length $6a$ results which corresponds to the circumference of the larger wheel. Since also the smaller wheel is unrolled we have a segment of length $6b$ which is not contiguous but intercepted by 5 gaps. Galileo now started a thought experiment: What happens if the number of edges is increased? Obviously, unrolling the larger wheel gives a segment of length $n \cdot a$, but the $(n-1)$ gaps between the pieces of length b get smaller. If n grows beyond all bounds we obviously arrive at circular wheels. Unrolling the larger circle results in its circumference DU, but also the smaller circle is unrolled and apparently also on the segment DU! But then both circles would have the same circumference which is impossible. Galileo's disputants are in a flap. Salviati had presented the problem and Sagredo says [Galilei 2014, p. 32]:

> *This business is truly very intricate, and no solution at all occurs to me; so tell us what occurs to you.*

Salviati comes up with an explanation based on indivisibles [Galilei 2014, p. 34f.]:

There are these [difficulties] indeed, and others; but let us remember that we are among infinities and indivisibles, the former incomprehensible to our finite understanding by reason of their largeness, and the latter by their smallness. Yet we see that human reason does not want to abstain from giddying itself about them. Taking some liberties on that account, I am going to produce a fantastic idea of mine which, if it concludes nothing necessarily, will at least by its novelty occasion some wonder. Or perhaps it will seem to you inopportune to digress at length from the road that we started on, and hence will be distasteful.

Since for every finite n there always remain gaps between the segments of length b and since the circle ($n \to \infty$) is nothing but a polygonal wheel with infinitely many infinitely small gaps there have to be infinitely many of such infinitely small gaps in the apparently contiguous segment on which the smaller circle unrolls. Hence DU has to be longer than the circumference of the smaller circle. In DU there are no such gaps, but in the unrolled circumference of the small wheel there are infinitely many!

Galilei and Indivisibles

But how can it happen that points are associated with lengths, since in Aristotle's wheel point-type gaps even lengthen a straight line segments? Salviati speaks [Galilei 2014, p. 35]:

Be it so, since that is the way you wish it. Let us begin from the first – how a single point can ever be understood to be equal to a line. The most that can be done at present is for me to try to put at rest, or at any rate to moderate, this improbability with an equal or greater one, as a marvel is sometimes put to rest by a miracle. [...]

Then a 'bowl' is considered as shown in figure 5.4.8. We follow [Baron 1987, p. 119f.]; in [Galilei 2014, p. 35] the bowl is drawn upside down. The bowl is the hemisphere ABF drawn inside a cylinder $ABED$ in which we additionally place a circular cone DCE. Imagine a plane GN intersecting the cylinder and in particular its axis CF in the point P.

Obviously,

$$CO^2 = PN^2,$$

since both distances are radii of the cylinder or the bowl, respectively. Applying the theorem of Pythagoras

$$PN^2 = CO^2 = CP^2 + PO^2 = PL^2 + PO^2,$$

where the last equation follows from $CBEF$ being a square and hence it holds $CP = PL$. But then it follows

$$PN^2 - PO^2 = PL^2.$$

This means that the area of the circular ring with width ON $(= PN - PO)$ equals the area of the circle with radius PL. We have assumed nothing concerning the location of the intersecting plane GN; hence our result remains valid for every position of the intersecting plane. Imagine moving the intersecting plane GN in position AB; then the circle with radius PL becomes a single point C, while the circular ring of width ON becomes the circle with radius CB. In this case we apparently have to admit that the point C somehow corresponds to the circumference of the circle with radius CB!

Salviati notes [Galilei 2014, p. 37]:

> [...] , *and so it appears that the circumference of an immense circle may be called equal to a single point!* [...] *Hence, along the line in which such speculations lead us, the circumferences of all circles, however unequal [in size], may be called equal to one another, and each of them [may be called] equal to a single point!*

The arguments employed here appear hair-raising today and can not stand up to our clear concept of limits. The area of the circular ring with width ON simply vanishes in AB as does the area of the circle with radius PL. It seems to me that here in particular Galileo's deep roots in the scholastic ideas of the Middle Ages with which he grew up showed.

We have already reported that Eberhard Knobloch [Knobloch 2004] has found an intellectual ribbon connecting Nicholas of Cusa and Gottfried Wilhelm Leibniz as far as the handling of indivisible quantities is concerned, and that he sees Galileo in the middle of this ribbon, cp. page 153. We can now at this point clearly see this connection between Galileo and the man from Cusa. As

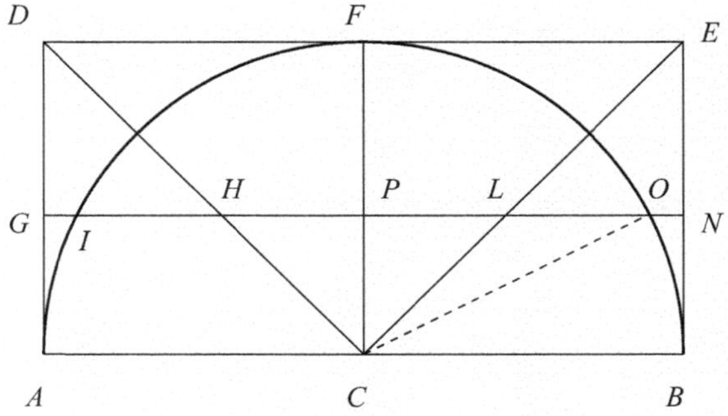

Fig. 5.4.8. Section of Galilei's 'bowl'

the man of Cusa, so also Galileo distinguished between quantities (quanti) and non-quantities (non quanti) and as Nicholas also Galileo postulated that non-quantities (meaning the infinitely large as well as the indivisibles) are not intelligible to the human mind. In moving from the finite to the infinite both entered a realm in which the well-known rules of finite mathematics cease to be valid.

The Cardinality of the Square Numbers

But the discussion about infinity continued and hence Galileo let Salviati speak [Galilei 2014, p. 39f.]:

> *These are some of those difficulties that derive from reasoning about infinities with our finite understanding, giving to them those attributes that we give to finite and bounded things. This, I think, is inconsistent, for I consider that the attributes of greater, lesser, and equal do not suit infinities, of which it cannot be said that one is greater, or less than, or equal to, another. In proof of this a certain argument once occured to me, which for clearer explanation I shall propound by interrogating Simplicio, who raised the difficulty. I assume that you know quite well which are square numbers, and which are not square.*

And now Galileo starts to discuss the problem of the cardinality of the set of square numbers and we feel a modern wind blowing. Compared with the natural numbers $1, 2, 3, 4, 5, 6, \ldots$ the square numbers $1, 4, 9, 16, 25, 36, \ldots$ appear to being thinned out. There are ten natural numbers between 1 and 10 (including the boundaries) but only three square numbers. There are ten natural numbers between 10 and 20 there are even only three square numbers and the distribution of square numbers gets 'thinner' the further we advance. Galileo came up with the idea to compare the infinite set of all natural numbers with the infinite set of all square numbers in that he numbered the square numbers.

$$\begin{array}{cccccccc} 1 & 2 & 3 & 4 & 5 & 6 & 7 & \cdots \\ \updownarrow & \updownarrow & \updownarrow & \updownarrow & \updownarrow & \updownarrow & \updownarrow & \cdots \\ 1 & 4 & 9 & 16 & 25 & 36 & 49 & \cdots \end{array}$$

Every square number gets a unique number and – inversely – every number is associated uniquely with a square number. Today we would say: Galileo has constructed a bijective mapping (one-to-one and onto) between both sets. But then there is one and only one possible conclusion: Both sets have the same cardinality. Salviati says [Galilei 2014, p. 41]:

> *I don't see how any other decision can be reached than to say that all numbers are infinitely many; all squares are infinitely many; all*

their roots are infinitely many; that the multitude of squares is not less than that of all numbers, nor is the latter greater than the former. And in final conclusion, the attributes of equal, greater, and less have no place in infinite, but only in bounded quantities.

5.5 Cavalieri, Guldin, Torricelli, and the High Art of Indivisibles

Among Galileo Galilei's students Bonaventura Francesco Cavalieri (1598–1647) became famous. He succeeded in perfecting the methods of indivisibles and to shape it so that they became a mighty tool in analysis.

He was born in Milan which then belonged to the Habsburg Empire. Already in 1615 he became a member of the order of Jesuati[3] (not Jesuits!).

In the following year he came to the Jesuit convent at Pisa and in contact with Galileo Galilei whose student he subsequently became. Cavalieri's genius had already attracted the attention of Cardinal Federico Borromeo in Milan and he informed Galileo of this contact. In Pisa Cavalieri learned mathematics from Benedetto Castelli who worked at the university and in whose lectures he was substituted by Cavalieri from time to time. The application to a

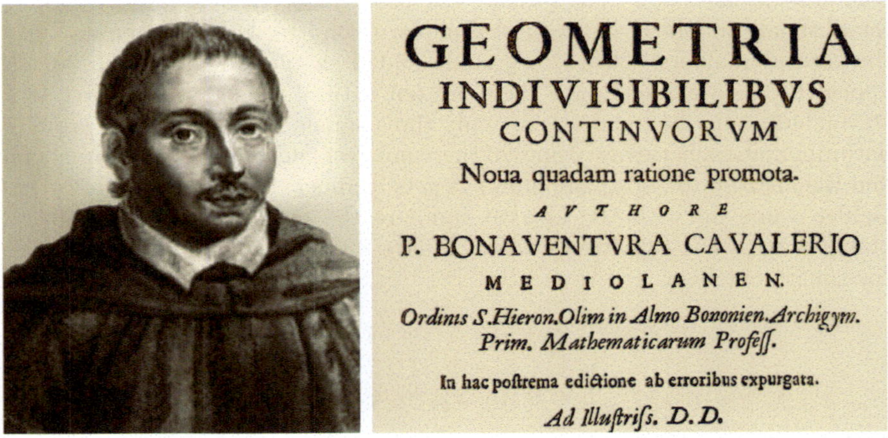

Fig. 5.5.1. Bonaventura Cavalieri and the title of his work regarding indivisibles of 1635

[3] The Jesuati order or the Apostolic Clerics of Saint Jerome had been founded as a lay movement about 1360 and was approved by Pope Urban V in 1367. In the year 1668 the Jesuati order was abolished by Pope Clement IX but the branch of female Jesuati (Jesuati sisters) remained for another 200 years. In Italy and in Toulouse the Jesuati maintained some 40 settlements.

vacant professorship at Bologna in 1619 failed due to Cavalieri's youth and when Castelli left for Rome Cavalieri was again not considered. So he went to the convent at Milan as a personal employee of Cardinal Borromeo where he taught theology until 1623 when he became Prior of Saint Peter in Lodi in Lombardy. After three years he changed to the convent at Parma where he stayed for another three years. Then, in 1629, his time had come: he got the mathematical chair at the University of Bologna.

At that time he had already accomplished groundbreaking work concerning the theory of indivisibles. In his work *Geometria indivisibilibus continuorum nova quadam ratione promota* which was published in 1635 he combined the Archimedean method of exhaustion with Kepler's indivisibles and went far beyond these methods. We mention 'Cavalieri's principle' in particular, from which two areas (volumes) are equal if they have the same height and if at every height every intersection of the two leads to lines (areas) of equal length (area). Cavalieri was heavily attacked by Jesuit circles, in particular by Paul Guldin. Guldin was born Habakuk Guldin in St Gallen christened as a Protestant. In 1597 he converted to the Catholic faith and changed his name to Paul Guldin. Shortly after he entered the Jesuits. His mathematical work *Centrobaryca* appeared in four volumes 1635, 1640, and 1641 in Vienna and among other results contains the famous 'Guldin's rule':

Guldin's rule: *The volume of a solid of revolution is computed from the product of the area and the distance the barycentre has to traverse.*

Example: To illustrate Guldin's rule we treat rotation of a triangle about the x-axis so that a circular cone emerges, see figure 5.5.3. The area of the triangle obviously is

$$A := \frac{1}{2}r \cdot h$$

and the distance travelled by the barycentre S at one full rotation is the circle with radius s. The barycentre in a triangle lies at

$$s = \frac{1}{3}r$$

and therefore the distance travelled will be

$$\ell := 2\pi s = \frac{2}{3}\pi \cdot r.$$

Following Guldin's rule the volume of the circular cone is

$$V := A \cdot \ell = \frac{1}{3}\pi \cdot r^2 \cdot h.$$

Due to the many attacks on Cavalieri's method of indivisibles he published a work *Exercitationes geometricae sex* in 1647 which became the most cited source of computations of areas and volumes in the 17th century.

Fig. 5.5.2. Paul Guldin (Oil painting in the Fachbibliothek of the University of Graz). This portrait shows great similarity with, or seems to be nearly identical to, a painting of Christophorus Clavius in the Collegium Romanum in Rome, and with an engraving after an oil painting by Francisco Villamena of 1606, and with the portrait of Clavius on the frontispiece of his *Opera Mathematica.* The differences – e.g. in the great circles of the armillary sphere, in holding the compass, and in the items on the table – let us suppose that the posthumously painted portrait of Guldin was copied from an earlier painting, for example from the copper engraving of Villamena's oil paintings. Such 'typed paintings' of Jesuits were widespread. Guldin had studied mathematics under Clavius at the Collegium Romanum and was ordered from there to go to the Jesuit University in Graz about 1618 where he became professor of mathematics [Gronau 2009]

Cavalieri not only corresponded intensely with Galileo Galilei but also with Marin Mersenne (1588–1648) in France and with Evangelista Torricelli (1608–1647). The Italian came in contact with the writings of Galileo and was immediately impressed and enthused. He became one of the most important physicists of the baroque period and influenced many scientists with his writings on mechanics and on air pressure; among them Blaise Pascal.

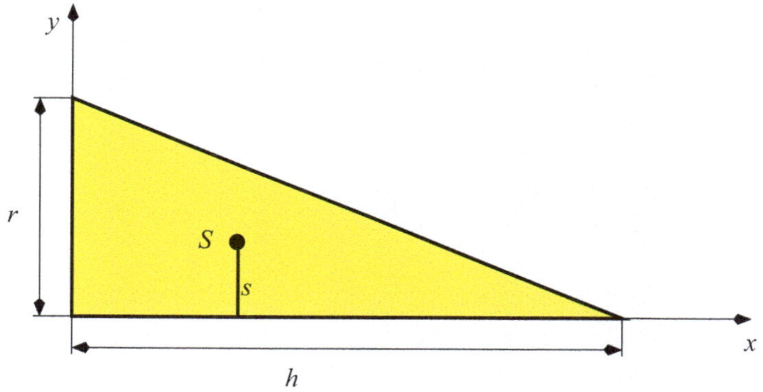

Fig. 5.5.3. Guldin's rule

Torricelli initially also critically opposed Cavalieri's method of indivisibles but then he recognised its value and even superseded them in that he not only used plane, but also curvilinear surfaces as indivisibles. A cylinder can be thought of being build from plane circular planes as Cavalieri proposed, but, as proposed by Torricelli, it can also be thought of being made up from cylindrical indivisibles as shown in figure 5.5.4. Such an assembly of a cylinder by cylindrical indivisibles had been used already by Kepler in 1615 [Knobloch 2005]. In 1644 Cavalieri published *Opera Geometrica*. His works were (and still are!) extremely difficult to understand – try reading the English translation of some extracts of Cavalieri's works in [Struik 1969, p. 209–219] – but Torricelli's work was accessible and explained Cavalieri's ideas in a way that others were able to follow. The reception of Cavalieri's ideas by John Wallis in England in particular can not be thought of without Torricelli.

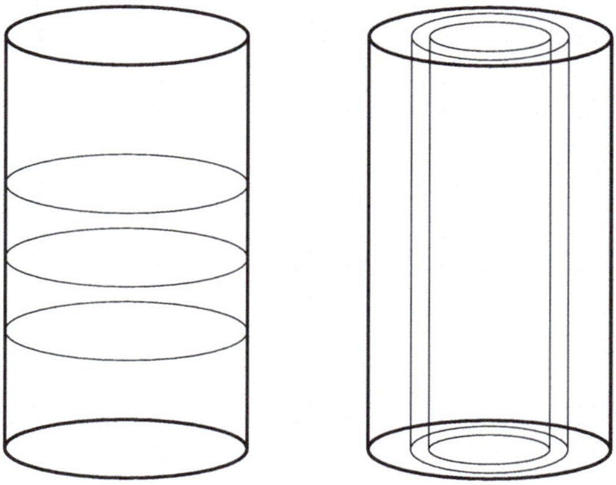

Fig. 5.5.4. Cavalieri's (left) and Torricelli's (right) indivisible in the cylinder

Fig. 5.5.5. Evangelista Torricelli and his mercury column for measuring air pressure

5.5.1 Cavalieri's Method of Indivisibles

We follow Margaret Baron [Baron 1987, p. 122–135] and the fundamental work [Andersen 1985] by Kirsti Andersen, and we develop the method of indivisibles of Cavalieri with an example concerning the computation under a curve $f(x) = x^n$ over the interval $[0, 1]$, where n is a natural number. Hence, in modern notation, we consider

$$\int_0^1 x^n \, dx.$$

Cavalieri built his method of indivisibles on two basic principles: The 'collective' and the 'distributive' principle. In the collective principle indivisibles of two different areas (volumes) are aggregated and then put in relation. If two areas F_1 and F_2 are given Cavalieri divides them into equidistantly distributed indivisibles (=lines) ℓ_i^1 and ℓ_i^2, respectively, where i is chosen from

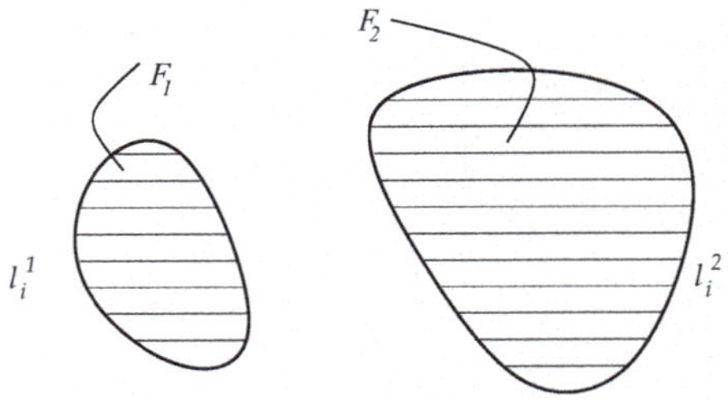

Fig. 5.5.6. Cavalieri's indivisibles in plane figures

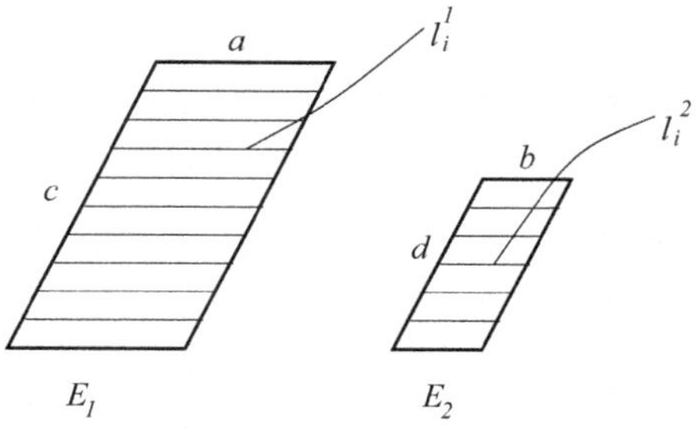

Fig. 5.5.7. Similar parallelograms with indivisibles

some set of indices[4]. Although the aggregation (or entirety) of indivisibles can *not* be expressed as an ordinary sum we nevertheless use the symbol $\sum$ for such an ordinary sum. The entirety of the indivisibles for F_1 and F_2, respectively, is therefore

$$\sum_i \ell_i^1, \quad \text{bzw.} \quad \sum_i \ell_i^2.$$

If then it holds

$$\frac{\sum_i \ell_i^1}{\sum_i \ell_i^2} = \frac{\alpha}{\beta}$$

for two numbers $\alpha, \beta > 0$, Cavalieri concludes that

$$\frac{F_1}{F_2} = \frac{\alpha}{\beta}$$

has to hold for their areas. The indivisbles remain still intangible and abstract but we will soon turn to their concrete realisation.

The distributive principle is expressed by the

Theorem of Cavalieri: *If E is a plane figure and R a spatial one with corresponding indivisibles ℓ (lines) and e (plane surfaces), respectively, then two plane figures E_1 and E_2 are coextensive, if $\sum_i \ell_i^1 = \sum_i \ell_i^2$ holds. Two spatial figures R_1 and R_2 are coextensive, if $\sum_i e_i^1 = \sum_i e_i^2$ holds.*

Besides the concept of coextensiveness (equality) of figures another important concept is that of similarity. Let us look at two similar parallelograms as shown in figure 5.5.7.

Each indivisible in E_1 has length a; the indivisibles in E_2 all have length b. Obviously it holds

[4] The superscript numbers 1 and 2 do *not* denote powers but are only labels. To avoid confusion we later write $(\ell_i)^2$, etc.

$$\frac{\ell_i^1}{\ell_i^2} = \frac{a}{b}.$$

But because of the similarity of the parallelograms we have

$$\frac{a}{b} = \frac{c}{d},$$

so that altogether we arrive at

$$\frac{\ell_i^1}{\ell_i^2} = \frac{a}{b} = \frac{c}{d}. \tag{5.5}$$

Now the entirety of indivisibles is considered:

$$\frac{E_1}{E_2} = \frac{\sum_i \ell_i^1}{\sum_i \ell_i^2} = \frac{\sum_i a}{\sum_i b} = \frac{c \cdot a}{d \cdot b} \overset{(5.5)}{=} \frac{a^2}{b^2}.$$

This result allows us to investigate plane figures with curvilinear boundaries with the help of Cavalieri's indivisibles and to compute their areas, because arbitrary similar figures can be enclosed in similar parallelograms as shown in figure 5.5.8. The quotient of the entirety of indivisibles then becomes

$$\frac{\sum_i \ell_i^1}{\sum_i \ell_i^2} = \frac{\sum_i a}{\sum_i b} = \frac{c \cdot a}{d \cdot b} = \frac{a^2}{b^2} \overset{(5.5)}{=} \left(\frac{\ell_i^1}{\ell_i^2}\right)^2.$$

Now consider a curve DHB in a reactangle $ABCD$ as in figure 5.5.9.

We look at indivisibles parallel to AB; indivisble EF is one of this flock which intersects the curve in the point H and the diagonal DB in the point G. We introduce the notations

$$\ell_i^1 = HF, \quad \ell_i^2 = GF.$$

Then it holds

$$\frac{\text{area } CBHD}{\text{area } ABCD} = \frac{\sum_i \ell_i^1}{\sum_i a}$$

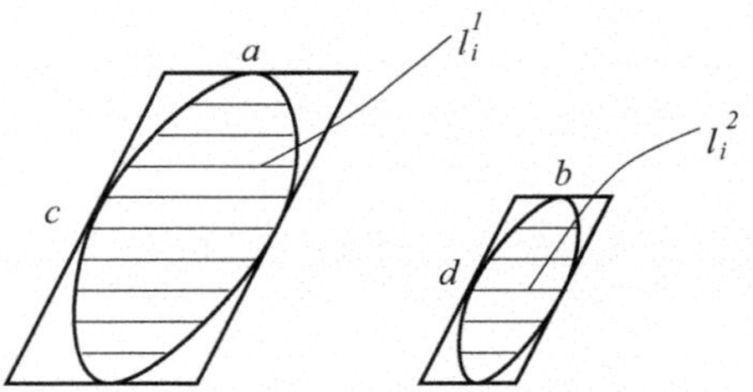

Fig. 5.5.8. Similar plane figures in parallelograms with indivisibles

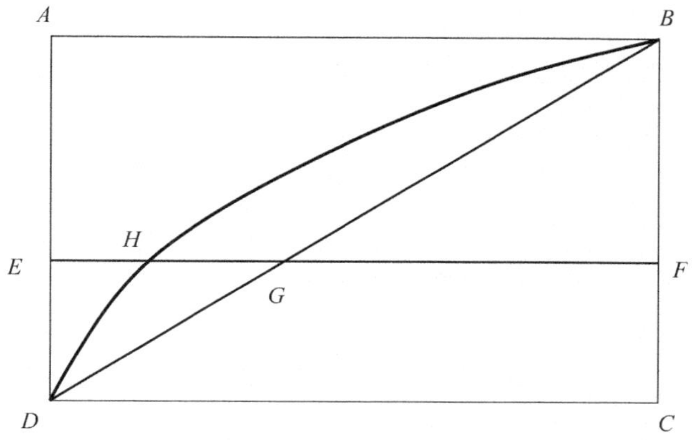

Fig. 5.5.9. Subdivision of indivisibles by a curve. The curve DHB divides the rectangle $ABCD$ into two parts. Imagining the rectangle consisting of indivisibles EF, then EH belongs to the one part of the rectangle and HF to the other

where $a = AB = CD$. If we now assume that the curve DHB can be written as a function

$$\ell_i^1 = f(\ell_i^2),$$

it follows

$$\frac{\sum_i \ell_i^1}{\sum_i a} = \frac{\sum_i f(\ell_i^2)}{\sum_i a}.$$

If f is a power function then the area $CBHD$ can be computed by certain powers of indivisibles ℓ_i^2 in the triangle BCD. This becomes trivial in case DHB being the diagonal BD. The area of the triangle BCD apparently is exactly one half of the area of the rectangle $ABCD$, i.e. we have $\ell_i^1 = \ell_i^2 =: \ell_i$ and

$$\frac{\sum_i \ell_i}{\sum_i a} = \frac{1}{2}.$$

This result is valid still, of course, if $ABCD$ will no longer be a rectangle but a parallelogram and we shall turn to this more general case. Cavalieri wanted to compute even higher powers of indivisibles, as

$$\frac{\sum_i (\ell_i)^2}{\sum_i a^2}, \quad \frac{\sum_i (\ell_i)^3}{\sum_i a^3},$$

and so on. To do so he needed an algebraic result sent to him by the French mathematician Jean Beaugrand (about 1590–1640):

$$(a+b)^n + (a-b)^n = 2\left(a^n + n \cdot C_2 \cdot a^{n-2} \cdot b^2 + n \cdot C_4 \cdot a^{n-4} \cdot b^4 + \ldots\right) \quad (5.6)$$

with certain numbers $C_2, C_4, \ldots$. In case of $n = 1$ it follows:

$$(a + b) + (a - b) = 2a.$$

For $n = 2$ the formula gives

$$(a+b)^2 + (a-b)^2 = 2(a^2 + b^2),$$

and so on. Let us take a closer look at the case $n = 5$ which yields

$$(a+b)^5 + (a-b)^5 = 2(a^5 + 10a^3b^2 + 5ab^4). \tag{5.7}$$

We assume furthermore that Cavalieri had already arrived at

$$\frac{\sum_i (\ell_i)^2}{\sum_i a^2} = \frac{1}{3}, \quad \frac{\sum_i (\ell_i)^3}{\sum_i a^3} = \frac{1}{4}, \quad \frac{\sum_i (\ell_i)^4}{\sum_i a^4} = \frac{1}{5} \tag{5.8}$$

by means of the same method we shall describe in detail soon. He is now looking for the value of $\sum_i (\ell_i)^5 / \sum_i a^5$ and considers a parallelogram $ACGQ$ as in figure 5.5.10.

The parallelogram is divided into two congruent triangles ACQ and CGQ by the diagonal QC. Further subdividing by joining the opposite centres of the sides as shown in the right part of figure 5.5.10 results in four congruent parallelograms $ABMD$, $BCHM$, $DMQN$, and $MHGN$. In a final step we further subdivide by means of the lines RV and $R'V'$ parallel to DH and at equal distance from DH, see figure 5.5.11.

Let the length of the segment RS be a and let b denote the length of ST. With the help of Beaugrand's formula (5.7) it follows for the paralleolgram $ACHD$:

$$\sum_{ACMD} (a+b)^5 + \sum_{CHM} (a-b)^5 = 2 \sum_{ABMD} a^5 + 20 \sum_{BCM} a^3b^2 + 10 \sum_{BCM} ab^4$$

and accordingly for $DHGQ$:

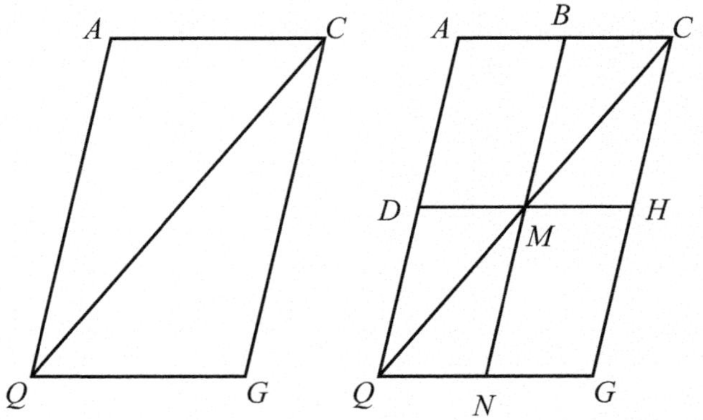

Fig. 5.5.10. Subdivision of a parallelogram

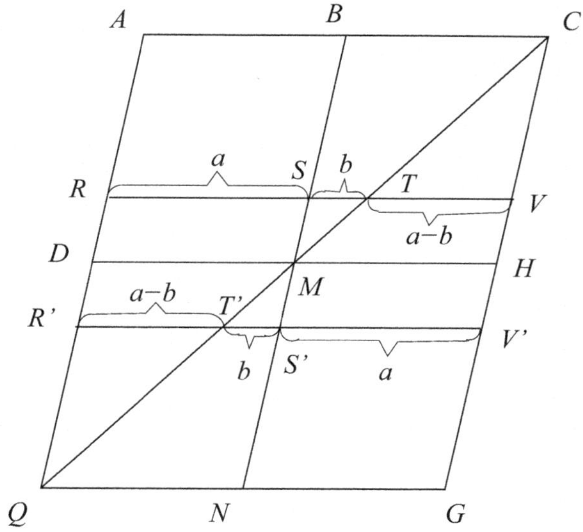

Fig. 5.5.11. Repeated subdivision of a parallelogram

$$\sum_{DHGQ}(a+b)^5 + \sum_{DMQ}(a-b)^5 = 2\sum_{MHGN} a^5 + 20\sum_{QMN} a^3b^2 + 10\sum_{QMN} ab^4.$$

It holds $ABMD = MHGN$ und $BMC = QMN$; hence, adding the 'summations' for $ACHD$ and $DHGQ$:

$$\sum_{ACGQ}\left((RT)^5 + (TV)^5\right) = 2\sum_{ABNQ} a^5 + 40\sum_{BCM} a^3b^2 + 20\sum_{BCM} ab^4. \quad (5.9)$$

Now $ACQ = CQG$, hence

$$\sum_{ACQ}(RT)^5 = \sum_{CGQ}(TV)^5$$

and

$$\sum_{ACGQ}\left((RT)^5 + (TV)^5\right) = 2\sum_{ACQ}(RT)^5. \quad (5.10)$$

We now use the results (5.8) which were previously derived by the very same method. We then get

$$\sum_{BCM} b^2 = \frac{1}{3}\sum_{BCHM} a^2 = \frac{1}{6}\sum_{BCGN} a^2 \quad (5.11)$$

and

$$\sum_{BCM} b^4 = \frac{1}{5}\sum_{BCHM} a^4 = \frac{1}{10}\sum_{BCGN} a^4 \quad . \quad (5.12)$$

Combining all partial results yields

$$\begin{aligned}\sum_{ACGQ} ((RT)^5 + (TV)^5) &= 2\sum_{ACQ}(RT)^5 \\ &\overset{(5.9),(5.11),(5.12)}{=} 2\sum_{ABNQ} a^5 + \frac{40}{6}\sum_{BCGN} a^5 + \frac{20}{10}\sum_{BCGN} a^5 \\ &= \frac{32}{3}\sum_{ABNQ} a^5 \overset{AC=2a}{=} \frac{1}{3}\sum_{ACGQ}(AC)^5\end{aligned}$$

for (5.10). It is therefore clear that

$$\sum_{ACQ}(RT)^5 = \frac{1}{6}\sum_{ACGQ}(AC)^5,$$

or, written slightly differently,

$$\frac{\sum_i (\ell_i)^5}{\sum_i a^5} = \frac{1}{6}.$$

What Cavalieri had achieved with his method is actually a table of integrals for the functions $f(x) = x^n$. From his results

$$\frac{\sum_i \ell_i}{\sum_i a} = \frac{1}{2}, \quad \frac{\sum_i (\ell_i)^2}{\sum_i a^2} = \frac{1}{3}, \quad \ldots, \quad \frac{\sum_i (\ell_i)^5}{\sum_i a^5} = \frac{1}{6}$$

Cavalieri extrapolated the general relation

$$\frac{\sum_i (\ell_i)^n}{\sum_i a^n} = \frac{1}{n+1}.$$

In modern notation he had proved

$$\int_0^1 x^n \, dx = \frac{1}{n+1},$$

i.e.: The area under the curves given by $f(x) = x^n$ on the interval $[0,1]$ is $1/(n+1)$.

Looking at Cavalieri's 'summations' today is breathtaking and hair raising. There were lines of thickness 0 airily 'summed' and put into ratios; hence it may be not surprising that opposition formed quickly against Cavalieri's method of indivisibles. Nevertheless were all of Cavalieri's results not only correct, but he had also presented a concrete method to carry out quadratures and, analogously, cubeages, i.e. computations of volumes.

The simplicity of Cavalieri's method of indivisibles can be seen in the case of the Archimedean spiral the quadrature of which had already been

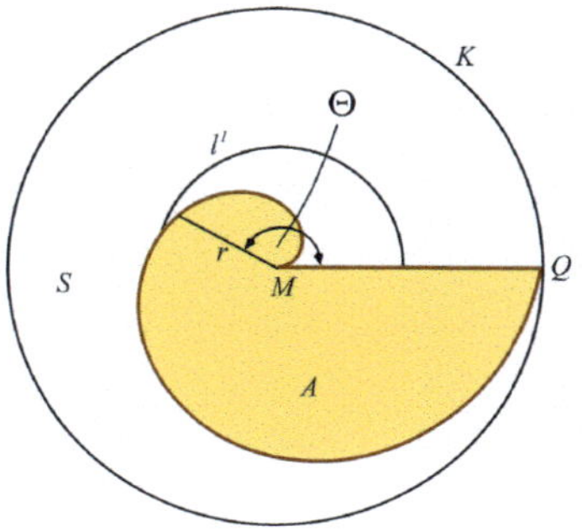

Fig. 5.5.12. Computing the area enclosed by the Archimedean spiral by means of indivisibles

accomplished by Archimedes himself, cp. page 81. The Archimedean spiral is described in polar form as $r = a \cdot \Theta$ where r is the lenth of the radius vector, $a > 0$ a parameter, and $\Theta \in [0, 2\pi]$. At $\Theta = 0$ it follows $r = 0$, at $\Theta = \pi/2 (\widehat{=} 90°)$ is $r = a \cdot \pi/2$, and so on. The spiral is shown in figure 5.5.12.

The area A is sought. The area S is the difference between the area of the circle and A. Cavalieri employed circular arcs in peripheral direction as indivisibles, namely ℓ_i^1 as the indivisible up to the spiral and ℓ_i^2 as indivisible of the cirlce K with radius R. Now Cavalieri imagines the circle being cut open at MQ and then bend open in form of a rectangle as shown in figure 5.5.13.

This results in

$$\frac{\text{area } S}{\text{area } K} = \frac{\sum_i \ell_i^1}{\sum_i \ell_i^2} = \frac{\sum_i GX}{\sum_i NX} = \frac{\text{area } BGDC}{\text{area } BNDC}$$

for the areas.

Now $R = 2\pi a$ and

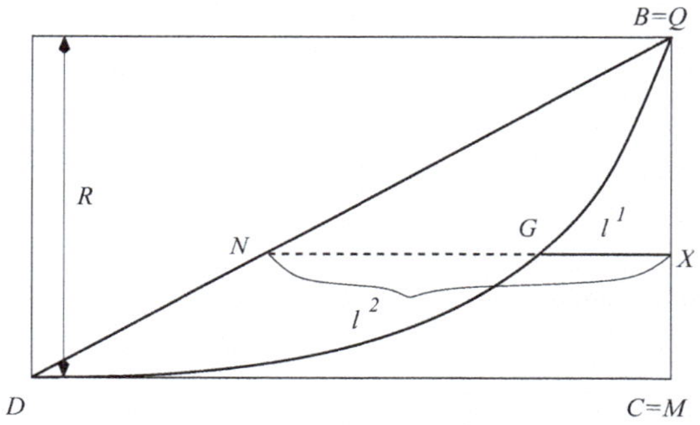

Fig. 5.5.13. The Archimedean spiral bend open

$$\ell^1 = r \cdot \Theta \overset{r=a\Theta}{=} \frac{r^2}{a} \overset{a=\frac{R}{2\pi}}{=} \frac{2\pi}{R} \cdot r^2.$$

In the bent-up state the spiral is nothing but a parabola and parabolas were already treated by Cavalieri! In the case of a parabola he had shown that

$$\frac{\text{area } BGDC}{\text{area } BCD} = \frac{2}{3},$$

and hence we get

$$\text{area } S = \frac{2}{3} \text{ area of circle } K, \quad \text{area } A = \frac{1}{3} \text{ area of circle } K.$$

5.5.2 The Criticism of Guldin

After Cavalieri had published *Geometria indivisibilibus continuorum nova quadam ratione promota* in 1635 Paul Guldin accused Cavalieri of having simply taken his method of indivisibles from Johannes Kepler. This unfounded charge of plagiarism was not a great blow to the method; however, Guldin's claim that the method would lead to paradoxes was! As an example Guldin studied a non-equilateral triangle AGH in his book *Centrobaryca*, see figure 5.5.14.

Drawing in equal distances parallel to AB the lines KM, JL, NR, and so on, results in the indivisibles KB, JC, NE, and so on left of the footpoint D of the height DH, and RF, LO, MI, and so on on the right. Now it holds

$$KB = MI, \quad JC = LO, \quad NE = RF$$

and so on for all indivisibles. Following Cavalieri it would follow

$$\sum_{ADH} JC = \sum_{HDG} LG,$$

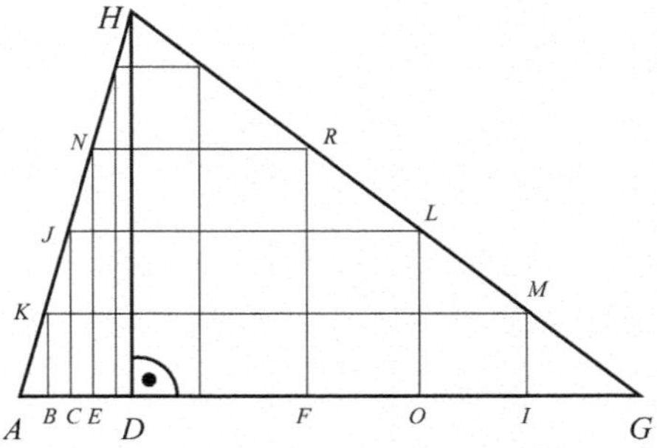

Fig. 5.5.14. Guldin's paradox

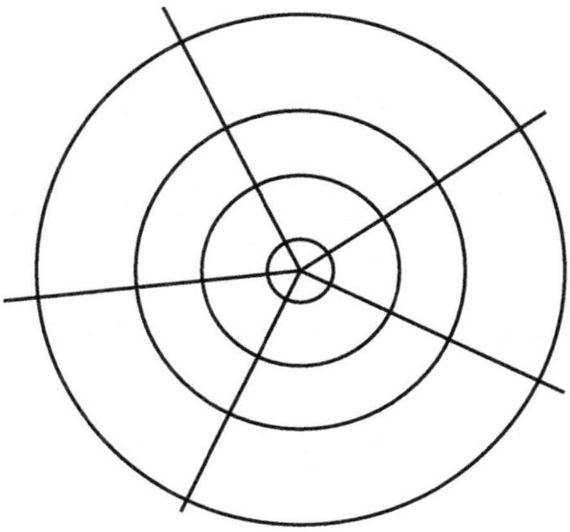

Fig. 5.5.15. Galilei's paradox

showing that the area ADH would be equal to the area HDG which would be ridiculous.

Cavalieri answered shrewdly in *Exercitationes geometricae sex.* He made it clear that the method of indivisibles was based on indivisibles being distributed evenly. The distribution of indivisibles in triangle AGH is completely different as in HDG, however.

5.5.3 The Criticism of Galilei

Galileo Galilei also saw paradoxes when Cavalieri's indivisibles were employed [Mancosu 1996, p. 119ff.]. One of his objections reminds us on the arguments of Christian scholastics against atomism. He considered a family of concentric circles which are intersected by their radii as in figure 5.5.15.

Every one of the circles is intersected by a radial ray in exactly one point; hence every circle consists of infinitely many points where every point of a larger circle can be assigned uniquely to a corresponding point of each smaller circles. Hence all circles share the same number of points. Since, as Cavalieri claims, all lines consist of indivisibles – the points, that is – all circles must have the same circumference and that would be ridiculous. Cavalieri's answer is interesting in two ways. Firstly he makes it clear that his theory is not able, and can not be used, to show that the continuum consists of points. Secondly, the same number of points does not mean that the lengths are equal. Or, as Mancuso has remarked in [Mancosu 1996, p. 122], equal cardinality does not allow any statement on a metric. Furthermore we have the same kind of issue here as in the paradox of Guldin, since the indivisibles in a larger circle and a smaller one are differently distributed.

5.5.4 Torricelli's Apparent Paradox

We have already noted that Evangelista Torricelli surpassed Cavalieri's method of indivisibles in that he allowed for crooked surfaces as indivisibles. In 1641 he made an apparently paradox discovery which quickly made him widely known in mathematical circles.

In the year 1642 he became Galileo's successor on the mathematical chair at Florence but was largely unknown as a geometer. Only two years later when he had published his *Opera Geometrica* he was well known. The reason for this lies in a result which he communicated to his friend Cavalieri already in 1641: Let the hyperbola $xy = a^2$ be bounded from below by a straight line ED in figure 5.5.16. If this truncated hyperbola rotates around the ordinate an infinitely long solid of revolution (an infinite 'Torricelli's trumpet') emerges which has finite volume! Torricelli believed that he was the first having computed the finite volume of an infinite solid, but Nicole Oresme had done that already in the 14th century, of course; see page 143, [Boyer 1959, p. 125f.].

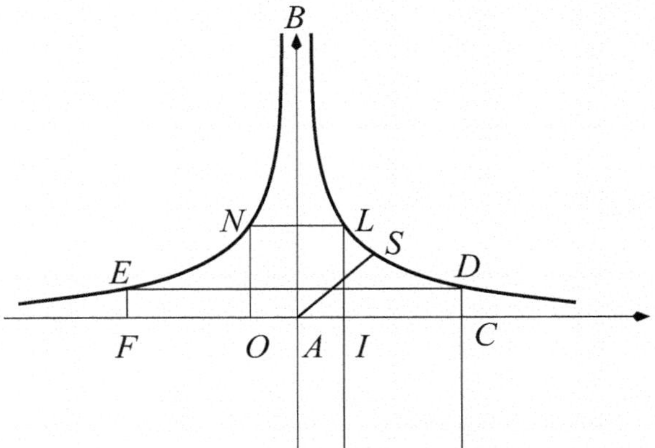

Fig. 5.5.16. Torricelli's apparent paradox

Figure 5.5.16 shows both branches of the hyperbola $xy = a^2$ with the truncation line ED. Rotating the truncated hyperbola around the ordinate axis HAB results in a solid of rotation which is infinitely extended in direction B. The most interesting fact concerning Torricelli's proof is his use of cylindrical indivisibles; in figure 5.5.16 these are the lateral surfaces of the cylinders defined by $ONLI$. For a better imagination of the three-dimensional situation consider figure 5.5.17.

Point S on the hyperbola marks the point of the shortest distance connecting A with the branches of the hyperbola and we define

$$AH = 2 \cdot AS.$$

The cylinder $ACGH$ being constructed below the hyperbola is also thought of being built up from indivisibles which are circular discs.

Torricelli prepended five lemmas to his main theorem and these he proved without the use of indivisibles [Mancosu 1996, p. 133]. In 'Lemma 5' it was being proved that the area of the surface of any cylinder $ONLI$ equals the area of a circle of radius AS, i.e. the surfaces of the cylinders $ONLI$ and $FEDC$ share the same area. Hence it is clear that every cylindrical indivisible $ONLI$ corresponds to a circular indivisible of the cylinder $ACGH$. This gives Torricelli's main result:

Theorem: *The volume of Torricelli's trumpet* EDB *equals the volume of the cylinder* $ACGH$.

It is very likely the Torricelli knew that an opposition would quickly develop against his proof by indivisibles the more so as the (correct!) result is counterintuitive. For this reason he has given a second proof exploiting a classical Greek method of exhaustion [Mancosu 1996, p. 135].

Mancosu and Vailati [Mancosu/Vailati 1991, p. 136] have pointed out the philosophical significance of Torricelli's result. In the eyes of Torricelli his trumpet is actually of infinite length, hence actual infinite. All thinkers

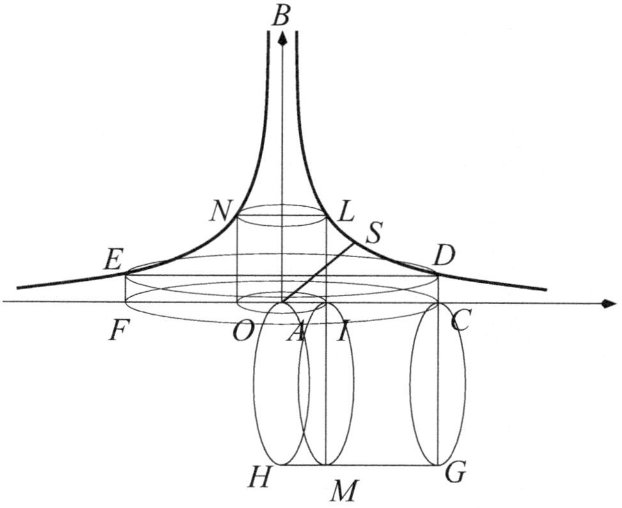

Fig. 5.5.17. Torricelli's apparent paradox in threedimensional view

Fig. 5.5.18. Christophorus Clavius in an engraving after an oil painting by Francisco Villamena of 1606, and Grégoire de Saint-Vincent. The painting of Clavius probably served as the template for the oil painting of Guldin (see figure 5.5.2)

rejecting infinity, as Pierre Gassendi (1592–1655) in France and Thomas Hobbes (1588–1679) in England, felt shocked by Torricelli's result. But also positively inclined philosophers who quite believed that they knew a few things about infinity rated Torricelli's theorem quite differently.

Torricelli's infinitely long trumpet incidentally has an infinite surface area and an infinitely large cross section! We can fill the trumpet with a finite amount of paint, but this finite amount of paint would not suffice to colour the inside surface of the trumpet. This *is* counterintuitive, isn't it? Here the paradox of the infinite clearly shows!

5.5.5 De Saint-Vincent and the Area under the Hyperbola

Grégoire de Saint-Vincent (1584–1667), latinised Gregorius a Sancto Vincentio, was born in Bruges and became a student of the great Christophorus Clavius (1538–1612)[5], who was called 'Euclid of the 16th century' and who became the architect of the Gregorian reform of the calendar. As was his teacher de Saint-Vincent became a Jesuit.

[5] As Eberhard Knobloch has shown in [Knobloch 1988] Clavius was *not* born in 1537 as one can still read, but in 1538. See also [Lattis 1994, p. 12].

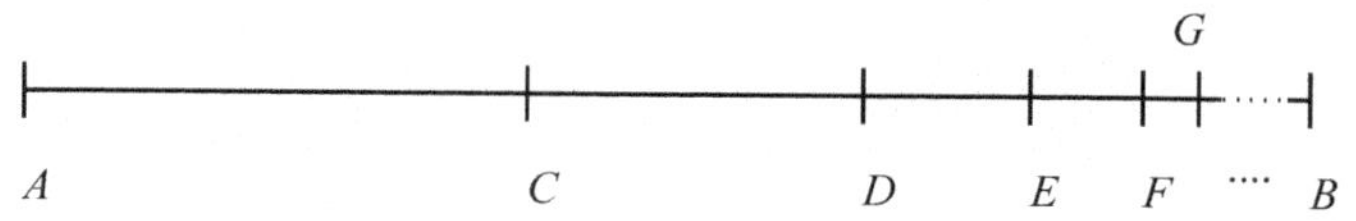

Fig. 5.5.19. Partitioning of a segment AB in geometric progression

He developed into an excellent mathematician and founded a mathematical school around him in Antwerp. He coined the word 'exhaustion' used for the Greek method of quadrature, cp. page 36. Furthermore he rigorously treated the geometric series with which he explained Zeno's paradox of Achill and the tortoise, see page 57. He worked on the foundations of analytic geometry and used methods of indivisibles independent of the one Cavalieri employed. He also designed methods using infinitesimal rectangles. In his work *Opus geometricum* of 1647 he tried to square the circle but failed, of course. He was the first to compute the area under the hyperbola and saw that it was given by the logarithm.

The Geometric Series of Saint-Vincent

De Saint-Vincent imagined a straight line segment AB with a parition as in figure 5.5.19, so that

$$\frac{AB}{CB} = \frac{CB}{DB} = \frac{DB}{EB} = \dots$$

holds.

Hence we have a geometric progression

$$AB, CB, DB, EB, \dots .$$

De Saint-Vincent realised [Baron 1987, p. 136] that since

$$\frac{AB - CB}{CB - DB} = \frac{CB - DB}{DB - EB} = \dots = \frac{AB}{CB}, \tag{5.13}$$

hence

$$\frac{AC}{CD} = \frac{CD}{DE} = \frac{DE}{EF} = \dots, \tag{5.14}$$

also the difference segments $AC, CD, DE, \dots$ are in geometric progression. But then the sequence

$$AC, CD, DE, \dots$$

defines a geometric series where, due to (5.14) and (5.13),

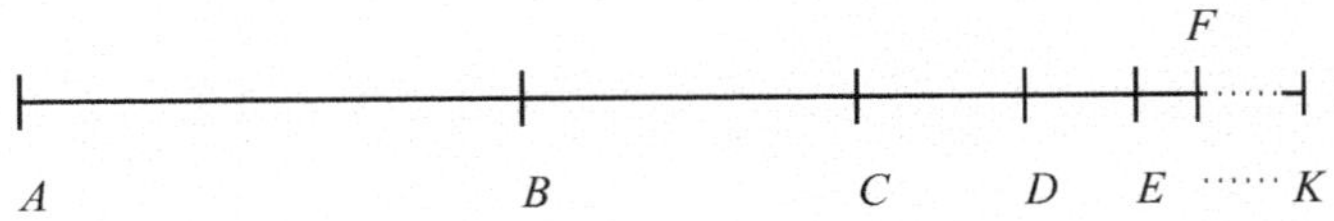

Fig. 5.5.20. Sum of a geometric series as length of a line segment

$$\frac{AC}{CD} = \frac{AB}{CB}$$

holds. Let us look now at the geometric series of the differences,

$$AC + CD + DE + \dots .$$

The distance defined by this series will come close to AB with an ever growing number of summands. Hence we can view AB as being the 'sum' of the geometric sequence $AC, CD, DE, \dots$.

In reverse we can consider an arbitrary geometric progression AB, BC, CD, ... as in figure 5.5.20 and then determine a point K so that

$$\frac{AB}{BC} = \frac{BC}{CD} = \dots = \frac{AK}{BK}$$

holds. The quantity AK therefore has to be seen as the 'sum' of the geometric series $AB + BC + CD + \dots$.

De Saint-Vincent apparently understood the concept of the 'sum' of an infinite series being the limit of the sequence of its partial sums. He also gave a geometric construction of the 'sum' by means of similar triangles as shown in figure 5.5.21.

If in the construction in figure 5.5.21

$$\frac{AM}{BN} = \frac{AB}{BC}$$

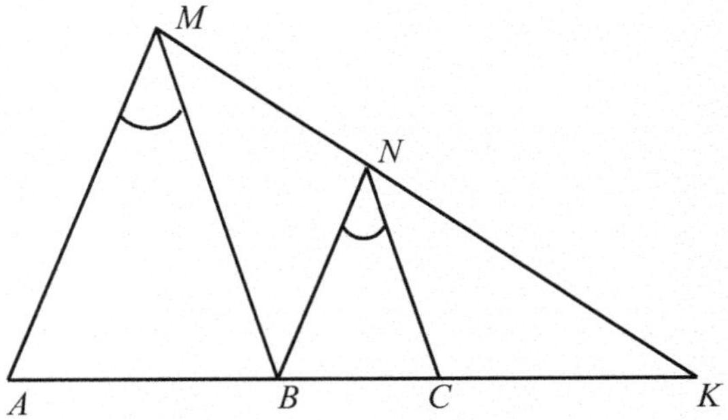

Fig. 5.5.21. Construction of the limit of a geometric series

holds then it follows from the similarity of the triangles drawn in the figure:

$$\frac{AM}{BN} = \frac{AK}{BK}$$

and

$$\frac{AB}{BC} = \frac{AK}{BK}.$$

Hence AK is the limit of the geometric series $AB + BC + \ldots$.

On the basis of his knowledge of the geometric series de Saint-Vincent started to resolve Zeno's paradox of Achilles and the tortoise. He realised that the time both are running is a geometric series and calculated their 'sums'. He also realised that this limit gives exactly the time at which Achilles will overtake the tortoise.

Horn Angles at Saint-Vincent

Grégoire de Saint-Vincent also could not resist to study horn (or corcnicular) angles. He considered two circles with different radii, midpoints on BH, and with a common tangent AB as in figure 5.5.22.

It seems intuitively clear that

$$\measuredangle ABC < \measuredangle ABE,$$

where the angles have the meaning of horn angles. If one now would compare the horn angles with the (genuine!) angle ABH as in [Baron 1987, p. 137] then it follows

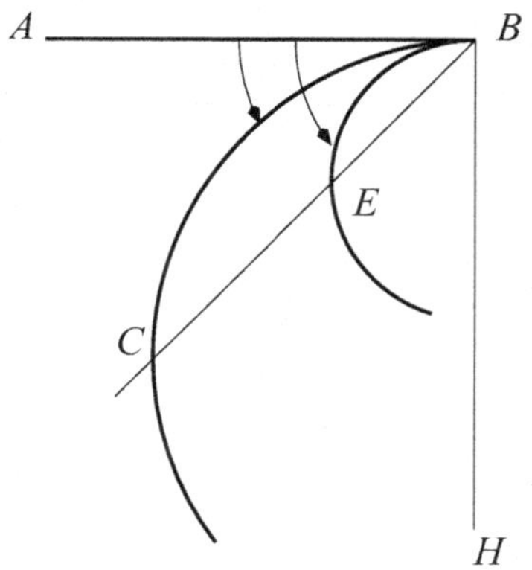

Fig. 5.5.22. Horn angles

$$\measuredangle ABH > \frac{\measuredangle ABH}{2} > \frac{\measuredangle ABH}{2^2} > \frac{\measuredangle ABH}{2^3} > \ldots > \frac{\measuredangle ABH}{2^n} > \ldots > \measuredangle ABC$$

for all n and, accordingly,

$$\measuredangle ABH > \ldots > \frac{\measuredangle ABH}{2^n} > \ldots > \measuredangle ABE.$$

But if n increases $\measuredangle ABH/2^n$ decreases and hence it becomes clear that

$$\measuredangle ABC = \measuredangle ABE = 0$$

holds for the horn angles. It is rather fascinating that Grégoire de Saint-Vincent realised this fact but that he could never accept it! He rather concluded that the laws of finite geometry could no longer be valid in the realm of the infinitely small and he doubted Euclid's famous principle: 'The whole is larger than the part.' Gottfried Wilhelm Leibniz will resume de Saint-Vincent's work later.

The Area Under the Hyperbola Following Saint-Vincent

De Saint-Vincent considered the hyperbola defined by

$$xy = 1,$$

hence the function $y = f(x) = 1/x$, on the interval $[a, b]$ with positive a and b [Edwards 1979, p. 154ff.]. Partitioning the intervall into n equal parts

$$a = x_0 < x_1 < x_2 < \ldots < x_{i-1} < x_i < \ldots < x_{n-1} < x_n = b,$$

upper and lower bounds of the area under the hyperbola from a to b can be computed by summing rectangles of width $x_i - x_{i-1}$ in two different ways – an anticipation of the later definition of a definite integral by Riemann in the 19th century.

Defining the heights of the rectangles as being always the smallest function value the i-th rectangular area is given by

$$(x_i - x_{i-1}) \cdot f(x_i) = \frac{x_i - x_{i-1}}{x_i},$$

since the hyperbola is a monotonically decreasing function and the smalles value is always assumed at the right boundary of the intervals. If we agree to always use the largest function value on each interval then this is assumed at the left boundary of each interval and it follows

$$(x_i - x_{i-1}) \cdot f(x_{i-1}) = \frac{x_i - x_{i-1}}{x_{i-1}}.$$

But it holds

$$x_i - x_{i-1} = \frac{b-a}{n},$$

so that an estimate for the true area $A_{a,b}$ between a and b is given by

$$\sum_{i=1}^{n} \frac{b-a}{nx_i} \leq A_{a,b} \leq \sum_{i=1}^{n} \frac{b-a}{nx_{i-1}}. \tag{5.15}$$

Now de Saint-Vincent considered the shifted interval $[ta, tb]$ where $t > 0$. A partitioning into n subintervals yields parts of length

$$\frac{tb - ta}{n}$$

and repeating the above idea to estimate the area under the hyperbola on $[ta, tb]$ results in

$$\sum_{i=1}^{n} \frac{tb-ta}{ntx_i} \leq A_{ta,tb} \leq \sum_{i=1}^{n} \frac{tb-ta}{ntx_{i-1}},$$

which, after reducing the factor t, yields (5.15) again. Hence Saint-Vincent has proven the property

$$A_{a,b} = A_{ta,tb} \tag{5.16}$$

for every positive t.

Following [Baron 1987, p. 139] de Saint-Vincent used a geometric progression to partition the abscissa starting at a positive x_1:

$$x_1, \quad x_2 := tx_1, \quad x_3 := t^2x_1, \quad x_4 := t^3x_1, \quad x_5 := t^4x_1, \ldots.$$

Then the differences $x_{i+1} - x_i$ on the abscissa are given by

$$x_1(t-1), \quad x_1t(t-1), \quad x_1t^2(t-1), \quad x_1t^3(t-1), \ldots.$$

The function value at the point x_1 is $y_1 := 1/x_1$. Considering the function values at the other points $x_i = t^{i-1}x_1$ yields

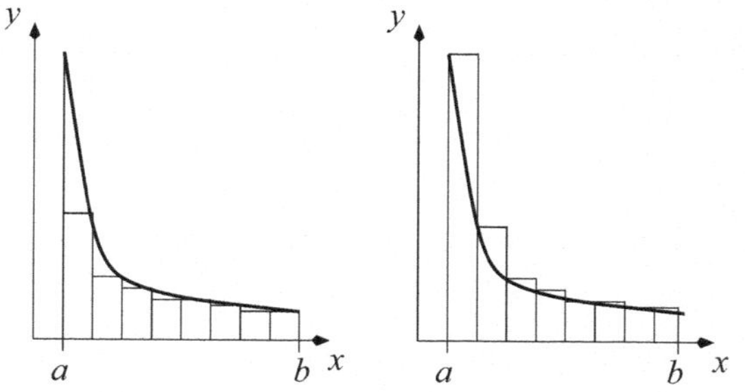

Fig. 5.5.23. Computing the area under the hyperbola

$$y_i = \frac{1}{x_i} = \frac{1}{t^{i-1}x_1} = \frac{y_1}{t^{i-1}}.$$

Hence the ordinates corresponding to the x_i are given by

$$y_1, \quad \frac{y_1}{t}, \quad \frac{y_1}{t^2}, \quad \frac{y_1}{t^3}, \quad \frac{y_1}{t^4}, \ldots.$$

Hence we can now compute the area of the rectangles which emerge by our partitioning the abscissa and the function values at the ordinates. The first rectangle has width $x_2 - x_1 = x_1(t-1)$ and height y_1, hence an area of $x_1 y_1(t-1)$. The second rectangle has width $x_3 - x_2 = x_1 t(t-1)$ and height y_1/t, and the area again is $x_1 y_1(t-1)$, and so on. The area of every rectangle therefore is

$$x_1 y_1 (t-1).$$

This shows: It always holds

$$\frac{x_{r+m}}{x_r} = \frac{t^{r+m-1}x_1}{t^{r-1}x_1} = t^m$$

and

$$\frac{x_{s+n}}{x_s} = \frac{t^{s+n-1}x_1}{t^{s-1}x_1} = t^n,$$

and the areas under the hyperbola from x_r until x_{r+m} and from x_s until x_{s+n}, respectively, satisfy

$$\frac{\int_{x_r}^{x_{r+m}} y\,dx}{\int_{x_s}^{x_{s+n}} y\,dx} = \frac{m \cdot x_1 \cdot y_1 \cdot (t-1)}{n \cdot x_1 \cdot y_1 \cdot (t-1)} = \frac{m}{n}.$$

But this is exactly the ratio of the logarithms

$$\frac{m}{n} = \frac{\log(x_{r+m}/x_r)}{\log(x_{s+n}/x_s)},$$

and hence the ratio of the areas under the hyperbolas is the same as that of the logarithms.

De Saint-Vincent's student Alphonse Antonio de Sarasa (1618–1667) found a particular area function from (5.16) when studying the book *Opus geometricum* written by his master. Following [Edwards 1979, p. 156f.] this area function is

$$L(x) := \begin{cases} A_{1,x} \; ; x \geq 1 \\ -A_{x,1} \; ; 0 < x < 1 \end{cases}.$$

One easily proves that

$$L(x \cdot y) = L(x) + L(y)$$

holds and this is nothing else but the law of logarithms which we nowadays call the functional equation of logarithms. In fact the function $x \mapsto L(x)$ is the natural logarithm as can be seen in [Edwards 1979, p. 157f.].

Isaac Newton will also compute the area under the hyperbola as Mercator before him. The idea of logarithms was in the air and in 1614 the first table of logarithms was published by the Scotsman Napier. We shall report on that important development later.

A wide class of functions was considered 'logarithms' in those days so that we have to be careful when we meet the notion of logarithms in the 17th century [Burn 2001]. A thorough investigation of the 17th century mathematics in which also the role of logarithms is illuminated was presented by Whiteside in [Whiteside 1960–62].

Analysis and Astronomy during the Renaissance

1494–1575	FRANCESCO MAUROLICO, polymath and translator of works of antiquity, solves difficult problems concerning barycentres of solids
1543	*De Revolutionibus Orbium Caelestium* by NICOLAUS COPERNICUS is published in Nuremberg
1548–1620	SIMON STEVIN, ingenious Dutch mathematician, physicist and engineer, used rigorous limits in his computations of barycentres
1565	FEDERICO COMMANDINO writes *Liber de centro gravitatis solidorum.* He also translates texts of antiquity and edits them
1569	Mercator's conformal map of the earth is published
1589	GALILEI becomes lector of mathematics at the University of Pisa
1589–1592	GALILEI works on the thermometer, the law of fall, and the law of the pendulum
1592	GALILEI gets the chair of mathematics at the University of Padua. Meanwhile he has become a follower of Copernicus
1597	KEPLER's *Mysterium Cosmographicum* appears in Tübingen
1600	WILLIAM GILBERT publishes *De magnete* in London
1604	*Ad Vitellionem paralipomena quibus astronomiae pars optica traditur*, a book on optics by KEPLER, is published
1604	A supernova can be observed in all of Europe
1609	GALILEI learns about the discovery of the telescope and recreates it
1609	KEPLER's *Astronomia nova* is published
1610	KEPLER's *Dioptrice* concerning the theory of the telescope is published
1610	GALILEI's *Sidereus nuncius* is published in which Galilei believed to have observed four new planets. GALILEI becomes court mathematician and court philosopher at the court of Cosimo II de'Medici. At the same time he becomes professor of mathematics in Pisa
1615	KEPLER's *Nova stereometria* published
1616	The cleric PAOLO ANTONIO FOSCARINI publishes his opinion that the Copernican system does not contradict the bible. The inquisition initiates a trial against him; his book is banned
1617	KEPLER gets to know logarithms in Prague
1619	KEPLER's *Harmonice mundi* published
1620	The table of logarithms by JOST BÜRGI published
1623	GALILEI's provoking writing *Saggiatore* is published
1627	KEPLER publishes the *Tabula Rudolphinae*
1630	GALILEI's *Dialogue concerning the two chief world systems* published in Italian language
1633	GALILEI has to revoke the Copernican system in a lawsuit of the inquisition. He is arrested in his house
1635	GALILEI's *Discourses & Mathematical Demonstrations Concerning Two New Sciences Pertaining to Mechanics & Local Motions* first published north of the Alps in a Latin translation
1638	The Italian text of *Discourses & Mathematical Demonstrations* by GALILEI is published in Leiden

6 At the Turn from the 16th to the 17th Century

T. Sonar, *3000 Years of Analysis*, https://doi.org/10.1007/978-3-030-58223-4_6

1556–1598	Philipp II King of Spain
1558	Elisabeth I becomes Queen of England
1560	Charles IX of France becomes King. The name 'Huguenots' for the Calvinistic oriented Protestants comes in use
1562	Edict of Saint-Germain: French Protestants gain legal recognition. Massacre of Vassy in which Hueguenots were slaughtered
1562/63	First religious war in France
1568–1648	Eighty Years' War between the Netherlands and Spain ends simultaneously with the Thirty Years' War and marks the birth of the 'Republic of the United Netherlands'
1572	St Bartholomew's Day massacre, approx 13 000 Huguenots are murdered in France
1577–1580	Englishman Francis Drake is the second man on earth circumnavigating the globe
1587	Execution of Mary Stuart
1588	The Spanish Armada is whitewashed by the English
1589	Begin of theatrical performances of William Shakespeare's works in London
1598	The Edict of Nantes terminates the religious wars and secures the position of the Huguenots
1603	Elisabeth I dies. The Scottish King James VI becomes King of England as James I. Peace between Spain and England
1605	Gunpowder Plot in England on 5th November
1610–1643	Louis XIII King of France
1622	War between England and Spain
1624–1642	Cardinal Richelieu becomes leading minister in France
1625	James I dies in England. His son becomes King Charles I
1629	Charles I dissolves the English Parliament and reigns without it for 11 years. Peace with Spain and France. His opponent and leader of the Revolutionary Army is Oliver Cromwell
1629	Edict of Alès: Huguenots loose political rights
1635	France enters the Thirty Years' War on the Swedish side openly supporting the Protestants
1642	Charles I has to flee London. The Civil War begins. Charles I takes winter quarters in Oxford
1648	Second English Civil War due to Scottish invasion. Cromwell triumphs
1649	Charles I beheaded. Monarchy abolished in England
1653	Oliver Cromwell becomes *Lord Protector* of England
1660	The English restoraton begins with the coronation of Charles II
1661–1715	Louis XIV King of France
1665–1683	Colbert Controller-General of the French finances
1665	Plague in London
1672–1679	War between France and Holland
1685	Edict of Nantes cancelled. 300 000 Huguenots leave France
1713	Papal bull 'Unigenitus' against the Jansenists.

6.1 Analysis in France before Leibniz

6.1.1 France at the turn of the 16th to the 17th Century

France was also not spared from religious distortions in the 16th century. The catholic court party under rule of the House of Guise stood irreconcilably against the leading Protestants of the country in order to gain influence to the still underaged King Francis II. The mother of the young king was Catherine de' Medici who attempted to balance both parties. She was supported by Michel de l'Hôpital who was appointed chancellor in 1560 and demanded religious tolerance.

From outside the country the Spanish King Philip II interfered. He was worried about the spread of 'heresy' in France and got supported by the Jesuits. Against all odds Catherine enacted the Edict of Saint-Germain on 17th January 1562 which guaranteed the French Protestants legal security of existence for the first time. Alas, the Edict was followed by a period of religious feuds and wars which lasted for more than 30 years between 1562 and 1598. The most violent and radiating confrontation between Catholics and Protestants happened in the night from 23rd to 24th August 1572: The night of Bartholomew's Day. What had happened? While Spaniard Philip II attacked France to grab French royal dignity himself the Protestant

Fig. 6.1.1. Philip II, King of Spain (1556–1598), [Painting by J. Pantoja de la Cruz, after Antonio Moro 1606] and Catherine de' Medici (1519–1589), wife of the French King Henry II and mother of his successors Francis II, Charles IX, and Henry III. As a ruler in place of her underaged sons Francis and Charles she enacted the Edict of Saint-Germain in 1560 to protect the Protestants, but then she caused a bloodbath among them in 1572 at Bartholomew's Day ([Painting attributed to François Clouet, about 1555] Victoria & Albert Museum, London)

armed forces had become mighty and important during the second French-Spanish war, so that the Admiral of France, Duke Gaspard de Coligny, already openly dreamt of a Protestant royal house. Thereupon Catherine de' Medici disgustedly turned to the Catholic side and tried having Coligny assassinated during the wedding ceremony of her daughter. However, the victim was only wounded. When the Protestants received the news and started to revolt, Catherine gave order to murder the leading Huguenots in Paris. The order acquired an independent existence though, so that all over France an indescribable series of murders took place which went down on history as the massacre at St Bartholomew's Day.

In the time following, a Catholic league was founded aiming at a reCatholisation of all of France. The situation escalated anew as a Protestant pretender of the throne appeared in 1584: Henry of Navarre, who ruled as Henry IV and converted to the Catholic faith in 1593.

Henry was now loved from no side, but he succeeded at least to superficially pacify the religious conflicting parties. He was an early absolute monarch and steered France towards the kingship of his successors Louis XIII and Louis XIV. When riots started about 1600 in the peasant population who suffered from the plague and marauding mercenary groups he reacted by lowering the peasant tax. However, the indirect taxes went up then. Henry also promoted trade and commerce and prepared mercantilism. He was assassinated but the way of French absolutism was paved for his son Louis XIII and his mighty minister Richelieu.

With the reign of Louis XIV began a great time in 1661 for France as far as culture, science, and politics was concerned. In the days of the Sun King Louis XIV a religious group emerged within the Catholics adhering to the doctrines of the Bishop of Ypern, Cornelius Jansen (1585–1638). Jansen's writing *Augustinus* was only published posthumously in 1640. In it Jansen expounded an alternative concept to the Catholicism of the Jesuits. He thereby referred to the Doctor of Church, Saint Augustine of Hippo (354–430) and his doctrine of grace. Jansen propagated the primeval sin of man and his salvation exclusively by Gods will. Jansen saw himself and his doctrine by all means as part of the Counter-Reformation but rejected the way of the Jesuits and only relied on God's grace, but not on the authority of the church and human will. The Port-Royal abbey close to Versailles became the centre of Jansenism, and among the ardent Jansenists were the mathematicians and philosophers Blaise Pascal (1623–1662) and Antoine Arnauld (1612–1694), and the great French tragedian Jean Racine (1639–1699).

Arguments with the Jesuits were preprogrammed this way, but only when the Jansenists started to attack Richelieu on the grounds that he had supported the Protestant side in the Thirty Years' War and conflicts with the French court arose. In 1709 Port-Royal Abbey was destroyed on the order of Louis XIV. Jansenism was banned in 1719 by the Vatican.

Fig. 6.1.2. Kings of France of the House of Bourbon: (upper left:) Louis XIII (1610–1643) [Artist: Peter Paul Rubens, about 1623], (upper right:) Louis XIV (1643–1715) [Artist: H. Rigaud 1701], (lower left:) Henry IV (1589–1610) [Artist: Frans Pourbus the Younger 1610]; (lower right:) Cardinal Richelieu [Artist: Philippe de Champaigne, about 1639], leading minister of Louis XIII fighting against the Huguenots and in favour of the king as an absolute monarch

Fig. 6.1.3. Cornelius Jansen [Painting: Evêque d'Ypres, 1st half of the 17th c.] and the title page of his main work *Augustinus* 1640. He was the originator of 'Jansenism', named after him and later forbidden by the pope. Jansenism was an exaggeration of Augustine's doctrine of grace and was fought against by the Jesuits

6.1.2 René Descartes

René Descartes was born on the 31st March 1596 in La Haye[1] in the Touraine. After an utterly eventful life as a superb philosopher, mathematician, physicist, bon vivant, mercenary, and wrangler he died on 11th February 1650, shortly before his 54th birthday, in Stockholm. His life was so eventful that it even gave rise to a novel [Davidenko 1993].

Antoine Arnauld and Blaise Pascal and also Descartes originated from families of liberal lawyers. The Descartes family belonged to the gentry; René's ancestors and even himself still wrote their family name as 'Des Cartes' [Specht 2001]. One year after his birth he lost his mother. Aged eight, in 1604, René attended the Jesuit boys' school Collège Royal in La Flèche; founded shortly before and already one of the best schools in all of Europe. As a child of wealthy parents René could occupy a single room and he was not woken up early in the morning as were all the other boys. Whether he was too dainty in childhood or whether his uncle being a padre we do not know. There is an anecdote that he was allowed to sleep long because of his brilliant

[1] To honour Descartes La Haye was renamed in 1802 'La Haye Descartes' and finally in 1967 'Descartes'.

Fig. 6.1.4. René Descartes ([Painting from the school of Frans Hals, 1648], Musée du Louvre Paris)

performance in mathematics, but this is nothing more than a well-invented story. It is however certain that Descartes was so good in mathematics that he sometimes embarrassed his teachers. At that time he also learned Latin of which he will become a master, and his love for the French language awoke. In the year 1610 the heart of Henry IV was buried in La Flèche and Descartes attended the ceremony since the padres had also taught their pupil's to love the school's founder Henry. Descartes also learned fencing and fighting; the Jesuits inspired him for astronomy and even reading 'forbidden texts' was possible [Specht 2001, p. 13].

Eventually Descartes left the school and his traces disappear between between 1612 and 1618. We only know that his father sent him to Holland in 1618 for military training. On the Protestant side a new flexible form of warfare had been developed by means of which even the apparently invincible Spanish troops had been beaten. That the Catholic Descartes was now fighting for the Protestants obviously did not bother him. In Holland he met Isaac Beeckmann, eight years his senior, who dreamt to mathematise physics. Both men recognise in themselves the same interests and inclinations and they became friends.

In 1619 we find Descartes in Copenhagen and travels further on to Danzig (now Gdańsk), Poland, Hungary, Austria, and Bohemia. In Ulm he met the mathematician and engineer Johann Faulhaber, member of the secret society of the Order of the Golden and Rosy Cross – the Rosicrucians. Something remarkable happened on the 10th November 1619: Descartes reported that he had discovered the foundations of a wonderful invention. We do not know what he meant by that but for certain it was concerned with the mathematisation of knowledge; perhaps he had already found the foundations of his later analytical geometry. Shortly after he had three strange dreams which unsettled him and led to a crisis. He left Ulm, fought as a volunteer in the troops of the Duke of Bavaria and wrote [Specht 2001, p. 20]:

> *I began to see the foundations of a wonderful discovery on 11th November 1620*
>
> (Am 11. November 1620 begann ich, die Grundlagen einer wunderbaren Erfindung zu erkennen.)

Between 1620 and 1625 he was criss-crossing Europe. From 1625 he was in Paris where he slept long and thought about his new philosophy, but in 1628 he emigrated to Holland where he stayed until 1649. He not only found peace and security there. In the Netherlands the free and dry wind of Protestantism is blowing. Research was free, freedom of religion was granted, and the universities and libraries counted among the best in Europe. In this climate a 'papist' like Descartes was barely noticed. He met interesting people: Constantijn Huygens (1596–1687), secretary to the Prince of Orange, and his enormously gifted son Christiaan (1629–1695). Descartes immediately recognised the talent of the latter and found him so sympathetic that he calls himself his father. And also Christiaan Huygens has apparently worshiped Descartes more than his biological father. Descartes also met the father of the mathematician Frans van Schooten (1615–1660). Later Frans will supply Descartes with important illustrations, but he will also translate, popularise, and spread Descartes' new geometry. Descartes dissected animals, showed a great interest in medicine, often changed his residence, and wrote his great works. With a maid he begot a child but the little girl died at age five. All this happened in Holland.

Although a Jesuit student and loyal to papacy Descartes developed the idea that one should be intellectually independent of authorities and should think with ones own head. Traditional doctrines are not valid *a priori* but should always be rethought and collated with ones own conscience. So Descartes came into opposition with ecclesiastical doctrine. The mathematician Gilles Personne de Roberval (1602–1675), known to be a difficult man, wrote in 1638 on Descartes [Specht 2001, p. 37f.]:

> *He deduced his private opinions quite with clarity [...] Whether they are true or not can only know the one who knows everything. Concerning ourself we have no proofs whatsoever, nor in favour and neither against; and perhaps even the author owns none. We think that he would be in some trouble if he had to prove his major premises.*

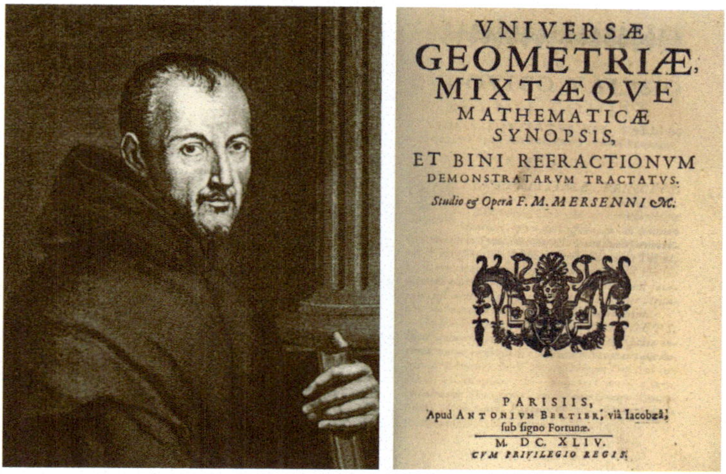

Fig. 6.1.5. Marin Mersenne and the title page of his work *Universae Geometriae mixtaeque Mathematicae synopsis* of 1644 (Bayerische Staatsbibliothek München, Res/4 Math.u. 72#Beibd.1, Titelblatt)

> (Er deduziert seine Privatmeinungen mit ziemlicher Klarheit [...] Ob sie wahr sind oder nicht, weiß ausschließlich der, der alles weiß. Was uns betrifft, so haben wir keinerlei Beweise, weder dafür noch dagegen; und vielleicht besitzt sogar der Autor selber keine. Denn wie wir glauben, befände er sich in ziemlicher Verlegenheit, wenn er seine Obersätze beweisen sollte.)

The 'deduction of his private opinions quite with clarity' was apparently freely applied by Descartes since there was barely a French philosopher or mathematician with whom he was not in conflict: Pascal (father and son), Fermat, and the theologian Pierre Gassendi. Only Marin Mersenne (1588–1648), his friend from school days, stayed his friend. Mersenne is best known today for the discovery of Mersenne's prime numbers. He became a theologian and did mathematical work in his spare time. About 1635 he founded the 'Academia Parisiensis', an association of men interested in mathematics who met at Mersenne's place. We can only conjecture that Descartes got along so well with Mersenne because he saw no competitor in the latter. Today we have to reassess the role of Mersenne. In Paris he maintained the connection between French mathematicians of the first class – Descartes in Holland, Roberval in Paris, Fermat in Toulouse – and with the colleagues in Italy: Galileo Galilei, Bonaventura Cavalieri, and Evangelista Torricelli. In the year 1644 Mersenne even visited Italy to get in personal contact with the Italians [Baron 1987, p. 149]. It is not our job here to penetrate deeper into Descartes' philosophy. His physics, characterised by vortex theories, was eventually superseded by Newton's physics.

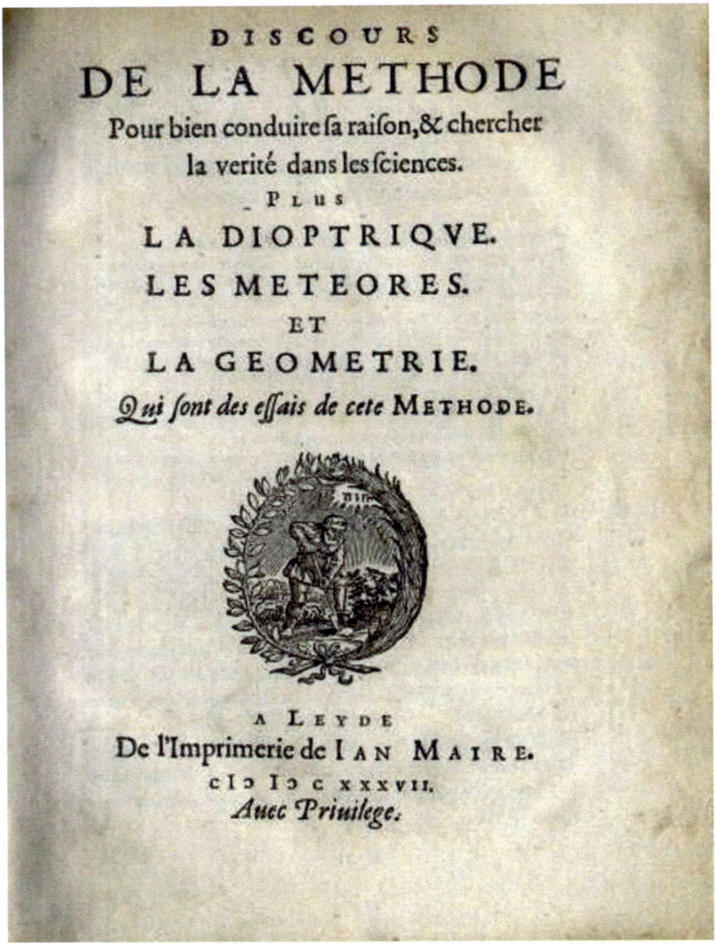

DISCOURS
DE LA METHODE
Pour bien conduire ſa raiſon, & chercher
la verité dans les ſciences.
PLUS
LA DIOPTRIQVE.
LES METEORES.
ET
LA GEOMETRIE.
Qui ſont des eſſais de cete METHODE.

A LEYDE
De l'Imprimerie de IAN MAIRE.
CIↃ IↃ C XXXVII.
Auec Priuilege.

Fig. 6.1.6. Title page of the book *Discours de la méthode* by René Descartes 1637 (Leeds University Library)

However, in 1637 there appeared a book which has revolutionised mathematics: *Discours de la méthode pour bien conduire sa raison, & chercher la verité dans les sciences, plus la Dioptrique, les Météores et la Géométrie qui sont des essais de cete méthode* (Discourse on the Method of Rightly Conducting One's Reason and of Seeking Truth in the Sciences, in addition the Refraction of Light, the meteors, and the Geometry as Trial Applications of This Method), in short: *Discours* [Descartes 2001]. Philosophers find here an epistemology, also the famous 'Cogito ergo sum' (I think, therefore I am) [Descartes 2001, p. 27f.], thoughts on ethics, and on metaphysics. In the about 100 pages long part on geometry [Descartes 1954], [Descartes 2001, pp. 177-259] we find the birth of what is now called 'analytical geometry'. He distinguished between two kind of 'functions', or better: curves, namely the mechanical curves and those describable by a functional relation, as simple as possible – in Descartes' mathematics these are the polynomials. Among the mechanical curves are the quadratrix (cp. page 43, and figures 2.1.16(a) and 2.1.16(b)), and all further curves which are constructible by means of 'mechanical processes'. Mathematically Descartes tried to stay away from the mechanical curves [Mancosu 1996, p. 81f.] because in their analysis he saw himself entangled in infinite processes of approximation. And indeed, if one would like to describe curves like the quadratrix by polynomials then infinite series come in inevitably (and therefore 'infinite polynomials'). In analysis Descartes came up with an ingenious (finite!) method to compute tangents at given curves. This method is known as the 'circle method'. In physics Descartes replaced Aristotle's world view by a 'causal' one, in which all processes in the surrounding nature follow from 'push and impact'. Descartes' world view hence is a 'mechanistical' one (cp. [Sorell 1999]), and this kind of thinking became a building block for theories in many empirical sciences. In this mechanistic world view animals are 'machines' and even man, after Descartes in contrast to animals, consisting of body and soul, has a mechanistic interpretation. Even today orthopaedists and physiotherapists look at us (rightly and successfully!) as if we were moving machines.

In 1614 a further philosophical work appears: *Meditationes de prima philosophia, in qua Dei existentia et animae immortalitas demonstratur* (Meditations on the first philosophy, in which God's existence and the immortality of the soul is proven). In these meditations [Descartes 2017] Descartes gave a proof of the existence of God and expounded his conviction that everything may be doubted by means of ones own intellect. The *Meditationes* belong to the most important classical works of philosophy today.

In the realm of philosophy he was now famous beyond all bounds. For quite some time he had corresponded with the Swedish Queen Christina (1626–1689); in autumn 1649 he followed an invitation to Stockholm.

The queen wanted to know more about philosophy and ordered Descartes to teach her in the early morning hours. The philosopher being a late riser all his life was now forced to rise at the crack of dawn in the Swedish winter and hurry to the palace. It is an old legend – almost folklore – that this contributed to his death. He died at the beginning of 1650 after having suffered, at least according to the official version, from pneumonia.

Fairly recently some doubts were shed on this legend. A rumour arose that Descartes' was poisoned and that he hence was assassinated [Pies 1996]. While the arguments presented in [Pies 1996] were not convincing the German professor of philosophy Theodor Ebert has published a scientifically rigorous study of the last days and the death of Descartes in 2009 [Ebert 2009]. Ebert included all of the extant letters concerning the death of the philosopher in his study and analyses the symptoms of the illness thoroughly. Having read Ebert's book I have no doubts that Descartes was indeed poisoned

Fig. 6.1.7. René Descartes explains his philosophy to Queen Christina [Detail of an oil painting by Pierre Louis Dumesnil, 1884, copied by Nils Forsberg]

and assassinated by the chaplain of the French embassy in Stockholm, padre François Viogué, first by means of the application of a host poisoned with arsenic (arsenic trioxid) during Holy Mass on 2nd February 1650. While this first poisoning was not lethal a second poisoned host, administerd again by Viogué on 8th February while he visited the ill man, killed Descartes, who died on 11th February 1650. Everything indicates that Descartes himself was aware of his poisoning since he ordered a substance to cause vomitting (tobacco resolved in wine) on 8th February (he wouldn't have done this if he would have believed to suffer from pneumonia). However, his servants strongly diluted the emetic in view of his ill health so that it had no effect.

Viogué was in Sweden as 'Apostolic missionary' for the Northern countries on behalf of Pope Innocent X and most likely prepared the conversion of Queen Christina to the Catholic faith; indeed Christina abdicated in 1654, left the crown to Charles X Gustav of Sweden and converted in 1655 in Innsbruck. After her conversion she called herself Maria Alexandra. Although Descartes was a 'good Catholic' on the surface it can be seen from his writings, which were on the 'Index librorum prohibitorum' (Index of forbidden books) from 1663 and could only be read by Catholics after an official permit of the responsible bishop, that he was a freethinker indeed. However, Descartes was convinced that a baptised Christian should stay in the faith assigned to him by baptism until the end. Viogué apparently feared that Descartes, being close to the queen, could prevent the conversion.

The Circle Method of Descartes

The circle method of Descartes, described in [Descartes 1954, p. 92ff.], is somewhat out of the ordinary as far as indivisibles or infinitesimals are concerned. Actually, Descartes was proud of *not* having to employ infinitely small quantities in his computation of tangents. The reason for Descartes' being a 'finitist' can be explained on ground of his philosophical inclination. The reader is urged to compare the study of Mancosu in [Mancosu 1996]. Descartes' rejection of mechanical curves has a lot to do with his finitism since he did not want to be entangled in non-transparent infinite operations and approximations. Nevertheless there can be no doubt that Descartes himself had some mastery of the method of indivisibles and infinitesimals.

Descartes' circle method is a purely algebraic-geometric one and was therefore already described in [Scriba/Schreiber 2000, S. 339 ff.]. This method nevertheless made a great impact on the development of analysis and therefore it belongs here; see also [Baron 1987, p. 165f.], [Edwards 1979, p. 125ff.], and [Stedall 2008, p. 74ff.].

Descartes aims at computing the slope of the tangent at a function $x \to f(x)$ in a point x_0. To this end he imagines a circle K with radius r and centre v on the abscissa as in figure 6.1.8.

A circle with centre at $(v, 0)$ and radius r is described by

$$y^2 + (x - v)^2 = r^2.$$

If the circle intersects the function f in at least one point then we may replace y by $f(x)$ in the equation above, hence

$$(f(x))^2 + (x - v)^2 = r^2. \tag{6.1}$$

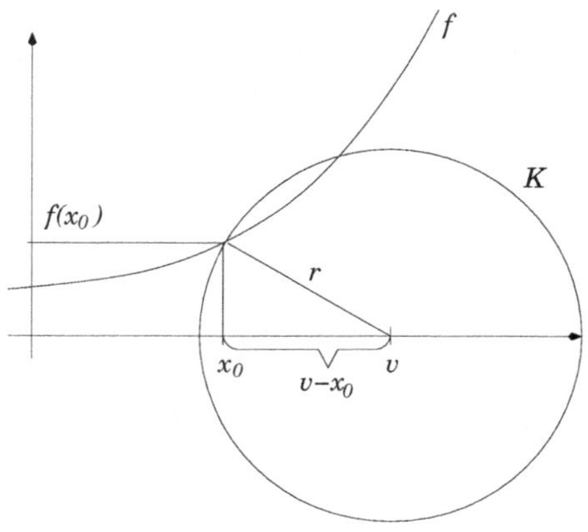

Fig. 6.1.8. The circle method of Descartes

Descartes assumes that the function f is a polynomial, hence f^2 is also a polynomial. This restriction to polynomials may seem quite strong in modern eyes but non-polynomial (transcendet) functions were mostly not yet discovered.

If the circle intersects the function in exactly two different points as in figure 6.1.8, then there would be two different solutions x_0 of equation (6.1). Is r so small that circle and function do not intersect at all, then there will be no solution of (6.1). But what happens in case of exactly *one* point of intersection? It is exactly this case that the circle K is tangential to f in the point x_0, i.e. the radius between the two points $(x_0, f(x_0))$ and $(v, 0)$ gives us the direction of the *normal* at f in x_0. But if we have the normal we can easily compute the *tangent* since normal and tangent are orthogonal to each other. What does exactly one point of intersection mean concerning equation (6.1)? The polynomial equation (6.1) has one double root x_0 in this case and the representation of the polynomial $(f(x))^2 + (x-v)^2 - r^2 = 0$ in linear factors will be

$$(f(x))^2 + (x-v)^2 - r^2 = (x-x_0)^2 \cdot \sum_i c_i x^i, \tag{6.2}$$

since a double root implies the factor $(x-x_0)^2$ and $\sum_i c_i x^i$ is a polynomial of lesser degree.

By equating coefficients a relation can be gained for v in dependence on x_0. The *normal* at f in x_0 then has the slope

$$-\frac{f(x_0)}{v-x_0}$$

and the *slope of the tangent* is

$$\frac{v-x_0}{f(x_0)}. \tag{6.3}$$

How these relations are derived can simply be seen in figure 6.1.9. Employing the notations in the figure the slope of the tangent is given by $\Delta y/\Delta x$. Now the gradient triangle is similar to the triangle defined by the three points $(x_0, 0)$, $(v, 0)$, and $(x_0, f(x_0))$ since their angles coincide. But then

$$\frac{\Delta y}{\Delta x} = \frac{v-x_0}{f(x_0)}.$$

Normal and tangent are mutually orthogonal. If the slope of the tangent is positive the slope of the normal will be negative and vice versa. In figure 6.1.9 we have already a gradient triangle giving the normal, namely the one defined by the points $(x_0, 0)$, $(v, 0)$, and $(x_0, f(x_0))$. The slope of the normal apparently is

$$-\frac{f(x_0)}{v-x_0}.$$

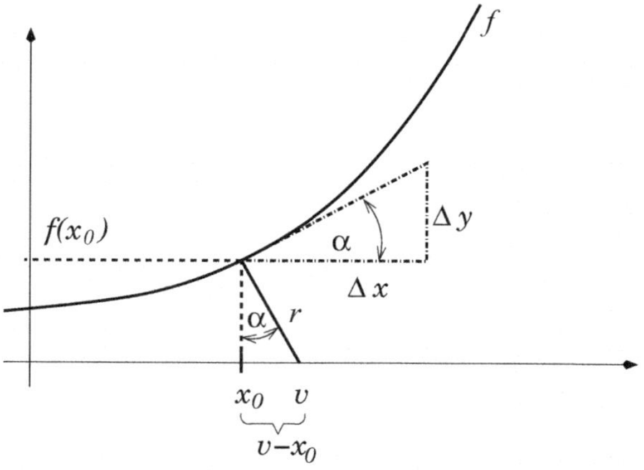

Fig. 6.1.9. Determining the equations of tangent and normal

It holds in general that if the slope of the tangent is m, the slope of the normal will be given by $-1/m$.

Example: We compute the remaining polynomial and the representation (6.2) in case of the function $y = f(x) = x^2$. Hence (6.1) becomes:

$$x^4 + (x - v)^2 = x^4 + x^2 - 2vx + v^2 = r^2.$$

The double root results in a representation $(x-x_0)^2 \cdot \sum_i c_i x^i$. Since we started with a polynomial of degree 4 and since $(x - x_0)^2$ apparently is a polynomial of degree 2, the remaining polynomial has to be of degree 2, hence

$$x^4 + x^2 - 2vx + v^2 - r^2 = (x - x_0)^2 \cdot (x^2 + ax + b)$$

with unknown coefficients a and b. Multiplying out and equating coefficients yields

$$\begin{aligned} a - 2x_0 &= 0 \\ b - 2ax_0 + x_0^2 &= 1 \\ ax_0^2 - 2bx_0 &= -2v, \end{aligned}$$

leading to $v = 2x_0^3 + x_0$. In the case of the *subnormal* $v - x$ which is the projection of the normal to the abscissa, this yields

$$v - x_0 = 2x_0^3.$$

Hence the slope of the tangent of $f(x) = x^2$ at the point x_0 follows from (6.3) as

$$\frac{v - x_0}{f(x_0)} = \frac{2x_0^3}{x_0^2} = 2x_0.$$

Since x_0 was an arbitrarily chosen point we can state in modern terminology that the tangent slope $f'(x)$ of $f(x) = x^2$ at some point x is given by

$$f'(x) = 2x.$$

6.1.3 Pierre de Fermat

Fermat was the most unobtrusive of the great French mathematicians of the 17th century – no known scandals, no life as a mercenary, and no sharp turning points in his life; at least as far as we know. He was, however, one of the most profound thinkers of his age. Having said that, he is nevertheless the most famous in our days since Andrew Wiles was able to prove the world-famous 'Fermat's Last Theorem' after decades of hard work and after 358 years of effort of hundreds of mathematicians in 1994. However, Fermat's Last Theorem belongs to the realms of number theory and not of analysis, see [Wußing 2008, p. 407], [Wußing 2009, p. 541ff.], and [Singh 1997]. Even without his Last Theorem Fermat was a mathematician of the first rank who stimulated important developments in analysis.

Until recently there was no doubt that Pierre de Fermat saw the light of day on 17th August 1601. As Klaus Barner has convincingly verified in [Barner 2001] and [Barner 2001a] a half-brother of our Fermat from the first marriage of his father was born on 17th August 1601 but died prematurely. The correct date of birth of *our* Fermat, who got the same name as his deceased half-brother by tradition, can only be some day at the end of 1607 or at the beginning of 1608.

Pierre's father was a wealthy wholesaler of rural products from the small French town of Beaumont-de-Lomagne close to the Pyrenees. There, at the place of his birth, Pierre grew up with two sisters and one brother and went to school; probably to one of the three elementary boys' schools. His name back then was just Pierre Fermat.

From 1623 to 1626 he studied civil rights at the University of Orleans. His lifelong friendship with Pierre de Carcavi (1600–1684) goes back to this student years. In autumn 1626 he settled as a lawyer in Bordeaux where he remained until the end of 1630. In this time he already made important discoveries concerning analysis; for example his reflections on maxima and minima. Already in 1626 Fermat is said to have come in contact with Jean Beaugrand (about 1590–1640), probably a pupil of François Viète (1540–1603), see [Alten et al. 2005, p. 266ff.], [Alten et al. 2014, p. 284ff.]. Beaugrand pointed him to a circle of mathematicians in Bordeaux so that Fermat chose Bordeaux in which to settle. We know little about Beaugrand. He was the teacher of Louis XIII in calligraphy, worked on topics of geostatics and on mathematics, and kept contact with Descartes, Mersenne, and, of course, with Fermat.

When Fermat's father died on 20th June 1628 he inherited a fortune. After the necessary four years working as a lawyer Fermat bought the office of a judge at the Parlement of Toulouse on 29th December for the enormous sum of 43500 Livres. A 'parlement' was a provincial appellate court in the *Ancien Régime* of France. Attached with that purchase was an enoblement; he was now allowed to call himself Pierre 'de' Fermat but he never made use of that. In December 1637 Fermat sold his office and bought the one of a deceased colleague in a higher chamber of Parlement. He kept this office until his death but moved up once again in 1652. In 1653 Fermat got seriously ill with plague which claimed many victims in the region in the 1650s. He was officially declared death but in fact survived.

In his life there still remained leisure to do mathematics and he became one of the great mathematicians of his day. His friendship with Beaugrand stayed on but Pierre de Carcavi (1600–1684) became a true friend as far as mathematics was concerned. As Fermat also Pierre de Carcavi was an amateur mathematician, but he had no university education. Carcavi became not known for his own mathematical work but for his correspondence with Fermat, Pascal, and Huygens.

In Toulouse Fermat had told his friend about his mathematical discoveries and when Carcavi moved to Paris in 1636 and became Royal Librarian he entered the circle around Marin Mersenne and a correspondence between Fermat and Mersenne began. At first both men corresponded on the laws of fall and on an error which Fermat thought he had detected in the writings of Galilei. But Fermat was not really interested in the applications of mathematics to physics. Already the very first letter contained two problems concerning the computation of maxima and minima. It is somewhat typical of Fermat that he asked Mersenne to introduce the Parisian mathematicians to

Fig. 6.1.10. Marble sculpture of Pierre de Fermat on a high pedestal in front of the roof of the great market hall of Beaumeont-de-Lomagne [Photo: Martin Barner] and a painting of him [unknown artist, 17th c.]

these problems. During the following correspondence Roberval and Mersenne have to admit that Fermat's problems could not be treated with the conventional methods and hence Fermat was asked to lay his methods open. In 1637 Fermat sent his works *Methodus ad disquirendam maximam et minimam et de tangentibus linearum curvarum* (Methods to determine maxima and minima and tangents of crooked curves), *Apollonii Pergaei libri duo de locis planis restituti*, Fermat's reconstruction of a text by Apollonius of Perga (about 262–about 190 BC) on conic sections, and *Ad locos planos et solidos isagoge* (Introduction to plane and spatial loci) to Mersenne. In this last publication unintentionally there was dynamite which ignited shortly after! Fermat had developed the foundations of an analytic geometry independently of Descartes. A short time after Mersenne received Fermat's manuscript Réne Descartes sent him the galley proofs of his *Discours*, containing Descartes' version of an analytic geometry. Of course Descartes was not willing to accept another inventor of the new methods! Additionally, the work by Fermat concerning the computation of maxima and minima and tangents was not particularly detailed and comprehensibly written so that the flood gates were open for criticism.

Why did two scientists arrive at the same theory at almost the same time? Both men shared the same mathematical background. Both men were interested in 'curves' – the conic sections of Apollonius, the curves in the writings of Pappus of Alexandria (about 300), the conchoid, the Archimedean spiral, and so on. Furthermore algebra was sufficiently developed by François Viète (1540–1603) and others (cp. [Alten et al. 2005], [Alten et al. 2014]) so that it could applied to geometry. From today's view Descartes deserves priority as far as analytic geometry is concerned since he thought about it much earlier than Fermat [Mahoney 1994, p. 73ff.].

As fuel to the conflict with Descartes Fermat had written a critique of Descartes' optical theories, in which he could not believe, for Mersenne. He was shown Descartes' *Dioptrique* by his friend Beaugrand in May 1637 and Mersenne asked for a comment. Fermat, who should later arrive at the correct law of refraction, immediately recognised the shortcomings of Descartes' arguments. In favour of Fermat it has to be said that he was neither aware of Descartes' eruptive temper nor of the attacks to which Descartes was exposed from other sides. Additionally, Beaugrand had got the *Dioptrique* surreptitiously and was by no means authorised to show it to Fermat. In any case a storm broke loose in 1638 [Mahoney 1994, p. 171f.]. Mersenne had kept Fermat's letter for quite some months – after all he knew Descartes – but when Descartes asked him to send all criticisms regarding the *Dioptrique* he also sent Fermat's comment. An unpleasant quarrel now developed between Fermat and Descartes in which Descartes not even shied away from personal insults. If Descartes had been able to destroy Fermat he would have done it. He now tried to discredit Fermat under all circumstances. In their correspondence he time and again posed problems to expose Fermat but the latter could even sharpen his own methods in solving these problems. As Mahoney remarked in [Mahoney 1994, p. 171] Descartes learned nothing from this correspondence, even if he was in the wrong. He was and remained stubborn like a donkey but Fermat gained new insights so that the struggle, quickly joined by Roberval and the older Pascal, turned out to be advantageous for the development of analysis.

From 1643 until 1654 the contact with the group in Paris ceased. Fermat had a demanding career and political crises in Toulouse added to these demands. Finally there was the plague. During these years Fermat in his not very lush leisure time worked on problems of number theory. Only in 1654 Étienne and Blaise Pascal turned to Fermat to discuss combinatorial problems concerning probability theory.

About this time a student of Descartes' asked Fermat for help in editing the correspondence between his teacher and Fermat. This gave Fermat the opportunity to rethink his now 20-years old arguments against Descartes' *Dioptrique*. He found the law of refraction of Snell by means of the principle that light always traverses between two points in least time (Fermat's principle). Although Fermat's principle did not find many followers under the mathematicians of his time it was an early application of a new theory within analysis, the variational calculus, which was fully developed only in the 18th century.

At the end of a biographical sketch of this great 'amateur mathematician' it seems appropriate and instructive to note his opinions of some of his contemporaries [Mahoney 1994, p. 15]:

Descartes: 'braggart'
Pascal: 'the greatest mathematician in all of Europe'
Mersenne: 'the learned councillor from Toulouse'
Wallis: 'the damned Frenchman'

The Quadrature of Higher Parabolas

Already the Pythagoreans and Archimedes knew the summation formulae for $\sum_{k=1}^{n} k$ and $\sum_{k=1}^{n} k^2$. Alhazen had already employed the formulae for $\sum_{k=1}^{n} k^3$ and $\sum_{k=1}^{n} k^4$. Johann Faulhaber whom René Descartes visited in Ulm was in possession of the formula for $\sum_{k=1}^{n} k^{13}$ and today these formulae are called 'Faulhaber polynomials'.

Fermat employed those formulae in the quadrature of 'higher parabolas' defined by

$$\frac{y}{a} = \left(\frac{x}{b}\right)^p$$

for positive $p = 1, 2, 3, \ldots$. Following [Baron 1987, p. 151f.] Fermat exploited the recursive relations

Fig. 6.1.11. Johann Faulhaber and a Detail from his *Perspektive & Geometrie & Würfel & Instrument* [Faulhaber/Remmelin, 1610]

$$2\sum_{k=1}^{n} k = n(n+1)$$

$$3\sum_{k=1}^{n} k(k+1) = n(n+1)(n+2)$$

$$4\sum_{k=1}^{n} k(k+1)(k+2) = n(n+1)(n+2)(n+3)$$

$$\vdots \quad \vdots \quad \vdots$$

to finally prove

$$\lim_{n\to\infty}\sum_{k=1}^{n}\frac{k^p}{n^{p+1}} = \frac{1}{1+p} \tag{6.4}$$

for the function $y = x^p$. Very likely Fermat had deduced from $(p = 2, 3, 4, \ldots)$

$$(p+1)\sum_{k=1}^{n} k(k+1)(k+2)\cdot\ldots\cdot(k+p-1) = n(n+1)(n+2)\cdot\ldots\cdot(n+p-1)(n+p)$$

that

$$(p+1)\sum_{k=1}^{n} k^p + \text{ lower order terms } = n^{p+1} + \text{ lower order terms}$$

holds true. Division by n^{p+1} and $p+1$ yields

$$\frac{1}{n^{p+1}}\sum_{k=1}^{n}k^p = \sum_{k=1}^{n}\frac{k^p}{n^{p+1}} + \frac{1}{n^{p+1}}\cdot \text{ lower order terms}$$
$$= \frac{1}{p+1} + \frac{1}{n^{p+1}}\cdot \text{ lower order terms}$$

and with increasing n the terms

$$\frac{1}{n^{p+1}}\cdot \text{ lower order terms}$$

will decrease and finally vanish in the limit $n \to \infty$ so that (6.4) follows.

Subdividing the interval $[0, 1]$ in n parts of length $1/n$ the corresponding function values $y = x^p$ follow to

$$y_k = \left(k \cdot \frac{1}{n}\right)^p, \quad k = 0, 1, 2, \ldots, n.$$

Hence an approximation of the area under the graph of the function $y = x^p$ has been achieved by means of sums of rectangular areas, 'rectangle sums', as shown in the left part of figure 6.1.12. The area of the k-th rectangle is

$$\frac{1}{n}\cdot\left(k \cdot \frac{1}{n}\right)^p = \frac{k^p}{n^{p+1}}.$$

Hence the rectangle sum

$$\sum_{k=1}^{n}\frac{k^p}{n^{p+1}}$$

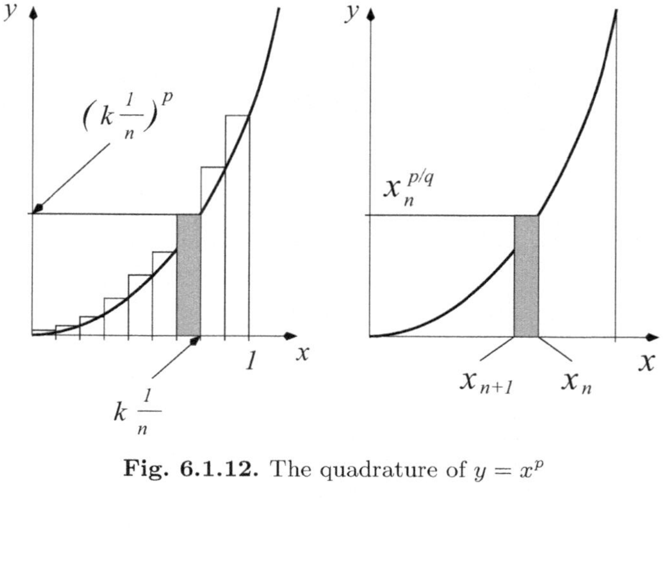

Fig. 6.1.12. The quadrature of $y = x^p$

is an approximattion of the true area and thanks to (6.4) it follows

$$\int_0^1 x^p\,dx = \frac{1}{p+1}.$$

Fermat was apparently not satisfied with this method and he thus turned to another idea. Instead of subdividing $[0,1]$ into equidistant subintervals Fermat now chose a geometric progression, i.e. a sequence of points $(x_n)_{n=0}^{\infty}$ having the property

$$x_n = at^n, \quad 0 < t < 1. \tag{6.5}$$

The first point x_0 of the subdivision is just a and increasing n the sequence of x_n come ever closer to 0. Fermat wanted to compute the area under the graph of the function

$$y = x^{\frac{p}{q}}, \quad p, q = 1, 2, 3, \ldots$$

on $[0,a]$ i.e. he wanted to compute

$$\int_0^a x^{p/q}\,dx.$$

Looking at a rectangle on this partition as in the right part of figure 6.1.12 the rectangle sum turns out to be

$$\begin{aligned} S &= \sum_{k=0}^{\infty} x_k^{p/q}(x_k - x_{k+1}) \overset{(6.5)}{=} \sum_{k=0}^{\infty} (at^k)^{p/q}(at^k - at^{k+1}) \\ &= a^{(p+q)/q}(1-t)\sum_{k=0}^{\infty} t^{k(p+q)/q}. \end{aligned}$$

Introducing a new variable $s := t^{(p+q)/q}$ yields

$$S = a^{(p+q)/q}(1-t)\sum_{k=0}^{\infty} s^k = a^{(p+q)/q}\frac{1-t}{1-s},$$

where we have employed the formula for the sum of a geometric series which, of course, Fermat knew. Again introducing a new variable $u := t^{1/q}$, then $t = u^q$ and $s = t^{p+q}$ and hence

$$S = a^{(p+q)/q}\frac{1-u^q}{1-u^{p+q}}.$$

Now Fermat employed the equation

$$(1-u)(1+u+u^2+\ldots+u^{k-1}) = 1-u^k,$$

so that our rectangle sum finally takes the form

$$S = a^{(p+q)/q} \frac{1 + u + u^2 + \ldots + u^{q-1}}{1 + u + u^2 + \ldots + u^{p+q-1}}.$$

The true area now follows from $t \to 1$. Since $u = t^{1/q}$ also u goes to 1 and for the limit follows

$$\int_0^a x^{p/q}\,dx = \frac{q}{p+q} a^{(p+q)/q}.$$

Fermat's Method of Pseudo-Equality

It is likely that Fermat was the first to carry a truly dynamical thought into analysis: Standing at an extremal point (maximum or minimum) a small change of the abscissa x would change the value $y = f(x)$ of the function only insignificantly, cp. [Stedall 2008, p. 72ff.]. Let us exploit this idea by means of a geometric problem:

Divide a straight line segment of length b at x in two parts x and $b - x$, so that the area of the rectangle with sides x and $b - x$ becomes maximal.

Obviously the area is given by

$$f(x) := x(b - x) = bx - x^2.$$

Now Fermat argues as follows: If x would already be the abscissa of the maximum of f then f would barely change its value if x would be perturbed by a 'small' e, hence if we would look at

$$f(x + e) = b(x + e) - (x + e)^2 = bx + be - x^2 - 2ex - e^2.$$

Fermat now equates $f(x)$ and $f(x + e)$ (both values are not really equal, hence the name 'pseudo-equality) and one gets

$$f(x) \sim f(x + e) \quad \Rightarrow \quad bx - x^2 \sim bx + be - x^2 - 2ex - e^2,$$

where we used $\sim$' to denote the pseudo-equality. Treating '$\sim$' in exactly the same way as '=' yields

$$2ex + e^2 \sim be.$$

Now e is very small, but not zero. Hence one may divide by e and gets

$$2x + e \sim b.$$

After this division Fermat neglected e as if e would be zero and arrives at the solution of the extremal problem:

$$x = \frac{b}{2}.$$

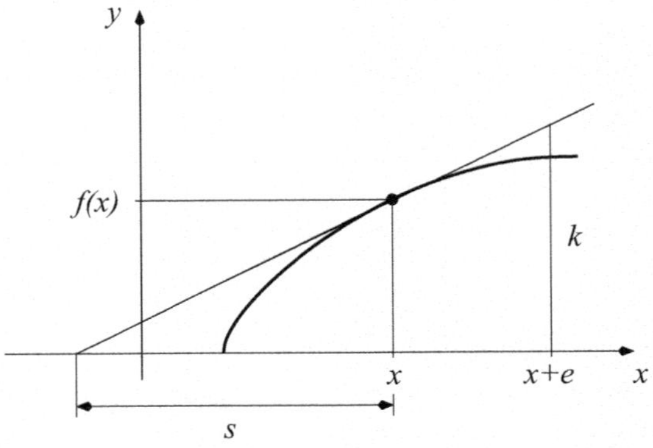

Fig. 6.1.13. Computation of the tangent

Indeed Fermat's method does lead to the desired result, but without a clear understanding of limits the treatment of e remains odd. $e \neq 0$; otherwise we would not have been allowed to divide, on the other hand $e = 0$ and can be neglected.

What did Fermat do in view of our modern standpoint? He has equated $f(x)$ and $f(x+e)$ and divided by e, i.e. he has computed

$$\frac{f(x+e) - f(x)}{e} \sim 0.$$

But this is Leibniz's difference quotient!

Fermat has also employed his method of pseudo-equality to compute tangents at curves.

From figure 6.1.13 we deduce:

$$\frac{s+e}{s} = \frac{k}{f(x)}.$$

The quantity s is called the subtangent of the function f. We now replace k by $f(x+e)$ by pseudo-equality, hoping that the difference $k - f(x+e)$ will be small if e will be small enough. Solving the last equation for the subtangent yields

$$s \sim \frac{ef(x)}{f(x+e) - f(x)}.$$

Division by e leads to

$$s \sim \frac{f(x)}{\frac{f(x+e)-f(x)}{e}}$$

and in the denominator we recognise the difference quotient. In modern notation it follows for $e \to 0$:

$$s = \frac{f(x)}{f'(x)},$$

showing the relation between the derivative (slope) f' and the subtangent. We follow Fermat's idea in treating the function $f(x) = x^2$. In this case we have

$$s \sim \frac{x^2}{\frac{(x+e)^2 - x^2}{e}} = \frac{x^2}{2x + e}$$

and e is neglected again to arrive at

$$s \sim \frac{x^2}{2x}.$$

Hence the slope of the tangent of $f(x) = x^2$ is given by $f'(x) = 2x$.

Descartes did not believe in the penetrating power of Fermat's computations of tangents. In particular he suspected Fermat being dependent on an equation of the form $y = f(x)$ to apply his technique. Hence he proposed to Fermat the task of computing the tangent of the Folium of Descartes

$$f(x, y) = x^3 + y^3 - nxry = 0,$$

[Baron 1987, p. 169f.]. But Fermat had no problems whatsoever! From the triangle in figure 6.1.13, shown in figure 6.1.14 with increment, the intercept theorem gives

$$\frac{y + \delta y}{s + e} = \frac{y}{s}.$$

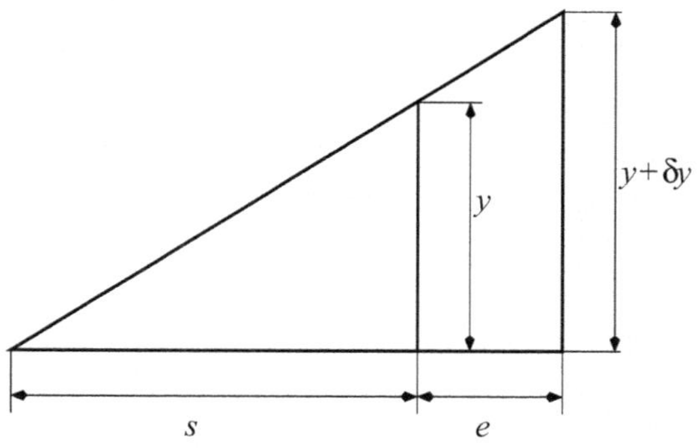

Fig. 6.1.14. The triangle of the subtangent s with increment

Hence the increment δy is given by

$$\delta y = \frac{ye}{s}$$

and it follows from Fermat's estimate, that

$$f(x,y) \sim f(x+e, y+\delta y) = f\left(x+e, y+\frac{ye}{s}\right).$$

To see where Fermat's computations finally led him we exploit the series expansion

$$f\left(x+e, y+\frac{ye}{s}\right) = f(x,y) + e\frac{\partial f}{\partial x} + \frac{ye}{s}\frac{\partial f}{\partial y} + \ldots = 0.$$

Negelcting the ...-terms and division by e leads to

$$\frac{\partial f}{\partial x} + \frac{y}{s}\frac{\partial f}{\partial y} = 0$$

and hence it follows for the slope of the tangent

$$\frac{y}{s} = -\frac{\frac{\partial f}{\partial x}}{\frac{\partial f}{\partial y}}.$$

This is nothing but the formula of implicit differentiation.

6.1.4 Blaise Pascal

Blaise Pascal was born in the year 1623 in Clermont which today is Clermont-Ferrand in the Auvergne, and died already in 1662, aged 39 and ravaged by disease.

The story of his life is fascinating; he was a prodigy and has left traces not only in mathematics but also in physics, philosophy, and theology. Nevertheless, Blaise Pascal is

> ... *a fascinating, but elusive personality ...,*
>
> (... eine faszinierende, aber schwer fassbare Persönlichkeit ...,)

as Loeffel wrote in his biography of Pascal in [Loeffel 1987].

His father Étienne Pascal and his mother Antoinette came from families of French officials. Étienne was a vice president at the high fiscal court and was a serious mathematician himself. He not only socialised with the great scientists of his day but made discoveries on his own as the limaçon of Pascal

$$(x^2 + y^2 - ax)^2 = b^2(x^2 + y^2),$$

a fourth-order curve which is named after him and *not* after his son Blaise. Belonging to the family were the two sisters Gilberte and Jacqueline; the latter entered Port-Royal Abbey in Paris, the stronghold of Jansenism, at the age of 26 in 1653. When the mother died in 1626 the father took his responsibility to educate his children which was quite uncommon in those days. The education closely followed the ideas of the great Michel de Montaigne (1533–1592), an unorthodox humanist and moral philosopher who is famous to this day due to his monumental philosophical work *Essais* [Montaigne 1998]. Following Montaigne learning material was no longer drilled into the pupils but the goal was to achieve a real insight into the material.

The father attached importance to good knowledge of Greek and Latin, but little Blaise already showed early signs of his excellent mathematical gifts. The father was scared that his prodigy would be on his way to develop into a one-sided personality, single-mindedly becoming an unworidly mathematician. To counter this development he locked away all books on mathematics but Blaise started to do maths with his own made up crude notations like 'rounds' and 'rods' [Loeffel 1987, p. 13f.]. It is said that he was able to work his way through to the 32nd theorem of Euclid's *Elements* with his own definitions and concepts! This again scared his father who now saw the isolation of his son in his own world of symbols and father's library of mathematics book was again opened. In 1631 Blaise travelled with his family to Paris and made contact with Gilles Personne de Roberval who had just stepped up as professor of mathematics at the Collège Royal. Hence Blaise became acquainted with the *Académie Parisienne* founded by Marin Mersenne. In this academy one discussed mathematics, physics, philosophy, and theology. Mersenne is best imagined as the central spider in a web encompassing all of Europe. He kept contact with scientists from France and Italy and mathematical news travelled fast via his 'intelligence service'.

Fig. 6.1.15. Two paintings of Blaise Pascal (1623–1662), left: [unknown artist], right: [Copy of a painting by François II Quesnel, about 1691]

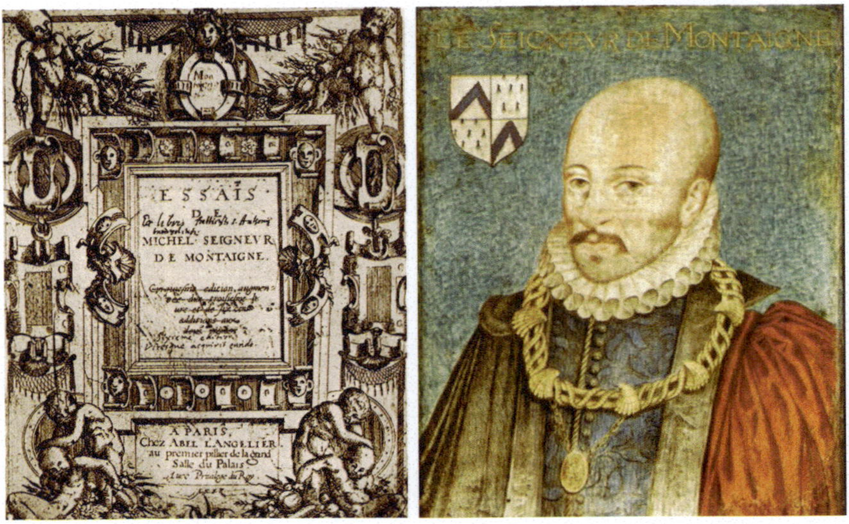

Fig. 6.1.16. Michel de Montaigne [(right) Painting by Thomas de Leu, about 1578 or later] and the title page of his *Essais*

Central dates of a true 'mathematisation' of natural sciences are the publishing dates of Descartes' *Discours de la méthode* 1637 and of Galilei's *Discorsi* one year later. Blaise Pascal got captured by this development. In the year 1638 Étienne Pascal fell from the grace of minister Cardinal Richelieu since he protested against the cancellation of state pensions. He fled Paris to avoid imprisonment but he was pardoned already in the next year and appointed royal taxation commissioner of Normandy. To make life easier for his father Blaise invented a calculating device in 1642; the famous 'Pascaline'. Their childhood diseases were overcome in 1645 and the Pascaline was then also sold to the public [Wußing 2008, p. 422f.].

However, mathematics did not miss out with the young Pascal. Already in 1639 the manuscript *Brouillon project d'une atteinte aux événements des rencontres du Cone avec un Plan* which was difficult to understand circled in Paris. The author was the architect and mathematician Girard (Gérard) Desargues (1591–1661) from the circle around Mersenne. In this manuscript Desargues developed the idea of projective geometry. Young Blaise immediately grasped the content and meaning of the manuscript and presented his first mathematical work *Essay pour les coniques* (Treatise on conic sections) in 1640. In a few years the work *Traité des coniques* on conic sections developed. We know that Leibniz truly admired this work. After having studied this work he called the author one of the greatest intellects of the century [Loeffel 1987, p. 45]. With his projective view on geometry young Pascal manoeuvred himself to an opposite position as far as the analytic geometry of René Descartes was concerned. Both men met on 23rd and

Fig. 6.1.17. René Descartes and Blaise Pascal on stamps [Monaco 1996, France 1962]

24th September 1647 where Roberval was present on their first meeting. The views of Pascal and Descartes turned out to be in opposition so that no genuine agreement could be reached. Father and son Pascal had criticised René Descartes' *Discours de la méthode* sharply and Blaise's projective views on geometry could not acknowledge the worth of an analytic geometry.

At the outset of 1646 Étienne Pascal suffered a severe accident slipping on black ice. He had to be cared for and this was done by two Jansenists, followers of the doctrine of the theologian Cornelius Jansen from the Netherlands. The centre of Jansenism is Port-Royal. Through the encounter with this Jansenist father and son Pascal entered the circle of the Jansenists. Blaise's sister Jacqueline will enter Port-Royal in 1653, the same year in which Pope Innocent X condemned some of Jansen's doctrine. The orientation of Blaise Pascal towards Jansenism is called his 'first conversion'. In the late autumn of 1646 Pascal got news of Torricelli's experiments with a barometer and father and son reproduced these experiments. Already in October 1647 a treatise on the vacuum flew from Pascal's quill.

As far as religion was concerned Blaise, being sickly since the day of his birth, developed into a man of conviction. When the young theologian Jacques Forton, called Frère Saint-Ange, appeared in Rouen to preach on a rationally oriented philosophy he was accused of seduction of young people by Pascal! Forton indeed had to revoke some of his theses but his intolerant attitude in religious matters will later turn against Pascal. Étienne Pascal died in September 1651 and Blaise was engulfed in broodings on the immortality of the soul. In July 1652 the so-called 'worldly period' of Pascal began. By means of a flattering letter to Christina, Queen of Sweden, he had apparently tried to offer himself as a replacement of the deceased Descartes, but this attempt failed. In examining the distribution of chances of winning games Pascal developed important fundamentals of probability theory which he discussed with Fermat. Thereby an important role is played by the discovery of 'Pascal's triangle'. The year 1654 turned out to be very fertile for Blaise as far as science was concerned, but his abhorrence towards the world also grew. Long-standing overwork, a sickly constitution, and his search for God led to a decisive event, his 'second conversion'. In the night from 23rd to 24th November 1654 Pascal suffered a 'mystical illumination'. He wrote the *Mémorial* which he sewed into the lining of his dress. In the English translation[2] of Elizabeth T. Knuth on `https://www.vofoundation.org/faith-and-science/memorial-blaise-pascal/` the *Mémorial* reads as:

The year of grace 1654,
Monday, 23 November, feast of St. Clement, pope and martyr, and others in the martyrology.
Vigil of St. Chrysogonus, martyr, and others.
From about half past ten at night until about half past midnight,

FIRE

GOD of Abraham, GOD of Isaac, GOD of Jacob
not of the philosophers and of the learned.
Certitude. Certitude. Feeling. Joy. Peace.
GOD of Jesus Christ.
My God and your God.
Your GOD will be my God.
Forgetfulness of the world and of everything, except GOD.
He is only found by the ways taught in the Gospel.
Grandeur of the human soul.
Righteous Father, the world has not known you, but I have known you.
Joy, joy, joy, tears of joy.
I have departed from him:
They have forsaken me, the fount of living water.
My God, will you leave me?
Let me not be separated from him forever.
This is eternal life, that they know you, the one true God, and the one that you sent, Jesus Christ.
Jesus Christ.
Jesus Christ.
I left him; I fled him, renounced, crucified.
Let me never be separated from him.
He is only kept securely by the ways taught in the Gospel:
Renunciation, total and sweet.
Complete submission to Jesus Christ and to my director.
Eternally in joy for a day's exercise on the earth.
May I not forget your words. Amen.

Affected by this night Pascal withdrew from the world, wanted to renounce all amenities what so ever and to avoid anything superfluous. The relation to his sister Jacqueline became closer and in 1655 Blaise Pascal became a hermit close to Port-Royal. From there he wrote his *Lettres à un Provincial* using the pseudonym Louis de Montalte. These letters, belonging to the treasure trove of French prose, were serious attacks on the Jesuits as enemies of the Jansenists; in particular they were attacks on the moral philosophy of the Jesuits. The provincial letters contained

[2] A good German translation can be found in [Beguin 1998, p. 111f.].

too much dynamite and hence they were put on the index in 1657 – now Pascal's religious intolerance turned against himself.

Nevertheless Pascal's interests in mathematics continued. He wrote mathematical manuscripts and books further on but he also intended to write a great apologia of Christianity. For this purpose he wrote down his thoughts – they were published under the title *Pensées* (thoughts) after his death and rank today under the great philosophical writings [Pascal 1997]. However, the great apologia which should have emerged from this thoughts was never completed.

Once again Pascal became absorbed in mathematics, namely under the influence of Cavalieri's writings on indivisibles. He succeeded in discovering his own techniques to compute areas and volumes. In June 1658 a public price task appeared directed to all famous mathematicians of the day. The author was Amos Dettonville, but this is just an anagram of Pascal's pseudonym Louis de Montalte! The title of the price task was *Première lettre circulaire relative à la cycloide.* The mathematicians were requested to solve some basic problems concerning the cycloide within a certain time. The price should have been 60 Spanish gold doubloons but Pascal did not have to disburse the award – he was not satisfied with any of the solutions sent to him. Pascal's mathematical literary style was quite conservative. Although some mathematical notations already existed due to the groundworks of François Viète (1540–1603) and others Pascal completely did without them. He formulated all of his results purely in a verbal way that certainly hampers our understanding.

Meanwhile the religious quarrels in Pascal's life increased further. He is asked to sign a writing condemning Jansenism but he refused. Then he turned away from mathematics. On 10th August 1660 he wrote to Pierre de Fermat [Loeffel 1987, p. 27f.]:

> *I think of mathematics as being the highest school of the mind, but at the same time I recognised it as being useless, so that I make no difference between a man being only a mathematician and a skilful craftsman ... In my case it adds that I am absorbed in studies which are so far away from the spirit of mathematics that I can barely remember there being such a spirit.*
>
> (Ich halte zwar die Mathematik als die höchste Schule des Geistes, gleichzeitig aber erkannte ich sie als nutzlos, daß ich wenig Unterschied mache zwischen einem Manne, der nur ein Mathematiker ist und einem geschickten Handwerker ... Bei mir kommt aber jetzt noch hinzu, daß ich in Studien vertieft bin, die so weit vom Geist der Mathematik entfernt sind, dass ich mich kaum mehr daran erinnere, daß es einen solchen gibt.)

Pascal now turned to help the needy and he received the royal patent for the world's first bus route being opened in Paris in 1662. In August of that same year Pascal drafted his last will and testament and shortly after he received the last sacraments. He died on the 19th August 1662 and was buried behind the chancel of the church of Saint-Étienne-du-Mont in Paris.

The Integration of x^p

Pascal's triangle was known in different cultures even before Pascal. It is all about the triangle of numbers

$$
\begin{array}{c|ccccccccccc}
n & & & & & & & & & & & \\
\hline
0 & & & & & & 1 & & & & & \\
1 & & & & & 1 & & 1 & & & & \\
2 & & & & 1 & & 2 & & 1 & & & \\
3 & & & 1 & & 3 & & 3 & & 1 & & \\
4 & & 1 & & 4 & & 6 & & 4 & & 1 & \\
5 & 1 & & 5 & & 10 & & 10 & & 5 & & 1
\end{array}
$$

which can be extended arbitrarily far at the bottom. An entry in this 'arithmetic' triangle is just the sum of the entries left and right of it in the row just above. The entries in Pascal's triangle are called binomial coefficients and can be defined by

$$\binom{n}{k} := \frac{n!}{k! \cdot (n-k)!} \tag{6.6}$$

for $k = 0, 1, 2, \ldots, n$. Here '!' denotes the factorial defined by

$$n! := \begin{cases} 1 \cdot 2 \cdot 3 \cdot 4 \cdots (n-1) \cdot n \; ; n = 1, 2, 3, \ldots \\ 1 \qquad\qquad\qquad\qquad\qquad ; n = 0 \end{cases} .$$

Hence Pascal's triangle can be rewritten in the form

$$
\begin{array}{c|ccccccccccc}
n & & & & & & \binom{n}{k} & & & & & \\
\hline
0 & & & & & & \binom{0}{0} & & & & & \\
1 & & & & & \binom{1}{0} & & \binom{1}{1} & & & & \\
2 & & & & \binom{2}{0} & & \binom{2}{1} & & \binom{2}{2} & & & \\
3 & & & \binom{3}{0} & & \binom{3}{1} & & \binom{3}{2} & & \binom{3}{3} & & \\
4 & & \binom{4}{0} & & \binom{4}{1} & & \binom{4}{2} & & \binom{4}{3} & & \binom{4}{4} & \\
5 & \binom{5}{0} & & \binom{5}{1} & & \binom{5}{2} & & \binom{5}{3} & & \binom{5}{4} & & \binom{5}{5}
\end{array}
$$

and now it is easily seen how the triangle can be extended further down. Pascal knew in about 1654 that the binomial coefficients can be computed by

$$\binom{n}{k} = \frac{n \cdot (n-1) \cdot (n-2) \cdots (n-k+1)}{k!}, \qquad n = 1, 2, 3, \ldots$$

which is equivalent to (6.6) in case of positive n. In his work *Potestatum numericarum summa* he even gave a 'proof' which does not meet modern standards [Baron 1987, p. 197f.]. The binomial coefficients are just the coefficients occurring in the computation of the so-called binomial $(a+b)^n$:

$$
\begin{aligned}
(a+b)^0 &= 0 \\
(a+b)^1 &= 1 \cdot a + 1 \cdot b \\
(a+b)^2 &= 1 \cdot a^2 + 2 \cdot ab + 1 \cdot b^2 \\
(a+b)^3 &= 1 \cdot a^3 + 3 \cdot a^2 b + 3 \cdot ab^2 + 1 \cdot b^3 \\
&\vdots \; \vdots \; \vdots
\end{aligned}
$$

In general the binomial theorem holds:

$$(a+b)^p = \sum_{k=0}^{p} \binom{p}{k} a^{p-k} b^k \tag{6.7}$$

where a and b denote two arbitrary numbers.

Pascal is now looking for the area under the curves $f(x) = x^p$ for $p = 1, 2, 3, \ldots$ from $x = 0$ up to a positive value $x = c$; in todays notation that is:

$$\int_0^c x^p \, dx.$$

We shall see in a moment how skilfully Pascal exploited the binomial theorem. However, we have to do some work first. To begin with we write

$$S_n^{(p)} := \sum_{k=1}^{n} k^p = 1^p + 2^p + 3^p + \ldots + n^p \tag{6.8}$$

for the sum of the p-th power of the first n natural numbers. Following the binomial theorem (6.7) it holds

$$\begin{aligned}(n+1)^{p+1} &= \sum_{k=0}^{p+1} \binom{p+1}{k} n^{p+1-k} \cdot 1^k \qquad (6.9)\\ &= n^{p+1} + \binom{p+1}{1} n^p + \binom{p+1}{2} n^{p-1} + \ldots + \binom{p+1}{p} n + 1,\end{aligned}$$

where we now insert the values $n = 0, 1, 2, \ldots, N$:

$$\begin{aligned}(0+1)^{p+1} &= 1\\ (1+1)^{p+1} &= 1^{p+1} + \binom{p+1}{1} 1^p + \binom{p+1}{2} 1^{p-1} + \ldots + \binom{p+1}{p} 1 + 1\\ (2+1)^{p+1} &= 2^{p+1} + \binom{p+1}{1} 2^p + \binom{p+1}{2} 2^{p-1} + \ldots + \binom{p+1}{p} 2 + 1\\ (3+1)^{p+1} &= 3^{p+1} + \binom{p+1}{1} 3^p + \binom{p+1}{2} 3^{p-1} + \ldots + \binom{p+1}{p} 3 + 1\\ &\;\vdots\\ (N+1)^{p+1} &= N^{p+1} + \binom{p+1}{1} N^p + \binom{p+1}{2} N^{p-1} + \ldots + \binom{p+1}{p} N + 1.\end{aligned}$$

Summing all terms on the left hand side we get

$$1^{p+1} + 2^{p+1} + 3^{p+1} + \ldots + (N+1)^{p+1} = S_{N+1}^{(p+1)},$$

where we have used definition (6.8). Summation of all terms on the right hand side yields:

$$\underbrace{(1+1+\ldots 1)}_{(N+1)\text{-mal}} + (1^{p+1} + 2^{p+1} + \ldots N^{p+1}) + \binom{p+1}{1}(1^p + 2^p + \ldots + N^p)$$
$$+ \binom{p+1}{2}(1^{p-1} + 2^{p-1} + \ldots N^{p-1}) + \ldots$$
$$\ldots + \binom{p+1}{p}(1+2+\ldots+N)$$
$$= (N+1) + S_N^{(p+1)} + \binom{p+1}{1} S_N^{(p)} + \binom{p+1}{2} S_N^{(p-1)}$$
$$+ \ldots + \binom{p+1}{p} S_N^{(1)}.$$

Since left and right hand sums have to be equal we have happily arrived at

$$S_{N+1}^{(p+1)} = (N+1) + S_N^{(p+1)} + \binom{p+1}{1} S_N^{(p)} + \binom{p+1}{2} S_N^{(p-1)} + \ldots + \binom{p+1}{p} S_N^{(1)}.$$

In view of (6.8) we see that

$$S_{N+1}^{(p+1)} - S_N^{(p+1)} = (N+1)^{p+1}$$

holds and so we have the important equation

$$(N+1)^{p+1} - 1 = N + \binom{p+1}{1} S_N^{(p)} + \binom{p+1}{2} S_N^{(p-1)} + \ldots + \binom{p+1}{p} S_N^{(1)} \quad (6.10)$$

being valid for all N.

We have formulated Pascal's result in our modern terminology but as we already have reported, Pascal never used the language of formulae available at his time. Bourbaki has commented on that in [Bourbaki 1971, p. 222f.]:

> *Pascal's language is exceptionally clear and precise, and although one does not understand why he refused to use algebraic notation – not only the one of Descartes but also the one introduced by Viète, one can only admire the tour de force he accomplished to which his mastery of language alone has enabled him.*
>
> (Pascals Sprache ist ganz besonders klar und präzise, und wenn man auch nicht begreift, warum er sich den Gebrauch algebraischer Bezeichnungsweisen – nicht nur der von Descartes, sondern auch der von Viète eingeführten – versagt, so kann man doch nur die Gewalttour, die er vollbringt und zu der ihn allein seine Beherrschung der Sprache befähigt, bewundern.)

We now use (6.10) to replace $S_N^{(p)}$ by the other $S_N^{(k)}$, $k < p$. After division of (6.10) by N^{p+1} we arrive at:

$$\frac{S_N^{(p)}}{N^{p+1}} = \frac{1}{p+1}\left[\left(1+\frac{1}{N}\right)^{p+1} - \frac{1}{N^{p+1}} - \frac{N}{N^{p+1}}\right.$$
$$\left. - \frac{\binom{p+1}{2}S_N^{(p-1)}}{N^{p+1}} - \frac{\binom{p+1}{3}S_N^{(p-2)}}{N^{p+1}} - \ldots - \frac{\binom{p+1}{p}S_N^{(1)}}{N^{p+1}}\right] \tag{6.11}$$

After this preliminary work Pascal started attacking the actual problem; the computation of the area under the function $f(x) = x^p$ on the interval $[0, c]$. Pascal certainly was *no* mathematician working with indivisibles like Cavalieri, although he admired the mathematics of the latter. To Pascal an indivisible is always an 'infinitely small rectangle' [Loeffel 1987, p. 100], hence an infinitesimal. Thus he did not view an indivisible ℓ as a line in an area as Cavalieri, but the infinitesimal rectangle $\ell \cdot dx$. In case of the problem of the computation of the area under $f(x) = x^p$ Pascal subdivided the interval $[0, c]$ into N equal parts of length

$$h := \frac{c}{N}.$$

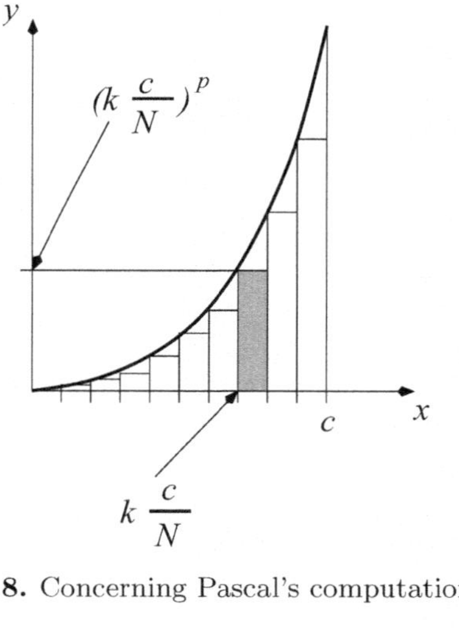

Fig. 6.1.18. Concerning Pascal's computation of area

The area of the single rectangles shown in figure 6.1.18 is given by

$$\frac{c}{N} \cdot \left(k\frac{c}{N}\right)^p, \quad k = 0, 1, 2, \ldots, N.$$

Hence the sum of all such rectangular areas is

$$\frac{c}{N} \cdot \sum_{k=0}^{N} \left(k\frac{c}{N}\right)^p = \frac{c^{p+1}}{N^{p+1}} \sum_{k=0}^{N} k^p = c^{p+1} \frac{S_N^{(p)}}{N^{p+1}}$$

and we can see now why Pascal invested so much preliminary work to compute $S_N^{(p)}/N^{p+1}$! To compute the true area under x^p Pascal has to compute the limit

$$\int_0^c x^p\,dx = \lim_{N\to\infty} \frac{c}{N} \cdot \sum_{k=0}^{N} \left(k\frac{c}{N}\right)^p.$$

But this is nothing but

$$\int_0^c x^p\,dx = c^{p+1} \lim_{N\to\infty} \frac{S_N^{(p)}}{N^{p+1}}.$$

In the view of (6.11) Pascal he reasoned, geometrically, that all $S_N^{(k)}$ with $k < p$ are negligible as compared to N^{p+1} if only N will be large enough. Hence

$$\lim_{N\to\infty} \frac{S_N^{(k)}}{N^{p+1}} = 0 \quad \text{für } k < p.$$

If N increases beyond all bounds all summands in (6.11) except $(1 + 1/N)^p$ are vanishing and the remaining one becomes just 1, because $1/N$ gets ever smaller if N increases. Hence Pascal has proven

$$\int_0^c x^p\,dx = \frac{c^{p+1}}{p+1}.$$

The Characteristic Triangle

In the treatise *Traité des sinus du quart de cercle* (Treatise on the ordinates in a quarter circle) Pascal for the first time employed a technique which later served Leibniz as the central seed of his differential calculus. It is all about the use of the characteristic triangle; a name given to it by Leibniz. We follow [Loeffel 1987, p. 108ff.] on our tour.

In a quarter circle as in figure 6.1.19 Pascal is concerned with the tangent at point D and with it forms the triangle $EE'K$ which Leibniz later called the characteristic triangle.

The triangles ODI and EKE' are similar, so that

$$DI : OD = EK : EE',$$

or

$$y : r = \Delta x : \Delta s,$$

hence

$$y \cdot \Delta s = r \cdot \Delta x. \tag{6.12}$$

Now Pascal computed the 'sum of the ordinates', meaning the sum

$$\sum DI \cdot \operatorname{arc}(DD')$$

between two values x_1 and x_2 of the abscissa. The arc DD' can be understood in the following sense. The characteristic triangle can be build not only at one point D on the quarter circle, but also at a point D' different from D. The arc DD' should then be understood as the circular arc between two neighbouring points D. Following Pascal the arc DD' can be replaced with arbitrary accuracy by the segment EE' if there are infinitely many summands, hence

$$\sum DI \cdot \text{Bogen}(DD') = \sum DI \cdot EE'$$

and using our notations in figure 6.1.19 this is

$$\sum DI \cdot \text{Bogen}(DD') = \sum DI \cdot EE' = \sum y \cdot \Delta s.$$

The last sum runs over the arc length s of the arc DD', in fact traversing from arc length $s := s_2$ corresponding to the abscissa x_2, up to $s := s_1$ corresponding to $x_1 < x_2$. Hence we traverse the arc in the mathematically positive sense.

Using (6.12) it finally follows

$$\sum_{s_2}^{s_1} y \cdot \Delta s = -\sum_{x_2}^{x_1} r \cdot \Delta x = r \cdot (x_2 - x_1). \tag{6.13}$$

The negative sign can be explained since a positive traverse of the arc from s_2 to s_1 implies a backward traverse on the abscissa from x_2 to $x_1 < x_2$.

Translating into Leibniz's (i.e. todays) symbols we get for the arc length

$$s = r \cdot \varphi$$

and hence it follows $ds = r \cdot d\varphi$ for the differentials. Furthermore we have

$$y = r \cdot \sin\varphi, \quad x = r \cdot \cos\varphi.$$

So the points x_1, x_2 can be written as

$$x_1 = r \cdot \cos\varphi_1, \quad x_2 = r \cdot \cos\varphi_2$$

and from (6.1.19) it follows

$$\int_{x_1}^{x_2} y \cdot ds = \int_{\varphi_1}^{\varphi_2} r \cdot \sin\varphi \cdot r \cdot d\varphi = r(x_2 - x_1) = r^2(\cos\varphi_1 - \cos\varphi_2).$$

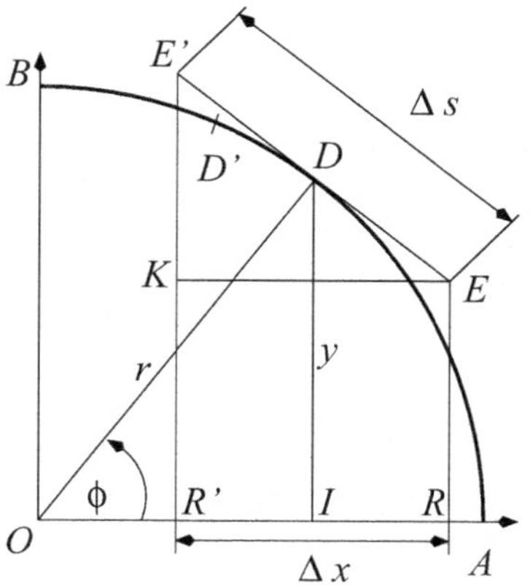

Fig. 6.1.19. The characteristic triangle at a quarter circle

Division by r^2 yields

$$\int_{\varphi_1}^{\varphi_2} \sin\varphi \, d\varphi = \cos\varphi_1 - \cos\varphi_2 = -\cos\varphi\Big|_{\varphi_1}^{\varphi_2}.$$

From our modern standpoint Pascal has computed the integral of the sine function by methods of elementary geometry.

Further Works Concerning Analysis

Pascal also tackled computations of barycentres and volumes. In his work *Traité des Trilignes Rectangles et de leurs Onglets* (Treatise on curve triangles and their 'adjunct bodies'[3]) which appeared as a part of the *Lettre de A. Dettonville à M. Carcavy* in 1658 there is even the rule of integration by parts in geometric form; of course not in the form of integration but as summation of infinitesimals [Loeffel 1987, p. 106]. His computation of area at the cycloid in the *Traité général de la Roulette* ('Roulette' is the name for the cycloid!) confirmed a result which was found before him by Roberval. Pascal reduced the geometric quantities of the cycloid to the geometry of the circle and hence could use already known facts of the circle's geometry. He has also successfully tackled the rectification of the cycloid. Concerning the computation of tangents there is no progress in Pascal's work. Loeffel writes in [Loeffel 1987, p. 118]:

> *The consequent circumvention of the tangent problem and the persistence in purely geometrical ideas is perhaps one reason that he failed a decisive breakthrough to a universal infinitesimal calculus. Nevertheless, Pascals genius can be seen more in the spontaneity and the uniqueness of his ideas as in the continual and consequent processing of a thought.*
>
> (Das konsequente Umgehen des Tangentenproblems und das Verharren in rein geometrischen Vorstellungen ist vielleicht ein Grund dafür, daß ihm der entscheidende Durchbruch zur Schaffung eines *universellen Infinitesimalkalküls* versagt blieb. Doch Pascals Genius ist mehr in der *Spontaneität* und Einmaligkeit seiner Ideen zu sehen, als in der kontinuierlichen und konsequenten Verarbeitung eines Gedankens.)

6.1.5 Gilles Personne de Roberval

Gilles Personne was born the son of an ordinary family of farmers in 1602 in Roberval in the Arrondissement Senlis, but he enjoyed a certain school education. Only later he called himself Gilles Personne *de Roberval* probably to suggest he came from nobility. He traveled France as an autodidact and worked as a teacher until he entered the circle of Marin Mersenne in Paris. There he became the only really professional mathematician of the group. In fact, his knowledge in the sciences were so excellent that he became professor at the Collège Gervais in 1632. Already in 1634 he became the successor of Petrus Ramus (1515–1572) at the Collège Royal (later renamed as Collège de France). He now had to apply to this post periodically and he did that with a very specific tactic.

[3] The word 'onglet' actually means 'claw' or 'hoof'.

Fig. 6.1.20. Jean Baptiste Colbert, minister of finace of Louis XIV, introduces members of the Royal Society of Sciences to the King, among them Gilles Personne de Roberval ([Painting by Henri Testelin about 1660, Detail] Musée du Château, Versailles)

He published none of his results but presented them at the periodic application speeches to outmanoeuvre the other competitors. This tactic together with an eruptive temperament brought Roberval in conflict with numerous other mathematicians. He accused Cavalieri and Torricelli of having stolen his ideas. With Descartes he fought a battle which drifted into personal accusations. When the Académie des Sciences was founded in 1666 Roberval became one of the first members. He worked on problems in physics – there is a Roberval balance – but he became known as the inventor of a calculus of indivisibles independently of Cavalieri.

The Area Under the Cycloid

A cycloid is that curve which arises from tracing the path of a fixed point on the circumference of a circle which unrolls on a plane as in figure 6.1.21.

Roberval was pointed to the problem of computing the area under the cycloid by Marin Mersenne. At first, the cycloid had to been described in mathematical terms. This Roberval achieved by decomposing the velocity of the point on the circle's circumference (radius a) into the two components of a rotation about the circle's centre and of a translation in y-direction. He then attacked the problem purely kinematically [Edwards 1979, p. 135f.]. By adding both components he arrived at the parametrical representation of the velocity of the point under consideration

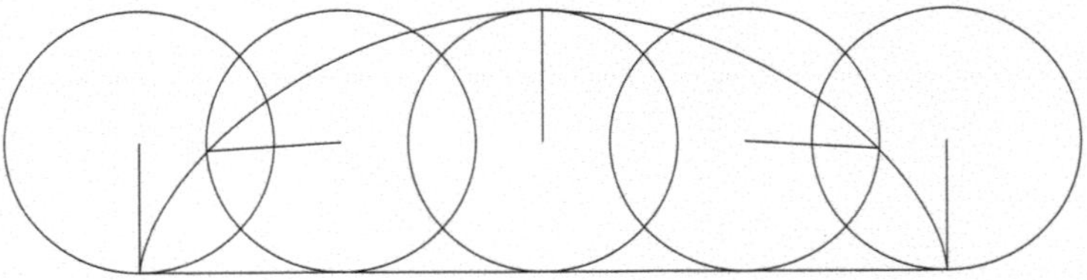

Fig. 6.1.21. The cycloid as roulette of a circle

and from there he advanced to the parametrical representation of the cycloid. The rotation angle Θ thereby served as the parameter:

$$x(\Theta) = a(\Theta - \sin\Theta) \tag{6.14}$$
$$y(\Theta) = a(1 - \cos\Theta). \tag{6.15}$$

He now determined the area under the cycloid during one period of revolution; i.e. from $\Theta = 0$ to $\Theta = 2\pi$. We follow [Baron 1987, p. 156] and take a look at figure 6.1.22.

There we see the half circle APB with centre O at time $t = 0$ formed by horizontal indivisibles. If the circle's centre moves through the rotation of the circle about the angle Θ from O to O', then it holds

$$AA' = OO'$$

and

$$PP' = QQ' = AA' = a\Theta.$$

The curve $AP'C$ is part of the cycloid and the points on the cycloid have coordinates (6.14) and (6.15). However, the point Q' lies on another curve which Roberval

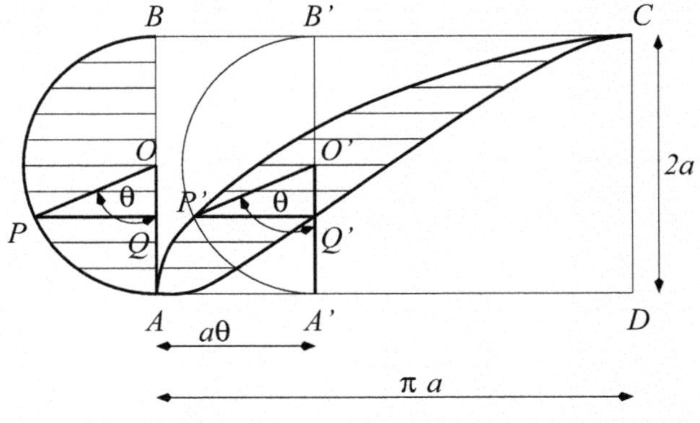

Fig. 6.1.22. The computation of the area under the cycloid

called the 'compagnon' of the cycloid. This 'compagnon' is the sine function since it obviously has the parametrical representation

$$x(\Theta) = a\Theta \tag{6.16}$$

$$y(\Theta) = a(1 - \cos\Theta). \tag{6.17}$$

Since $P'Q' = PQ$ and since such an equality is valid for all indivisibles the area $AP'CQ'A$ between the cycloid and the 'compagnon' is exactly the area of the half circle APB. The area of the rectangle $ABCD$ is exactly twice the area of the circle, $2\pi a^2$. If we now could compute the area $AQ'CD$ under the 'compagnon' we had hit the target.

Inserting (6.16) (in the form of $\Theta = x/a$) into (6.17) then the function

$$y(x) = a\left(1 - \cos\frac{x}{a}\right),$$

emerges which describes the 'compagnon'. But this function is completely symmetric on $[0, \pi a]$ so that the graph of the 'compagnon' divides the rectangle $ABCD$ into two exactly equal areas, cp. figure 6.1.23. The length of the indivisibles from the segment AD upward to the graph of the 'compagnon' is $a(1 - \cos\Theta)$ after (6.17). Likewise the length of BC downward to the graph is $2a - a(1 - \cos\Theta) = a(1 + \cos\Theta)$. Since the cosine changes sign on $[0, \pi]$ at $\Theta = \pi/2$ the lower indivisble corresponding to Θ can be found in the upper part at $\pi - \Theta$ since $a(1 + \cos(\pi - \Theta)) = a(1 - \cos\Theta)$. Expoiting this argument of symmetry Roberval reached his aim: The area $AP'CD$ under the (half) cycloid is

$$\text{rectangular area } ABCDA - \text{area } AP'CQ'A,$$

hence

$$a\pi \cdot 2a - \frac{1}{2}\pi a^2 = \frac{3}{2}\pi a^2.$$

This result of Roberval was communicated to the mathematicians of France by Mersenne. Thereupon also Descartes and Fermat squared the cycloid [Baron 1987, p. 157–160].

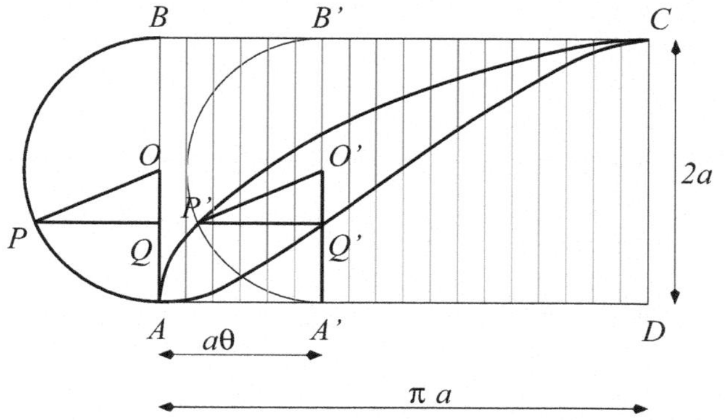

Fig. 6.1.23. Computing the area under the 'compagnon'

The Quadrature of x^p

Even against the background of the great variety of methods in his days Roberval's method to compute the area under the functions $f(x) = x^p$ in the *Traité des indivisibles* is remarkable. Although Roberval spoke of indivisibles he always computed with infinitesimals. These are strips which all have always the breadth 1. He divided the area under the straight line $y = x$ under $[0, b]$ as in figure 6.1.24.

If $[0, b]$ is subdivided into n parts, then, following Roberval

$$b = 1 \cdot n,$$

since every part has breadth 1. Of course, this is difficult to digest since the '1' can not be our real number 1 but a somewhat strange 'Roberval unit'. Yet if n increases there always has to hold $1 \cdot n = b$, rendering the use of the symbol '1' strange! Hence in the use of n by Roberval we also have to be careful, [Baron 1987, p. 154f.]. Roberval was convinced that Cavalieri actually meant little line segments when he spoke about a line consisting of indivisible points [Boyer 1959, p. 141f.]. This interpretation is wrong, but shows that Roberval actually interpreted the calculus of indivisibles as being an infinitesimal calculus.

As did Fermat so Roberval started his computation of areas by means of summation formulae like

$$\sum_{k=1}^{n} k = \frac{1}{2}n(n+1), \quad \sum_{k=1}^{n} k^2 = \frac{1}{3}n(n+1)(n+2),$$

and so on. In the case of the function $f(x) = x$ in figure 6.1.24 Roberval used

$$\sum_{k=1}^{n} k = \frac{n^2}{2} + \frac{n}{2}$$

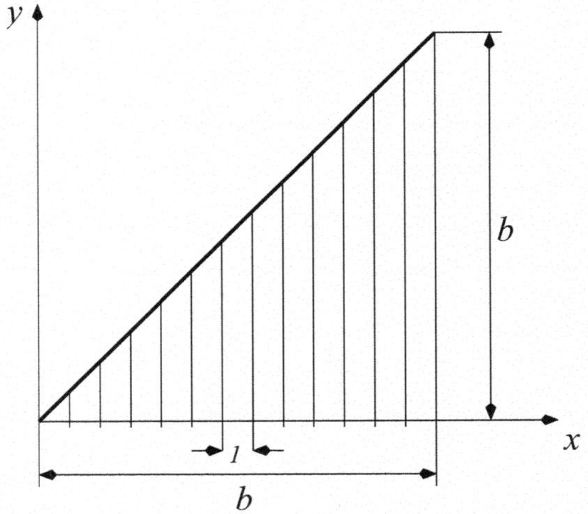

Fig. 6.1.24. Roberval's infinitesimals at a triangle

and argued as follows: If n increases beyond bounds then the term $n/2$ does not contribute to the area since this term does not belong to an area but to a line. Hence the area sought is (every rectangle has area $k \cdot 1$)

$$\sum k = \frac{n^2}{2}$$

and, since $1 \cdot n = b$, it follows

$$\int_0^b x\,dx = \frac{b^2}{2}.$$

In much the same way Roberval proceeds in the case of the area under $f(x) = x^2$ on $[0, b]$. Here he used

$$\sum_{k=1}^{n} k^2 = \frac{n^3}{3} + \frac{n^2}{2} + \frac{n}{6}$$

and now neither $n^2/2$ nor $n/6$ contribute since neither a square nor a line can have a relation to a cube. Fairly generally Roberval concluded that the difference

$$\sum_{k=1}^{n} k^p - \frac{n^{p+1}}{p+1}$$

can be neglected as long as n can be made arbitrarily large. In this case it follows

$$\lim_{n\to\infty} \sum_{k=1}^{n} \frac{k^p}{n^p} = \frac{1}{p+1},$$

and if $\sum k^p$ denotes the sum of all infinitesimals parallel to the y-axis under the curve

$$\frac{y}{a} = \left(\frac{x}{b}\right)^p,$$

then it apparently follows

$$\int_0^b y\,dx = \frac{ab}{p+1},$$

cp. figure 6.1.25.

6.2 Analysis Prior to Leibniz in the Netherlands

Writing on analysis in the Netherlands we do *not* mean Holland or the Netherlands as we know it today but the area for which the English have coined the name 'low countries'. In the late 15th century the Netherlands came under Habsburg rule, in particular under the rule of Emperor Charles V. In the wake of the Protestant Reformation large parts of the population converted to Protestantism which led to persecution and oppression by Emperor Charles and his son, Philip II of Spain. The 'Spanish Netherlands' included the territories of the present Netherlands, Belgium, and Luxembourg. Eventually the repressions and the attempts to re-Catholicise the country led to an uprising. The seven northern provinces of the Netherlands, Holland, Zeeland, Groningen, Utrecht, Friesland, Gelderland, and Overijssel, joined in 1579 to constitute the 'Union of Utrecht' from which arose the 'Republic of the

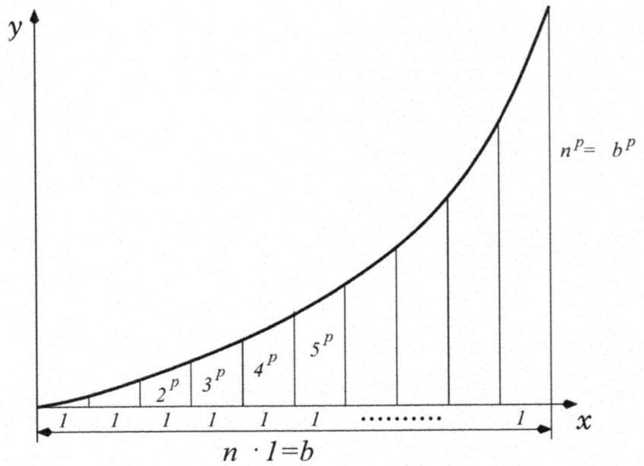

Fig. 6.1.25. The integration of $f(x) = (x/b)^p$

seven United Netherlands' (Dutch Republic) in 1581. In the so-called 'Eighty Years' War' or 'Dutch War of Independence' the Union of Utrecht won the independence from Spain between 1568 and 1648. The Peace of Westphalia in 1648 after the Thirty Years' War finally became the natal birth of the 'Republic of the United Netherlands'. The southern provinces, still being under Spanish rule, were separated and formed a territory which later became Belgium.

Due to the vicinity of France it seems natural that French mathematics spread fast in the Netherlands.

6.2.1 Frans van Schooten

Frans van Schooten (1615–1660) counts as the mathematical teacher of the Dutch people. In Leiden he met Descartes who visited van Schooten's father (Frans van Schooten, senior). Van Schooten got insight into the galley proofs of *La Géométrie*, and in Paris he got in contact with Roberval via Mersenne. In Paris he also became acquainted with the works of Fermat which at that time were discussed in Mersenne's circle. Back in Leiden van Schooten became professor of mathematics and developed into a great protagonist of the new geometry of Descartes. He succeeded in attracting a group of young and gifted Dutchmen; for a short time in 1646 among them the later super star of the Netherlands, Christiaan Huygens.

In 1646 van Schooten edited and published the *Opera mathematica* of François Viète in which he replaced Viète's cumbersome notation with the modern one of Descartes. He translated Descartes' *La Géométrie* into Latin and published it in 1649, augmented with a wealth of remarks. In these remarks van Schooten emphasised the importance of Descartes' circle method (cp. section 6.1.2). This Latin edition of Descartes' work has found wide distribution and has stimulated the general interest in computation of tangents. Van Schooten's pupils have promoted the development of analysis by their independent works.

Fig. 6.2.1. The partition of the Spanish Netherlands in the Peace of Westphalia 1648 into the independent Netherlands and the territory remaining with Spain. The attack on England 1588 was planned to be carried out from the territory of the Spanish Netherlands with troops which should be brought there by the Armada. The territory of the Spanish Netherlands from 1648 on corresponds almost exactly to the modern territory of Belgium [Map: H. Wesemüller-Kock]

6.2.2 René François Walther de Sluse

René François Walther de Sluse (1622–1685) (Slusius) came from the region around Liège, studied in Lyon, and remained for 10 years in Italy where he worked with a friend and pupil of Torricelli. In Italy he learned Cavalieri's method of indivisibles and got acquainted with the mathematical works of Torricelli. De Sluse developed his own method to compute tangents about 1655 which was published only in 1673 when de Sluse had to worry about his priority due to Newton's work. To illustrate his method [Baron 1987, p. 215f.] we consider functions

$$f(x,y) = \sum_{i=1}^{n}\sum_{j=1}^{m} a_{ij}x^iy^j =: \sum a_{ij}x^iy^j = 0.$$

If (x_1, y_1) is a point on the curve infinitesimally adjacent to (x, y) then, following Fermat, it holds

$$f(x_1, y_1) - f(x, y) = 0,$$

hence

$$\sum a_{ij}\left(x_1^iy_1^j - x^iy^j\right) = \sum a_{ij}\left(x_1^i(y_1^j - y^j) + y^j(x_1^i - x^i)\right). \tag{6.18}$$

From the known relations

$$\frac{x_1^i - x^i}{x_1 - x} = x_1^{i-1} + x_1^{i-2}x + x_1^{i-3}x^2 + \ldots + x^{i-1}$$
$$\frac{y_1^j - y^j}{y_1 - y} = y_1^{j-1} + y_1^{j-2}y + y_1^{j-3}y^2 + \ldots + y^{j-1}$$

it follows

$$x_1^i(y_1^j - y^j) = x_1^i(y_1 - y)\left(y_1^{j-1} + y_1^{j-2}y + y_1^{j-3}y^2 + \ldots y^{j-1}\right)$$
$$y_1^j(x_1^i - x^i) = y^j(x_1 - x)\left(x_1^{i-1} + x_1^{i-2}x + x_1^{i-3}x^2 + \ldots x^{i-1}\right).$$

Inserting this into (6.18) yields

$$(y_1 - y)\sum a_{ij}\left[x_1^i\left(y_1^{j-1} + y_1^{j-2}y + \ldots + y^{j-1}\right)\right]$$
$$+(x_1 - x)\sum a_{ij}\left[y^j\left(x_1^{i-1} + x_1^{i-2}x + \ldots + x^{i-1}\right)\right] = 0$$

Solving for the quotient $(y_1 - y)/(x_1 - x)$ it follows

$$\frac{y_1 - y}{x_1 - x} = -\frac{\sum a_{ij}[y^j(x_1^{i-1} + x_1^{i-2}x + \ldots + x^{i-1})]}{\sum a_{ij}[x_1^i(y_1^{j-1} + y_1^{j-2}y + \ldots + y^{j-1})]}. \tag{6.19}$$

We now let (x_1, y_1) approach (x, y). From figure 6.2.2 it holds in the limit, that

$$\lim_{(x_1,y_1)\to(x,y)} \frac{y_1 - y}{x_1 - x} = \frac{y}{s}$$

and if we set in the right hand side of (6.19) somewhat brutally $x_1 = x$ and $y_1 = y$ we get

$$\frac{y}{s} = -\frac{\sum a_{ij}y^j \cdot i \cdot x^{i-1}}{\sum a_{ij}x^i \cdot j \cdot y^{j-1}}.$$

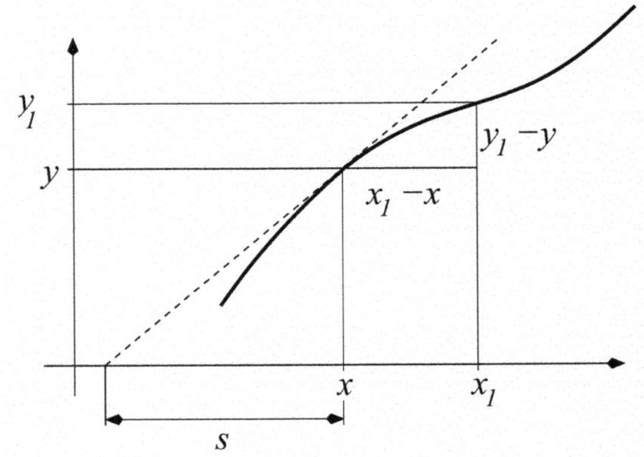

Fig. 6.2.2. The limit $(x_1, y_1) \to (x, y)$

On the left hand side is the slope of the tangent. Nominator and denominator on the right side are the derivatives of f with respect to x and y, respectively. Hence de Sluse has found

$$\frac{dy}{dx} = -\frac{\frac{\partial f}{\partial x}}{\frac{\partial f}{\partial y}},$$

and that is the formula of implicit differentiation.

6.2.3 Johannes van Waveren Hudde

Johannes (van Waveren) Hudde (1628–1704), born in Amsterdam, studied law at the University of Leiden and there got into the circle around van Schooten.

Hudde's mathematical achievements, as far as they have come down on us, come from the period between 1654 and 1663. After that period he got engaged in politics in his birth town and acted 21 times as mayor of Amsterdam until his death. This says something about his quality as a politician! Hudde was very interested in algebra [Lüneburg 2008, p. 55f.], but when he became acquainted with Fermat's method to compute tangents he tried to simplify it at least for polynomials. Hudde gave an explanation of his method in a letter of 1659; this letter will only be published in 1713 in the context of the priority dispute between Leibniz and Newton [Baron 1987, p. 217], [Sonar 2018]. The method is known as 'Hudde's rule' today. An example of this rule appeared in 1657 in *Exercitationes mathematicae* by van Schooten. In the edition of the Latin *La Géométrie* of Descartes in 1699 van Schooten added *De maximis et minimis*, a short tract of Hudde.

Hudde treated polynomials

$$f(x) = a_0 + a_1x + a_2x^2 + a_3x^3 + \ldots + a_nx^n$$

and looked for roots at which f is extremal, i.e. points $x = \alpha$ at which $f(\alpha) = f'(\alpha) = 0$. Hence the first derivative has a root in α and f therefore has a double root. Hudde wrote [Baron 1987, p. 218]:

> *If, in an equation two roots are equal and, if it be multiplied by any arithmetical progression, i.e., the first term by the first term of the progression, the second by the second term of the progression, and so on: I say that the equation found by the sums of these products shall have a root in common with the original equation.*

Choosing an arithmetic progression

$$p, p+q, p+2q, p+3q, \ldots,$$

Hudde computes the products

$$\begin{aligned} &p \cdot a_0 \\ &(p+q) \cdot a_1x \\ &(p+2q) \cdot a_2x^2 \\ &\vdots \\ &(p+nq) \cdot a_nx^n \end{aligned}$$

Fig. 6.2.3. Johannes Hudde, mayor and mathematician [Painting: Michiel van Musscher, probably end of 17th c.]

and sums up to arrive at

$$0 = \underbrace{pa_0 + pa_1x + pa_2x^2 + \ldots + pa_nx^n}_{=p\cdot f(x)}$$
$$+qa_1x + 2qa_2x^2 + 3qa_3x^3 + \ldots + nqa_nx^n,$$

hence

$$p \cdot f(x) + q \cdot \underbrace{\left(a_1x + 2a_2x^2 + 3a_3x^3 + \ldots + na_nx^n\right)}_{x\cdot f'(x)} = 0.$$

In modern notation Hudde has derived the equation

$$p \cdot f(x) + q \cdot x \cdot f'(x) = 0. \tag{6.20}$$

It is now evident that (6.20) shares a root with $f(x)$, since $f(\alpha) = f'(\alpha) = 0$ and hence $p \cdot f(\alpha) + q \cdot \alpha \cdot f'(\alpha) = 0$.

Starting from this algebraic considerations Hudde developed 'Hudde's rule' [Baron 1987, p. 218]:

> *Pull all the terms on one side (=0). Remove x, y, from the divisors. Arrange in descending powers of y and multiply each term by the corresponding term of any arithmetic progression whatsoever. Repeat this process for the terms containing x. Divide the first sum of products by the second. Multiply the quotient by $-x$ and this will give the subtangent.*

To illustrate Hudde's rule we start with

$$f(x,y) = 0$$

and think of the terms in y and x as being ordered as required. After (6.20) multiplication with an arithmetic progression $p, p+q, p+2q, p+3q, \ldots$ yields for the y-terms

$$p \cdot f(x,y) - q \cdot y \cdot \frac{\partial f}{\partial y}(x,y),$$

and for the x-terms after multiplication with $r, r+s, r+2s, r+3s, \ldots$

$$r \cdot f(x,y) - s \cdot x \cdot \frac{\partial f}{\partial x}(x,y).$$

Baron [Baron 1987, p. 218] has pointed out that Hudde always chose $q = s = -1$ in his examples, i.e. Hudde's rule yields

$$s = -x \cdot \frac{p \cdot f(x,y) + y\frac{\partial f}{\partial y}(x,y)}{r \cdot f(x,y) + x\frac{\partial f}{\partial x}(x,y)}$$

for the subtangent s. Since $f(x,y) = 0$ it follows

$$s = -x \cdot \frac{\frac{\partial f}{\partial y}(x,y)}{\frac{\partial f}{\partial x}(x,y)},$$

and this is the formula for the subtangent by means of implicit differentiation.

The importance of Hudde's rule does not so much lie in the method as such, but in the 'mechanisation' of the computation of tangents. Hudde's rule is an explicit algorithm which can be applied even without deeper knowledge.

6.2.4 Christiaan Huygens

Christiaan Huygens (1629–1695) was born the son of the influential diplomat and artist Constantijn Huygens (1596–1687) in The Hague.

His social interaction with famous men of his day was a normality since his father entertained guests as diverse as the painter Rubens or the mathematician and philosopher René Descartes. Constantijn Huygens was in touch with Mersenne as he was with the English court so that for the adult Christiaan Huygens all doors were open both in Paris and in London. Until the age of 16 young Huygens enjoyed private tutoring. Among other topics he learned geometry, building models, and playing the lute. We may safely assume that his education was directly influenced by Descartes who was adored by Huygens as well as by his father. Huygens studied law at the University of Leiden from 1645 to 1647 and was educated in mathematics by van Schooten. From 1647 to 1649 he continued his studies in Breda. He took part in a diplomatic mission to Denmark already in 1649 and hoped to visit Descartes in Sweden which did not materialise due to bad weather.

De Saint-Vincent had claimed to have solved the problem of squaring the circle. In his very first publication *Cyclometria* of 1651 young Huygens determined an error in the proof and in addition wrote a larger work *De circuli magnitudine inventa* in 1654. But Huygens was also educated as a practician and the new inventions of telescopes and microscopes fascinated him. He turned to grinding lenses and developed methods to produce better ones than those of his contemporaries. He constructed a telescope with his own ground and polished lenses and detected the first of the moons of Saturn in 1655; a discovery which he proudly reported to Paris. From there he got news of the works of Fermat and Pascal concerning the theory of probability and published the first printed booklet concerning this theory.

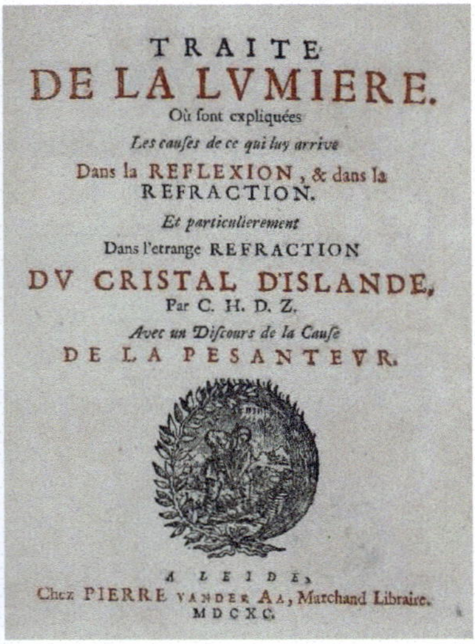

TRAITE
DE LA LVMIERE.
Où sont expliquées
Les causes de ce qui luy arrive
Dans la REFLEXION, & dans la
REFRACTION.
Et particulierement
Dans l'etrange REFRACTION
DV CRISTAL D'ISLANDE,
Par C. H. D. Z.
Avec un Discours de la Cause
DE LA PESANTEVR.

A LEIDE,
Chez PIERRE VANDER AA, Marchand Libraire.
MDCXC.

Fig. 6.2.4. Christiaan Huygens [Painting by Caspar Netscher, 1671] and the title page of his book on the nature of light 1690

Already in 1656 he discovered the true shape of the ring of Saturn and was able to explain the different forms of the ring in certain phases. Against some resistance Huyghens' theory had to be accepted shortly after; apparently Huyghens had the better telescope! Since accurate measurement of time is essential in astronomy it takes no wonder that Huygens was also concerned with the construction of accurate clocks. Today he is ranked as the inventor of the pendulum clock which he got patented in 1656. In his days and even long after it was the most accurate device for measuring time. Through Pascal's award challenge concerning the cycloid, published under his pseudonym Amos Dettonville, he was led to the construction of a cycloidal pendulum clock. In this device the pendulum moves on a cycloid and hence is an isochronous pendulum, i.e. it always shows the same period of oscillation independently of its amplitude. Huygens was convinced that the famous problem of the determination of longitude [Sobel 1995] could be solved by means of a pendulum clock, and some experiments were conducted on sea. His most famous book *Horologium oscillatorium sive de motu pendulorum* on the physics of the pendulum clock contains the theory of the movement of the pendulum and can be understood as a work serving as an introduction to Newton's *Principia.* Huygens discovered the law of centrifugal force at uniform circular motion and hence delivered the forward pass for Newton and Hooke concerning the $1/r^2$-law of gravitation which Huygens himself formulated.

In 1660 Huygens moved to Paris where he attended many meetings of the scholars; meeting, among others, Roberval, Pascal, and Desargues. A year later he traveled to London where he presented his superior telescope and attended experiments conducted by Robert Boyle (1627–1692); in particular Boyle demonstrating his vacuum pump. Huygens is very impressed by the mathematical works of John Wallis (1616–1703) and other English mathematicians and he kept contact with London further on. In 1663 he was received into the Royal Society and in 1666 state secretary Jean-Baptiste Colbert (1619–1683) invited him into the Académie Royale des Sciences in Paris which Colbert himself had founded. Huygens transferred his residence to Paris. In Paris works on physics originated; some of them falsifying some of Descartes' theories. Huygens always stayed in close contact with the Royal Society in London, and when he fell seriously ill in 1670 he let his unpublished manuscript on mechanics be sent to the Royal Society. Having overcome his illness in Holland he met young Gottfried Wilhelm Leibniz in Paris in 1672 and became his teacher and mentor. Huygens was an excellent mathematician who understood the tangent methods and the techniques used of his day. But when Leibniz made his breakthrough to the differential and integral calculus Huygens could not acquire a taste for the new methods of his pupil. He kept attached to classical geometry all of his life and was able to solve even the most complex problems by means of geometrical constructions which Leibniz could solve in a few lines.

Further times of illness forced Huygens back to The Hague again where he continued his scientific works. In 1683 Colbert died; in 1687 Huygens' father followed. A return to Paris without the protection of his patron Colbert seemed impossible and so Huygens went to England in 1689 where he met with Newton, Boyle, and other members of the Royal Society. Huygens adored Newton, although he thought the theory of gravitation of the latter as being absurd. But also Newton held Huygens dear and called him 'summus Hugenius': The very great Huygens! Huygens passed away in 1695 in the town of his birth, The Hague.

6.3 Analysis Before Newton in England

6.3.1 The Discovery of Logarithms

In calculus and analysis logarithms play an exceptional and valuable role as the inverse functions of the power functions. If

$$x = a^y, \quad a > 0,$$

then the logarithm to base a is defined as

$$y = \log_a x.$$

The logarithms to bases e, $\ln x = \log_{\mathsf{e}} x$ ('logarithmus naturalis', natural logarithmus) and 10, $\log_{10} x$ (common or decadic or decimal logarithm, Briggsian logarithm) have to be particularly highlighted. Since all logarithms satisfy the functional equation

$$\log_a(x \cdot z) = \log_a x + \log_a z$$

the logarithms were used to simplify tedious calculations in earlier times. Multiplication/division is reduced to addition/subtraction and since $\log_a b^x = x \cdot \log_a b$ the treatment of powers is significantly simplified also. Until far into the 20th century the logarithmic slide rule was employed for computations and an engineer could barely live without one. The principle of the slide rule was based on logarithmic scales so that multiplications/divisions could be carried out simply by adding/subtracting of segments on the slide rule.

Although the actual invention of the logarithms could be seen as a purely English issue (the Swiss Jost Bürgi (1552–1632) published his logarithms in 1620 as *Progreß Tabulen* [Clark 2015] but he did not prevail [Goldstine 1977, p. 20ff.]), there was an important precursor, namely the German Augustinian monk Michael Stifel (1487?–1567) [Stifel 2007], [Alten et al. 2005, Section 4.6.3] [Alten et al. 2014, p. 254ff.]. Stifel in his work *Arithmetica Integra* [Stifel 2007] of 1544 wrote down the arithmetic scale (=sequence)

$$\ldots, -3, -2, -1, 0, 1, 2, 3, \ldots$$

and beneath the corresponding geometric scale

$$\ldots, 2^{-3}, 2^{-2}, 2^{-1}, 2^0, 2^1, 2^2, 2^3, \ldots.$$

He noticed that multiplication on the geometric scale could be carried out by addition on the arithmetic scale. If one wants to compute $4 \cdot 16$ for example, one has to add the corresponding entries on the arithmetic scale, $2 + 4$, and then find the solution 64 on the geometric scale beneath the 6 on the arithmetic scale. However, Stifel's scales allow only the multiplication of powers of two and hence Stifel has not followed his idea further on in the direction of logarithms [Hofmann 1968]. This intellectual leap was only dared in Scotland and England where Stifel's book was read.

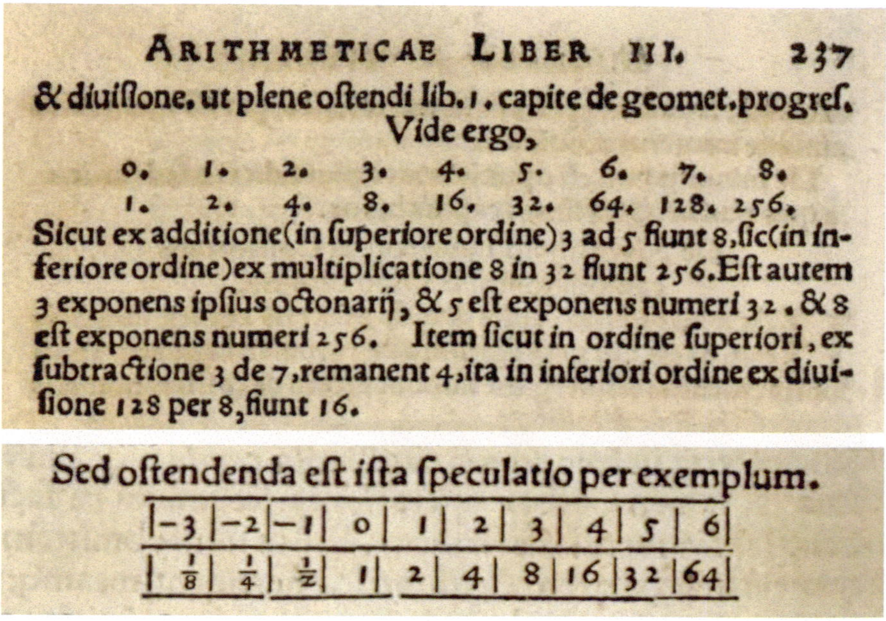

ARITHMETICAE LIBER III. 237

& diuiſione, ut plene oſtendi lib. 1. capite de geomet. progreſ. Vide ergo,

0.	1.	2.	3.	4.	5.	6.	7.	8.
1.	2.	4.	8.	16.	32.	64.	128.	256.

Sicut ex additione (in ſuperiore ordine) 3 ad 5 fiunt 8, ſic (in inferiore ordine) ex multiplicatione 8 in 32 fiunt 256. Eſt autem 3 exponens ipſius octonarij, & 5 eſt exponens numeri 32. & 8 eſt exponens numeri 256. Item ſicut in ordine ſuperiori, ex ſubtractione 3 de 7, remanent 4, ita in inferiori ordine ex diuiſione 128 per 8, fiunt 16.

Sed oſtendenda eſt iſta ſpeculatio per exemplum.

−3	−2	−1	0	1	2	3	4	5	6
$\frac{1}{8}$	$\frac{1}{4}$	$\frac{1}{2}$	1	2	4	8	16	32	64

Fig. 6.3.1. Stifel's scales from the *'Arithmetica Integra'* of 1544

6.3.2 England at the Turn from the 16th to the 17th Century

King Henry VIII died in 1547 and not only left behind the foundations of a sea fleet and modern defence fortifications on the English coast, but also a country torn apart and deeply divided by religion. Since the pope refused his permit to the divorce of the first marriage with Catherine of Aragon Henry had parted his country from the Catholic Church in a long process and had founded a reformed Protestant state church, the Anglican Church. From his first marriage there sprang his daughter Mary, born in 1516. His second wife, Anne Boleyn, gave birth to a little girl in 1533. This girl will turn England into the leading sea power as Queen Elizabeth I. After Anne was pronounced guilty of adultery with five men in an unpleasant trial she was beheaded. The way was paved for a third marriage, this time with Jane Seymour in 1536. From this marriage sprang the long awaited male heir to the throne, Edward, but his mother sadly died one week after having given birth. Three further marriages followed of which the first was annulled (Anna of Cleves 1541), the second being terminated by another execution of the wife (Catherine Howard 1542), and the third with Catherine Parr was terminated by the death of the king.

Henry himself had fixed the line of succession and Edward came first, then Mary, and Elizabeth came third. But Edward was a child so that the Duke of Hertford took the power, cared for Edward being crowned as King Edward VI, and reigned as protector of the boy king. But Edward died of consumption already in 1553, being 16 years of age. After intense but unsuccessful attempts by the Protestant

Fig. 6.3.2. Henry VIII ([Painter: Hans Holbein the Younger, 1539/40] Walker Art Gallery, Liverpool) and his daughter Elizabeth I ([Painting probably by William Scrots, about 1546] Royal Collection, Windsor Castle, Windsor Berkshire), dominating English rulers in the 16th century

side to prevent the devoted Catholic Mary she became the legitimate Queen Mary I of England and Ireland. She was animated by a serious hatred of everything protestant and became finally known in history as *Bloody Mary*. However, when she announced her plans to marry Philip II of Spain she lost the approval of many English Catholics who feared the loss of English independence. An attempt to overthrow Mary and enthrone Elizabeth instead failed and put Elizabeth in grave danger since the hatred of her sister was now directed towards her. Only due to her wise behaviour could Elizabeth save her life but she had to spend two months in the Tower. Thereafter she remained under observation. There can be no doubt that her upright attitude during this crisis and her loyal adherence to the Protestant faith served for the admiration she experienced in wide parts of the population. In addition, many English people were horrified by the extent of cruelties taking place after the marriage of Mary and Philip II of Spain. Both wanted to re-catholicise the country quickly and thoroughly. In 1555 Mary was thought to be pregnant but it turned out to have been a phantom pregnancy only. Philip thereupon turned away from Mary. She now suffered from oedema and premonition of death. Shortly before her death she was reconciled with her half sister Elizabeth, died in 1558 and 25 years old Elizabeth became the new queen.

Fig. 6.3.3. Englishmen fighting against the Spanish Armada, 8th August 1588 ([Painting by Phillip James de Loutherbourg, 1796] National Maritime Museum, Greenwich Hospital Collection, London)

England became the leading sea power on earth under the reign of Elizabeth and not without reason the Elizabethan period is called *England's Golden Age* or *England's Heroic Age*. The highly gifted and educated young woman succeeded in creating a sense of atmosphere of departure in the country. Protestant emigrants now came back to England in fairly large numbers. A new fleet was planned and built. We all know of the famous English pirates of those days, among them Francis Drake, whose hustle and bustle against the Spanish galleons in the Pacific and the Atlantic Ocean was officially condemned by Elizabeth; but unofficially they sailed on behalf of the queen and had to share their plunder with the crown. In 1588 the attack of the Spanish Armada against England in the Channel resulted in a disaster for the Spanish who had to hand over their rank as leading sea power to the English.

In the Arts it was the great time of William Shakespeare and the theatre. In the sciences the *Golden Age* also led to a heyday. The mysterious John Dee (1527–1608/9) [Woolley 2001], [French 1972] traveled on the continent, made friends with Gerhard Mercator and got acquainted with the latter's new map projection [Scriba/Schreiber 2000], particularly suited for sailing, which he introduced in England. His preface to the edition of Euclid's *Elements* [Scriba/Schreiber 2000, section 2.3] by Henry Billingsley 1570 [Dee 1570] became famous where Dee presented all branches of the mathematics of his days in a diagram. As advisor to the queen he coined the notion of the 'British Empire' and if he had to send secret messages he signed them with '007', stimulating author Ian Fleming centuries later to call his secret agent James Bond also 007. The brilliant mathematician Thomas

Harriot (1560–1621) [Shirley 1983], [Alten et al. 2014, section 5.2] worked on problems of navigation and algebra. Recent research [Stedall 2003] has revealed that Harriot was an algebraist of the first rank who most likely influenced mathematicians on the Continent like René Descartes. William Gilbert (1540–1603), physician to the Queen, published his groundbreaking work *De Magnete, Magneticisque corporibus, et de magno magnete tellure; Physiologia nova, plurimis & argumentis, & experimentis demonstrata* concerning the theory of magnetism in 1600 in London and therewith laid the ground to English 'natural philosophy'. The book has become so famous that it is still in print in English translation [Gilbert 1958] and was in parts an outcome of a collaboration of several mathematicians at Gresham College, London, [Pumfrey 2002, p. 175], among them Edward Wright and Henry Briggs (1561–1631) who contributed important works concerning the theory of navigation [Sonar 2001].

The universities of Oxford and Cambridge were in a sad and pitiful condition [Hill 1997]. Aristotelean physics was taught, the geocentric structure of the cosmos, and Galen's medicine. Not a word could be heard on Copernicus and the heliocentric view of the world, not a word on sophisticated geometry and trigonometry and not a word on problems of navigation. Mathematicians in these days were hence attracted to London where Thomas Gresham had donated a college to compensate for the deficits of Oxford and Cambridge. But also the rural population was attracted to the towns. While the population of London counted approximately 100 000 at Elizabeth's inauguration this number had doubled in 1600 [Suerbaum 2003, p. 314]. In 1596 Henry Briggs became the first *Gresham professor of Geometry* before in 1619 he became the first *Savilian professor of Geometry* at Oxford due to a reform movement initiated by Henry Savile. He became famous by his works concerning logarithms.

Decisive preliminary work was done by a Scotsman in Merchiston close to Edinburgh who invented an early form of logarithms.

6.3.3 John Napier and His Logarithms

John Napier (1550–1617) [Havil 2014] was the eighth Laird (not to be confused with 'Lord'!) of Merchiston. As with Stifel he was an apocalyptic and published *A Plaine Discovery of the Whole Revelation of St. John* [Rice/González-Valesco/Corrigan 2017, pp. 97-390] in 1593 in which he meant to prove the pope was the Antichrist of the Revelation of John with (pseudo-)mathematical methods. He possessed and knew Stifel's *Arithmetica Integra* and he recognised that Stifel's scales would be much too coarse in the process of successive computations. If we denote the numbers on the geometric scale (cp. figure 6.3.1)

i	-1	0	1	2	3	4	5	6
g_i	1/2	1	2	4	8	16	32	64

by $g_i := 2^i$, i.e.

$$g_0 = 2^0 = 1, \quad g_1 = 2, \quad g_2 = 4, \quad \text{etc.},$$

then the quotient of two successive numbers is

Fig. 6.3.4. John Napier. Painting by an unknown artist. It was a present of Napier's grandchild to the University of Edinburgh 1616

$$\frac{g_{i+1}}{g_i} = 2.$$

Hence the geometric scale gets ever larger gaps as we move further on the scale. This problem can only be overcome by a scale where

$$\frac{g_{i+1}}{g_i} \approx 1$$

holds. Napier quite wisely chose $g_{i+1}/g_i = 0.9999999 = 1 - 10^{-7}$. He started with a table of the first 100 numbers of the form

$$10^7(1 - 10^{-7})^n, \quad n = 0, 1, 2, 3, \ldots, 100.$$

Napier called the number n the *logarithm* of the number $10^7(1 - 10^{-7})^n$. Now why was the choice of the number 10^7 wise? In the days of Napier the use of the decimal point just came into use and only few people used it. Hence all calculations had to be performable so that no decimal places occurred. The reference value of all computations in those days was the 'whole sine'. In trigonometrical calculations in right-angled triangles the whole sine always was the hypotenuse and Napier eventually aimed to apply his logarithm to computations in the realm of geometry and navigation (spherical trigonometry). If the whole sine is chosen significantly smaller than 10^7 computations neglecting the decimal places will go wrong; choosing it much larger will result in unnecessary work. Obviously Napier knew for sure what he did!

In 1614 he published his logarithmic table *Mirifici Logarithmorum Canonis Descriptio* (The description of the wonderful tables of logarithms) [Rice/González-Valesco/Corrigan 2017, pp. 475-647] which explained the use of the table but revealed nothing about how they were computed. That only became public after Napier's death when his son published *Mirifici Logarithmorum Canonis Constructio* (The construction of the wonderful table of logarithms) [Rice/González-Valesco/Corrigan 2017, pp. 751-808] in 1619.

In all questions concerning logarithms, whether Napier's or Briggs', we refer the reader strongly to the internet resources of one of the leading experts in this field, Denis Roegel, and his internet page `https://members.loria.fr/Roegel/`.

The Construction of Napier's Logarithms

The first number in the table ($n = 0$) is obvious: 10^7. For the second we have to compute $10^7(1-10^{-7})^1 = 10^7 - 10^7 10^{-7} = 10^7 - 1$. For the third we actually would have to compute $10^7(1-10^{-7})^2$ but Napier saw that all numbers in the table can be computed just by subtraction. This can be seen as follows: Assume that $10^7(1-10^{-7})^k$ is already computed. Then

$$\begin{aligned} 10^7(1-10^{-7})^{k+1} &= \underbrace{10^7(1-10^{-7})^k}_{\text{already computed!}} \cdot (1-10^{-7}) \\ &= \underbrace{10^7(1-10^{-7})^k}_{\text{already computed!}} - \underbrace{10^7(1-10^{-7})^k}_{\text{already computed!}} \cdot 10^{-7}, \end{aligned}$$

but multiplication by 10^{-7} is nothing but the shift of the decimal point seven places to the left. Hence the first table in Napier's *Descriptio* reads as in figure 6.3.5.

n	$10^7(1-10^{-7})^n$.
0	10000000.0000000
	-1.0000000
1	9999999.0000000
	-0.9999999
2	9999998.0000001
	....
100	9999900.0004950

Fig. 6.3.5. Napier's first table

Apparently 100 is Napier's logarithm of 99999000004950. Hence Napier's logarithm of a number x is nothing but the number of multiplications of 10^7 with $1 - 10^{-7}$ until x results. We write

$$y = \text{NapLog}x \quad :\Leftrightarrow \quad x = 10^7(1 - 10^{-7})^y. \tag{6.21}$$

If two logarithms $y, \tilde{y}$ are given,

$$x = 10^7(1 - 10^{-7})^y,$$
$$\tilde{x} = 10^7(1 - 10^{-7})^{\tilde{y}},$$

then their quotient will be

$$\frac{x}{\tilde{x}} = (1 - 10^{-7})^{y-\tilde{y}}.$$

The differences of logarithms are hence only depending on the ratio of x and $\tilde{x}$. That is where the name 'logarithm' comes from, built from the Greek words *logos* and *arithmos* and meaning something like 'quotient number'.

It is interesting to calculate how many steps John Napier had to compute to reduce the starting number 10^7 to half its value $5 \cdot 10^6$. We thus ask for the number of steps so that $(1 - 10^{-7})^n = \frac{1}{2}$ holds. Employing our Briggsian logarithm (base 10) it follows

$$n \log_{10}(1 - 10^{-7}) = -\log_{10} 2,$$

hence

$$n = -\frac{\log_{10} 2}{\log_{10}(1 - 10^{-7})} \approx 6931471.$$

Such an enormous number of steps would have doomed the computation of a practical useful logarithmic table to failure. But Napier recognised that

$$10^7(1 - 10^{-7})^{100} \approx 10^7(1 - 10^{-5})$$

holds! Hence he proceeded with a second table containing the numbers $10^7(1 - 10^{-5})^n, n = 0, 1, 2, \ldots, 50$. From the first and second table a third one can be constructed consisting of 21 rows and 69 columns. From this third table Napier finally interpolated his logarithms [Edwards 1979].

Napier's Kinematic Model

There are two obvious drawbacks to the computation of a logarithmic table with the method just described. Firstly, a computed logarithm depends on all of the logarithms previously computed. If Napier made an error (and in fact he made a few) then all logarithms following would be wrong. The second drawback lies in the chosen starting point itself. It holds

$$\text{NapLog}10^7 = 0$$

and $\text{NapLog}x_1 > \text{NapLog}x_2$ if $x_1 < x_2$. This is indeed not one of the logarithms we know today. But there is something even weirder: We have described a *discrete* version of Napier's logarithm only but we would need a *function* to work reasonably.

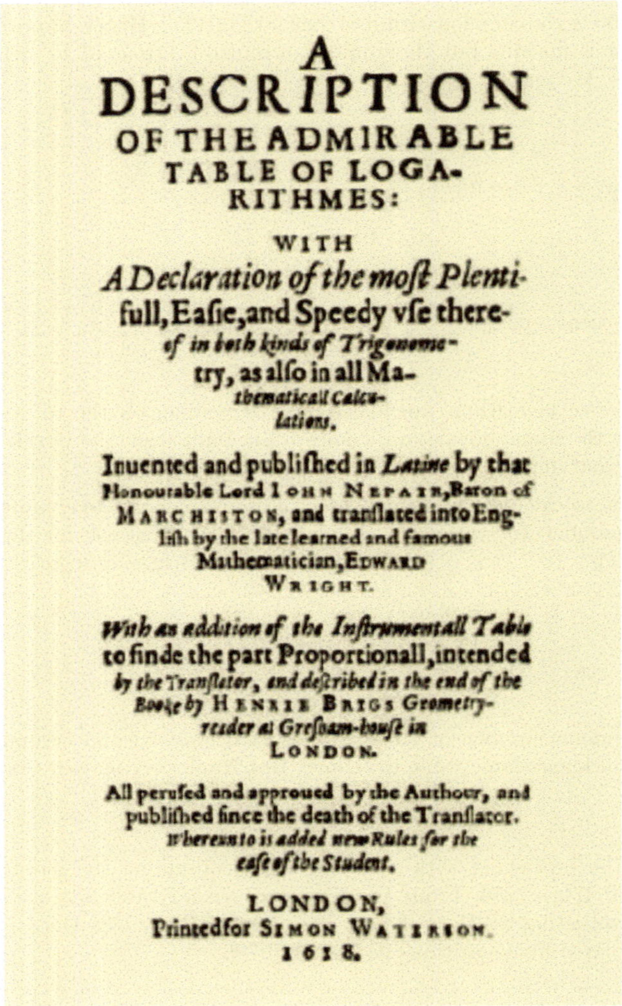

A
DESCRIPTION
OF THE ADMIRABLE
TABLE OF LOGA-
RITHMES:
WITH
A Declaration of the most Plenti-
full, Easie, and Speedy vse there-
of in both kinds of Trigonome-
try, as also in all Ma-
thematicall Calcu-
lations.

Inuented and published in Latine by that
Honourable Lord IOHN NEPAIR, Baron of
MARCHISTON, and translated into Eng-
lish by the late learned and famous
Mathematician, EDWARD
WRIGHT.

With an addition of the Instrumentall Table
to finde the part Proportionall, intended
by the Translator, and described in the end of the
Booke by HENRIE BRIGS Geometry-
reader at Gresham-house in
LONDON.

All perused and approued by the Authour, and
published since the death of the Translator.
Whereunto is added new Rules for the
ease of the Student.

LONDON,
Printed for SIMON WATERSON.
1618.

Fig. 6.3.6. Napier's *Descriptio* in an English translation of 1618 (translation by Edward Wright)

This interpretation as a *continuous* function was already given by Napier himself without having the modern notion of *function* at hand.

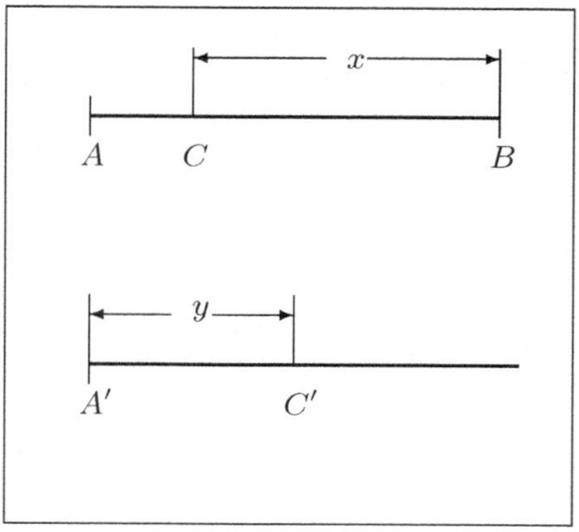

Fig. 6.3.7. Napier's kinematic model

John Napier exploited the idea of his logarithms by means of a kinematic model. He imagined a line segment AB on which a point C at A starts to move at time $t = 0$. Its velocity equals the distance CB everywhere, i.e. at $t = 0$ the velocity is $v_C(0) = AB$ and then the velocity gradually decreases, $v_C(t) = CB = x$. At the same time ($t = 0$) a point C' starts at A' to move on a segment extending to infinity to the right with constant velocity $v_{C'}(t) = AB$. Choosing the length of AB as $AB = 10^7$ one can show that this kinematic model exactly gives Napier's logarithms, i.e. $y = \text{NapLog}\, x$ [Phillips 2000, p. 60]. The kinematic model even served Napier as the very *definition* of his logarithms.

The kinematic model serves us to give a modern analysis of Napier's logarithms. The velocity of the point C obviously is the change of distance with respect to time. The distance apparently is $AC = AB - CB = 10^7 - x$ and the velocity equals the distance $CB = x$, hence

$$\frac{d}{dt}(10^7 - x) = x.$$

In the case of y things are even simpler since the velocity is constant. We have

$$\frac{dy}{dt} = AB = 10^7.$$

From the first equation it follows $\frac{dx}{dt} = -x$ which has the solution $x(t) = K\mathrm{e}^{-t}$ what, if one can not see it immediately, can be deduced by taking the derivative. Since $x(0) = AB = 10^7$ it follows $K = 10^7$, hence

$$x(t) = 10^7\mathrm{e}^{-t}. \tag{6.22}$$

The equation for y is even more simple to solve giving $y(t) = 10^7 t + k$ where the constant k evaluates to $k = 0$ since $y(0) = 0$, hence

$$y(t) = 10^7 t. \tag{6.23}$$

If we now apply the natural logarithm $\ln = \log_{\mathrm{e}}$ to (6.22) it follows

$$\ln x = \ln 10^7 - t \underbrace{\ln \mathrm{e}}_{=1} = \ln 10^7 - t,$$

hence

$$t = \ln 10^7 - \ln x = \ln\left(\frac{10^7}{x}\right).$$

Inserting this in (6.23) finally results in

$$y = \text{NapLog}\, x = 10^7 \ln\left(\frac{10^7}{x}\right). \tag{6.24}$$

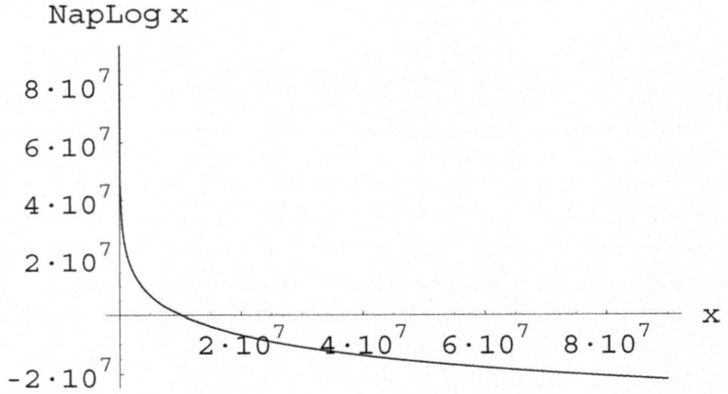

Fig. 6.3.8. The function NapLog $x = 10^7(\ln 10^7 - \ln x)$

The representation (6.24) as a function allows us to derive the functional equation of Napier's logarithm. One arrives at

$$\text{NapLog}\,(x_1 \cdot x_2) = \text{NapLog}\, x_1 + \text{NapLog}\, x_2 - \text{NapLog}\, 1.$$

And to make matters worse not even a 'neat' functional equation is valid but the value NapLog 1 has to be always subtracted.

Now it can also easily be proven that

$$\text{NapLog}\, x^n = n \cdot \text{NapLog}\, x + (1-n) \cdot \text{NapLog}\, 1$$

and the reader is asked to carry out the proof as an exercise.

From equation (6.24) we can even reveal the base of Napier's logarithm. It is part of the folklore of the history of mathematics that Napier's logarithm has often been mixed up with the natural logarithm. However, this is wrong. The base a is the number for which $y = \text{NapLog}\,(a) = 1$ holds. From (6.24) it follows

$$a = 10^7 \cdot \frac{1}{e^{10^{-7}}} = 10^7 \cdot \left(\frac{1}{e}\right)^{0.0000001}.$$

The Early Meaning of Napier's Logarithms

To judge the importance of the logarithms directly after Napier's first book was published in 1614 we refer to Johannes Kepler (1571–1630) who was one of the first users of Napier's logarithms. Without this computational tool his astronomical calculations would have been possible only with an enormous amount of work and time [Horsburgh 1982], if at all!

If one wanted to compute the product of two large numbers by repatriating to addition before Napier one employed the method of prosthaphaeresis [Toeplitz 2007, p. 86]. This method is based on the addition theorems

$$\begin{array}{r|l} & \cos(x+y) = \cos x \cdot \cos y - \sin x \cdot \sin y \\ + & \cos(x-y) = \cos x \cdot \cos y + \sin x \cdot \sin y \\ \hline & \cos(x+y) + \cos(x-y) = 2 \cdot \cos x \cdot \cos y, \end{array}$$

hence

$$\cos x \cdot \cos y = \frac{1}{2}\cos(x+y) + \frac{1}{2}\cos(x-y).$$

If two numbers A and B were to be multiplied one had to look up the angles x und y for which $\cos x = A$ and $\cos y = B$ in a sine table (which, of course, is also a cosine table). Now one looked up the cosines of $x + y$ and $x - y$ in the same table and computed the product $A \cdot B$ by means of the above formula. We leave the assessment of this method to Otto Toeplitz [Toeplitz 2007, p. 86]:

> *This is not bad for the purposes of astronomy and navigation, where sines and cosines are often multiplied, but it was slow; something simpler was needed.*

6.3.4 Henry Briggs and His Logarithms

Henry Briggs (1561–1631) was a Yorkshireman who studied at Cambridge and left the university in 1596 for London where he became the first professor of geometry at Gresham College. We have already pointed out the miserable state of the two English universities and the role played by Gresham College in the education of young, curious men and mariners. In a remarkabley short amount of time Briggs became the centre of a circle of Copernicans, among them the physician and natural philosopher[4] William Gilbert, the 'applied' mathematician and navigator Edward Wright, the instrument maker William Barlow, and the great populariser of scientific knowledge of his days, Thomas Blundeville [Hill 1997, p. 36]. Bevor Briggs learned about logarithms from Napier's *Descriptio* in 1614 he achieved important work in navigation; among others numerous astronomical tables

[4] The notion of 'natural philosophy', having emerged in England, denotes the natural sciences in a comprehensive and 'hard' sense. It has nothing whatsoever to do with the 19th century German philosophical school of thought called 'Naturphilosophie' to which many dreamers and despisers of natural sciences belonged.

in Edward Wright's *Certain Errors in Navigation*, widespread in many editions as from 1599. Concerning the determination of latitude by means of a new instrument, described for the first time in Gilbert's *De Magnete* of 1600, Briggs made relevant contributions [Sonar 2001, Chapter 2].

In the historical context his religious inclination is characteristic for this whole English period. He was a strict Presbyterian who supported the Puritans actively already during his days in Cambridge [Hill 1997, p. 52]. As well as the Catholics living in England and the members of the Anglican Church new religious currents had formed in the reformed camp. The Presbyterians refused the episcopalian constitution of the Anglican Church and were in resistance to the so-called independents (congregationalists). The notion of 'Puritans' emerged in mid 16th century as a mockery for strictly Calvinistic Protestants and the Presbyterians can be seen as a current within Puritanism. Strict Puritans refused any imagery; hence it takes no wonder that there are no paintings of Briggs. Out of the conflict between a kingship ogling with Catholicism and the Puritans the English Civil War will emerge in the middle of the 17th century. From the friendship between Briggs and the Puritan James Usher, later the Archbishop of Armagh in Ireland, stems a correspondence not only on religious themes but also revealing some details of Briggs' scientific works. Among other information we learn that he studied Kepler's cosmology and of his first encounter with Napier's *Descriptio* [Sonar 2001, pp. 28–29].

Briggs was also involved in the destinies of seafaring. He was a member of the Virginia Company, was very interested in the search for the Northwest passage, and he even draw a map of North America – the first in which California is wrongly depicted as an island.

Briggs' life changed when he got hold of Napier's *Descriptio* in 1614. He immediately recognised not only the value but also the problems of Napier's logarithms. Already in the summer of 1615 he undertook the exhausting journey to Merchiston Castle where he was warmly received by Napier [Gibson 1914]. Briggs suggested the construction of logarithms for which $\log 1 = 0$ holds. Napier is said to have agreed immediately and perhaps he himself had noted the shortcomings of his logarithms. Briggs' idea is based on the consequent use of base 10:

$$\begin{aligned} 10^0 &= 1 &\leftrightarrow\quad& \log_{10} 1 = 0\\ 10^1 &= 10 &\leftrightarrow\quad& \log_{10} 10 = 1\\ 10^2 &= 100 &\leftrightarrow\quad& \log_{10} 100 = 2 \end{aligned}$$

and so introduced the decadic logarithm which we denote simply by log instead of $\log_{10}$ from now on.

It is worthwhile and instructive to follow the construction of the Briggsian logarithms. For one we see here the birth of one of the founding pillars of early analysis by means of elementary mathematics, namely the calculus of finite differences, secondly the existing literature is somewhat sparse and in some cases like [Goldstine 1977] difficult to understand. A positive exception is [Phillips 2000, p. 65–71] and we shall follow the exposition given therein.

The Construction Idea of Briggsian Logarithms

In expanding the above table we can write

$$10^{\frac{1}{2}} = \sqrt{10} \quad \leftrightarrow \quad \log \sqrt{10} = \frac{1}{2}$$
$$10^{\frac{1}{4}} = \sqrt{\sqrt{10}} \quad \leftrightarrow \quad \log \sqrt{\sqrt{10}} = \frac{1}{4}$$
$$10^{\frac{1}{8}} = \sqrt{\sqrt{\sqrt{10}}} \quad \leftrightarrow \quad \log \sqrt{\sqrt{\sqrt{10}}} = \frac{1}{8}$$
$$10^{\frac{1}{16}} = \sqrt{\sqrt{\sqrt{\sqrt{10}}}} \quad \leftrightarrow \quad \log \sqrt{\sqrt{\sqrt{\sqrt{10}}}} = \frac{1}{16}$$

and so on. That inspired Briggs to construct a logarithmic table on the basis of successive root extractions.

Extracting the root of 10 n successive times, $\sqrt{\ldots\sqrt{10}}$, then this is equivalent to $10^{\frac{1}{2^n}}$. In case of 'large' n the term $10^{\frac{1}{2^n}}$ should be close to 1, since

$$\lim_{n\to\infty} 10^{\frac{1}{2^n}} = 1.$$

By means of numerical experiments (Briggs must have worked many days and nights on this!) he recognised that the logarithm of $1 + x$ divided by x is constant for 'small' values of x. What does that mean? He probably made a list as the one shown in table 6.1.

We want to postpone the discussion of the way Briggs computed the many roots (without any mechanical tools, let alone electronic ones), but he must have computed much further and with many more digits, to see that

$$\lim_{x\to 0} \frac{\log(1+x)}{x} = K \tag{6.25}$$

with a constant K which Briggs calculated to be

$$K \approx 0.4342944819032518. \tag{6.26}$$

We know today that two logarithms to different bases can be transformed into each other by

$$\log_b x = \log_b a \cdot \log_a x.$$

Thus we can write $\log(1+x) = \log \mathsf{e} \cdot \ln(1+x)$; remembering that log always means $\log_{10}$. Additionally we today know a series representation of the function named after Nicholas Mercator (1620–1687), namely

$$\ln(1+x) = x - \frac{x^2}{2} + \frac{x^3}{3} - \frac{x^4}{4} + \frac{x^5}{5} - \frac{x^6}{6} + \ldots, \tag{6.27}$$

which is easily verified using the techniques of Taylor series'[5]. Hence it holds $\frac{\ln(1+x)}{x} = 1 - \frac{x}{2} + \frac{x^2}{3} - \ldots$ and thus $\lim_{x\to 0} \frac{\ln(1+x)}{x} = 1$ follows. Inserting our transformation rule for logarithms yields $\lim_{x\to 0} \frac{\log(1+x)}{\log \mathsf{e}}/x = 1$, and, since $\log \mathsf{e}$ does not depend on x,

[5] In section 6.3.8 we describe Mercator's role in the development and understanding of logarithms.

$1+x$	$\log(1+x)$	$\frac{\log(1+x)}{x}$
10	1	0.111
$\sqrt{10} \approx 3.162278$	$\frac{1}{2}$	0.231
$\sqrt{\sqrt{10}} \approx 1.778279$	$\frac{1}{4}$	0.321
$\sqrt{\sqrt{\sqrt{10}}} \approx 1.333521$	$\frac{1}{8}$	0.375
$\sqrt{\sqrt{\sqrt{\sqrt{10}}}} \approx 1.154782$	$\frac{1}{16}$	0.404
$\sqrt{\sqrt{\sqrt{\sqrt{\sqrt{10}}}}} \approx 1.074608$	$\frac{1}{32}$	0.419
$\sqrt{\sqrt{\sqrt{\sqrt{\sqrt{\sqrt{10}}}}}} \approx 1.036633$	$\frac{1}{64}$	0.427
$\sqrt{\sqrt{\sqrt{\sqrt{\sqrt{\sqrt{\sqrt{10}}}}}}} \approx 1.018152$	$\frac{1}{128}$	0.430
$\sqrt{\sqrt{\sqrt{\sqrt{\sqrt{\sqrt{\sqrt{\sqrt{10}}}}}}}} \approx 1.009035$	$\frac{1}{256}$	0.432
$\sqrt{\sqrt{\sqrt{\sqrt{\sqrt{\sqrt{\sqrt{\sqrt{\sqrt{10}}}}}}}}} \approx 1.004507$	$\frac{1}{512}$	0.433
$\sqrt{\sqrt{\sqrt{\sqrt{\sqrt{\sqrt{\sqrt{\sqrt{\sqrt{\sqrt{10}}}}}}}}}} \approx 1.002251$	$\frac{1}{1024}$	0.434

Table 6.1. Successively extracting the roots of 10

$$\lim_{x \to 0} \frac{\log(1+x)}{x} = \log \mathsf{e}.$$

Hence the Briggsian constant is just $K = \log \mathsf{e}$, but that Briggs could not know, of course.

But if he once knew the value of K to 16 digits accuracy he was able to compute the logarithm of $1+x$ for small x thanks to (6.25):

$$\log(1+x) \approx Kx.$$

We are now in the position to describe Briggs' construction of his logarithmic table. Briggs computed only logarithms of prime numbers since every natural number can be represented uniquely by its prime number decomposition. For example it holds $60 = 2^2 \cdot 3 \cdot 5$ and therefore $\log 60 = 2\log 2 + \log 3 + \log 5$. For each prime number p Briggs extracts successive roots until the n-th root can be written as

$$p^{\frac{1}{2^n}} = 1 + x,$$

where $x \approx 10^{-16}$. It then follows

$$\log\left(p^{\frac{1}{2^n}}\right) = \frac{1}{2^n}\log p = \log(1+x) \approx Kx,$$

and hence the logarithm of p is

$$\log p \approx 2^n Kx.$$

We can easily check that Briggs must have taken approximately fifty (!) successive roots to make the construction described above possible. How did he do it? Surely *not* by means of an iterative algorithm like the method of Hero [Alten et al. 2005, section 1.3.6], [Alten et al. 2014, p. 42f.] since then he would have had no chance to ever finish! In answering the question of how he did it we clearly see a genius at work. He employed the calculus of finite differences.

The Successive Extraction of Roots

Henry Briggs counts as one of the inventors of the calculus of differences [Goldstine 1977, p. 13]; the other is his fellow countryman Thomas Harriot whom we have to discuss later. Both used very similar techniques of interpolation but whether the two men ever met and Briggs took a method of Harriot can not be revealed [Beery/Stedall 2009].

The bottleneck of his algorithm surely is the extraction of many successive roots and Briggs must have thought about it intensely. How can an algorithm be economical enough? During his numerical experiments he must have laboriously extracted the first six or seven roots (probably by a direct method by hand), like we did in table 6.2 for p=3; but he computed 30 digits! Briggs must have known that the computation following these lines would never have resulted in a usable table of logarithms. At least he was well into his fifties when he first came in contact with logarithms! But now he showed a mastership which is common in the work of truly brilliant mathematicians: He sees a pattern in table 6.2! Forget the digit left of the decimal point since the logarithm does not depend on the decimal point. This can be seen for example from $\log_{10} 5.1 = \log_{10} \frac{51}{10} = \log_{10} 51 - \log_{10} 10 = \log_{10} 51 - 1$

$$\begin{aligned}
3^1 &= 3.000000000\\
\sqrt{3} = 3^{1/2} &\approx 1.732050808\\
\sqrt{\sqrt{3}} = 3^{1/4} &\approx 1.316074013\\
\sqrt{\sqrt{\sqrt{3}}} = 3^{1/8} &\approx 1.147202690\\
\sqrt{\sqrt{\sqrt{\sqrt{3}}}} = 3^{1/16} &\approx 1.071075483\\
\sqrt{\sqrt{\sqrt{\sqrt{\sqrt{3}}}}} = 3^{1/32} &\approx 1.034927767\\
3^{1/64} &\approx 1.017313996\\
3^{1/128} &\approx 1.008619847
\end{aligned}$$

Table 6.2. The first seven successive roots of 3

and users of a slide rule know that one has to think about the decimal point only *after* the computation.

Without the leading digit the table 6.3 results from table 6.2. In the second column Briggs recognised that every entry is approximately half of its predecessor. This can not be seen from the second entry since (0.)732050808 is considerably larger than the half of its predecessor (1.)000000000, but somewhat further down the accordance gets much better. To see how good the idea of halving really is Briggs computed the *first Briggsian difference*

$$B_1^n := \frac{1}{2}Z(n) - Z(n+1).$$

Therewith we compute

$$\begin{aligned} B_1^0 &= \frac{1}{2}Z(0) - Z(1) = 0000000000 - 732050808 = 267949192 \\ B_1^1 &= \frac{1}{2}Z(1) - Z(2) = \frac{1}{2}732050808 - 316074013 = 49951391 \\ &\vdots \quad \vdots \quad \vdots, \end{aligned}$$

which we want to note in table 6.4. Viewing the column of the first Briggsian differences in table 6.4 Briggs noted that every entry is roughly one quarter of the preceeding one! To check this idea he computed the *second Briggsian difference*

$$B_2^n := \frac{1}{4}B_1^n - B_1^{n+1},$$

which we write down in table 6.5. It is now obvious how it proceeds further on. The second Briggsian differences are each one eighth of the preceeding value and give rise to the definition of the *third Briggsian difference*

$$B_3^n := \frac{1}{8}B_2^n - B_2^{n+1},$$

shown in table 6.6.

We want to consider even one further difference, the *fourth Briggsian difference*

n	Roots	$Z(n)$
0	3^1	0000000000
1	$3^{1/2}$	732050808
2	$3^{1/4}$	316074013
3	$3^{1/8}$	147202690
4	$3^{1/16}$	71075483
5	$3^{1/32}$	34927767
6	$3^{1/64}$	17313996
7	$3^{1/128}$	8619847

Table 6.3. The first seven successive roots of 3 without the leading digit

$$B_4^n := \frac{1}{16} B_3^n - B_3^{n+1}.$$

This results in table 6.7 and now Henry Briggs has finally reached his aim! The hour of birth of the calculus of differences is marked in the final column by a box surrounding a zero! All further entries in this column will be zero since the numbers decrease from top to bottom. Hence also $B_4^4 = B_4^5 = 0$ follows. To make things clear once more we have named the differences in table 6.8 explicitly. In bold font we see the differences which could not have been computed with our data since futher roots woud have been necesssary. But now we can fill the logarithmic table backwards from right to left without ever having to extract a root! In the column of the third differences the entry B_3^5 is unknown, but we know, that

$$B_4^4 = \frac{1}{2^4} B_3^4 - B_3^5$$

has to hold. From this equation we can compute B_3^5:

$$B_3^5 = \frac{1}{2^4} B_3^4 - B_4^4 = \frac{1}{16} 5 - 0 = 0,$$

since 5/16 falls below our accuracy (we fall below 1). Looking one column to the left the difference B_2^6 is unknown, but we know that

$$B_3^5 = \frac{1}{2^3} B_2^5 - B_2^6$$

has to hold. Hence B_2^6 can be computed from

$$B_2^6 = \frac{1}{2^3} B_2^5 - B_3^5 = \frac{1}{8} 321 - 0 = 40,$$

where $321/8 = 40.125$ and 0.125 falls below our accuracy. But now also B_1^7 is computable, since from

n	$Z(n)$	B_1
0	0000000000	
		267949192
1	732050808	
		49951391
2	316074013	
		10834317
3	147202690	
		2525862
4	71075483	
		609975
5	34927767	
		149888
6	17313996	
		37151
7	8619847	

Table 6.4. The first Briggsian differences

n	$Z(n)$	B_1	B_2
0	0000000000		
		267949192	
1	732050808		17035907
		49951391	
2	316074013		1653531
		10834317	
3	147202690		182717
		2525862	
4	71075483		21491
		609975	
5	34927767		2606
		149888	
6	17313996		321
		37151	
7	8619847		

Table 6.5. The first and second Briggsian differences

n	$Z(n)$	B_1	B_2	B_3
0	0000000000			
		267949192		
1	732050808		17035907	
		49951391		475957
2	316074013		1653531	
		10834317		23974
3	147202690		182717	
		2525862		1349
4	71075483		21491	
		609975		80
5	34927767		2606	
		149888		5
6	17313996		321	
		37151		
7	8619847			

Table 6.6. The first, second, and third Briggsian differences

$$B_2^6 = \frac{1}{2^2}B_1^6 - B_1^7$$

it follows

$$B_1^7 = \frac{1}{2^2}B_1^6 - B_2^6 = \frac{1}{4}37151 - 40 = 9248,$$

since $\frac{1}{4}37151 - 40 = 9247.75$, rounded up to 9248. Hence we have arrived at the column for Z and can now compute $Z(8)$ from

$$B_1^7 = \frac{1}{2}Z(7) - Z(8)$$

n	$Z(n)$	B_1	B_2	B_3	B_4
0	0000000000				
		267949192			
1	732050808		17035907		
		49951391		475957	
2	316074013		1653531		5773
		10834317		23974	
3	147202690		182717		149
		2525862		1349	
4	71075483		21491		4
		609975		80	
5	34927767		2606		$\boxed{0}$
		149888		5	
6	17313996		321		
		37151			
7	8619847				

Table 6.7. The first, second, third, and fourth Briggsian differences

to give

$$Z(8) = \frac{1}{2}Z(7) - B_1^7 = \frac{1}{2}8619847 - 9248 = 4300676,$$

since $\frac{1}{2}8619847 - 9248 = 4300675.5$.

We have now reached table 6.9. The entries in italics are those which we have computed backwards from $B_4^4 = 0$.

We now proceed to the entry $B_4^5 = 0$ and compute backwards until $Z(9)$ is determined, and so on. For every prime number considered Henry Briggs had to compute only a few roots by hand; the rest of the work was done by his brilliant technique of the calculus of differences.

Was Briggs' Difference Calculus Stolen From Bürgi?

We have already reported on the Swiss mathematician, maker of fine instruments and clocks, Jost Bürgi, starting on page 185, and on a recently found manuscript on page 187. In 2013 the German historian of mathematics Menso Folkerts realised from a manuscript in the University Library of Wroclaw in Poland that this was a work of Bürgi's which was supposed to be lost. This manuscript is the *Fundamentum Astronomiæ* and it contained an ingenious algorithm, called 'artificium' (skilfull/ artful method), to efficiently compute sine tables of all subdividing angles [Folkerts 2014], [Folkerts/Launert/Thom 2016], [Launert 2015, p. 46ff.], [Waldvogel 2016]. Although the technique used has nothing whatsoever in common with the technique of the Briggsian differences which we have described in detail, there were rumours about some similarities in the use of differences of higher order in Briggs' *Trigonometria Britannica* and the algorithm in Bürgi's *Fundamentum Astronomiæ*. Bürgi was a good friend of the astronomer Nicolaus Reimers Baer (Reimarus Ursus), called

n	$Z(n)$	B_1	B_2	B_3	B_4
0	0000000000				
		$267949192 = B_1^0$			
1	732050808		$17035907 = B_2^0$		
		$49951391 = B_1^1$		$475957 = B_3^0$	
2	316074013		$1653531 = B_2^1$		$5773 = B_4^0$
		$10834317 = B_1^2$		$23974 = B_3^1$	
3	147202690		$182717 = B_2^2$		$149 = B_4^1$
		$2525862 = B_1^3$		$1349 = B_3^2$	
4	71075483		$21491 = B_2^3$		$4 = B_4^2$
		$609975 = B_1^4$		$80 = B_3^3$	
5	34927767		$2606 = B_2^4$		$\boxed{0 = B_4^3}$
		$149888 = B_1^5$		$5 = B_3^4$	
6	17313996		$321 = B_2^5$		$\mathbf{0 = B_4^4}$
		$37151 = B_1^6$		$\mathbf{B_3^5}$	
7	8619847		$\mathbf{B_2^6}$		$\mathbf{0 = B_4^5}$
		$\mathbf{B_1^7}$		$\mathbf{B_3^6}$	
8	$\mathbf{Z(8)}$		$\mathbf{B_2^7}$		
		$\mathbf{B_1^8}$			
9	$\mathbf{Z(9)}$				

Table 6.8. Successive Briggsian differences

Ursus (the bear), and it is one of Ursus' publications which was thought to shed a new light on the connection of the difference techniques of Briggs' with Bürgi's 'artificium'. In the University Library of Leiden a copy of Ursus' book *Fundamentum Astronomicum* of 1588 can be found which was thoroughly annotated by different hands. We follow the report given by Launert [Launert 2015, p. 55].

The first owner of this book was one Daniel Moller of Mollenberg (1544–1600) of whom we know little. It came into the possession of Leiden University in 1690 from the bookshelves of Isaac Vossius (also anglicised as Voss) (1618–1689), a Leiden-born Dutch scholar and collector of manuscripts and books. He became the court librarian of Queen Christina of Sweden in 1648 and left Sweden in 1654. He became a member of the Royal Society in 1664 and went to England in 1670 where he most likely acquired Ursus' book and where he died in 1689.

Most of the annotations in Ursus' book are in English. It is well possible that some remarks stem from Voss' hand but Voss was a philologist and librarian and many annotations are concerned with medical recipes. Hans van de Velde, librarian at the University Library of Leiden, has suggested that these remarks and annotations stem from John Bainbridge, English astronomer, who worked as a physician in Ashby and London before he studied astronomy, and were written later than 1618, probably later than 1621. In 1619 he became the first Savilian Professor of Astronomy at Oxford University and hence was a direct colleague of Henry Briggs who was the first Savilian Professor of Geometry. Both men were buried in Merton College Chapel. That Bainbridge annotated in English instead

n	$Z(n)$	B_1	B_2	B_3	B_4
0	0000000000				
		$267949192 = B_1^0$			
1	732050808		$17035907 = B_2^0$		
		$49951391 = B_1^1$		$475957 = B_3^0$	
2	316074013		$1653531 = B_2^1$		$5773 = B_4^0$
		$10834317 = B_1^2$		$23974 = B_3^1$	
3	147202690		$182717 = B_2^2$		$149 = B_4^1$
		$2525862 = B_1^3$		$1349 = B_3^2$	
4	71075483		$21491 = B_2^3$		$4 = B_4^2$
		$609975 = B_1^4$		$80 = B_3^3$	
5	34927767		$2606 = B_2^4$		$0 = B_4^3$
		$149888 = B_1^5$		$5 = B_3^4$	
6	17313996		$321 = B_2^5$		$0 = B_4^4$
		$37151 = B_1^6$		$\mathit{0} = B_3^5$	
7	8619847		$\mathit{40} = B_2^6$		$\mathbf{0} = \mathbf{B_4^5}$
		$\mathit{9241} = B_1^7$		$\mathbf{B_3^6}$	
8	*4300676*		$\mathbf{B_2^7}$		
		$\mathbf{B_1^8}$			
9	$\mathbf{Z(9)}$				

Table 6.9. The backward computation of roots from the Briggsian differences

of Latin seems natural since students mocked his weak Latin abilities with the following mocking verse [Fauvel/Flood/Wilson 2000, p. 83]:

Dr. Bambridge
Came from Cambridge
To read *de Polis et Axis*;
Let him go back again
Like as a dunce he came,
And learn a new syntaxis.

Be it as it may, the main author of the annotations also used the insert sheets of Ursus' book for computations, and in fact a table relating to Bürgi's *artificium* can be found. In the middle of the page the annotator has written 'H. Briggs', probably to mention the name of the person who had the idea of using multiple differences, as Launert suggests [Launert 2015, p. 55f.].

In an attempt to declare Bürgi the sole inventor of the difference method, Fritz Staudacher, in his latest edition of his book [Staudacher 2018, p. 220ff.] on Bürgi, construed a hair-raising conspiracy theory. In this theory John Dee, working as an English spy on the Continent, made contact with Bürgi during a week in June 1589 in which Dee stayed in Kassel with his family (on page 223 of [Staudacher 2018] many other possible conspirators are listed, but Dee appears to be the main suspect). Somehow Dee is suspected having taken the idea of the 'artificium' to England where Briggs got to know it and only then was able to apply his difference method to construct his logarithms. One of Staudacher's evidences can be found in the second volume of Anton von Braunmühl's book *Vorlesungen über Geschichte der Trigonometrie* (Lectures on the History of Trigonometry) [Braunmühl 1900–1903] in which the author claimed [Braunmühl 1900–1903, Teil II, p. 28]:

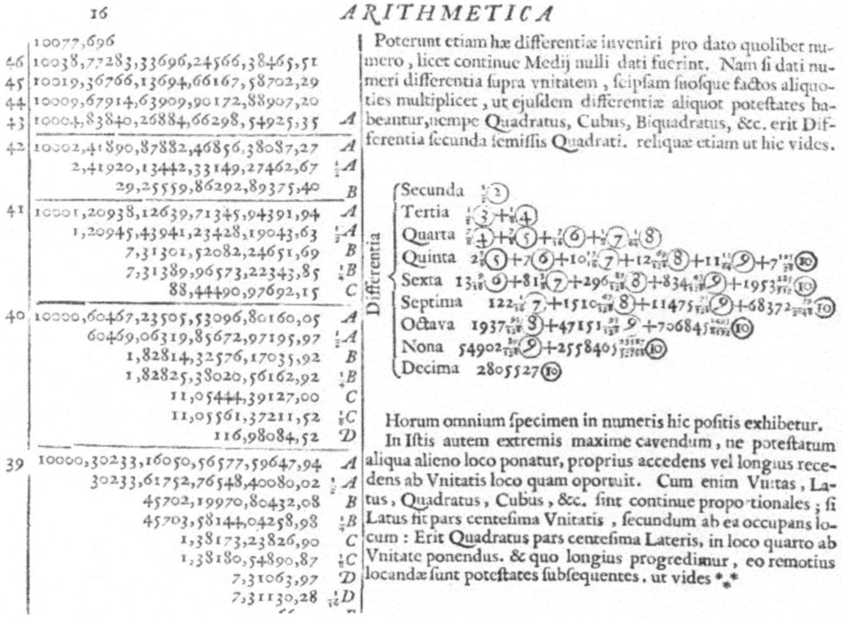

16 *ARITHMETICA*

	10077,696	
46	10038,77283,33696,24566,38465,51	
45	10019,36766,13694,66167,58702,29	
44	10009,67914,63909,90172,88907,20	
43	10004,83840,26884,66298,54925,35	A
42	10002,41890,87882,46856,38087,27	A
	2,41920,13442,33149,27462,67	½A
	29,25559,86292,89375,40	B
41	10001,20938,12639,71345,94391,94	A
	1,20945,43941,23428,19043,63	½A
	7,31301,52082,24651,69	B
	7,31389,96573,22343,85	¼B
	88,44490,97692,15	C
40	10000,60467,23505,53096,80160,05	A
	60469,06319,85672,97195,97	½A
	1,82814,32576,17035,92	B
	1,82825,38020,56162,92	¼B
	11,05444,39127,00	C
	11,05561,37211,52	⅛C
	116,98084,52	D
39	10000,30233,16050,56577,59647,94	A
	30233,61752,76548,40080,02	½A
	45702,19970,80432,08	B
	45703,58144,04258,98	¼B
	1,38173,23826,90	C
	1,38180,54890,87	⅛C
	7,31063,97	D
	7,31130,28	$\frac{1}{16}$D

Poterunt etiam hæ differentiæ inveniri pro dato quolibet numero, licet continue Medij nulli dati fuerint. Nam si dati numeri differentia supra vnitatem, seipsam suosque factos aliquoties multiplicet, ut ejusdem differentiæ aliquot potestates habeantur, nempe Quadratus, Cubus, Biquadratus, &c. erit Differentia secunda semissis Quadrati. reliquæ etiam ut hic vides.

Differentia:
- Secunda ½(2)
- Tertia ½(3)+⅛(4)
- Quarta [illegible](4)+[illegible](5)+$\frac{7}{16}$(6)+[illegible](7)+$\frac{1}{64}$(8)
- Quinta 2[illegible](5)+7(6)+10$\frac{11}{16}$(7)+12$\frac{69}{128}$(8)+11$\frac{11}{64}$(9)+7[illegible](10)
- Sexta 13$\frac{9}{16}$(6)+81[illegible](7)+296[illegible](8)+834[illegible](9)+1953[illegible](10)
- Septima 122$\frac{1}{16}$(7)+1510[illegible](8)+11475[illegible](9)+68372[illegible](10)
- Octava 1937[illegible](8)+47151[illegible](9)+706845[illegible](10)
- Nona 54902[illegible](9)+2558405[illegible](10)
- Decima 2805527(10)

Horum omnium specimen in numeris hic positis exhibetur.

In Istis autem extremis maxime cavendum, ne potestatum aliqua alieno loco ponatur, proprius accedens vel longius recedens ab Vnitatis loco quam oportuit. Cum enim Vnitas, Latus, Quadratus, Cubus, &c. sint continue proportionales; si Latus sit pars centesima Vnitatis, secundum ab ea occupans locum: Erit Quadratus pars centesima Lateris, in loco quarto ab Vnitate ponendus. & quo longius progredimur, eo remotius locandæ sunt potestates subsequentes. ut vides ***

Fig. 6.3.9. The computation of Briggsian differences in his *Arithmetica Logarithmica* of 1624

'Asking how Briggs came to his method we are, by the manner in which he arranged his difference tables, involuntarily reminded of the scheme which Reimers had published in his Fundamentum astronomicum and which he annotated with mysterious remarks [...]. *Either Briggs, as seems most likely to us, in tracing the meaning of these remarks has seen through Bürgi's method, or further details had come to him from some unknown source.'*

(Fragt man, wie Briggs zu seiner Methode gekommen ist, so wird man aus der Art und Weise, wie er seine Differenzentabellen anordnet, unwillkürlich an das Schema erinnert, das einst Reimers mit jenen mysteriösen Andeutungen versehen in seinem Fundamentum astronomicum veröffentlichte [...]. Entweder hat Briggs, was uns das wahrscheinlichste dünkt, dem Sinne dieser Andeutungen nachspürend, das Wesen von Bürgis Methode durchschaut, oder es kamen ihm aus irgend einer unbekannten Quelle nähere Mitteilungen über dieselbe zu.)

Von Braunmühl, however, was *not* writing about the tables in the *Arithmetica Logarithmica*, but on a table of sines which Briggs had included in his *Trigonometrica Britannica*, published in Gouda as late as 1633, hence three years after Briggs's death.

Peter Ullrich meanwhile has fully refuted Staudacher's hair-raising theory[6]. The *artificium* is definitely something different from the method of Briggs. In his book

[6] Private communication 2019. A publication is in preparation.

Staudacher returns four times to the visit of Dee in Kassel in 1586, each time accusing Dee of spying on Bürgi's *artificium*. What Staudacher does not mention is the first visit of Dee in Kassel having taken place in July 1586. Since it is accepted that Bürgi had the idea leading to the *artificium* sometime in 1586 or 1587 it may well be that it was Dee who took Briggsian ideas to Bürgi and hence enabled Bürgi to come up with his *artificium*. In [Staudacher 2018, p. 221] Staudacher is suspicious that Briggs gave his difference method without explaining how he constructed it. But the same is true of Bürgi, of course.

The Early Invention of the Binomial Theorem

Briggs had always to extract roots of $(1+x)$. In our notation introduced above we compute successive roots by $(1+Z(j)) = \sqrt{1+Z(j-1)}$. It is easy to show that Briggs had found a special case of the binomial theorem as a kind of side-effect of his work [Phillips 2000, p. 71], namely

$$(1+x)^{\frac{1}{2}} = 1 + \frac{1}{2}x - \frac{1}{8}x^2 + \frac{1}{16}x^3 - \frac{5}{128}x^4 + \dots.$$

Today we attribute the general binomial theorem to Newton who kept a copy of Briggs' logarithmic table [Briggs 1976] in his library.

6.3.5 England in the 17th Century

Queen Elizabeth I died on the 24th March 1603 aged 69. In 1603 the reign of the Stuarts started in England which will last until 1714. It is a time of upheaval, in politics as in mathematics. James VI (1566–1625), King of Scotland, who reigned from 1603 to 1625 was the son of Mary Stuart who was the Scottish Queen and was beheaded under Elizabeth I in 1587. James was raised as a Protestant. When a plan to free his mother from English imprisonment with French help was defeated by a Protestant Lord in 1583 James began to negotiate with Elizabeth to secure his claim to the throne. The Treaty of Berwick in 1586 united England and Scotland in a strategic alliance and finally was the foundation for a smooth succession of James to the throne [Haan/Niedhart 2002, p. 149f.]. He became James I in 1603. He strived for absolute power but his foreign politics can only be called weak. Under his personal union of England and Scotland the influence of the lower gentry in Parliament was raised strongly. At the same time a religious current towards Puritanism swept through the country which created a strong contrast to the Church of England due to claims to clean the Christian doctrine and the religious service from any Catholic influence. Many Puritans left the country and emigrated to the new English settlements in North America. In the year 1640 we already find approximately 25000 settlers in New England. This time of religious tensions falls in the reign of James' successor Charles I (1600–1649) who ruled from 1625 until his death. Charles showed strong absolutistic tendencies and was very much inclined to Catholicism. During the days of the 3rd Parliament from 1627 to 1629 a political polarisation of the nation took place between *court* and *country* [Haan/Niedhart 2002, p. 163]. Due to the war against Spain and France the king consistently demanded higher taxes.

Fig. 6.3.10. England's rulers after the death of Elizabeth I: James I [Painting: Paul von Somer] (© Museo Nacional del Prado, Madrid), Karl I. ([Painting: Anthonis van Dyck, 1636, Studio version of an original which was often copied] Royal Collection Windsor Castle), Oliver Cromwell [Painting by Robert Walker, 1650] (Sotheby's London, Ökumenisches Heiligenlexikon), Charles II [Painting by Peter Lely 1675] (Collection of Euston Hall, Suffolk, Belton House, Lincolnshire)

The parliamentarians answered with the Petition of Rights in 1628 in which they in turn demanded parliamentary rights on the basis of the Magna Charta.

From the point of view of the parliamentarians the king had violated applicable law already at the beginning of the war. Although the king had signed the Petition of Rights he did not feel responsible to act accordingly. This led to further resentment and to the Protestation of the Commons in 1629. This protestation was also made public. Thereupon Charles resolved Parliament without further ado and reigned for eleven years without it in an absolutist manner. He only reinstated Parliament in 1640 to demand money for the violent introduction of the episcopal church and the prayer book in Scotland. Then he overhastily resolved Parliament again. But when the Scottish invaded North England and peace became possible only in return of daily payments the king reinstated Parliament again; but now radical reformers held the majority. Charles thereupon tried to incorporate the moderate members of Parliament into the government and to arrest the five most radical Parliamentarians, but this attempt failed. The king now had to fear for life and limb and fled London on 10th January 1642. It is this situation which finally led to the Civil War.

In the eyes of philosopher Thomas Hobbes (1588–1679) the period between 1640 and 1660 was the climax of this era. The fights for power in state and society provided him with 'an overview of all kinds of injustices and follies which the world could ever afford' [Haan/Niedhart 2002, p. 167]. The famous classical dictum used by Hobbes: 'Man to Man is an arrant Wolfe'[7] can be understood only in view of the riots of the Civil War. His main philosophical work *Leviathan* was published 1651 in the English language and declared man to be a wretched warrior against

[7] This dictum goes back to Plautus (approx. 254–184 BC) who in his play 'Asinaria' wrote: 'Lupus est homo homini, non homo, quom qualis sit non novit' ('A man is a wolf, not a man, to another man which he hasn't met yet').

every other man, since in the state of nature the only concerns are survival and the personal benefit. Therefore it should be the task of the state to care for security and protection. Man therefore gives all power to this one instance, the state in the form of a leviathan, and they pay the price by the surrender of their free will.

Between Hobbes and John Wallis a 'war' concerning the quadrature of the circle started to develop because Hobbes thought he had succeeded in the task. However, Hobbes as a mathematician was no match for Wallis [Jesseph 1999]. Although Hobbes had to leave the country in 1640 due to his royalist attitudes he departed ever further from the position of the king due to his fierce criticism of the Church. He was attacked by Anglicans and Presbyterians alike and felt inclined to the Independents which were favoured by the military leader of the Parliamentarians, Oliver Cromwell.

Fig. 6.3.11. Philosopher and dilettante mathematician Thomas Hobbes [Engraving by W. Humphrys, 1839] and the title page of his main work *Leviathan*

From 1642 on the court of Charles I resided at Oxford, but Oxford fell to the Parliamentarians in 1646[8].

Charles surrendered to the Scots but he was handed over and beheaded in 1649. Monarchy was abandoned in England and Cromwell developed his military reign until his death on the 3rd September 1658.

Since his son and successor failed to get the support of the army a Rump Parliament formed in London which summoned the son of the executed king from abroad. After an act of grace concerning all doings of the past years and the guarantee of freedom of conscience he was crowned in 1660 as Charles II. The revolution was terminated – the restoration began.

After Cromwell had conquered Oxford a new time began also in sciences. After the takeover of some Colleges by Cromwell's followers, men with new scientific interests could prevail, in particular in mathematics and the natural sciences. 'The Baconic ideal of increasing the knowledge by observation gained pace' [Maurer 2002, p. 208].

In the year 1643 Isaac Newton was born. He will write that he stood on the shoulders of giants and could therefore look further ahead[9]. Without doubts the *English* giants were John Wallis in Oxford and Isaac Barrow in Cambridge.

6.3.6 John Wallis and the Arithmetic of the Infinite

John Wallis was born in Ashford in the east of the county of Kent on 23rd November 1616 as the oldest son of an Anglican minister.

During his school days at different schools he already presented himself as a very good pupil and was excellent in particular in Latin, Greek, and Hebrew, and showed a great appetite for learning. Much later he wrote [Scott 1981, p. 3]):

> *It was always my affection, even from a child, in all pieces of Learning or Knowledge, not merely to learn by rote, which is soon forgotten, but to know the grounds or reasons of what I learn; to inform my Judgement as well as furnish my Memory, and thereby make a better Impression on both.*

However, mathematics did not belong to the areas of interest of young Wallis. Only at Christmas 1630 his interest in mathematics was raised when a younger brother had to learn elementary arithmetic to become a merchant. As John Wallis wrote he learned *Common Arithmeticke* in less than two weeks. We know much about his development because as an 80 years old man he wrote an autobiography which was printed and commented upon in [Scriba 1970]. He started his studies in 1632 at

[8] In the north of Oxford are two streets, named 'South Parade' and 'North Parade', respectively. Attentive tourists are usually confused because South Parade is north of North Parade. The naming of the streets goes back to the siege of Oxford during the Civil War where South Parade was the most southern outpost of the Parliamentarians, while North Parade was the most northern street held by royal troops.

[9] See [Merton 1985] for a witty discussion on the origins of this phrase.

Emmanuel College, Cambridge, where he immersed himself in logic but also studied medicine under Doctor Glisson. In the academic year 1636/37 he became BA; four years later he finished his studies with the title of a Master. It is instructive to compare this time period with the 'modern' one after the so-called 'Bologna reform' where the time to acquire an MA or MSc is two years!

Wallis' genius could now clearly be seen and he should become a fellow of his college. Following the rules there should be exactly one fellow for every county and unfortunately the county of Kent was already represented. Hence an early bond between him and his college was not to be. Instead he became fellow of Queens' College, Cambridge, but left soon to take up a position as private chaplain and spiritual advisor. For the time being he was chaplain to Sir Richard Darley of Buttercamp, Yorkshire. One year later he could be found outside Yorkshire as chaplain to of Lady Vere, widow of Lord Horatio Vere.

This is also the time of an event which should contribute to the fame of Wallis. He was shown an encrypted letter which came in the possession of the parliamentary side at the conquest of Chichester on 29th December 1642. He decrypted the message in an unbelievably short amount of time and he was now employed as a cryptologist by the Parliamentarians. As a reward for his services he got earnings from the parish of St Gabriel in London. Already in 1644 he became secretary of Westminster Assembly (*Assembly of Divines at Westminster*) and married Susanna Glyde with whom he had several children; only one son and two daughters lived to adulthood.

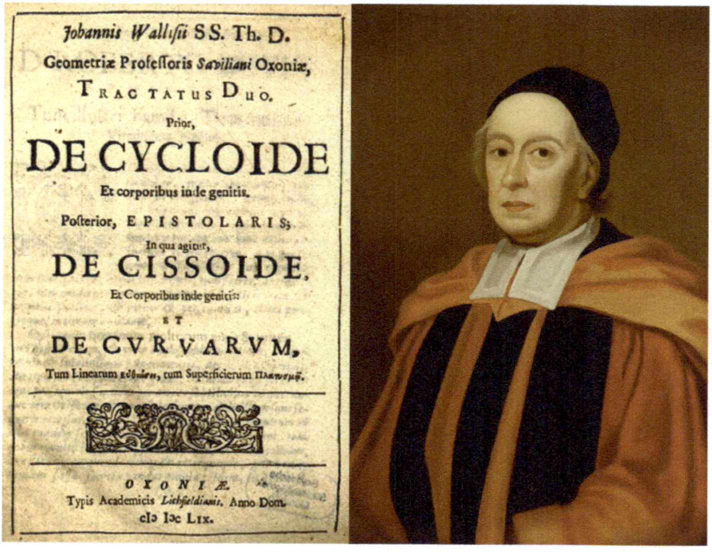

Fig. 6.3.12. John Wallis (right: [Pastel by HWK 2015 after a painting by Godfrey Kneller 1701]) and the title page of his treatise *De Cycloide* (left: Bayerische Staatsbibliothek München, 4 Math.p.391, title page)

Wallis was not yet thirty years old and his career had been steep, but he found the access to higher mathematics only through the founding of the Royal Society.

Wallis and the Establishing of the Royal Society

As we have already noted in section 6.3.2 *Gresham College* was founded in London as a kind of modern educational institution to contrast the weak universities Oxford and Cambridge. Everywhere in Europe we find the founding of extramural academies in the 17th century; the earliest in Naples in the 16th century and then the famous *Accademia dei Lincei* in 1603. The *New Learning*, characterising the advent of sciences freed from Aristotelianism, was strongly promoted in England by Francis Bacon (1561–1626) whose saying 'knowledge is power' received world-wide fame. Wallis became interested in the lectures at Gresham College and was a member of a group of men associated with the college: Dr Wilkins, later Bishop of Chester, Dr Goddard, Gresham professor of physics, and others. The impetus to form an own group of scientific enthusiasts meeting regularly came from Theodore Haak (1605–1690), a German Calvinist scholar who had settled in England. The group met in Goddard's rooms or in Gresham College. In 1648 some members of the group left for Oxford and formed the *Philosophical Society of Oxford* in 1651. The London branch met in Gresham College until 1658. Eventually this group was transformed into the Royal Society by royal grace in 1662. We must not think that the Royal Society in its infant years was already the recognised institution it became during the 19th century; we should rather think the opposite! At the start many dilettantes with ideas concerning idiosyncratic natural philosophy were among the circle of members and so we may not wonder that the young society became the

Fig. 6.3.13. Grandville's drawing of Swift's Laputians [Jonathan Swift, Gullivers Reisen (Gulliver's Travels), Reise nach Laputa (A Voyage to Laputa), 2nd ed., Stuttgart 1843]

Fig. 6.3.14. William Oughtred, author of the textbook *Clavis mathematicae* ([Engraving after a painting by Wenzel Hollar, 17th c.] University of Toronto, Wenceslas Hollar Digital Collection); Francis Bacon, philosopher and statesman, fighter against the scholastic doctrines and supporter of the natural sciences ('knowledge is power') [Engraving, William Rawley, 1627]

target of satire of a writer as great as Jonathan Swift (1667–1745). In his worldwide success *Gulliver's Travels*, published in 1726 for the first time, he described strange creatures in the second chapter of the travel to Laputa which is devoted to the mockery of the members of the Royal Society. Richard Steele, the founder of the literary journal *The Tatler*, wrote in October 1710 [Scott 1981, p. 11]):

> *I have made some observations in this matter so long that when I meet with a young fellow that is an humble admirer of the sciences, but more dull than the rest of the company, I conclude him to be a Fellow of the Royal Society.*

There can be no doubt that Wallis was a serious engine of the new society. He threw himself into works concerning all branches of the new sciences, conducted physical experiments, undertook astronomical observations, and investigated the earth's gravitation. However, his greatest achievements lie in mathematics.

Wallis' Mathematics at Oxford

Today there are no doubts that Wallis earned his appointment of the Savilian chair of geometry in Oxford in 1649 due to his achievements for, and his loyalty to the parliamentarian side of Cromwell in the Civil War. His predecessor, loyal royalist

Peter Turner, had been sacked. Although the circumstances of the vocation were more than questionable not many scholars appointed for political reasons have ever earned a mathematical chair by their mathematical brilliance as Wallis. In 1647 the book *Clavis mathematicae* by William Oughtred (1573–1660) fell into his hands. The *Clavis* was a textbook on elementary algebra, first printed in 1631 and already hopelessly outdated at the time of publishing [Stedall 2002, p. 55]. It was a small book of 88 pages with the full title *Arithmeticae in numeris et speciebus institutio: quae tum logisticae, tum analyticae, atque adeo totius mathematicae quasi clavis est* (Introduction into the calculation with numbers and letters: which is, so to speak, the key to arithmetic, then to analysis, and even to all of mathematics), but the running title was *Clavis mathematicae* and this title finally prevailed. Although the book was outdated and even in England overtaken by the works of Thomas Harriot [Stedall 2003] it has survived its author by 40 years. The engine behind ever new editions was Wallis who remained an ardent admirer of this work. Wallis mastered the contents of the *Clavis* in only a few weeks and after that created his own mathematics.

The two books which shaped Wallis' mathematical thinking most besides the *Clavis* were Descartes' *La Géométrie* in the Latin edition *Geometria* by Frans van Schooten of 1649 and Torricelli's *Opera geometrica* of 1644. Influenced by Descartes Wallis wrote *De sectionibus conicis* in 1652 in which he investigated conic sections not geometrically but by means of their algebraic equations. He was not the first to do so, but in this days Wallis did not belong to the circle of scientists reading the mathematical manuscripts of Pierre de Fermat [Stedall 2004, p. xiii]. The book by Torricelli actually was only a second-hand source of information since it described Cavalieri's method of indivisibles. Torricelli's book was clearly organised and intelligible, while Cavalieri's writing was hard to understand. Fortunately Cavalieri's book *Geometria indivisibilibus continuorum nova quadam ratione promota* of 1635 was hard to come by in England and so Wallis read Torricelli's book and became fascinated by the possibilities of the method of indivisibles. Already in the same year in which he wrote *De sectionibus conicis* Wallis wrote his *Arithmetica infinitorum* [Stedall 2004] which became a most important work in the development of analysis. It was dedicated to William Oughtred. The printing of both books began in 1655 and they finally were published in 1656 in the *Opera mathematica*.

Although some of the results in the *Arithmetica Infinitorum* were known prior to Wallis he can be seen as the first to follow a *systematic* approach to problems of quadrature. There are two new techniques which are continuously employed by Wallis: Induction and interpolation; both must not be confused with what we mean today by induction and interpolation! Wallis' induction is a heuristic one and *Proposition 1* at the beginning of the *Arithmetica Infinitorum* may serve as an example [Stedall 2004, p. 13]:

> *If there is proposed a series of quantities in arithmetic proportion (or as the natural sequence of numbers) continually increasing, beginning from a point or 0 (that is, nought, or nothing), thus as 0, 1, 2, 3, 4, etc., let it be proposed to inquire what is the ratio of the sum of all of them, to the sum of the same number of terms equal to the greatest.*

In modern notation the task is to compute the quotients

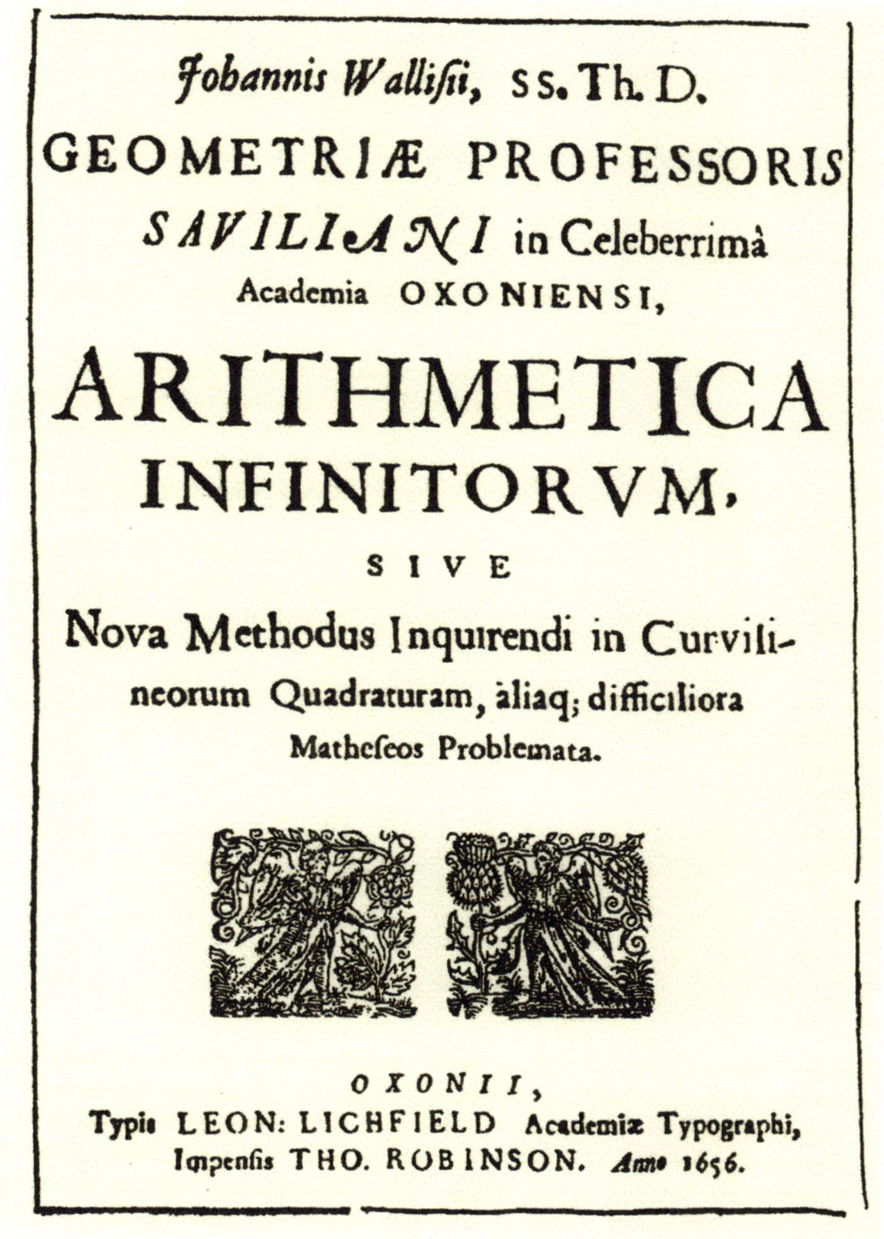

Johannis Wallisii, SS. Th. D.
GEOMETRIÆ PROFESSORIS
SAVILIANI in Celeberrimâ
Academia OXONIENSI,
ARITHMETICA
INFINITORUM,
SIVE
Nova Methodus Inquirendi in Curvilineorum Quadraturam, àliaq; difficiliora Matheseos Problemata.

OXONII,
Typis LEON: LICHFIELD Academiæ Typographi,
Impensis THO. ROBINSON. *Anno* 1656.

Fig. 6.3.15. Title page of *Arithmetica Infinitorum* by John Wallis 1656

$$\frac{0+1+2+3+\ldots+n}{n+n+n+n+\ldots+n}$$

depending on n. Wallis' method of induction proceeds as follows: Write down the result for some n and then deduce a general law. Scott writes [Scott 1981, p. 30]:

> *Wallis used the term Induction in the Baconian sense – namely, a generalisation from a number of cases which would prove true universally.*

Wallis calculated:

$$\frac{0+1}{1+1}=\frac{1}{2} \qquad \frac{0+1+2}{2+2+2}=\frac{1}{2}$$

$$\frac{0+1+2+3}{3+3+3+3}=\frac{1}{2} \qquad \frac{0+1+2+3+4}{4+4+4+4+4}=\frac{1}{2}$$

$$\frac{0+1+2+3+4+5}{5+5+5+5+5+5}=\frac{1}{2} \qquad \frac{0+1+2+3+4+5+6}{6+6+6+6+6+6+6}=\frac{1}{2}$$

and then concluded [Stedall 2004, p. 14]:

> *And in the same way, however far we proceed, it will always produce the same ratio of one half.*

Apparently this 'argument' was seen sufficient by Wallis. With the help of such sum formulae, proven by 'induction', Wallis succeeded to transform problems of quadrature from the prevailing geometrical treatment to an arithmetical one. Therein lies in fact the progress which Wallis carried into the English analysis [Stedall 2005, p. 25].

As an example we want to discuss the quadrature of the parabola and follow [Stedall 2005]. In Proposition 19 Wallis computed the ratio

$$\frac{0^2+1^2+2^2+\ldots n^2}{n^2+n^2+n^2+\ldots+n^2}$$

which depends on n by his method of induction and computed

$$\begin{aligned}\frac{0+1}{1+1}&=\frac{3}{6}=\frac{1}{3}+\frac{1}{6}\\ \frac{0+1+4}{4+4+4}&=\frac{5}{12}=\frac{1}{3}+\frac{1}{12}\\ \frac{0+1+4+9}{9+9+9+9}&=\frac{7}{18}=\frac{1}{3}+\frac{1}{18}\\ \frac{0+1+4+9+16}{16+16+16+16}&=\frac{9}{24}=\frac{1}{3}+\frac{1}{24}.\end{aligned}$$

He deduced that for ever increasing n this ratio will become exactly 1/3 because the deviation from 1/3 gets ever smaller.

Endowed with this result Wallis wants to compute the ratio of the area ATO (concave gusset) to the area of the rectangle $ATOD$ in case of the parabola shown

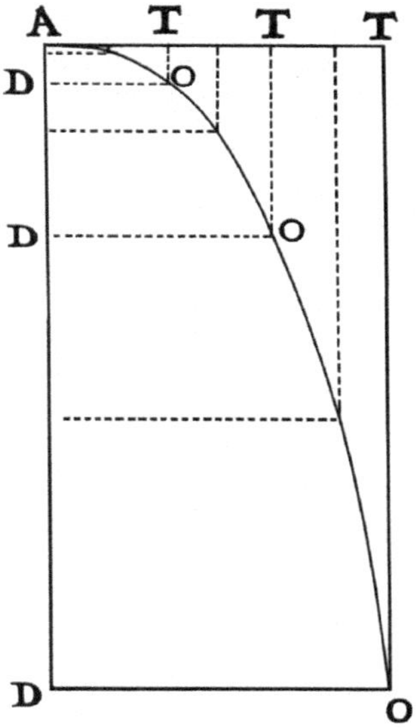

Fig. 6.3.16. Quadrature of the parabola after Wallis

in figure 6.3.16. Following Cavalieri's method of indivisibles he has to 'sum' the lines TO. Each TO has length $(OD)^2$ because they are bounded by the parabola. We now imagine a small distance[10] a between the indivisbles TO; hence Wallis has to compute the ratio

$$\frac{0^2 + a^2 + (2a)^2 + (3a)^2 + \ldots + (na)^2}{(na)^2 + (na)^2 + (na)^2 + \ldots + (na)^2}$$

for ever increasing n and ever decreasing a. From the induction above it immediately follows that this ratio is 1/3.

This result was already known but Wallis' 'proof' is an arithmetic one for the first time which convinced him that his method actually led to true results. We see here nothing less than the transformation from Cavalieri's *geometria indivisibilium* to Wallis's *Arithmetica infinitorum*. So assured he now turned to the quadrature of monomials of higher order.

In this way he 'saw' that

$$\lim_{n\to\infty} \frac{0^k + 1^k + 2^k + \ldots + n^k}{n^k + n^k + n^k + \ldots + n^k} = \frac{1}{k+1} \tag{6.28}$$

holds for every natural number k; in particular it follows for the parabola $f(x) = x^2$,

$$\lim_{n\to\infty} \frac{0^2 + 1^2 + 2^2 + 3^2 + \ldots + n^2}{n^2 + n^2 + n^2 + n^2 + \ldots + n^2} = \frac{1}{3} = \frac{\text{area } ATO}{\text{area } ATOD}.$$

[10] It can be seen here beautifully that Wallis did not care about the difference of *indivisibles* and *infinitesimals*. He used both techniques without ever mentioning it.

If the area of the gusset ATO in figure 6.3.16 is thus the third part of the rectangular area then the area AOD has to be two thirds of the rectangular area. But this area is just the area under the inverse function of $f(x) = x^2$, hence the function $y = \sqrt{x}$. By 'summing' the indivisibles DO having length $\sqrt{DO}$ Wallis arrived at the relation

$$\lim_{n\to\infty} \frac{\sqrt{0}+\sqrt{1}+\sqrt{2}+\sqrt{3}+\ldots+\sqrt{n}}{\sqrt{n}+\sqrt{n}+\sqrt{n}+\sqrt{n}+\ldots\sqrt{n}} = \frac{\text{area } ATOD - \text{area } ATO}{\text{area } ATOD} = \frac{2}{3}.$$

Wallis now rearranged:

$$\frac{2}{3} = \frac{1}{\frac{3}{2}} = \frac{1}{1+\frac{1}{2}}$$

and compared with (6.28), leading him to the insight that the root is nothing but a fractional power, namely

$$\sqrt{n} = n^{\frac{1}{2}}.$$

Analogously he noted that $\sqrt[3]{n} = n^{1/3}$ and that $n^0 = 1$ has to hold.

So far as Wallis' first cornerstone, induction, is concerned. The interpolation he used as an argument as follows: If (6.28) holds for all $k \in \mathbb{N}$, then such a relation as

$$\lim_{n\to\infty} \frac{0^p+1^p+2^p+\ldots+n^p}{n^p+n^p+n^p+\ldots+n^p} = \frac{1}{p+1}$$

has to hold also *for all rational numbers* p! Therewith he succeeded to give an arithmetical proof of

$$\int_0^1 x^{\frac{p}{q}}\, dx = \frac{q}{p+q}.$$

This result was also known but Wallis' way to get to it was brand new. A further gem was found by Wallis when he applied his 'interpolation' to the squaring of the circle, namely

$$\frac{4}{\pi} = \frac{1\cdot3\cdot3\cdot5\cdot5\cdot7\cdot7\cdot9\cdot9\cdot\ldots}{2\cdot4\cdot4\cdot6\cdot6\cdot8\cdot8\cdot10\cdot10\cdot\ldots}.$$

The noble Irishman William Brouncker (1620–1684), first president of the Royal Society, got his education at Oxford and collaborated closely with Wallis. He was so impressed by this formula that he worked on his own representation in form of a continued fraction which he found in 1655.

$$\frac{4}{\pi} = 1 + \frac{1^2}{2+\frac{3^2}{2+\frac{5^2}{2+\frac{7^2}{2+\frac{9^2}{\ddots}}}}}.$$

This formula found its way into Wallis' *Arithmetica Infinitorum* with a clear reference to Brouncker.

6.3.7 Isaac Barrow and the Love of Geometry

Besides Wallis in Oxford it was Isaac Barrow (1630–1677) in Cambridge on whose shoulders Isaac Newton, successor of Barrow's on the Lucasian Chair, stood and eventually could found modern analysis.

Fig. 6.3.17. William Brouncker [Unknown artist, 17th c., copy of a painting by Peter Lely], first president of the Royal Society; Isaac Barrow [Painting by Domenico Tempesti, 1690], first professor of mathematics on the chair donated by Henry Lucas at Trinity College, Cambridge. He was Isaac Newton's teacher

Isaac's father Thomas Barrow could not follow his intentions to become a scholar. Although the family could look back at renowned physicians and other graduates from the University of Cambridge Thomas was repelled by his father's hardness, left his family early, and became a merchant. Affected by this experience Thomas did everything to render a career in the sciences for his son Isaac possible [Feingold 1990]. He payed double fees to the headmaster of Charterhouse to ensure special care for Isaac, but this apparently failed since Isaac became a bully. Things worsened so much that the father in a devout ejaculation appealed to God that if He decided to take away one of his children it should be Isaac! After two or three years Isaac had to leave school and was introduced to Felsted School where John Wallis was a pupil a decade before Barrow. When Thomas Barrow had to face bankruptcy due to the collapse of the trade relation with Ireland headmaster Martin Holbeach received Isaac in his own home and arranged a position as tutor to Thomas Fairfax. When the young noble man Fairfax fell desperately in love with a girl and married her he lost any support of his family and Isaac Barrow lost his job. Thereupon Holbach wanted to bring Barrow back to Felsted to make him his successor as headmaster. Isaac Barrow refused and in 1646 went instead to Trinity College Cambridge with a former classmate.

In Cambridge James Duport became his tutor and he studied Greek, Latin, Hebrew, Spanish, Italian, but also chronology, geography, and theology under the tutelage of Duport. A serious study of mathematics was begun only in 1648 or 1649, probably under John Smith of Queens' College who was teaching Descartes' *Géométrie.* Eventually Trinity College made a quantum leap in mathematics through the arrival of the fellows Nathaniel Rowles and Charles Robotham in 1645; later both of them became university professors. Barrow finished his studies in 1649 with the BA. He became fellow of Trinity College and studied mathematics intensely. Without doubt Barrow was a royalist although he proclaimed to the Parliamentarian side in Cromwell's days. He withdrew later and was saved from being expelled from university the second time by the Master of the College. So he was able to finish his MA in 1654. He took the oath to study church history but is also interested in medicine. Through his studies in church history he was eventually led to astronomy. In 1655 a simplified version of Euclid's *Elements* was published by Barrow, documenting his permanent interests in mathematics. An attempt to get a lectureship was rejected with flimsy arguments, but Barrow's political attitude was probably the true reason behind this rejection.

With the financial help of his university Barrow travelled to France in 1655 where he stayed in Paris for ten months. He was disappointed by the local university which he had imagined to be more impressive, but at least he met Roberval. In February 1656 he travelled further to Florence, staying there for eight months. A visit to Rome was not to be since the plague was rampant there. In Florence he worked in the library of the House of Medici and became an expert on coins. He also met Italian mathematicians; among others Galilei's last pupil, Vincenzo Viviani (1622–1703), and Carlo Renaldini (1615–1679) who wrote a book on algebra at this time. Here Barrow was introduced to Italian mathematics, in particular to the methods of indivisibles.

Fig. 6.3.18. Vincenzo Viviani [Painting by Domenico Tempesti, about 1690]; Galileo Galilei and his pupil Vincenzo Viviani ([Painting by Tito Lessi 1892] in the Istituto e Museo di Storia della Scienza, Florence)

From Florence the journey continued by ship to Turkey, but Barrow's ship was attacked by pirates and he landed in Smyrna (today: Izmir) where he stayed for seven months before travelling to Constantinople where he stayed for more than a year, accommodated in the house of the English ambassador. He dedicated himself to the study of theology and of the Greek church. He missed to send regular reports to his university concerning his travels, nevertheless he got permission from Cambridge to extend his journey. Only in 1658 he began his return journey. The ship berthed in Venice when a fire broke out and destroyed all of Barrow's possessions on board. Overland across Germany and Holland the journey continued back to Cambridge where he arrived in September 1659. The restoration had changed the political climate in England; Charles II is the new king and Barrow got the professorship of Greek. His teacher Duport was forced out of office but was asked to again occupy his professorship. But he declined, so that Barrow now received an annual salary of £40. He was, however, not allowed to take any other positions within the university and to draw income from Trinity College. Barrow started a campaign to enable fellows to draw income from their College memberships. The financial frame stayed tight, though, and hence Barrow applied in 1662 to the professorship of geometry at Gresham College which had become vacant. He is successful! As Gresham professor he is allowed to keep his professorship in Cambridge; at Gresham College he taught geometry for one hour a week in English, and for one hour in Latin, and got an additional annual fee of £50.

At one of the first meetings of the still young Royal Society Barrow was elected as a fellow on 20th May 1663. However, he did not care much about the Society; he did not pay his dues and was dismissed. Meanwhile Henry Lucas, member of Parliament for Cambridge in the House of Commons, had donated landed property to establish a professorship at Trinity College. The Lucasian Chair of mathematics seemed to be created for Barrow and in 1663 he gave up his professorship of Greek and became the first Lucasian professor of mathematics. Due to the plague the University of Oxford was closed in 1665 and re-opened only in April 1666 until a second outbreak of the plague. Hence the university could open only at Easter 1667. Besides lectures on mathematics Barrow lectured on optics in 1668/69 and Isaac Newton attended these and got in private contact with his teacher. The mathematical period of Barrow can be seen as the years between 1663 and 1669. In writing the book *Lectiones Opticae* Barrow got help from Newton. The book was published in 1669 by John Collins who was in the service of the Royal Society. The mathematical works were also published by Collins as *Lectiones Geometricae* in 1670, and *Lectiones Mathematicae* in 1683 [Barrow 1973].

In the year 1669 Barrow withdraw from the Lucasian Chair. No doubt that Barrow had recognised Newton's genius and also no doubt that he admired Newton's scientific abilities. Hans Wußing [Wußing 1984, p. 21] gives us a tradition of a contemporary:

> *The Doctor* [Barrow] *had a tremendous esteem of his pupil and used to say many a time that he truly understood something of mathematics, but that he calculates like a child in comparison to Newton.*
>
> (Der Doktor [Barrow] hatte eine gewaltige Hochachtung vor seinem Schüler und pflegte des öfteren zu sagen, daß er wahrhaftig Einiges an Mathematik verstehe, daß er aber im Vergleich zu Newton wie ein Kind rechne.)

Anyway, the story that Barrow withdrew from the Lucasian Chair in favour of Newton is very likely purely fictional. In Barrow's mathematical works a true mastership of geometry and the use of infinitely small quantities can be found. He had even moved as far as to a geometric version of the fundamental theorem of calculus, so that he certainly did not 'calculate like a child in comparison to Newton'. Much more plausible are the arguments presented by Mordechai Feingold in [Feingold 1990, p. 80–83]; Barrow was in a crisis, concerned for the salvation of his soul, and hence consequently took the only logical step: he withdraw from his professorship and at the same time from mathematics. Feingold cites an early biographer of Barrow's [Feingold 1990, p. 80f.]:

> *He was afraid, as a clergyman, of spending too much time upon Mathematics; for ... he had vowed in his ordination to serve God in the Gospel of his Son, and he could not make a bible out of his Euclid, or a pulpit out of his mathematical chair.*

Barrow renounced a large part of his income in taking this step; on the other hand was he not willing to take a nicely endowed position within the church, although he had been elected in 1670 in Salisbury as royal chaplain to Charles II. It was the same Charles II who made Barrow Master of Trinity College in February 1673 so that Barrow came back to Cambridge. In April 1677 he undertook a journey to London where he contracted a feverish illness. It is said that had tried to cure himself by fasting and the use of opium, but he died on 4th May 1677 being 47 years old. He was buried in Westminster Abbey.

Barrows Mathematics

In the *Lectiones Geometricae* of the year 1670 Barrow discussed problems of the division of time and space and in doing so he is close to Oresme and Galilei. Barrow's mathematics on the one hand seems generally ancient, on the other hand it comprises very modern ideas [Mahoney 1990]. The continuum is arbitrarily divisible in space as in time. In the infinitely small, said Barrow, time consists of 'timelets' as space consists of 'linelets'. Today we interpret these quantities as infinitely small straight line segments. Barrow imagined all curves consisting of infinitely small straight line segments. Hence all higher derivatives would locally be zero and this would consequently lead to nilpotent infinitesimals, i.e. if o is an infinitesimal quantity, then it follows

$$o^n = 0, \quad n > 1.$$

In a later chapter where we discuss modern concepts we shall come back to these ideas.

In the computation of tangents Barrow used a method which originated in the unpublished works of Fermat [Edwards 1979, p. 132]. It is all about the computation of tangents at functions implicitly given by

$$f(x, y) = 0.$$

Barrow looked at two infinitesimally adjacent points M and N as shown in figure 6.3.19.

The coordinates of M are (x, y), the ones of N are $(x + e, y + a)$. Since M and N are both points lying on the curve it holds

$$f(x + e, y + a) = f(x, y) = 0.$$

The arc MN is not viewed as an arc by Barrow but as a 'linelet', i.e. a small straight line segment. In the equation above he therefore sets higher powers of a and e to zero, 'for these terms have no value'. We follow Edwards [Edwards 1979, p. 133] and demonstrate the method at the folium of Descartes

$$f(x, y) = x^3 + y^3 - 3xy = 0.$$

From $f(x + e, y + a) = f(x, y) = 0$ it follows

$$(x + e)^3 + (y + a)^3 - 3(x + e)(y + a) = x^3 + y^3 - 3xy = 0,$$

or

$$3x^2e + \underbrace{3xe^2}_{=0} + \underbrace{e^3}_{=0} + 3y^2a + \underbrace{3ya^2}_{=0} + \underbrace{a^3}_{=0} - 3xa - 3ye - \underbrace{3ae}_{=0} = 0,$$

where all powers of a and e with exponents larger than 1 (and therefore also the product $a \cdot e$, of course) are neglected. Then it follows

$$3x^2e + 3y^2a - 3xa - 3ye = 0$$

and hence the slope of the tangent is given by

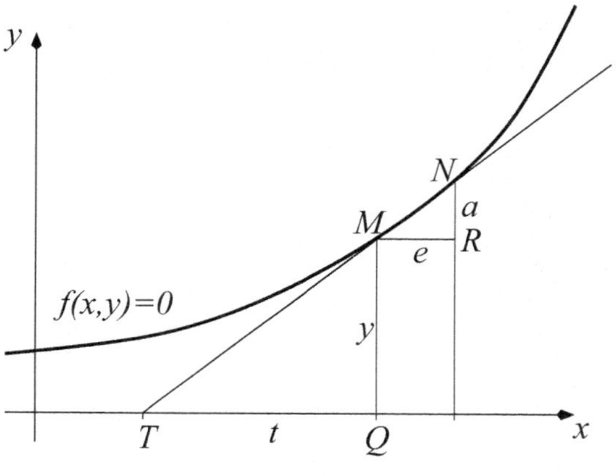

Fig. 6.3.19. Barrow's computation of tangents

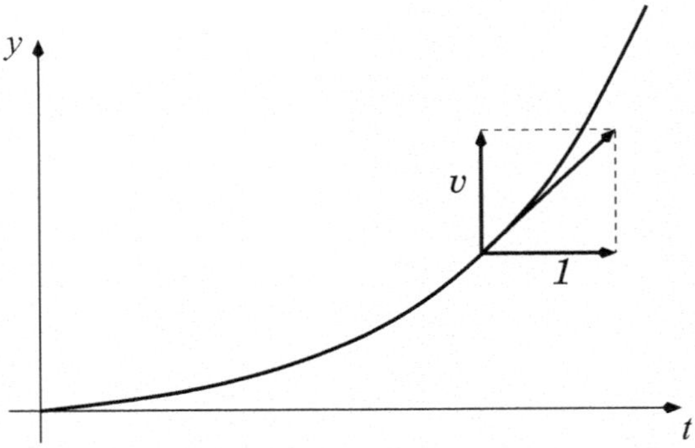

Fig. 6.3.20. Motion in the distance-time diagram

$$\frac{a}{e} = \frac{y - x^2}{y^2 - x}.$$

Particularly impressive are Barrow's geometrical reflections which led him to the fundamental theorem of calculus. In 1916 J. M. Child published a translation of parts of the *Lectiones Geometricae* [Child 1916]. In the preface he elevated Barrow to have been the actual and only discoverer of the calculus. But Childs took it much too far. Yes, one can discern the fundamental theorem from Barrow's geometrical constructions, but Child did that from the position of a modern man with modern mathematical education. In truth Barrow failed to draw this deep insight from his own works.

If a point moves with speed $t \mapsto v(t)$ along a straight line then the area under the velocity-time curve gives the distance travelled. This can be seen from arguments employing indivisibles. Representing the same movement in a distance-time diagram then the velocity *vector* at a point of the curve has components 1 in t-direction and v in y-direction, cp. figure 6.3.20.

If the velocity is given by $v = t^n$ then it follows $y = \frac{t^{n+1}}{n+1}$ for the distance. The facts are that the area under $y = x^n$ is given by $x^{n+1}/(n+1)$, and that the tangent at $y = x^{n+1}/(n+1)$ has slope x^n, were historically discovered in many different ways. Here, using motion as an example, it became clear for the first time that, and how, these things belong together. Edwards called this insight, which before Barrow was also bestowed on Torricelli, *'embryonic formulation of the fundamental theorem of calculus'*. Let us now take a look at how Barrow grasped this connection geometrically.

Figure 6.3.21 shows a monotonically increasing function $y = f(x)$ in a representation which looks unusual in modern eyes. Additionally the figure shows the function $z = A(x)$ which is the area under f over $[0, x]$. Let D be the point $(x_0, 0)$ on the x-axis and T a point for which

$$DT = \frac{DF}{DE} = \frac{A(x_0)}{f(x_0)}$$

holds. Then Barrow proves that the straight line TF touches the curve $z = A(x)$ only in the point F.

The slope of TF is

$$\frac{DF}{DT} = \frac{A(x_0)}{\frac{A(x_0)}{f(x_0)}} = f(x_0).$$

Let I be a point on the curve at $x = x_1 < x_0$ and K the point of intersection of TF with IL being parallel to the x-axis. Barrow now showed that the point K always lies to the right of I. To this end he deduces from

$$\frac{LF}{LK} = \frac{DF}{DT} = DE,$$

i.e.

$$LF = LK \cdot DE,$$

that

$$LF = DF - PI = A(x_0) - A(x_1) < \underbrace{DP \cdot DE}_{\text{rectangular area}}$$

holds. But then we have shown $LF = LK{\cdot}DE < DP{\cdot}DE$ or $LK < DP = LI$. The case $x_1 > x_0$ can be treated analogously.

6.3.8 The Discovery of the Series of the Logarithms by Nicholas Mercator

Nicholas Mercator was born in 1620 in what then was the Danish town of Eutin as Nicolaus Kauffman. His father was school master in the town

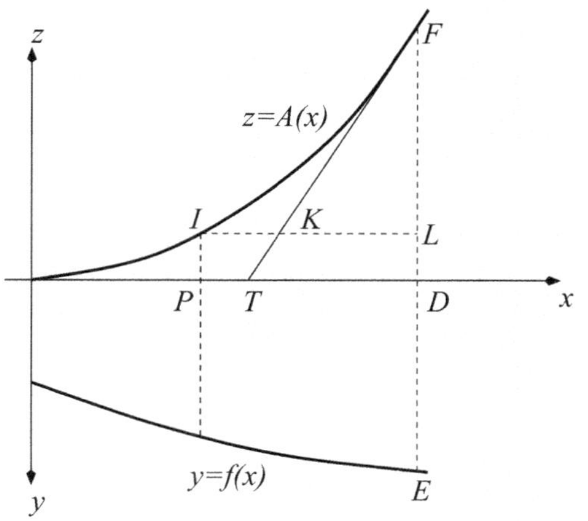

Fig. 6.3.21. Barrow's way to the fundamental theorem

Oldenburg in Holstein from 1623 to his death in 1638, and we can assume that little Nicholas received his school education from his father. In the famous short biographies of the time, John Aubrey's *Brief Lives* [Aubrey 1982, p. 200], we find the note that Lutheran reformer Philip Melanchthon was the brother of Mercator's great-grandmother; that Mercator was of small stature, with curly black hair and dark eyes, and of incredible inventiveness. In 1632 we find him at the University of Rostock at which he finished his studies in 1641 to go to Leiden. Already in 1642 he was back in Rostock where he was in a position at the faculty of philosophy. But in 1648 he went to the University of Copenhagen where he wrote books on geography (*Cosmographia*, 1651), astronomy (*Astronomia sphaerica*, 1651), and spherical trigonometry (*Trigonometria sphaericorum logarithmica*, 1651). Following the customs of his days he latinised his name (German Kaufmann = English merchant = Latin mercator) and called himself Nicholas Mercator. In astronomy he was open to Kepler's planetary theory. In his work *Rationes mathematicae subductae* of 1653 an instructive example can be found. Mercator distinguished rational and irrational numbers by comparing them with notes and tones in music. Rational ratios lead to harmonic tones while irrational ratios lead to discords. At this point Mercator noted that Kepler's planetary theory led to rational ratios of the orbits while observations suggested irrational ratios. Mercator was also interested in a calendar reform and in *De emendatione annua* he drafted a calendar with 12 months and months varying from 29 to 31 days.

Even before 1660 Mercator turned to England. Any presumptions that he was invited to England by Cromwell to initiate a much needed calendar reform can not be proven. He failed to get a position at one of the two English universities so that that he had to earn his living as a private tutor in London. However, his mathematical abilities got about quickly in England. He corresponded with Oughtred and Pell and after a pause in publication of more than 10 years he published his ideas concerning the structure of the cosmos in 1664 as *Hypothesis astronomica nova*, where Kepler's planetary theory served as foundation.

A major problem of the time was the exact determination of longitude on the sea. Latitude is invariant under the rotation of the earth and can be determined by observation of the sun at midday or by targeting the polar star at night. Matters are much worse with longitude since longitude moves with the earth's rotation. The fascinating story of the determination of longitude was recently described by Dava Sobel but Mercator was apparently forgotten! It was known that the problem of the determination of longitude could be solved by very accurate clocks on board of ships. The time difference of the true time at midday (highest position of the sun) and the time shown by the clock (always the local time of the haven where the journey started) can be used to compute longitude.

Christiaan Huygens had already invented his pendulum clock but its accuracy was not sufficient for long sea journeys; additionally a storm at sea would perturb (or even stop) the pendulum. Mercator also designed a pendulum clock for use at sea which brought him the membership in the Royal Society in 1666. There he got in contact with Robert Hooke who struggled at this time with Huygens on the priority of the invention of the balance wheel. Mercator's clock, most likely not accurate and robust enough for use at sea, had no fortune. Aubrey [Aubrey 1982, p. 200] reports that Mercator wanted to sell his clock to Charles II who had praised the work. However, the king did not pay a penny! The clock came in the hands of a courtier who sold it to a clock maker named Knibb. Since Knibb did not understand the mechanism of the clock he sold it to the clock maker Fromanteel who once built it. The selling price is said to have been 5 pounds and Aubrey remarks not without malice that Fromanteel offered the clock in 1638 for 200 pounds.

In 1668 Mercator published his most important contribution to analysis, *Logarithmotechnia* [Mercator 1975], in which the famous series

$$\ln(1+x) = x - \frac{x^2}{2} + \frac{x^3}{3} - \frac{x^4}{4} + \frac{x^5}{5} - \dots$$

appeared in purely verbal form, i.e. without the use of formulae.

The *Logarithmotechnia* consists of three parts. Following [Hofmann 1939] the first two parts were published separately already in 1667 and are concerned with the calculation of logarithms. The number range between 1 and 10 is divided into 10 million parts by the introduction of geometric means, called 'ratiunculae' by Mercator. The logarithm of a number between 1 and 10 hence is the number of ratiunculae between 1 and this number. By means of a very skilful algorithm concerning successive squaring he was able to compute the number of ratiunculae corresponding to a given number and hence the logarithm. In the third part of the *Logarithmotechnia* the function $1/(1+x)$ is represented by polynomial long division as

$$\frac{1}{1+x} = 1 - x + x^2 - x^3 + \dots.$$

Entirely in line with Wallis and with his methods of the *Arithmetica Infinitorum* Mercator represented the area under the hyperbola $1/(1+x)$ as a 'sum' of indivisibles, cp. figure 6.3.23. As to how exactly the indivisibles have to be 'summed' in view of the mononomials $1, x, x^2, \dots$ appearing in the series of the logarithm Mercator said only little. Hofmann [Hofmann 1939, Remark 9 on p.44] has concluded that the modalities of this behaviour may hint to Mercator having not known Cavalieri's original work but had got his knowledge from lectures at Rostock, Copenhagen, or Danzig (Gdansk). However, reading Wallis immediately shows that Mercator had used his methods.

In modern notation Mercator had arrived at

LOGARITHMO-TECHNIA:
SIVE
Methodus conſtruendi
LOGARITHMOS
Nova, accurata, & facilis;
SCRIPTO
Antehàc Communicata, Anno Sc. 1667.
Nonis *Auguſti*: Cui nunc accedit.
Vera Quadratura Hyperbolæ,
&
Inventio *Summæ* Logarithmorum.

AUCTORE *NICOLAO MERCATORE*
Holſato, è Societate Regia.

HUIC ETIAM JUNGITUR
MICHAELIS ANGELI RICCII Exercitatio
Geometrica de Maximis & Minimis; hîc ob Argumenti
præſtantiam & Exemplarium raritatem recuſa.

LONDINI,
Typis *Guilielmi Godbid*, & Impenſis *Moſis Pitt* Bibliopolæ, in
vico vulgò vocato *Little Britain*. Anno M.DC.LXVIII.

Fig. 6.3.22. Title page of the *Logarithmotechnia*, 1668

$$\int_0^x \frac{d\xi}{1+\xi} = x - \frac{x^2}{2} + \frac{x^3}{3} - \frac{x^4}{4} + \ldots, \tag{6.29}$$

and in a daring flight of thought he declared that this is expressible as the logarithm of $1 + x$. The *Logarithmotechnia* ends with an also only verbally given description of a computation using Cavalieri's indivisibles, leading to

$$\int_0^x \log(1+\xi)\, d\xi = \frac{x^2}{1\cdot 2} - \frac{x^3}{2\cdot 3} + \frac{x^4}{3\cdot 4} - \ldots.$$

The third part of the *Logarithmotechnia* was praised by Wallis in a book review immediately after it was published. Wallis even improved on Mercator's representation by introducing formulae. He recognised that Mercator's series (6.29) is valid only for $|x| < 1$ and completed Mercator's considerations by giving the series

$$\int_0^x \frac{d\xi}{1-\xi} = x + \frac{x^2}{2} + \frac{x^3}{3} + \frac{x^4}{4} + \ldots, \qquad 0 < x < 1.$$

In a short work *Some Illustration of the Logarithmotechnia of M Mercator, who communicated it to the publisher, as follows*, published also in 1668 in the *Philosophical Transactions*, Mercator called the logarithm defined by his series *logarithmus naturalis*. Mercator also recognised that the natural logarithm can be gained from Briggs' logarithm by multiplication with $K = \log \mathsf{e} = \frac{1}{\ln 10}$, cp. (6.26).

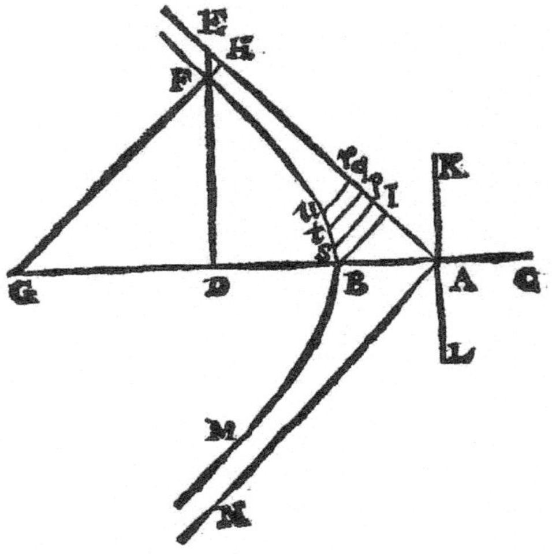

Fig. 6.3.23. Indivisibles building the area under the hyperbola [Mercator 1975]

Mercator's series (6.29) plays a very important role in the early mathematical studies of Newton. A certain conclusion in the works concerning logarithms was only reached by Robert Cotes (1682–1716) who, building on considerations of Edmund Halley (1656–1742), found the functional relation (or functional equation) which we today write as

$$\log_b a^x = x \log_b a,$$

and who *defined* the logarithmic function through this functional relation. For the first time logarithms were separated from geometrical considerations concerning hyperbolas.

As translator of the Danish work *Algebra ofte Stelkonst* by Kinckhuysen into Latin Mercator got in contact with Newton in 1669. The two men corresponded on the motion of the moon. In Newton's library a copy of Mercator's two-volume work *Institutiones astronomicae* could be found which contained many annotations in Newton's hand [Harrison 1978, catalogue no. 1072, p. 191].

Despite Mercator's recognised achievements he failed to take roots at an English university. Even in 1676 Hooke proposed him for a position as *Mathematical Master* at Christ Hospital but Mercator also failed. Then he was invited by the founding member of the Académie des Sciences and minister of finance Jean-Baptiste Colbert to design the waterworks at Versailles. So he went to France in 1682. But Mercator quarrelled with Colbert and the project failed. Mercator died in Paris on 14th January 1687.

6.3.9 The First Rectifications: Harriot and Neile

The computation of the length of curves or of pieces of curves is one of the basic tasks of analysis. The circumference of a circle with radius r was fairly early known to be $2\pi r$, and hence the circle is historically the first curvilinear figure the length of which could be computed. The actual history of *rectification* begins, however, only with Thomas Harriot (1560–1621), who is also known as an eminent algebraic [Stedall 2002], [Stedall 2003].

Thomas Harriot

Thomas Harriot (1560–1621) was born in 1560 in or about Oxford in an humble parental home. We know very little about the first twenty years of his life but he must have been an excellent pupil since in 1577 we find in the matriculation register of the University of Oxford the admission entry to Saint Mary Hall [Shirley 1983, p. 51]:

Fig. 6.3.24. Thomas Harriot ([unknown painter, 1602] Trinity College, Oxford); Sir Walter Raleigh [Painting attributed to the French School] (Kunsthistorisches Museum Wien, Austria / Bridgeman Images)

1577
20 Dec. *S. Mary H.* *Hariet, Thomas;* *Oxon.,* *pleb. f.,* *17.*

In Trinity College, Oxford, the portrait shown in figure 6.3.24 is presented which is said to show Harriot. This, however, is not certain, especially since the dates on the portrait do not fit, cp. [Batho 2000].

Harriot came in contact with Walter Raleigh who organised the Elisabethan fleet, and hence became acquainted with the burning problems of navigation at sea. Harriot's duties included the computation of navigational tables (declination tables), but also the education of navigators. In [Pepper 1974] manuscripts are discussed which Harriot had compiled for the Guiana expedition of Raleigh in 1595. These manuscripts clearly show that Harriot not only was an excellent theoretician, but could also give practical instructions concerning the use of navigational instruments of the time.

The problem which finally led Harriot to the first rectification of a curve was the problem of a nautical chart which was useful in practice. Until the 16th century Portonal charts were used in which it was assumed that the charted area was flat. If the charted area became large these charts showed extreme errors rendering them useless for longer sea travels. In the case of the Mediterranean there were charts which included rays of compass roses at different places leading to a pattern of rhombs. Sailing on these rays was dangerous though, since a fixed bearing did not correspond to a straight ray on the chart! Every navigator wants to transfer his reading of the compass directly to his sea chart, i.e. the true course on the chart should in fact be a straight line if the bearing is held constant.

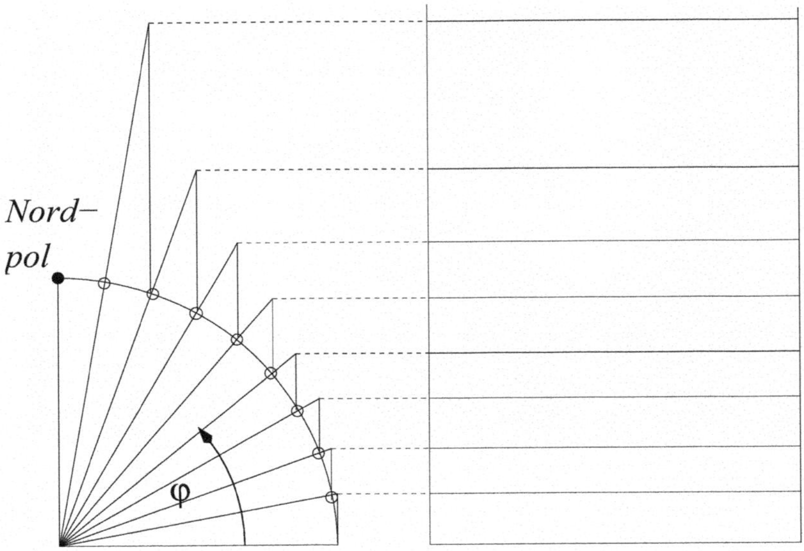

Fig. 6.3.25. The principle of the Mercator mapping

The very first truly conformal chart on which a fixed bearing intersecting latitudes under a fixed angle was a straight line was made possible by the work of Gerhard Mercator (1512–1594). Gerhard worked in the Belgian town of Louvain in the workshop of Gemma Frisius (1508–1555), manufacturing fine navigational instruments and globes. Mercator was known for his excellent work in producing globes.

Gerhard Mercator – not to be confused with Nicholas Mercator! – conceived a cylindrical projection where the earth is put in a cylinder so that sphere and cylinder touch each other in the equator of the earth. In the projection of the earth's surface onto the surface of the cylinder each piece of arc is strechted by a certain factor. The closer one gets to the poles the larger the distortions on the cylinder. Figure 6.3.25 shows the basic principle of the Mercator projection. The latitudes φ are subdivided into 9 parts corresponding to $\Delta\varphi = 10°$ as in the figure. The distance of two adjacent latitudes on the chart become ever greater the closer one gets to the poles.

Today we see immediately the mathematics behind this projection which we illustrate in figure 6.3.26. Let us consider a point on the earth's surface on the latitude φ and map an infinitesimal piece $d\varphi$ so that this piece would appear on the chart with length $d\eta$. From the figure one immediately sees $\cos\varphi = d\varphi/d\eta$, hence

$$d\eta = \frac{d\varphi}{\cos\varphi} = \sec\varphi \cdot d\varphi, \tag{6.30}$$

where, as common among cartographers, we have used the secant function, defined by $\sec\varphi := 1/\cos\varphi$. This central equation was not known by

Mercator, of course; he might have constructed his charts and globes by means of an approximative method, in that he used a certain partition of latitudes as in figure 6.3.25. It was the famous and notorious John Dee (1527–1608/1609) who brought the knowledge of the Mercator projection to England. Dee undertook extensive travels on the Continent and befriended Mercator in Louvain. When he came back to England after 1550 he brought with him

> *Two Globes of Gerardus Mercators best making [...] also divers other instruments [...] of Gerhardus Mercator his own making for me purposedly,*

as he wrote in retrospect in an autobiographic note in 1592 [Skelton 1962]. Dee stood in the service of the *Muscovy Company* as an instructor to skippers and navigators, and he had direct contact with Thomas Harriot [Woolley 2001].

However, the Mercator projection became widely known only through the publication of the book *Certaine Errors in Navigation, Arising either of the ordinarie erroneous making or using of the sea Chart, Compasse, Crosse staffe, and Tables of declination of the Sunne, and fixed Starrs detected and corrected* by Edward Wright in 1599 in London. Wright had tabulated the Mercator projection where the distances between two lines of constant latitudes were chosen as 10 minutes of arcs. In the second edition of his book which soon became famous among navigators he even chose one minute of arc. Hence it is not surprising that the Mercator chart was often called *Mr Wright's chart* in England [Skelton 1962, p. 165].

The straight connections of a location A with another location B on a Mercator chart correspond to a bearing which intersects every line of constant latitude under a fixed angle α. On the sphere of the earth this is a spiral-shaped curve, the so-called *loxodrome*, defined by the equation

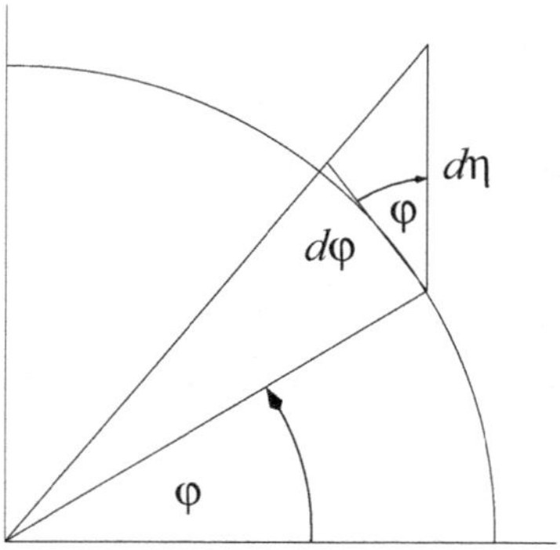

Fig. 6.3.26. Mathematical background of the Mercator projection

$$\begin{pmatrix} x(t) \\ y(t) \\ z(t) \end{pmatrix} = \frac{1}{\cosh(t \cdot \tan\alpha)} \begin{pmatrix} \cos t \\ \sin t \\ \sinh(t \cdot \tan\alpha) \end{pmatrix}.$$

Since the rays of the compass roses on the old Portolan charts created rhombus-shaped patterns one used to call the *loxodrome* the *rhumb line* in England. Figure 6.3.27 (left) shows a loxodrome intersecting the lines of constant latitude under an angle of $\alpha = 5^\circ$. Loxodromes approach the poles in ever narrowing spirals. For navigators the knowledge of the length of a section of a loxodrome is important since that is the the actual distance to travel between two points A and B. Also of essential importance is the difference in latitude from A to B. Edward Wright had calculated this difference by addition of small pieces, and also Harriot had used such summations when he began his work on the loxodrome. Pepper has shown in [Pepper 1968] that Harriot very likely had achieved the rectification of the loxodrome by means of infinitesimal methods already in 1614.

We consider a loxodrome starting at the equator as in figure 6.3.27 (right) and which intersects all lines of constant latitude under the angle α. The pieces $A'B$, $B'C$, and so on are sections on the lines of constant latitudes. The partition is chosen so that $AB = BC = \cdots = d$ holds with some constant d. The triangles $AA'B$, $BB'C$, and so on are assumed to be so small that approximately

$$AA' = BB' = \cdots = d \cdot \cos\alpha$$
$$A'B = B'C = \cdots = d \cdot \sin\alpha$$

holds. The sections AA', BB', and so on are direct differences of latitudes so that on a journey on the loxodrome the difference in latitudes travelled will be

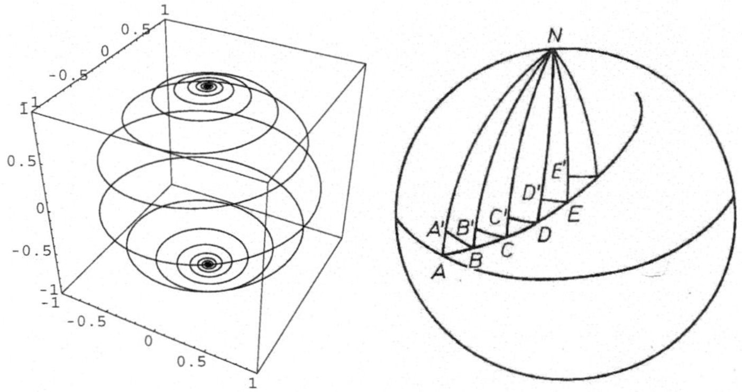

Fig. 6.3.27. (left) A loxodrome. (right) Rectification of the loxodrome [Pepper 1968]

$$\cos\alpha \sum d,$$

where the number of terms in the sum corresponds to the pieces of the partition travelled. The case of the difference of longitude is a different matter since we have to take into account the stretching factor (6.30) of the Mercator projection. Hence for the difference in longitude one gets

$$\sin\alpha \sum_i d \cdot \sec\varphi_i,$$

where φ_i denotes latitudes travelled on the pieces of the partition. This equation was already known to Wright and he had laboriously added the values of the secant to compute his tables. In our notation it follows for the difference L in longitude travelled from latitude 0 (equator) to latitude β:

$$L(\beta) = k \lim_{i\to\infty} \sum_i \sec\varphi_i,$$

where we have introduced a scaling constant k so that we will not have to care about null sequences of values of ds. This equation is but nothing else than

$$L(\beta) = k \int_0^\beta \sec\varphi \, d\varphi = k \ln\tan\left(\frac{\pi}{4} + \frac{\beta}{2}\right). \tag{6.31}$$

It seems unbelievable that Harriot knew this equation but Pepper has conclusively proven this fact from Harriot's manuscripts. Harriot developed the fundamental equation

$$\tan\left(\frac{\pi}{4} - \frac{\Theta_n}{2}\right) = \tan^n\left(\frac{\pi}{4} - \frac{\Theta_1}{2}\right)$$

valid for a sequence of angles $\Theta_1, \Theta_2, \ldots$, and tabulated the values (6.31) by means of an ingenious computation of the logarithm of the tangent function, where besides an independent form of a logarithmic table he also used interpolation to keep the computing effort small. The actual relation between the integral to be computed and the logarithm of the tangent function is found by Harriot by a subtle investigation of the length of a spiral and the area enclosed by the spiral. Full of admiration Pepper writes in [Pepper 1968] that these computations concerning the spiral alone, based on infinitesimal methods, belong to the true highlights in the history of rectification and of quadrature.

Thus Harriot has rectified the loxodrome already in 1614! Today we compute the length of the loxodrome as

$$s = \frac{1}{\cos\alpha} \int_0^t \frac{d\xi}{\cosh(\xi \cdot \tan\alpha)} = \frac{2}{\sin\alpha} \left(\arctan \mathrm{e}^{t\cdot\tan\alpha} - \arctan 1\right).$$

William Neile

Very little is known of William Neile (1637–1670). He came to fame by the rectification of the semi-cubic parabola which today is called Neile's parabola in 1657. In the older literature this rectification was called the first, but Thomas Harriot outstripped Neile as the first rectifier by some fifty years, since he succeeded in rectifying the loxodrome.

Neile was the grandson of the Archbishop of York in whose palace he was born. He studied at Wadham College, Oxford, from 1652 and came in contact with Seth Ward, then Savilian Professor of Astronomy, and John Wilkins, a founding member of the Royal Society. Both of these men recognised Neile's mathematical genius and gave him support [Scott 1981, p. 206]. In 1663 he became a member of the Royal Society and in 1669 presented his highly acclaimed work *De Motu* to the society. The great hopes placed in him were suddenly destroyed by his early death when he was just 33 years old.

The rectification of the semi-cubic parabola, in our notation

$$9y^2 = 4kx^3,$$

was achieved by Neile in a geometric fashion by 'summing' indivisibles. By comparison of terms in his 'sums' of infinitely small quantities he found familiar constructs and hence was able to compute the length of the function's curve. It then was easy for Wallis to turn Neile's proof into a purely arithmetical one and William Brouncker added an even further leading result [Stedall 2004]. All three results were published by Wallis in his treatise *De Cycloide* of 1659. We can illuminate Neile's idea by treating the semi-cubic parabola

$$y^2 = x^3$$

since any constant factors do not matter. Following figure 6.3.28 and applying Pythagoras' theorem we can express the arc length of Neile's parabola as

$$s \approx \sum_{i=1}^{n} s_i$$

where

$$s_i := \sqrt{(x_i - x_{i-1})^2 + (y_i - y_{i-1})^2},$$

if we assume a partition $0 = x_0 < x_1 < x_2 < \ldots < x_n =: a$.

Hence Neile knew that he had to 'sum' infinitesimal small quantities of the square root function and therefore looked at it more closely. The area under the square root function $z = \sqrt{x}$ between $x = 0$ and $x = x_i$ is, in today's notation,

$$A_i := \int_0^{x_i} \sqrt{x}\, dx,$$

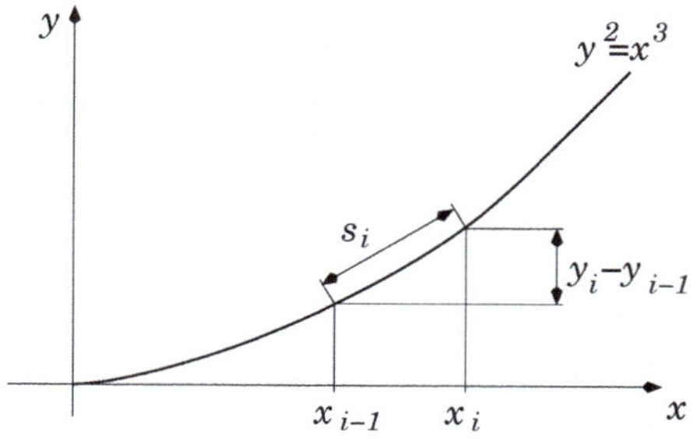

Fig. 6.3.28. Neile's parabola

and this is just a sum as it also appears in the arc length. In figure 6.3.29 the represented area is just $A_i - A_{i-1}$. Neile knew, of course, Wallis' results

$$\int_0^a x^{\frac{p}{q}}\, dx = \frac{q}{p+q} a^{\frac{p+q}{q}},$$

and thus the representation of the area in the form

$$A_i = \int_0^{x_i} x^{\frac{1}{2}}\, dx = \frac{2}{3} x_i^{\frac{3}{2}}$$

is possible. By insertion we now arrive at

$$y_i - y_{i-1} = x_i^{\frac{3}{2}} - x_{i-1}^{\frac{3}{2}} = \frac{3}{2}(A_i - A_{i-1})$$

and if the area $A_i - A_{i-1}$ is approximated by a rectangle it follows

$$y_i - y_{i-1} \approx \frac{3}{2} z_i (x_i - x_{i-1}). \tag{6.32}$$

To utilise this 'equation' successfully we recast our formula of the arc length a little bit:

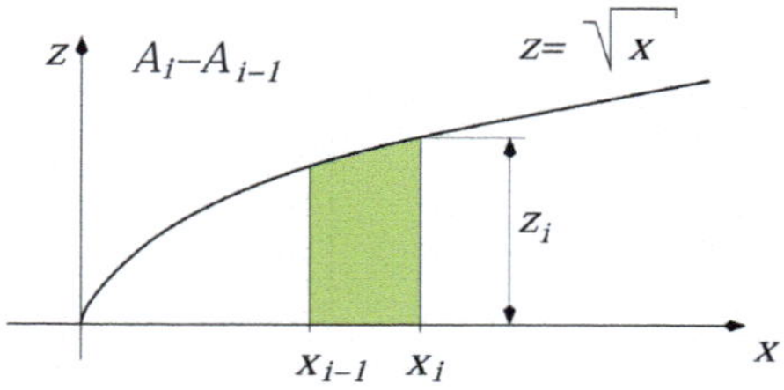

Fig. 6.3.29. The function $z = \sqrt{x}$

$$
\begin{aligned}
s &\approx \sum_{i=1}^{n}[(x_i - x_{i-1})^2 + (y_i - y_{i-1})^2]^{\frac{1}{2}} \\
&= \sum_{i=1}^{n}\left[(x_i - x_{i-1})^2\left\{1 + \left(\frac{y_i - y_{i-1}}{x_i - x_{i-1}}\right)^2\right\}\right]^{\frac{1}{2}} \\
&= \sum_{i=1}^{n}\left[1 + \left(\frac{y_i - y_{i-1}}{x_i - x_{i-1}}\right)^2\right]^{\frac{1}{2}}(x_i - x_{i-1}).
\end{aligned}
$$

If we now insert (6.32) it follows

$$
s \approx \sum_{i=1}^{n}\left[1 + \frac{9}{4}z_i^2\right]^{\frac{1}{2}}(x_i - x_{i-1}),
$$

and since $z_i = \sqrt{x_i}$ this is just

$$
s \approx \sum_{i=1}^{n}\left[1 + \frac{9}{4}x_i\right]^{\frac{1}{2}}(x_i - x_{i-1}) = \sum_{i=1}^{n}\frac{3}{2}\left[\frac{4}{9} + x_i\right]^{\frac{1}{2}}(x_i - x_{i-1}).
$$

But this means that s is approximately the area under the curve $y(x) = \frac{3}{2}\sqrt{\frac{4}{9} + x}$ between $x = 0$ and $x = a$. This consideration is depicted in figure 6.3.30. Shifting the whole function by 4/9 to the right shows the arc length being the area under the function $y = \frac{3}{2}\sqrt{x}$ between $x = \frac{4}{9}$ and $\frac{4}{9} + a$. Hence we can give Neile's result in modern notation as

$$
s = \int_{\frac{4}{9}}^{\frac{4}{9}+a}\frac{3}{2}x^{\frac{1}{2}}\,dx = \frac{(9a+4)^{\frac{3}{2}} - 8}{27}.
$$

6.3.10 James Gregory

There can be no doubts that the Scotsman James Gregory (1638–1675) was a great genius in 17th century Great Britain who was only surpassed by Isaac Newton [Turnbull 1940].

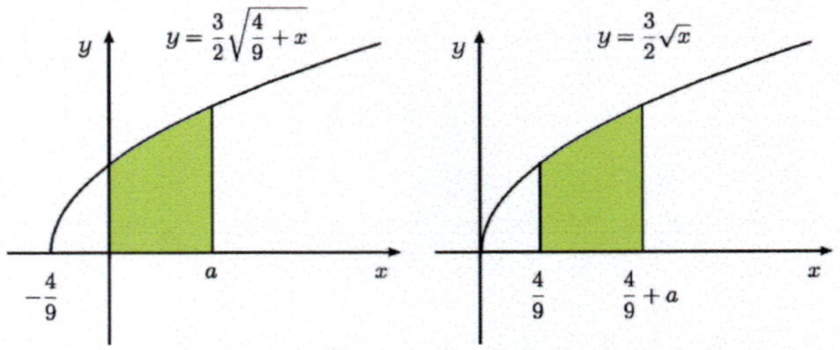

Fig. 6.3.30. The arc length of Neile's parabola as area under a square root function

Gregory studied in Italy from 1664 to 1668 where he was introduced to the methods of indivisibles of Cavalieri and Torricelli by Stefano degli Angeli (1623–1697) [Turnbull 1939, pp. 1–15]. Gregory inhabited professorships at the University of St Andrews and at the University of Edinburgh and, as did Newton , worked on a reflecting telescope. This telescope was called the 'Gregorian telescope' and was actually built by Robert Hooke in 1673. As an astronomer he developed a method to calculate the distance of the earth from the sun by a transit of Venus.

While in Padua he published in 1667 in *Vera Circuli et Hyperbolae Quadratura* on convergent infinite series and on the computation of the area under the hyperbola. Apparently he was also aware of the fundamental theorem of calculus stating that quadrature is the inverse operation to computing tangents. Even the series which later was named Taylor series after Brook Taylor (1685–1731) can be found in his work. The series expansions of the functions $\sin x, \cos x, \arccos x, \arcsin x, \arctan x$ were also found by Gregory. Gregory was an admirer of Newton and corresponded with him. However, his early death prevented that his genius could unfold. Some of his mathematical works can be found in [Baron 1987, p. 228ff.], in particular his form of the fundamental theorem.

6.4 Analysis in India

Our journey through the history of analysis was pretty much eurocentrered up to now except of our report on the very beginnings in Egypt and Mesopotamia. Hence the obvious question appears whether 'analysis' did develop also in non-European countries. The most likely candidate at first

Fig. 6.3.31. James Gregory [unknown artist] and his reflecting telescope of about 1735 in the Putman Gallery, Harvard Science Center [Photo: Sage Ross, 2009]

glance is China since so many discoveries and invention were made in China first. But the Chinese apparently had not much interest in developing an analysis and the few known works concerning infinite series originated only from the middle of the 18th century; and they were probably initiated by imported ideas from Europe [Martzloff 2006, p. 353f.]. It is India where we can find much more material [Baron 1987, p. 61f.].

In Hindu mathematics computational aspects were always in the foreground. Hence it never came to the development of a culture of proofs as happened in the Greek culture. In analysis it is interesting that even Brahmagupta (598–668) was concerned with the computation of zeros in 628. He came up with the algebraic laws

$$a - a = 0, \quad a + 0 = 0, \quad a - 0 = 0, \quad 0 \cdot a = 0, \quad a \cdot 0 = 0.$$

But Brahmagupta also allowed division by 0 in that he imagined 0 being an infinitesimal becoming actually zero. So he wrote $a/0$ without assigning a value to the quotient.

Bhāskara II (1114–1185) about 1150 formulated the idea

$$a/0 = \infty, \quad \infty + a = \infty$$

and Ganeśa becomes even more articulate in 1558 [Baron 1987, p. 62]:

> *$a/0$ is an indefinite and unlimited or infinite quantity: since it cannot be determined how great it is. It is unaltered by the addition or subtraction of finite quantities.*

Arithmetic and geometric series were known in India already in the 4th century BC and about 200 BC Pingala seems to have known the binomial coefficients. Between 300 and 1350 we find problems in which squares and third powers of natural numbers had to be summed. Already in the 15th century the series expansions of $\sin x, \cos x, \arctan x$ were known, and in astronomy numerical quadrature methods were applied which were derived in Europe only two centuries later.

In 1671 James Gregory discovered the series of the arcus tangent, but this series was known to Hindu mathematicians already in the 15th century and was known as 'Talakulattura's series'. To derive it the Indians used a circular arc Apq with radius OA as shown in figure 6.4.1.

The radii Op, Oq meet the tangent of the circle in the points P and Q. Let pm be the perpendicular of p on OQ and $OA = Op = Oq = 1$. Then it follows from the sine theorem

$$\frac{PQ}{OP} = \frac{\sin \angle QOP}{\sin \angle OQP}.$$

Since

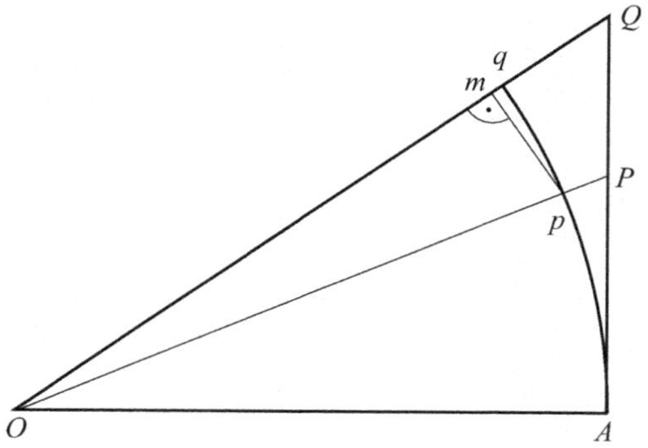

Fig. 6.4.1. Derivation of the series of the arcus tangent in India, part 1

$$\sin\measuredangle OQP = \frac{OA}{OQ} = \frac{1}{OQ},$$
$$\sin\measuredangle QOP = \frac{pm}{Op} = pm,$$

it therefore follows

$$\frac{PQ}{OP} = \frac{pm}{\frac{1}{OQ}} = pm \cdot OQ,$$

and hence

$$pm = \frac{PQ}{OP \cdot OQ}.$$

If PQ is 'small' then $pm \approx \text{arc } pq = \frac{PQ}{OP^2}$ and with the theorem of Pythagoras it then follows

$$pm \approx \frac{PQ}{1 + AP^2}.$$

We now turn to figure 6.4.2 and require $\angle BOA < \frac{\pi}{4}$.

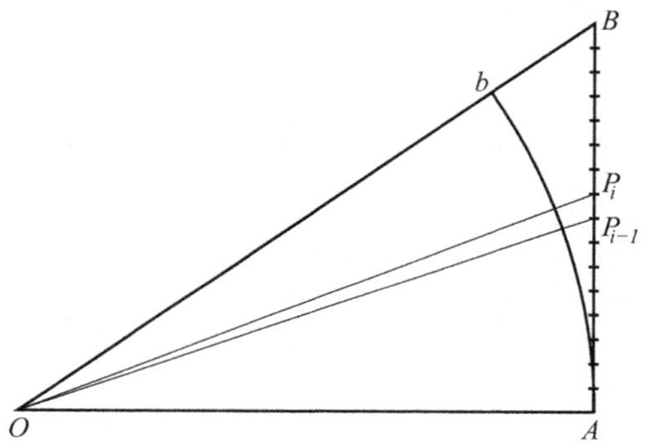

Fig. 6.4.2. Derivation of the series of the arcus tangent in India, part 2

We define $t := \tan\measuredangle AOB = AB$ and partition the segment AB into n equidistant parts, so that the points $A = P_0, P_1, P_2, \ldots, P_n = B$ emerge. Let the point b denote the point of intersection of OB with the circular arc. Then it holds

$$\begin{aligned}\text{arc } Ab &= \lim_{n\to\infty} \sum_{r=0}^{n-1} \frac{P_r P_{r+1}}{1 + AP_r^2} \\ &= \lim_{n\to\infty} \sum_{r=0}^{n-1} \frac{\frac{t}{n}}{1 + \left(\frac{rt}{n}\right)^2} \\ &= \lim_{n\to\infty} \sum_{r=0}^{n-1} \frac{t}{n} \left(1 - \left(\frac{rt}{n}\right)^2 + \left(\frac{rt}{n}\right)^4 - + \ldots\right).\end{aligned}$$

The next step is remarkable. The Hindu mathematicians knew about 1500 that

$$\lim_{n\to\infty} \frac{\sum_{r=0}^{n-1} r^p}{n^{p+1}} = \frac{1}{1+p}$$

holds; long before Wallis and other mathematicians in Europe. Applying this results above it follows

$$\text{arc } Ab = t - \frac{t^3}{3} + \frac{t^5}{5} - \frac{t^7}{7} + - \ldots$$

and that is the series of the function $\arctan t$. To rigorously justify the application we have to show the exchangeability of two limit processes:

$$\lim_{n\to\infty} \sum_{r=0}^{n-1} \sum_{p=0}^{\infty} (-1)^p t^{2p+1} \frac{r^{2p}}{n^{2p+1}} \stackrel{!}{=} \sum_{p=0}^{\infty} (-1)^p t^{2p+1} \lim_{n\to\infty} \sum_{r=0}^{n-1} \frac{r^{2p}}{n^{2p+1}}.$$

We may safely assume that the Indian mathematicians were not aware of these convergence problems. Setting $t = 1$, i.e. $\arctan 1 = \pi/4$, then Leibniz's famous series (cp. page 414)

$$\frac{\pi}{4} = 1 - \frac{1}{3} + \frac{1}{5} - \frac{1}{7} + - \ldots$$

follows. Hindu mathematics certainly came into the Arabian world by trade relations as did the Indian place-value notation with digits $0, 1, \ldots, 9$. This is a remarkable example of the accomplishments the Hindu mathematicians achieved long before the Europeans. A detailed overview of the achievements of Indian mathematics can be found in [Plofker 2007].

Development of Analysis in the 16th/17th Century

about 1500	The series expansion of the arcus tangent is known in India
1544	MICHAEL STIFEL publishes arithmetic and geometric scales in his *Arithmetica Integra*
1558	In India GANEŚA recognises the infinity when dividing by zero
1596	HENRY BRIGGS becomes the first Gresham Professor of Geometry
1600	WILLIAM GILBERT founds English natural philosophy with his *De Magnete*
1614	JOHN NAPIER publishes the first logarithms in *Mirifici Logarithmorum Canonis Descriptio*
about 1614	THOMAS HARRIOT rectifies the loxodrome
1619	HENRY BRIGGS becomes first Savilian Professor of Geometry at Oxford University
1624	*Arithmetica Logarithmorum* by HENRY BRIGGS is published in London. Logarithms are established
1629–1695	CHRISTIAAN HUYGENS. The Dutchman becomes the greatest natural scientist of his country. He ranks as teacher of Leibniz and was highly esteemed by Newton
1635	MARIN MERSENNE founds the 'free academy' in Paris
vor 1636	GILLES PERSONNE DE ROBERVAL succeeds in the quadrature of the cycloid
1637	RENÉ DESCARTES publishes *Discours de la méthode ...*, the birth of analytical geometry. 'Circle method' for the determination of tangents at curves
1637	PIERRE DE FERMAT sends his mathematical works to *Mersenne* in Paris. Among these are ingenious works for the calculation of extrema
1640	BLAISE PASCAL publishes his first mathematical work on conic sections
1649	The Dutchman FRANS VAN SCHOOTEN JR translates *La Géométrie* of Descartes into Latin. This translation becomes influential
1654–1663	The DutchmanJOHANN VAN WAVEREN HUDDE develops 'Hudde's rule'
1656	*Arithmetica Infinitorum* by JOHN WALLIS is published in Oxford
1657	WILLIAM NEILE rectifies the semi-cubic parabola
1658	BLAISE PASCAL publishes his famous price contest in Paris under the pseudonym of 'Amos Dettonville'
1659	BLAISE PASCAL publishes *Traité des sinus du quart de cercle.* Here Leibniz will find the idea of the characteristic triangle
1668	NICOLAUS MERCATOR publishes *Logarithmotechnia* in London, in which Mercator'sche series of the logarithm appears
1673	RENÉ FRANÇOIS WALTHER DE SLUSE publishes an own method to determine tangents in Holland which he knew already about 1655

7 Newton and Leibniz – Giants and Opponents

T. Sonar, *3000 Years of Analysis*, https://doi.org/10.1007/978-3-030-58223-4_7

1618–1648	Thirty Years' War
1620	The pilgrim fathers, Puritans from England, settle at Cape Cod in North America
1633	Galilei has to withdraw his confession to the Copernican world system before the inquisition
1643–1715	Louis XIV King of France
1643	Isaac Newton born on 4th January (new system) in Woolsthorpe close to Grantham
1644	Blaise Pascal builds the first mechanical computing machine still extant
1646	Gottfried Wilhelm Leibniz born on 1st July in Leipzig
1649	King Charles I of England beheaded. Commonwealth introduced by Cromwell
1660–1685	Charles II King of England
1658–1705	Leopold I Holy Roman Emperor
1658	The first League of the Rhine forms against the Emperor
1662	Foundation of the Royal Society in London
1665–1667	Second Anglo-Dutch War
1665	Outbreak of the plague in London
1666	Great Fire of London destroys large part of the city. Reconstruction led by Christopher Wren und Robert Hooke Foundation of the Paris Academy of Sciences
1672	Leibniz invents the Leibniz wheel as an element of mechanical computing machines
1672–1678	War of conquest by Louis XIV against the Netherlands
1683–1699	Great Turkish War
1685–1688	The Catholic James II King of England
1688	Protestants invite William of Orange who arrives in London at the end of December. James flees to France
1688–1713	Frederick III Elector of Brandenburg. From 1701 King in Prussia as Frederick I
1689–1725	Peter the Great Tsar of All Russia
1702–1714	Anne, Queen of Ireland, England, and Scotland
1702–1713	English participation in the War of the Spanish Succession
1703	Foundation of St Petersburg
1705–1711	Joseph I Holy Roman Emperor
1709	Ehrenfried Walter von Tschirnhaus and Johann Friedrich Böttger invent the European white hard porcelain
1711–1740	Charles VI Holy Roman Emperor
1714–1727	The Elector of Hanover, George Louis becomes King of England as George I
1725	Opening of the Academy of Sciences in St Petersburg

7.1 Isaac Newton

On the first day of Christmas in the year 1642 (old system) two obstetricians hurry from Woolsthorpe in Lincolnshire to North-Witham. They want to fetch strengthening medicine for the early-born Isaac Newton, son of the widow Hannah Newton. After a pregnancy of only seven months the midget has barely a chance to survive since he fits into a quart mug. As the old Newton reported the obstetricians were not so much in hurry since they both were convinced that the infant would be dead when they would come back [Westfall 2006, p. 49]. Who could have thought in this situation that the premature baby would not only become 84 years old, but also one the greatest mathematicians and physicist of all times?

7.1.1 Childhood and Youth

First to the date of birth: On the Continent the new Gregorian calendar was already accepted to a great extent; not so in Protestant England. Newton is hence a Christmas child of the year 1642 only if we follow the Julian calendar. Following the new system his birthday is 4th January 1643. Newton's father Isaac was a yeoman; one of the many free peasants with small estate, Woolsthorpe Manor.

When little Isaac was thrown into this life his father had already died during the pregnancy of his wife Hannah, née Ayscough. The Ayscoughs were much better off than the Newtons and we can view this marriage as

Fig. 7.1.1. Woolsthorpe Manor: Newton's birthplace

Newton's successful attempt to rise in the social status of his time. Hannah Newton could certainly not throw money around but a truly poor family the Newtons were not. With his father missing, the relation between little Isaac and his mother must have developed into something special. When his mother decided to get married with the 63 years old minister Barnabas Smith, moving with him to the neighbouring community of North-Witham and thereby leaving her little son with his grandmother in Woolsthorpe, the world of the child must have been collapsed. Many an author has since tried to explain Newton's behaviour as an adult on grounds of this emotional catastrophe. Most prominently in this respect ranks Frank Edward Manuel's book [Manuel 1968]. It is known that Newton developed murder phantasies towards his stepfather and even towards his mother. When twenty years old he admitted a 'sin no. 13' in a notebook [Manuel 1968, p. 26]:

> *Threatning my father and mother Smith to burne them and the house over them.*

Throughout his life Newton ranked a difficult character. In the priority dispute with Leibniz he presented himself from his very worst side, cp. [Sonar 2018]. Physicist Stephen Hawking, one of the successors of Newton on the Lucasian Chair in Cambridge characterised him as follows [Hawking 1988, p. 191]:

> *Isaac Newton was not a pleasant man. His relations with other academics were notorious, with most of his later life spent embroiled in heated disputes.*

It seems reasonable to follow Manuel and to connect Newton's disconcerting character with his horrible experience of separation from his mother at an early age. His stepfather Barnabas Smith died in August 1653 and Newton's mother returned to Woolsthorpe; however, she was accompanied by three half siblings of Isaac: Benjamin, Mary, and Hannah Smith. Newton was ten years old and had already lived seven years motherless. Although Barnabas Smith had apparently never looked after his stepson between 200 and 300 books on theology which formerly belonged to Smith found their way to the self-built bookshelves in Newton's room in Woolsthorpe as a 'heritage' of the deceased stepfather. We can not be exactly sure whether Newton's lifelong obsession concerning theological matters began with this books but it seems very likely. Two years were left to Newton to enjoy the presence of his mother, although these years certainly were not free of tensions due to his three little half siblings, until he entered the *free grammar school* in Grantham, about 9 km north of Woolsthorpe. Before that he had attended an elementary school, but the school in Grantham was already a place of higher education. Although the school enjoyed an excellent reputation we should point out the content of teaching: Latin, a little bit of Greek, and the study of the bible; barely any mathematics [Westfall 2006, p. 56]. This corresponded to the demands of the time. The frail Newton is said to have been beaten up by his crude schoolmates. Although he returned the beating full of rage after school he decided to take revenge also on the intellectual level and quickly advanced to the top of his class. The knowledge of Latin which Newton acquired at school in Grantham served him very well during his life; not only in writing his important works, but also in the correspondence with scientists on the Continent.

In Grantham the young pupil lived in the house of apothecary Clark whose wife was a friend of Newton's mother. With children of same age, in particular with boys, Newton always had problems. Surely the long time staying with his grandmother had made an impact. Moreover, as the top of his class he was subject to the usual teasings and other children refused to play with him due to his faster perception. It is said that the adolescent had a delicate flirtation with Clark's stepdaughter, a Miss Storer. This is the first and last story relating Newton to any female being; his mother being an exception, of course.

During his days in Grantham Newton showed great mechanical skills in designing and constructing machines and we may even be tempted to find here already the young experimental physicist. His room filled up with tools for which he spent all the money he got from his mother. For the girls he built doll furniture, but also a model of a windmill the original of which was just under construction in Grantham. He complemented his model with an impeller driven by a mouse. He also built a four-wheeled carriage for himself driven by a crank handle. Using a self-constructed foldable lantern he lit his way to school in winter and he also mounted it on a kite which he let fly at night to scare the inhabitants of Grantham. During a storm he conducted experiments concerning the force of the wind in that he jumped with the wind and against it, and measured the distances which the wind had moved him. He also measured the length of the shadows of sticks in the sunlight and constructed sundials. He was said having been so good at it that he was not only able to read off the solstices from the length of the shadows but also to determine the individual days of a month by it. Life in the home of an apothecary awakened Newton's interest in chemicals and drugs. His love of (al)chemistry will last all of his lifetime and he will spend much more time in alchemy than in mathematics, physics, and astronomy [Westfall 2006, p. 63].

At the end of 1659, shortly before Newton's 17th birthday, his schooldays ended and he returned to his mother in Woolsthorpe, who doubtlessly had hoped for some male help at the farm and certainly also for Isaac inheriting the small manor. The hopes of mother Hannah were dashed, however. Although a servant received the order to train Newton in farmer's work the latter was elsewhere with his thoughts. When he was supposed to tend sheep he instead built a small water mill while the sheep trampled down the corn of the neighbour. Newton even bribed the servant on market days to let him go and do the sales without him, meanwhile reading books or tinkering

Fig. 7.1.2. Trinity College, Cambridge, in an engraving of 1690 (Newton's rooms were located at the first floor right of the main entrance)

around with new little machines. When going to Grantham the library of the apothecary, freely opened for his use, was always his first destination. On one occasion he walked having a horse on the leash behind him when he was so immersed in his thoughts that he did not realise that the horse detached from the leash and went away. He arrived at home with the leash in hand; the horse was gone. Yes, Isaac Newton certainly was a complete failure as far as farming was concerned.

At this point the mother's brother, clergyman William Ayscough, intervened in favour of Newton. Ayscough must have looked benevolently at his nephew and must have realised that the lad should go to university. He was supported by the Grantham schoolmaster Mr. Stokes who unambiguously pointed out to Newton's mother that Isaac working in farming would simply be a waste of talent. Stokes went so far to exempt Newton from the school fees and to offer free accommodation in his own house. In autumn 1660 Isaac Newton was back in school; this time with the clear aim of training for studying at a university.

7.1.2 Student in Cambridge

Newton entered the University of Cambridge in June 1661. Since he was not the son of a wealthy big landowner or a nobleman he was enrolled as a 'subsizar'; a servant for the fellows and older and richer students. This needed not be! His mother Hannah had a yearly income of about £700 [Westfall 2006, p. 72] and therefore would have easily been able to fully fund her son at university. Apparently the mother, who could barely write, was deeply

Fig. 7.1.3. Isaac Newton (Statue in the University Museum of Natural History, Oxford) [Photo: Thomas Sonar]

disappointed by her son having shown to be a complete failure in farming and having instead turned to (in her eyes) unprofitable academic studies.

His mother merely provided her son with £10 annually. Newton enrolled at Trinity College where his uncle William Ayscough had studied before him. Another reason for the choice of Trinity may have been the brother of the apothecary Clark's wife in Grantham, Humphrey Babington. Babington was an influential fellow of Trinity College and was very fond of Newton.

At Trinity Newton shared rooms with John Wickins with whom a kind of friendship developed, if the notion of 'friendship' had a meaning at all for a maverick like Newton. But what did Newton had to learn at Cambridge University? Today Cambridge counts as one of the leading universities in the world; however, in Newton's days the situation was very deplorable. Aristotle still was the central authority and hence one had to learn Aristotelian logic, rhetoric, natural philosophy, and physics. Many wealthy students enrolled at Cambridge to indulge into dog races, escalating binge drinking, and to enjoy affairs with young women. If the fathers of such students were of the opinion that it became necessary for their offsprings to enter real life a leaving certificate was bought for a lot of money. It fits the overall picture that many lecturers were paid by the Colleges without ever having to give a single lecture.

But Newton was different. He bought books not actually needed in the curriculum and started in self studies: he read books by Descartes, Robert Boyle, Thomas Hobbes, Henry More, Galilei's *Dialogue Concerning the Two Chief World Systems*, and books by other authors considered modern in those days. Before he went to university he read a book on logic on the recommendation of his uncle which the latter had to study 30 years before at Cambridge. At university Newton had to discover that the same book still belonged to the compulsory reading but that he already knew more logic than his tutor.

In 1662 Newton apparently suffered a religious crisis in which he compiled a list of all of his committed sins. In a notebook we find Newton's true occupation in 1663: Theorems concerning conic sections following Pappus, remarks concerning geometrical theorems by Viète, van Schooten, and Oughtred, theorems concerning arithmetic by John Wallis, methods of grinding lenses, questions of natural philosophy, theology, and alchemy. He was also very keen on optical questions. Where do the colours come from? In trying to answer this question he even risked his health. Recognising that pressure on the eyeball results in the sensation of colours he pushed a hatpin behind his eye and exerted pressure close to the optic nerve, see figure 7.1.4.

He did not loose his eyesight but he could not see for some time and had to stay in his darkened room.

In 1663 the Lucasian Chair was donated at Cambridge and the first inhabitant, Isaac Barrow, began his lectures in 1664. Contrary to folklore Barrow was not Newton's tutor, but Westfall suspected in [Westfall 2006, p. 99] that in all of Cambridge it could have been only Barrow to have lent Newton books by Wallis. A new crisis loomed in 1664 concerning the awarding of stipends. Newton had to postpone all of his private studies and to start caring for the neglected university curriculum. In fact, Newton succeeded to be elected and became a 'scholar'. Together with this election came annual grants from the college as well as an annual stipend. Apparently even more important for Newton were the secured next four years in which he could continue his studies.

He was now able to throw himself into his studies with all his power and he did that with dedication; one can even speak of obsession. He often forgot

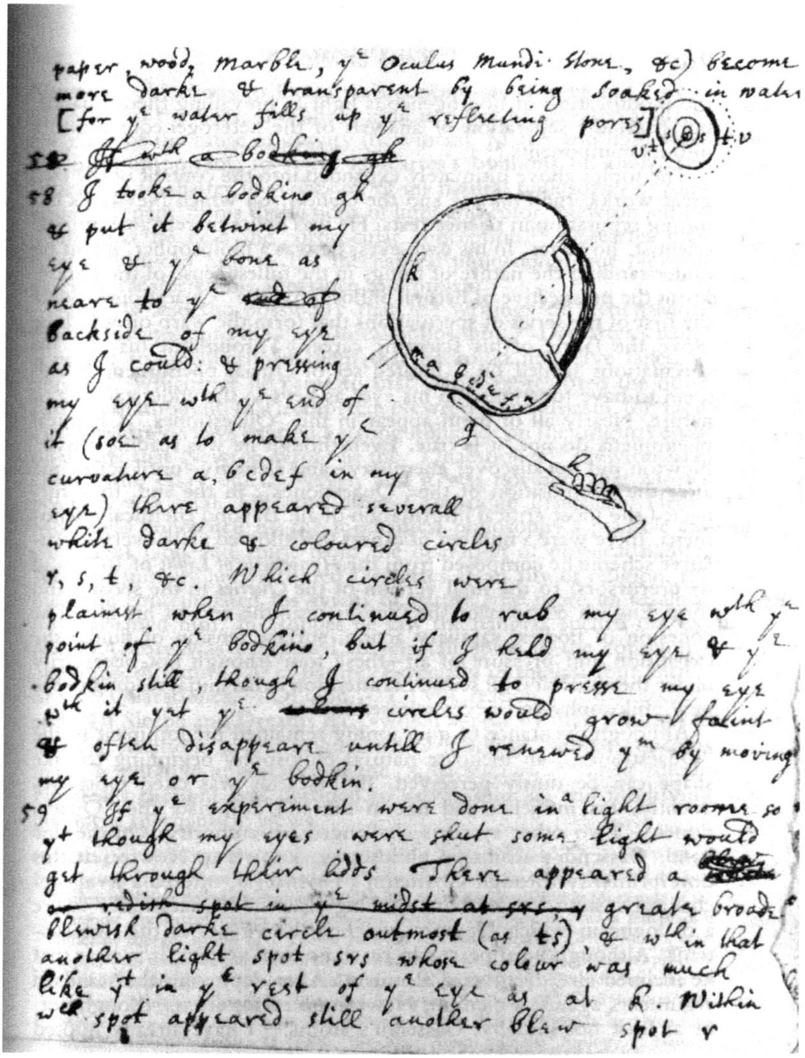
paper, wood, marble, ye Oculus Mundi Stone, &c) become
more darke & transparent by being soaked in water
[for ye water fills up ye reflecting pores]
58 ~~If wth a bodkin gh~~
58 I tooke a bodkin gh
& put it betwixt my
eye & ye bone as
neare to ye ~~end of~~
backside of my eye
as I could: & pressing
my eye wth ye end of
it (soe as to make ye
curvature a, bcdef in my
eye) there appeared severall
white darke & coloured circles
r, s, t, &c. Which circles were
plainest when I continued to rub my eye wth ye
point of ye bodkin, but if I held my eye & ye
bodkin still, though I continued to presse my eye
wth it yet ye circles would grow faint
& often disappeare untill I renewed ym by moving
my eye or ye bodkin.
59 If ye experiment were done in a light roome so
yt though my eyes were shut some light would
get through their lidds There appeared a
~~reddish spot in ye midst at~~ greate broade
bluish darke circle outmost (as ts), & wthin that
another light spot srs whose colour was much
like yt in ye rest of ye eye as at k. Within
wch spot appeared still another blew spot r

Fig. 7.1.4. Newton's recording concerning the experiment with his own eye (Reproduced by kind permission of the Syndics of Cambridge University Library, Ms. Add. 3995 p. 15 Bound notebook of 174 leaves)

to eat, so that the cat living with Newton and his roommate Wickins got ever fatter since it ate all of Newton's food. As little as he ate so little did he sleep. If he was gripped by a calculation he did not think of going to bed but did not rest until he was in command of the solution. In his later years his servants still called Newton half an hour early to dinner. When he then came out of his study and by chance found an interesting book or manuscript the food remained untouched. It was not uncommon that Newton ate cold porridge or eggs in milk served warm the evening before for breakfast. There is also no doubt that Newton's psychosis which he took from Woolsthorpe to Cambridge got stronger in the periods in which he was working hard on scientific problems, since he then behaved strangely to Wickins and others [Westfall 2006, p. 104].

Following an anecdote, when Newton in his old age was asked how he discovered the law of gravitation he answered: 'By permanent thinking'. Whether this anecdote is made up or not, it shows a characteristic feature of Newton: work, work, work, and once again work. This having said, we have to bear in mind that Newton by no means repeated the thoughts of others. He had already some time ago started to move into uncharted intellectual territory which he entered as the first human being ever. After having studied the works of Wallis, Descartes (in van Schooten's second Latin edition), and Viète he came into the possession of the binomial theorem some time between winter 1664 and summer 1665. In the midst of the labour pains of his new analysis he was asked in 1665 to take his exams for a Bachelor degree which he passed by the skin of his teeth. In the summer of 1665 the plague broke out. The university was closed and the students were sent home. By the 7th August 1665 Newton went back to his mother in Woolsthorpe and moved back to Cambridge only on the 20th March 1666. When the plague flared up again the university was again closed; again Newton went back to his mother and returned to Cambridge at the end of April 1667. The years from 1664 to 1666 are usually called Newton's 'anni mirabiles'. Newton himself wrote about 50 years later [Westfall 2006, p. 143]:

> *In the beginning of the year 1665 I found the Method of approximating series & the Rule for reducing any dignity of any Binomial into such a series. The same year in May I found the method of Tangents of Gregory & Slusius, & in November had the direct method of fluxions & the next year in January had the Theory of Colours & in May following I had entrance into ye inverse method of fluxions. And the same year I began to think of gravity extending to ye orb of the Moon & (having found out how to estimate the force with w^ch [a] globe revolving within a sphere presses the surface of the sphere) from Keplers rule of the periodical times of the Planets being in sesquialterate proportion of their distances from the center of their Orbs, I deduced that the forces wch keep the Planets in their Orbs must [be] reciprocally as the squares of their distances from the centers about wch they revolve: & therebye compared the force requisite to keep the Moon in her Orb with the force of gravity at the surface of the earth, & found them answer pretty nearly. All this was in the two plague years of 1665–1666. For in those days I was in the prime of my age for invention & minded Mathematickes & Philosophy more then at any time since.*

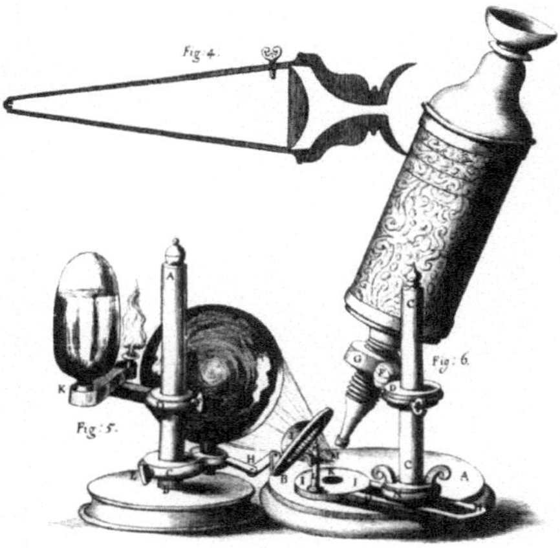

Fig. 7.1.5. Engraving showing a microscope in Hooke's *Micrographia* of 1665

In 1666 Newton turned away from mathematics for the time being and addressed himself to physics; with the same energy and exclusiveness, of course. The often circulated story of Newton having found the law of gravitation when an apple fell on his head while sitting under an apple tree in Woolsthorpe and thinking, is a well-invented anecdote only. He did *not* have finished ideas in his mind for about twenty years leading eventually to the publication of his *Principia*, but his notebooks, the *Quaestiones*, clearly show that his physical conceptions of the world developed only from 1666 on. He was interested in optics; in particular in the emergence of colours. In 1665 Robert Hooke's (1635–1703) *Micrographia* had been published in which Hooke, curator of the physical collection of the Royal Society, wrote on the microscope and discussed a theory of colours which was refuted by Newton.

He bought a prism and started experimenting with it. In his work *A new Theory about Light and Colours* sent to the Royal Society in February 1672 he wrote in retrospect [Turnbull 1959–77, Vol. I, p. 92]:

> *Sir,*
> *To perform my late promise to you, I shall without further ceremony acquaint you, that in the beginning of the Year 1666 (at which time I applyed my self to the grinding of Optick glasses of other figures than Sperical,) I procured me a Triangular glass-Prisme, to try therewith the celebrated Phœnomena of Colours. And in order thereto having darkened my chamber, and made a small hole in my window-shuts, to let in a convenient quantity of the Suns light, I placed my Prisme at its entrance, that it might be thereby refracted to the opposite wall. It was at first a very pleasing divertisement, to view the vivid and intense colours produced thereby; but after a while applying my self to consider them more circumspectly, I became surprised to see them in*

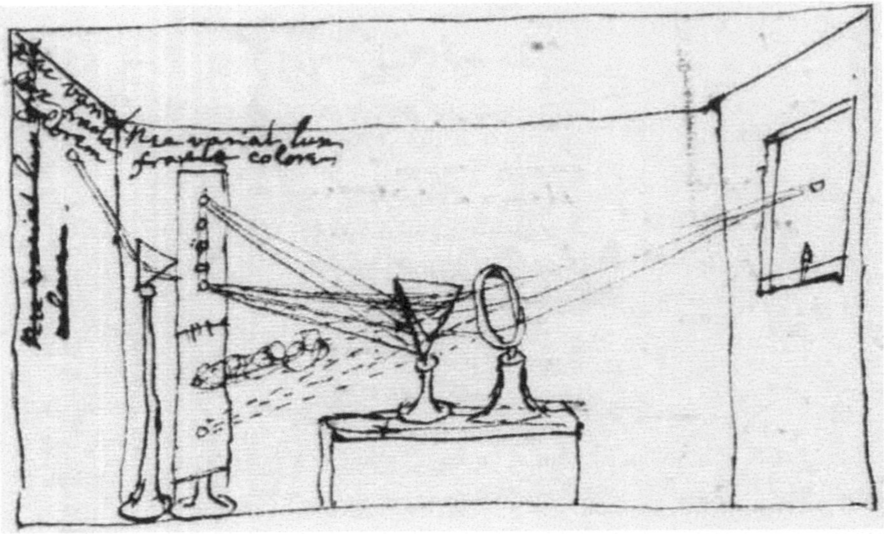

Fig. 7.1.6. Newton's sketch of the experimentum crucis (© Courtesy of the Warden and Scholars of New College, Oxford/Bridgeman Images)

> *an oblong form; which, according to the received laws of Refraction, I expected should have been circular.*
> *They were terminated at the sides with streight lines, but at the ends, the decay of light was so gradual, that it was difficult to determine justly, what was their figure; yet they seemed semicircular.*
> *Comparing the length of this coloured Spectrum with its breadth, I found it about five times greater; a disproportion so extravagant, that it excited me to a more then ordinary curiosity of examining, from whence it might proceed.*

After further descriptions of many experiments, mainly to refute the hypotheses of others, Newton finally described his 'experimentum crucis' [Turnbull 1959–77, Vol. I, p. 92–102], [Rosenberger 1987, p. 63f.]. Thereby he was able to prove that white light consists of a mixture of coloured light.

In 1667 Newton became 'minor fellow' of his college in Cambridge and shortly after, having been promoted to 'Master', he became 'major fellow'. His future seemed to be secured. In becoming 'minor fellow' he was sworn to the 39 articles of the Church of England and he had to make a solemn confession to remain celibate. Meanwhile Isaac Barrow had recognised the mathematical talents of his pupil and he wanted to make it known. He sent Newton's manuscript *De Analysi per Aequationes Numero Terminorum Infinitas* to John Collins (1625–1683) in 1669 who was the librarian and mathematical impresario of the Royal Society in London. Barrow had been sent Mercator's *Logarithmotechnia* and had studied it keenly, whereby he saw that Newton

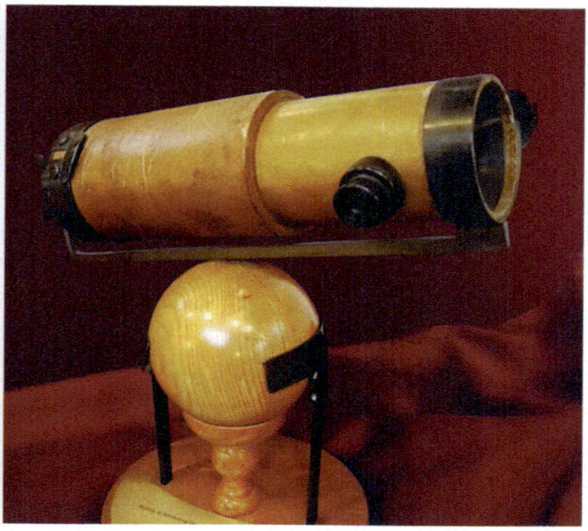

Fig. 7.1.7. Isaac Newton [Painting by Godfrey Kneller 1689] and his reflecting telescope [Photo: Andrew Dunn] (Whipple Museum of the History of Science, Cambridge)

had achieved the same. With Newton's permit Collins showed the manuscript to the president of the Royal Society, Lord William Brouncker (1620–1684), but then suddenly Newton required his manuscript to be send back to him. It is certain, however, that Sluse and Gregory learnt through Collins of Newton's work.

7.1.3 The Lucasian Professor

For quite some time Isaac Barrow was thinking of a career within the Church of England. In 1669 he renounced his chair at Cambridge and recommended his pupil Isaac Newton as his successor. Hence the only 27 years old Newton became professor on the Lucasian Chair. His first lecture at Cambridge was not concerned with mathematics, but with optics. Since the lenses in these days always delivered distorted pictures independently of how good and exact they were ground, Newton developed a quite new principle of the telescope: the reflecting telescope. Today one can buy 'Newtonian' reflecting telescopes to watch the sky all around the world.

In 1672 Newton became a member of the Royal Society because of his invention of the reflecting telescope. In the same year he presented his work on the theory of colours, containing the results of his experiments of the past few years, to the Royal Society. This work was very well received; only Robert Hooke brusquely refuted Newton's theory. A new hostility started to develop: In 1675 Hooke will even claim that Newton had stolen some of his own results in the theory of optics. Again Newton's psychosis disrupted. He will stay away from the meetings of the Royal Society until Robert Hooke's death in March 1703! Although both men still exchanged polite letters we can be sure that Hooke became Newton's archenemy. Christiaan Huygens also criticised Newton's theory since he believed in the wave character of light while in Newton's view light consisted of particles (corpuscles). This controversy was resolved only at the beginning of the 20th century with the wave-particle dualism discovered by French physicist Louis de Broglie (1892–1987). It seems very likely that Newton would have had published his book *Opticks* much earlier but it came out only in 1704 after Hooke's death in 1703. Bound with it was *De quadratura curvarum*, a work on the calculus of fluxions, which must have bewildered some of it's readers. But Newton had *nothing* published on his analysis until 1704 and got under pressure by the works of Leibniz and others on the Continent to reveal some of his results to the public.

In 1679 Newton corresponded with Hooke on the law of gravitation. In fact, Hooke had told Newton that attraction between two bodies would behave like the inverse square of the distance of these bodies, but Hooke had no proof and also could not offer one later on. In contrast, Newton had *proven* this law from Kepler's laws and his own law of centrifugal forces. Obviously Hooke claimed the law of gravitation being entirely his discovery; but eventually Hooke's boastings annoyed Edmond Halley (1656–1742) so that he urged Newton to publish the result.

Halley, a learned mathematician and ingenious astronomer, had early travelled to observe a transit of Mercury and to survey positions of stars. In 1684 in one of the new and popular coffee houses he had a discussion with Christopher Wren and Robert Hooke concerning a proof of Kepler's laws but the three men could not find one. So he travelled to Cambridge in the same year to meet Newton who happened to have the proof in his drawer. In 1703 Halley was appointed the Savilian Chair of Geometry at Oxford. He worked on the determination of orbits of comets and on the determination of longitude at sea. After Royal Astronomer John Flamsteed (1646–1719) had died Halley became his successor in Greenwich.

Three years after Halley had visited Newton in Cambridge Newtons opus magnum, *Philosophiae Naturalis Principia Mathematica*, in short: *Principia*, was published in 1687. This is the natal hour of modern physics. Since the Royal Society was short of money the printing was financed by Edmond Halley privately. In three 'books' of the *Principia* the laws of mechanics, the mechanics of motion in viscous fluids, the law of gravitation, and celestial mechanics can be found. It says all about the immense importance of the *Principia* that even today, more than 330 years after its first publication, it can still be bought in a number of different reprints and translations of which I recommend [Newton 1999] most.

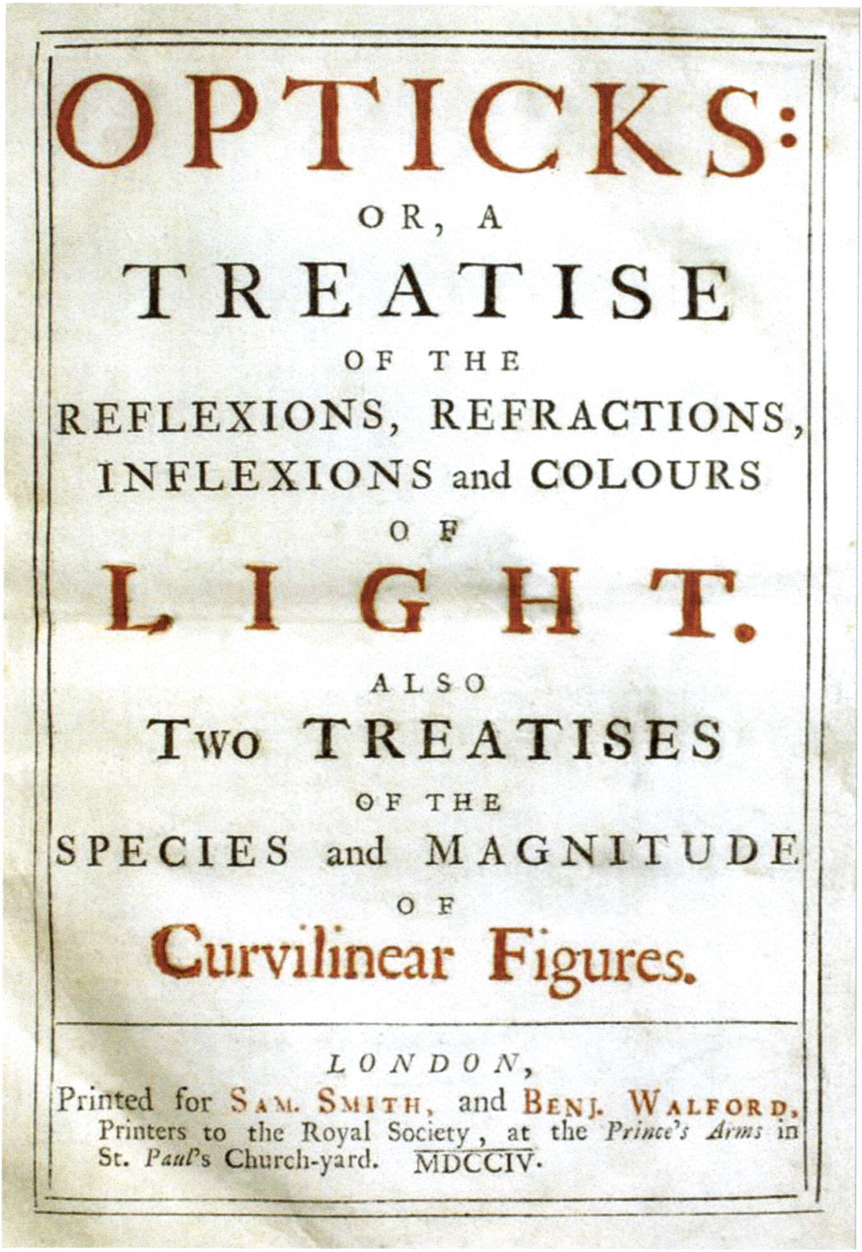

OPTICKS:
OR, A
TREATISE
OF THE
REFLEXIONS, REFRACTIONS,
INFLEXIONS and COLOURS
OF
LIGHT.
ALSO
Two TREATISES
OF THE
SPECIES and MAGNITUDE
OF
Curvilinear Figures.

LONDON,
Printed for SAM. SMITH, and BENJ. WALFORD,
Printers to the Royal Society, at the *Prince's Arms* in
St. *Paul's* Church-yard. MDCCIV.

Fig. 7.1.8. Newton's *Opticks*, Title page of the first edition 1704

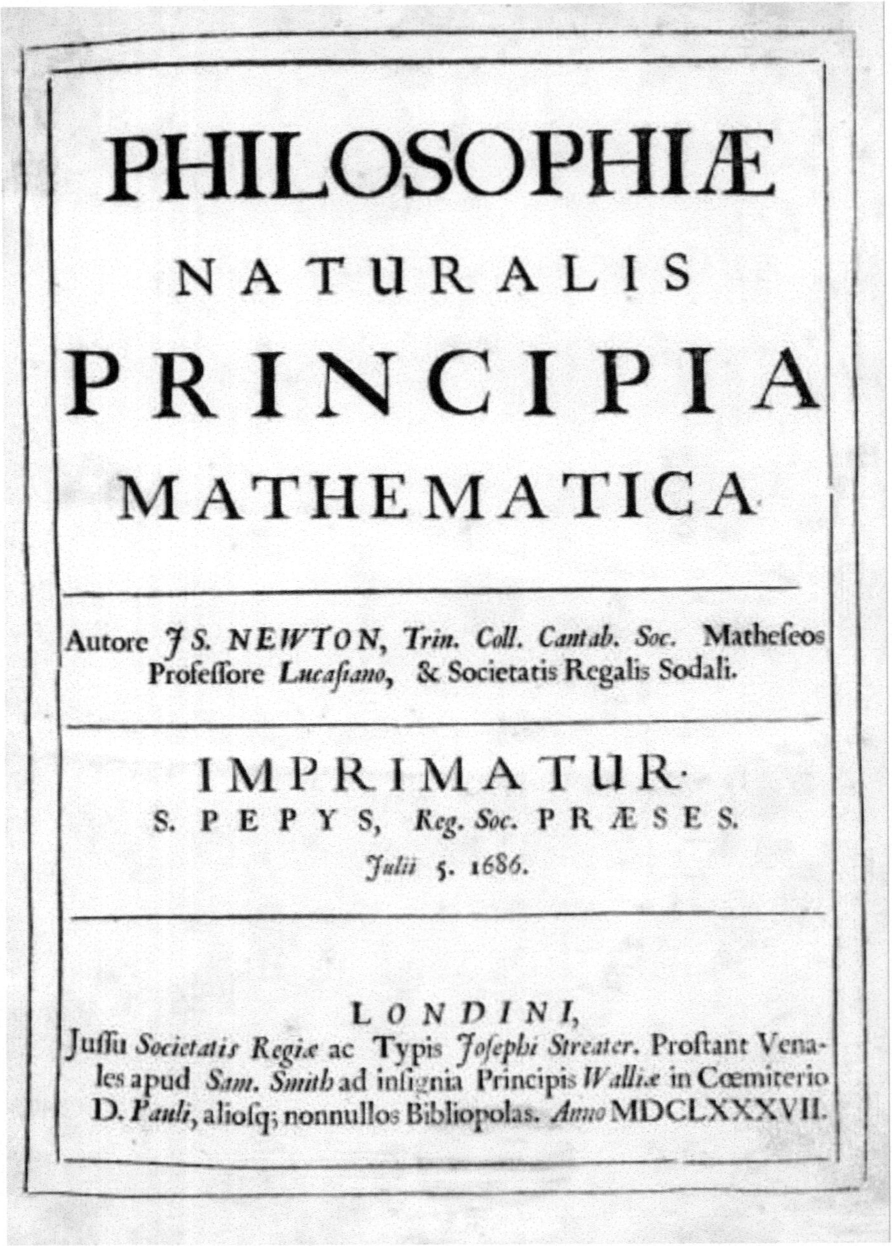

PHILOSOPHIÆ
NATURALIS
PRINCIPIA
MATHEMATICA

Autore *JS.* NEWTON, *Trin. Coll. Cantab. Soc.* Matheseos Professore *Lucasiano*, & Societatis Regalis Sodali.

IMPRIMATUR.
S. PEPYS, *Reg. Soc.* PRÆSES.
Julii 5. 1686.

LONDINI,
Jussu *Societatis Regiæ* ac Typis *Josephi Streater*. Prostant Venales apud *Sam. Smith* ad insignia Principis *Walliæ* in Cœmiterio D. *Pauli*, aliosq; nonnullos Bibliopolas. *Anno* MDCLXXXVII.

Fig. 7.1.9. Title page of Newtons *Principia* 1687

Fig. 7.1.10. Edmond Halley and the comet named after him. Bust in the Museum Royal Greenwich Observatory, London [Photo: K.-D. Keller 2006]; Photo of the comet ([Kuiper Airborne Observatory] Photo No. AC86-0720-2, C141 aircraft April 8/9, 1986)

One may guess (and in fact I did many decades ago before I bought a copy) that Newton's *Principia* swarms from applications of the calculus of fluxions. This guess is dead wrong, however! Except for two lemmas in which 'last quantities' are discussed the whole work is based on classical geometric arguments. One feels almost reminded of Archimedes who arrived at his results by means of his method of indivisibles but then transferred it into the language of classic geometry. The complete penetration of physics by the new analysis was achieved only by Leonhard Euler in the 17th century.

In 1685 James II became King of England. He had converted to the Catholic faith and tried by and by to place Catholics in important positions. When he gave order to Trinity College to award a Benedictine monk without any examinations with a university degree Newton rebelled against it. He encouraged the vice chancellor of Cambridge University to stick strictly to the rules which did not permit such a request; the vice chancellor followed Newton's advise and was removed from his post. But Newton was not willing to give in and prepared documents for the defence of the university. In November 1688 William of Orange, who had been invited by numerous Protestant leaders, landed with his troops in England. James fled to France and Newton became a member of Parliament representing the University of Cambridge on the 15th January 1689.

Whether it was the time as parliamentarian which he spent in London or whether Newton was simply disappointed by his lonely life in Cambridge we do not know; he wanted to leave Cambridge. An opportunity came in 1696 when a new Warden of the Mint was sought in London. Newton got the offer, took the opportunity, and became very successful! He also had to track down forgers and was soon feared by the numerous clippers and forgers active in

London. In 1699 he even became Master of the Mint so that he gave up all his positions he had held in Cambridge in 1701 and settled in London.

7.1.4 Alchemy, Religion, and the Great Crisis

Up to now we have get to know Newton as the great natural scientist and mathematician. It is lesser known that he devoted much more work to chemical experiments and religious studies as to mathematics and physics! His occupation with religion certainly began with the estate of his stepfather Barnabas Smith consisting of a theological library. Throughout his life Newton has collected books on theology and studied them. He became convinced that he had to reject the doctrine of trinity and became an Arian. The doctrine of Arius (about 260–336) of Alexandria stated that God alone is the father and that his son was God's creation but could not be put on a level with God. Arianism was seen as heresy and hence Newton had to keep his religious convictions strongly to himself. In a recent monumental survey [Iliffe 2017] Rob Iliffe has shown in detail that, and how, Newton's religious thinking has influenced his way of thinking about science.

Fig. 7.1.11. Sir Isaac Newton [Painting attributed to the English School, about 1720]

The occupation with chemistry began about the year 1669 since in this year expenses for glasses and chemicals can be found in his books. He must have studied the ancient literature concerning alchemical studies for decades and ran his own laboratory for many years. As Dobbs [Dobbs 1991] has convincingly shown that alchemy has also influenced Newton's physics; so in the postulation of the ether or in the corpuscular theory of light. It belongs to the Newton legends that it was alchemy which plunged him into a deep crisis at the beginning of the 1690s. Following the legend (see, for example, [Rosenberger 1987, p. 278], Newton is said to have gone to church some time in 1692 and had left his doggie in one of his rooms with a burning candle. The doggie is then said to have upset the candle resulting in important manuscript on alchemy burning away. Other interpretations speak of a fire in his laboratory. Shocked by this event Newton is said to have gone mad. To Samuel Pepys (1633–1703), author of the now famous 'secret diary' and president of the Royal Society between 1684 and 1686, Newton wrote a strange letter dated 13th September 1693 [Turnbull 1959–77, Vol. III, p. 279]:

> *Sir,*
>
> *Some time after Mr Millington had delivered your message, he pressed me to see you the next time I went to London. I was averse; but upon his pressing consented, before I considered what I did, for I am extremely troubled at the embroilment I am in, and have neither ate nor slept well this twelve month, nor have my former consistency of mind. I never designed to get anything by your interest, nor by King James's favour, but am now sensible that I must withdraw from your acquaintance, and see neither you nor the rest of my friends any more, if I may but leave them quietly. I beg your pardon for saying I would see you again, and rest your most humble and most obedient servant,*
>
> *Is. Newton*

Pepys did not know what to think of this letter and carefully and cautiously asked Mr Millington of Magdalen College in Cambridge, but Millington's information was somewhat blank. Then Pepys wrote openly to Millington [Brewster 1855, Vol. II, p. 143]:

> *For I was loth at first dash to tell you that I had lately received a letter from him so surprising to me for the inconsistency of every part of it, as to be put into great disorder by it, from the concernment I have for him, lest it should arise from that which of all mankind I should least dread from him and most lament for, – I mean a discomposure in head, or mind, or both. Let me, therefore, beg you, Sir, having now told you the true ground of the trouble I lately gave you, to let me*

know the very truth of the matter, as far at least as comes within your knowledge.

Millington's answer of 10th October was [Brewster 1855, Vol. II, p. 144f.]:

> ... *he* [Newton] *told me that he had writt to you a very odd letter, at which he was much concerned; added, that it was in a distemper that much seized his head, and that kept him awake for above five nights together, which upon occasion he desired I would represent to you, and beg your pardon, he being very much ashamed he should be so rude to a person for whom he hath so great an honour. He is now very well, and, though I fear he is under some small degree of melancholy, yet I think there is no reason to suspect it hath at all touched his understanding, and I hope never will; and so I am sure all ought to wish that love learning or the honour of our nation, ...*

The philosopher John Locke (1632–1704), being befriended with Newton, got an even stranger letter from Newton [Turnbull 1959–77, p. 280]:

> *Sr*
>
> *Being of opinion that you endeavoured to embroil me wth woemen & by other means I was so much affected with it as that when one told me you were sickly & would not live I answered twere better if you were dead. I desire you to forgive me this uncharitableness. For I am now satisfied that what you have done is just & I beg your pardon for my having hard thoughts of you for it & for representing that you*

Fig. 7.1.12. John Locke [Painting by Godfrey Kneller, 1697] und Samuel Pepys [Painting by John Hayls, 1666]

struck at ye root of morality in a principle you laid down in your book of Ideas & designed to pursue in another book & that I took you for a Hobbist. I beg your pardon also for saying or thinking that there was a designe to sell me an office, or to embroile me, I am

your most humble & most

unfortunate Servant

Is. Newton

The idea that Newton's strange mental condition was caused by a fire in his laboratory or on his desk is hard to believe, to say the least. Therefore two other variants developed in recent days. One variant is based on the fact that Newton was physically and mentally exhausted at the beginning of the 1690s so that his psychoses led to phases of manic depressions [Lieb/Hershman 1983]. Following this variant the great crisis was just a particularly harsh manic depression. The other variant is concerned with an assumed poisoning by heavy metals, in particular mercury [Keynes 1980], [Johnson/Wolbarsht 1979], [Spargo/Pounds 1979]. In fact, Newton had phases in which he slept besides the fire in his laboratory for weeks while dangerous mixtures of heavy metals simmered continuously. The symptoms therefore could be symptoms of serious lead or mercury poisoning or poisoning with other heavy metals. Hairs of Newton were found and chemically analysed [Spargo/Pounds 1979]. Indeed a high pollution with heavy metals could be found but the time when the hair was cut off is not known and it seems certain that *all* experimental scientists in these days were exposed to high doses of heavy metals. Since Newton was buried in Westminster Abbey and exhumations are not allowed there the secrets concerning the poisoning variant will never be clarified.

Fig. 7.1.13. Fire in Newton's laboratory [Engraving, Paris 1874]

Fig. 7.1.14. Nicolas Fatio de Duillier and Giovanni Domenico Cassini (in front of the Paris observatory)

It seems much more likely that Newton's crisis was triggered when his close young friend Nicolas Fatio de Duillier (1664–1753) broke up their intimate friendship. Fatio was a Swiss mathematician who went to Paris when he was 18 years old to work with Giovanni Domenico Cassini at the observatory where he did excellent research. Having met Jacob Bernoulli and Christiaan Huygens in 1686 Fatio became interested in the new methods of infinitesimal mathematics. A year later he travelled to London where he met Wallis and Locke and became a member of the Royal Society in 1688. He did work on the theory of gravitation where he tried to reconcile a theory of Huygens' with the theory of Newton. Fatio also worked on his own theory of gravitation. Fatio was impressed by Newton and, conversely, Newton became attracted by Fatio. A curious friendship developed. When Fatio fell ill with influenza Newton wrote letters full of anxiety in a unique intensity. Fatio attempted to get a new edition of Newton's *Principia* off the ground in 1691 but he failed. In 1693 the relation between Newton and Fatio suddenly cooled off and that was most likely the reason for Newton's fatal state of mind. The inglorious role Fatio played in the history of the priority dispute between Newton and Leibniz we have described in [Sonar 2018].

7.1.5 Newton as President of the Royal Society

Newton was elected in 1703 the President of the Royal Society for the first time and was re-elected year by year until his death. We have to report on the priority dispute with Leibniz later on, in which Newton did not play fair. He was knighted in 1705 by Queen Anne and hence became the first scientist ever honoured in that way. On 1st July 1725 Newton headed a meeting of the Royal Society in which the educator of Louis XV, Abbé Alari, was honoured. A report concerning this visit allows once again some insight into Newton's character [Rosenberger 1987, p. 387]:

> [...] *Since the Abbé was very well versed in the reading of Greek and Latin authors he appealed to the learned man and was invited for dinner. Newton was stingy, the meal was abominable, the drinks he offered his guest were gifts from Palma and Madeira. After dinner he led his guest to the Royal Society and let him sit to his right. Shortly after the meeting had started Newton fell asleep.*[...]
>
> ([...] Da der Abbé in der Lectüre griechischer und lateinischer Autoren sehr bewandert war, so gefiel er dem alten Gelehrten und wurde zum Dinner behalten. Newton war geizig, die Mahlzeit abscheulich, die Getränke, die er seinem Gaste vorsetzte, nur geschenkte Weine von Palma und Madeira. Nach seinem Dinner führte er den Gast in die Royal Society und liess ihn zu seiner Rechten sitzen. Gleich nach Beginn der Sitzung schlief Newton ein. [...])

On 2nd March 1727 he headed a meeting of the Royal Society for the last time. On 20th March Sir Isaac Newton died and found his last resting place in Westminster Abbey. His tomb monument was erected in 1731 by his heirs. Its inscription reads[1]:

Fig. 7.1.15. Isaac Newton, honoured on a British One Pound Note

[1] http://www.westminster-abbey.org/our-history/people/sir-isaac-newton

Here is buried Isaac Newton, Knight, who by a strength of mind almost divine, and mathematical principles peculiarly his own, explored the course and figures of the planets, the paths of comets, the tides of the sea, the dissimilarities in rays of light, and, what no other scholar has previously imagined, the properties of the colours thus produced. Diligent, sagacious and faithful, in his expositions of nature, antiquity and the holy Scriptures, he vindicated by his philosophy the majesty of God mighty and good, and expressed the simplicity of the Gospel in his manners. Mortals rejoice that there has existed such and so great an ornament of the human race! He was born on 25th December 1642, and died on 20th March 1726.

(H. S. E. ISAACUS NEWTON Eques Auratus, / Qui, animi vi prope divinâ, / Planetarum Motus, Figuras, / Cometarum semitas, Oceanique Aestus. Suâ Mathesi facem praeferente / Primus demonstravit: / Radiorum Lucis dissimilitudines, / Colorumque inde nascentium proprietates, / Quas nemo antea vel suspicatus erat, pervestigavit. / Naturae, Antiquitatis, S. Scripturae, / Sedulus, sagax, fidus

Fig. 7.1.16. Newton's tomb monument in Westminster Abbey [Photo: Klaus-Dieter Keller, 2006]

> Interpres / Dei O. M. Majestatem Philosophiâ asseruit, / Evangelij Simplicitatem Moribus expressit. / Sibi gratulentur Mortales, / Tale tantumque exstitisse / HUMANI GENERIS DECUS. / NAT. XXV DEC. A.D. MDCXLII. OBIIT. XX. MAR. MDCCXXVI)

The English poet Alexander Pope (1688–1744) got to the heart of Newton's importance:

> *Nature and nature's laws lay hid in night;*
> *God said: 'Let Newton be!' and all was light.*

7.1.6 The Binomial Theorem

After having studied John Wallis' book *Arithmetica infinitorum* in 1664 Newton was inspired to a new discovery in 1665 which should develop into a central element and a work horse of analysis. In his first letter to Leibniz (epistola prior) 1676 he reported thereon [Edwards 1979, p. 178]:

> *Extractions of roots are much shortened by this theorem,*

$$(P+PQ)^{m/n} = P^{m/n} + \frac{m}{n}AQ + \frac{m-n}{2n}BQ + \frac{m-2n}{3n}CQ + \frac{m-3n}{4n}DQ + etc.$$

> *where $P+PQ$ signifies the quantity whose root or even any power, or the root of a power, is to be found; P signifies the first term of that quantity, Q the remaining terms divided by the first, and m/n the numerical index of the power of $P+PQ$, whether that power is integral or (so to speak) fractional, whether positive or negative.*

The quantities $A, B, C, \cdots$ denote the corresponding preceding term, hence $A = P^{m/n}$, $B = (m/n)AQ$, and so on. In this form the binomial theorem is hardly digestible today. We write this theorem as

$$(1+x)^\alpha = \sum_{k=0}^{\infty} \binom{\alpha}{k} x^k$$

for arbitrary $\alpha \in \mathbb{R}$ and with the binomial coefficients $\binom{\alpha}{k}$, which, in case of integer α, can be read off Pascal's triangle. In that case the series above is a finite series, i.e. terminates with $k = \alpha$. Setting $P = 1$ Newton's binomial theorem can be put as

$$(1+Q)^{m/n} = 1 + \frac{m}{n}Q + \frac{m}{n}\frac{m-n}{2n}Q^2 + \frac{m}{n}\frac{m-n}{2n}\frac{m-2n}{3n}Q^3 + \dots.$$

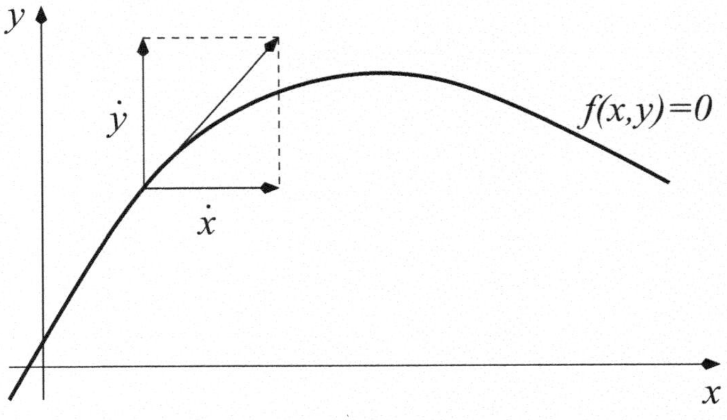

Fig. 7.1.17. Fluxions (velocity components) in the motion alongside a curve

We have already discussed that the binomial theorem for the case of $\alpha = 1/2$ was already known to Henry Briggs when he computed his logarithms, but Newton went a large step further, of course. We note in passing that Newton never gave a proof of his important theorem. For him there apparently was enough numerical and empirical evidence for the theorem to be valid [Stedall 2008, p. 191]. Questions concerning convergence of the series and hence restrictions of the quantity Q were not discussed by Newton. In the further history of analysis Newton turned out to be a true master in the manipulation of infinite series. His turn to series and his conviction that series yield important tools in the development of analysis was surely initiated by his early discovery of the binomial theorem.

7.1.7 The Calculus of Fluxions

If one would like to make a sharp point and characterise just one principal difference between Newton and Leibniz then Newton could be called a physicist while Leibniz could be called a mathematician. Newton thought in terms of motion and velocities when he attempted to compute tangents of curves of the form $f(x, y) = 0$. In the eyes of Newton the curve $f(x, y) = 0$ itself 'results' from the points of intersection of two moving lines which we can interpret as being the velocity components in x- and y-direction.

From about 1690 on Newton denoted the velocity in x-direction by $\dot{x}$ and the one in y-direction by $\dot{y}$. Before that he wrote p and q. It is a kind of irony of history that today we discuss Newton's analysis in Leibniz's notation but Newton's designations could not stand up against Leibniz's genial notations. The connection between the two is given by

$$\dot{x} = \frac{dx}{dt}, \quad \dot{y} = \frac{dy}{dt},$$

where in fact one should think of t as being the time.

Since the curve $f(x, y) = 0$ comes into existence by a 'fluent' motion Newton called the quantities x and y the 'fluents', and $\dot{x}$ and $\dot{y}$ the 'fluxions'. From the fluxions at time t there immediately follows the slope of the tangent of the curve (cp. figure 7.1.17) as

$$\frac{\dot{y}}{\dot{x}} = \frac{\frac{dy}{dt}}{\frac{dx}{dt}} = \frac{dy}{dx}.$$

Newton summarised his works on the calculus of fluxions in a manuscript dated October 1666. This manuscript was left unpublished but there were copies available to the English mathematicians. This 'October tract' [Whiteside 1967–1981, Vol.I, p. 400–448] can be found in Newton's *Mathematical Papers*, collected, edited, translated into English, and issued by Derek Thomas Whiteside (1932–2008) in eight volumes [Whiteside 1967–1981]. In this tract Newton asked for the ratio of the fluxions $\dot{x}$ and $\dot{y}$ in case of the curve

$$f(x, y) = \sum_{i,j} a_{ij} x^i y^j = 0,$$

where finite sums are considered. Let us read Newton himself where we have to read 'the' instead of '$\mathrm{y^e}$', and 'which' instead of '$\mathrm{w^{ch}}$' [Edwards 1979, p. 193]:

> *Set all y^e termes on one side of y^e equation that they become equal to nothing. And first multiply each terme by so many times $\dot{x}/x$ as x hath dimensions in that terme. Secondly multiply each terme by so many times $\dot{y}/y$ as y hath dimensions in it [...] the summe of all these products shall bee equall to nothing. W^{ch} Equation gives y^e relation of y^e velocitys.*

Following Newton's instructions step by step we have to equate all terms to zero, i.e.

$$\sum_{i,j} a_{ij} x^i y^j = 0.$$

Now every term has to be multiplied by $\dot{x}/x$ as often as the 'dimension' of x is in this equation. Newton used 'dimension' for 'power'; hence the dimension of x is simply i. Likewise we have to multiply j times by $\dot{y}/y$. The sum of all these products has to be equated to zero. Hence we get

$$\sum_{ij} \left(\frac{i\dot{x}}{x} + \frac{j\dot{y}}{y} \right) a_{ij} x^i y^j = 0. \tag{7.1}$$

Looking closer at this sum and tidying up a bit results in

$$\sum_{ij} a_{ij} i x^{i-1} \dot{x} y^j + \sum_{ij} a_{ij} j y^{j-1} \dot{y} x^i = 0,$$

but it holds $\frac{\partial f}{\partial x} = \sum_{i,j} a_{ij} i x^{i-1} y^j$ and $\frac{\partial f}{\partial y} = \sum_{i,j} a_{ij} j y^{j-1} x^i$, so that Newton's equation is nothing but

$$\dot{x}\frac{\partial f}{\partial x} + \dot{y}\frac{\partial f}{\partial y} = 0,$$

or

$$\frac{\dot{y}}{\dot{x}} = \frac{dy}{dx} = -\frac{\frac{\partial f}{\partial x}}{\frac{\partial f}{\partial y}},$$

and we arrive at the rule of implicit differentiation.

But how did Newton prove this result? He thought in terms of motion! He introduced an infinitesimally small time interval o, and since $\dot{x}$ denotes the velocity in x-direction a point moves in time o from position x to $x + o\dot{x}$. Therefore he inserted $x + o\dot{x}$ and $y + o\dot{y}$ for x and y in the equation $f(x, y) = 0$:

$$\sum_{i,j} a_{ij} (x + o\dot{x})^i (y + o\dot{y})^j = 0.$$

Then Newton applied his binomial theorem to every factor:

$$\begin{aligned}\sum_{i,j} a_{ij} x^i y^j &+ \sum_{i,j} a_{ij} x^i (j y^{j-1} o\dot{y} + \text{ terms in } o^2) \\ &+ \sum_{i,j} a_{ij} y^j (i x^{i-1} o\dot{x} + \text{ terms in } o^2) \\ &+ \sum_{i,j} a_{ij} (i x^{i-1} o\dot{x} + \ldots)(j y^{j-1} o\dot{y} + \ldots) = 0.\end{aligned}$$

According to the premises it holds $\sum_{i,j} a_{ij} x^i y^j = 0$ and Newton set all further terms containing o^2 and higher powers to zero,

> *... because they are infinitely lesse y^n those in wch o is but of one dimension.*

Therefore the only remaining term is

$$\sum_{i,j} a_{ij} (i x^{i-1} y^j o\dot{x} + j x^i y^{j-1} o\dot{y}) = 0.$$

Division by o finally yields the result

$$\sum_{i,j} a_{ij} (i x^{i-1} y^j \dot{x} + j x^i y^{j-1} \dot{y}) = 0.$$

7.1.8 The Fundamental Theorem

Modern analysis could only begin with the thorough knowledge that differentiation and integration are inverse operations. Barrow had this result implicitly but it was left to Newton and Leibniz to clearly acknowledge the central place of the fundamental theorem.

Having solved the problem of computing the slope $dy/dx = \dot{y}/\dot{x}$ of a curve $f(x, y) = 0$ it seems natural today to wish for a solution of the inverse problem, i.e. the computation of a curve with given slope of the tangent. In Newton's notation a function $y(x)$ is sought from an equation of the form

$$\frac{\dot{y}}{\dot{x}} = \phi(x).$$

This case was called 'anti-differentiation' by Newton, while the more complicated case of

$$g(x, \dot{y}/\dot{x}) = 0$$

calls for the solution of an ordinary differential equation. The historically first rigorous statement of the fundamental theorem can be found in Newton's October tract. If A denotes the area under a function $y = f(x)$ then Newton's form of the fundamental theorem is (in the notation of Leibniz)

$$\frac{dA}{dx} = y.$$

It should not come as a surprise that Newton derived also this result from considerations of motion. As in figure 7.1.18 we denote the area under the function $q = f(x)$ by y.

We think of this area as being generated by a horizontal motion of the segment bc being of variable length. The velocity of the motion in x-direction

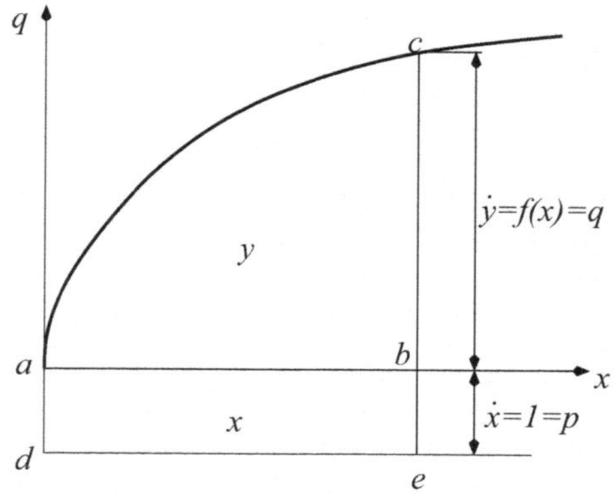

Fig. 7.1.18. Deriving the fundamental theorem

is assumed to be $\dot{x} = 1$. Simultaneous with bc the segment ad moves with velocity $\dot{x} = 1$ and hence form a rectangle $adeb$ having area x. Now Newton argues as follows [Edwards 1979, p. 196], [Whiteside 1967–1981, Vol. I p. 427]:

> *supposing y^e line cbe by parallel motion to describe y^e two [areas] x and y; The velocity wth wch they increase will bee, as be to bc: y^t is, y^e motion by wch x increaseth being $be = p = 1$, y^e motion by wch y increaseth will be $bc = q$. which therefore may bee found by prop: 7th. viz: $\frac{-\mathcal{X}y}{\mathcal{X}x} = q = bc$.*

Hence Newton grasped the rate of change of the area y being $q = f(x)$ at $\dot{x} = 1$ rather intuitively, i.e.

$$\frac{\dot{y}}{\dot{x}} = \dot{y} = f(x).$$

In Newton's notation we have $\mathcal{X} = f(x, y) = f(x, y(x)) = 0$, $\mathcal{X} = x\frac{\partial f}{\partial x}$, $\mathcal{X} = y\frac{\partial f}{\partial y}$, and it holds $\frac{\partial f}{\partial x} + \frac{\partial f}{\partial y}\frac{dy}{dx} = 0$. Hence

$$q = -\frac{\mathcal{X}y}{\mathcal{X}x} = \frac{dy}{dx},$$

and for the area under the curve between a and b it holds

$$\int q\,dx = y,$$

thus in Leibniz's notation:

$$\int \frac{dy}{dx}\,dx = \int dy = y.$$

Now it becomes clear why the slope of the tangent at the curve

$$x \mapsto \frac{x^{n+1}}{n+1}$$

is just x^n and that the area under x^n conversely is just $\frac{x^{n+1}}{n+1}$, since if we put

$$y = \frac{x^{n+1}}{n+1}$$

as the area then it follows from (7.1) that

$$\frac{\dot{y}}{\dot{x}} = x^n$$

and vice versa. Putting $\dot{x} = p = 1$ and $\dot{y} = q$, then

$$q = x^n$$

is the curve of figure 7.1.18. Newton never cared about constants of integration; he always assumed his curves passing through the origin.

7.1.9 Chain Rule and Substitutions

By means of differentiation and anti-differentiation Newton was able to derive important rules of his new analysis. Let us consider with [Edwards 1979, p. 197] the function

$$y = \sqrt{(1+x^n)^3} = (1+x^n)^{\frac{3}{2}},$$

in which Newton substituted $z = 1 + x^n$. Following (7.1) the fluxion of z is $\dot{z} = nx^{n-1}\dot{x}$. We have

$$y^2 = (1+x^n)^3 = z^3$$

and the fluxions hence satisfy

$$2y\dot{y} = 3z^2\dot{z}.$$

Solving for $\dot{y}/\dot{x}$ and inserting $z^2 = (1+x^n)^2$ and $y = (1+x^n)^{3/2}$ one gets

$$\frac{\dot{y}}{\dot{x}} = \frac{dy}{dx} = \frac{3}{2}n(1+x^n)^{1/2}x^{n-1},$$

as is easily confirmed by means of Leibniz's chain rule. The books [Baron 1987] and [Edwards 1979] contain further examples to illustrate this technique. Edwards justly writes [Edwards 1979, p. 196] that the chain rule is 'built in' Newton's calculus of fluxions. We tend to see this rather as a drawback of Newton's analysis since this important rule does not appear as an explicit *rule* but 'arises' depending on the individual case.

7.1.10 Computation with Series

In 1666 Newton wrote *De Analysi per Aequationes Numero Terminorum Infinitas* (On the analysis with equations in which the number of terms is infinite) which was published only in 1711. However, the publication circulated before in the form of a manuscript. Among other things he developed a technique to approximately determining roots in this work which we call 'Newton's method' or 'Newton-Raphson method' today. If a root of the equation

$$f(x) = \sum_{i=0}^{k} a_i x^i = 0$$

is sought let us assume we have found an approximation x_n of a root (by educated guess, say). Then insert $x_* := x_n + p$ into the equation to get

$$0 = \sum_{i=0}^{k} a_i x_*^i = \sum_{i=0}^{k} a_i (x_n + p)^n.$$

Again the binomial theorem comes into play and we get

$$0 = \sum_{i=0}^{k} a_i \left(x_n^i + i x_n^{i-1} p + \ldots\right) = \underbrace{\sum_{i=0}^{k} a_i x_n^i}_{=f(x_n)} + p \underbrace{\sum_{i=0}^{k} i a_i x_n^{i-1}}_{=f'(x_n)} + \ldots,$$

hence

$$0 = f(x_n) + p f'(x_n) + \ldots.$$

Neglecting terms of higher order it follows for p:

$$p \approx -\frac{f(x_n)}{f'(x_n)}$$

and thus the (hopefully) improved approximation of the root is given as

$$x_* \approx x_n + p = x_n - \frac{f(x_n)}{f'(x_n)} =: x_{n+1}.$$

Newton nowhere gave a derivation of his method by means of geometry although this seems to be obvious. He also extended his method to the case of the determination of roots for functions defined implicitly by $f(x,y) = 0$.

In essence Newton required this method for the inversion of series, e.g. when he solved the series

$$z = x - \frac{1}{2}x^2 + \frac{1}{3}x^3 - \frac{1}{4}x^4 + \frac{1}{5}x^5 - + \ldots$$

(describing the area under the hyperbola $y = (1+x)^{-1}$) for x. Newton neglected all terms with powers 6 and higher and got

$$\frac{1}{5}x^5 - \frac{1}{4}x^4 + \frac{1}{3}x^3 - \frac{1}{2}x^2 + x - z = 0. \tag{7.2}$$

The first approximation consists in a brute force linearisation $x \approx z$. According to Newton's method Newton inserted $x = z + p$ into (7.2) which yields

$$0 = \left(-\frac{1}{2}z^2 + \frac{1}{3}z^3 - \frac{1}{4}z^4 + \frac{1}{5}z^5\right) + p(1 - z + z^2 - z^3 + z^4) \\ + p^2\left(-\frac{1}{2} + z - \frac{3}{2}z^2 + 2z^3\right) + \ldots$$

Neglecting all terms with p^2 and higher powers it follows

$$p \approx \frac{\frac{1}{2}z^2 - \frac{1}{3}z^3 + \frac{1}{4}z^4 - \frac{1}{5}z^5}{1 - z + z^2 - z^3 + z^4} = \frac{1}{2}z^2 + \ldots,$$

hence a new approximation is given as

$$x \approx z + \frac{1}{2}z^2.$$

This is now again inserted, and so on. Eventually Newton arrived at

$$\begin{aligned} x &= z + \frac{1}{2}z^2 + \frac{1}{6}z^3 + \frac{1}{24}z^4 + \frac{1}{120}z^5 + \dots \\ &= z + \frac{1}{2!}z^2 + \frac{1}{3!}z^3 + \frac{1}{4!}z^4 + \frac{1}{5!}z^5 + \dots \end{aligned}$$

and that is the series representation of $x = \mathrm{e}^z - 1$, since $z = \ln(1+x)$ is fully equivalent to $x = \mathrm{e}^z - 1$.

By applying this technique Newton was able to derive the series of $\sin x$ and $\cos x$, cp. [Edwards 1979, p. 205ff.].

7.1.11 Integration by Substitution

In his work *De methodis serierum et fluxionum* [Whiteside 1967–1981, Vol. III, p. 32–372], cpl. [Hairer/Nørsett/Wanner1987, p. 388] of 1671[2] Newton was concerned with area-preserving transformations in 'Problem 8' (p. 119–209). To this end he looked at two functions $v = f(x)$ and $y = g(z)$, cp. [Edwards 1979, p. 210], defining the curves FDH and GEI as shown in figure 7.1.19.

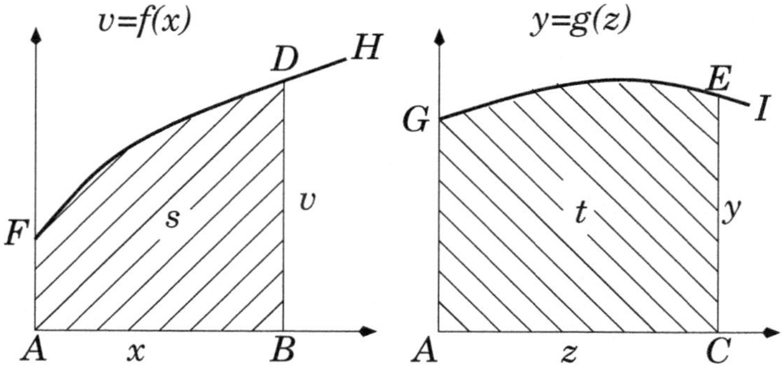

Fig. 7.1.19. Figure concerning integration by substitution

As was the case with the fundamental theorem Newton imagined the areas s and t as being generated by horizontal motion of the line segments BD

[2] This work did not appear during Newton's lifetime. Posthumously John Colson published an English translation 1736 in London under the title *The Method of Fluxions and Infinite Series; With its Application to the Geometry of Curve Lines.* The first publication in Latin appeared only in 1779 and was edited by Samuel Horsley. Since then the work is known under the title *Methodus fluxionum et serierum infinitarum.*

and EC. The change of area s with time is therefore given by the length v, multiplied by the velocity $\dot{x}$ in horizontal direction. Analogously the change of t with time can be interpreted. Hence, Newton got

$$\frac{\dot{s}}{\dot{t}} = \frac{v\dot{x}}{y\dot{z}}.$$

Putting $\dot{x} = 1$ then $\dot{s} = v$. Since $\dot{t} = y\dot{z}$ it follows $y = \dot{t}/\dot{z}$. Assuming further equality of areas $s = t$, then it follows $\dot{s} = \dot{t} = v$ and thus

$$y = \frac{v}{\dot{z}}. \tag{7.3}$$

Assuming further a mutual functional connection between x and z, e.g.

$$z = \phi(x), \quad x = \psi(z), \tag{7.4}$$

then (7.3) *defines* that function $y = g(z)$ for which the equality of areas holds, namely

$$y = \frac{v}{\dot{z}} = \frac{f(x)}{\phi'(x)\dot{x}} = \frac{f(\psi(z))}{\phi'(\psi(z))\dot{x}} \overset{\dot{x}=1}{=} \frac{f(\psi(z))}{\phi'(\psi(z))}.$$

Since (7.4) it holds $x = \psi(z) = \psi(\phi(x))$ and therefore $\dot{x} = 1 = \psi'(z)\phi'(x)$, hence $\phi'(x) = \phi'(\psi(z)) = (\psi'(z))^{-1}$. Thus $y = g(z)$ has been found:

$$y = f(\psi(z))\psi'(z),$$

and so Newton had proven the transformation formula

$$\int f(x)\,dx = \int f(\psi(z))\psi'(z)\,dz.$$

7.1.12 Newtons Last Works Concerning Analysis

In 1704 Newtons second Opus Magnus *Opticks* [Newton 1979] was published. Bound with this was an appendix including the work *De quadratura curvarum* (not included in [Newton 1979]), stemming from the time between 1691 and 1693, representing Newton's mature Analysis [Edwards 1979, p. 226]. A German translation can be found in [Leibniz/Newton 1998]. It is Newton's most technically advanced work in which he tried to give his calculus of fluxions and fluents a sound foundation. Many problems are posted and solved, and many theorems are stated. We cite here only one example from *De quadratura* [Edwards 1979, p. 229]:

Let

$$\begin{aligned} R &= e + fx^\eta + gx^{2\eta} + hx^{3\eta} + \ldots \\ S &= a + bx^\eta + cx^{2\eta} + dx^{3\eta} + \ldots \\ r &= \theta/\eta, s = r + \lambda, t = s + \lambda, v = t + \lambda, \ldots. \end{aligned}$$

Then

$$\int x^{\theta-1}R^{\lambda-1}S\,dx = x^{\theta}R^{\lambda}\left[\frac{a/\eta}{re} + \frac{b/\eta - sfA}{(r+1)e}x^{2\eta} \right.$$
$$+\frac{c/\eta - (s+1)fB - tgA}{(r+2)e}x^{2\eta}$$
$$\left. +\frac{d/\eta - (s+2)fC - (t+1)gB - vhA}{(r+3)e}x^{3\eta} + \dots\right].$$

Here A, B, C, ... are coefficients of the respectively previous power of x, hence

$$A = \frac{a/\eta}{re}, B = \frac{b/\eta - sfA}{(r+1)e}, C = \frac{c/\eta - (s+1)fB - tgA}{(r+2)e}, \dots.$$

What looks very technical here is nothing else than the integral formula for rational functions. As the French mathematician Jacques Hadamard (1865–1963) remarked in 1947 (cited from [Edwards 1979, p. 229]):

> De Quadratura *brings the integration of rational functions to a state hardly inferior to what it is now.*

7.1.13 Newton and Differential Equations

We can not delve into the history of differential equations in this book since a new book would be necessary devoted to it. The ordinary differential equations – and, much more so – the partial differential equations comprise very active and large areas within analysis and are vital for the applications. But since Newton as well as Leibniz immediately started solving differential equations after they invented their new calculus we want to shed a little light at least.

A particularly simply differential equation is

$$y' = y$$

concerning a function $x \mapsto y(x)$, but we can already see the principle: an (ordinary) differential equation is an equation for a sought-after function, in which derivatives of this function appear. In the case of the simple differential equation above the solution can be immediately seen: all functions $y(x) = C \cdot \mathrm{e}^x$ where C is an arbitrary constant are solutions (the function $y \equiv 0$ is also a solution, of course). If one wants to single out a solution we have to prescribe an 'initial condition'. The initial condition $y(0) = 5$, for example, would lead to the (uniquely defined) solution $y(x) = 5 \cdot \mathrm{e}^x$.

In *De methodis serierum et fluxionum* and *Methodus fluxionum et serierum infinitarum*, dating from 1671, Newton applied his doctrine of series to solve the differential equation (cp. [Hairer/Nørsett/Wanner1987, p. 4f.])

$$\frac{\dot{y}}{\dot{x}} = 1 - 3x + y + xx + xy$$

under the initial condition $y(0) = 0$. We rewrite the differential equation as

$$y' = 1 - 3x + y + x^2 + xy.$$

Newton sought the solution as a series

$$y(x) = a_0 + a_1 x + a_2 x^2 + a_3 x^3 + a_4 x^4 + \dots$$

and immediately recognised from the initial condition $y(0) = 0$ that $a_0 = 0$, i.e. the series starts with

$$y(x) = 0 + \dots.$$

Now $y = 0$ is to be inserted into the differential equation yielding $y' = 1 + \dots$. Integration leads to

$$y(x) = x + \dots,$$

which again is inserted into the differential equation yielding $y' = 1 - 3x + x + \dots = 1 - 2x + \dots$ and integration reveals

$$y(x) = x - x^2 + \dots.$$

The procedure goes on until Newton finishes with

$$y(x) = x - x^2 + \frac{1}{3}x^3 - \frac{1}{6}x^4 + \frac{1}{30}x^5 - \frac{1}{45}x^6 + - \dots.$$

7.2 Gottfried Wilhelm Leibniz

7.2.1 Childhood, Youth, and Studies

Gottfried Wilhelm Leibniz (1646–1716) was born near the end of the Thirty Years' War the son of the university professor and actuary Friedrich Leibnütz and his third wife Catharina Schmuck in Leipzig. Hence the family belonged to the upper bourgeoisie, and an ancestor, captain Paul von Leubnitz, had even earned a title of nobility and a coat of arms for services rendered in 1600 [Finster/van den Heuvel 1990, p. 7]. The title was not hereditary; nevertheless Leibniz later used it on occasion and wrote 'Gottfried Wilhelm von Leibniz', although he was not entitled to. He probably did it to impress correspondents.

We know only little about Leibniz's childhood; most of the information coming from the memory of the adult Leibniz. He must have been a bookworm since he wrote [Müller/Krönert 1969, p. 4]:

When I grew up I acquired an exceptional pleasure in reading stories, and the German books I could lay my hands on I did not put down before I had read them completely.

(Als ich heranwuchs, fand ich am Lesen von Geschichten ein außerordentliches Vergnügen, und die deutschen Bücher, deren ich habhaft werden konnte, legte ich nicht eher aus der Hand, als bis ich sie ganz gelesen hatte.)

In 1654 Leibniz self-taught himself Latin by means of an illustrated edition of a book by Livy. In retrospect he reported [Müller/Krönert 1969, p. 4]:

In the [book by] *Livy I often got stuck;* [...]. *But because it was an old edition including woodcuts I eagerly looked at them, read here and there the words below them, carefree of the dark parts, and what I did not understand at all I skipped. Having done that repeatedly and having browsed the whole book and starting anew after some time, I understood much more. I was very delighted thereupon and hence proceeded without any dictionary until most of it became clear to me and I penetrated even deeper into the meaning.*

Fig. 7.2.1. Gottfried Wilhelm Leibniz [Painting by B. Chr. Franke, about 1700] (Herzog Anton Ulrich-Museum, Brunswick)

(In dem Livius dagegen blieb ich öfter stecken; [...]. Weil es aber eine alte Ausgabe mit Holzschnitten war, so betrachtete ich diese eifrig, las hier und da die darunterstehenden Worte, um die dunklen Stellen wenig bekümmert, und das, was ich gar nicht verstand, übersprang ich. Als ich dies öfter getan, das ganze Buch durchgeblättert hatte und nach einiger Zeit die Sache von vorn begann, verstand ich viel mehr davon. Darüber hoch erfreut, fuhr ich so ohne irgendein Wörterbuch fort, bis mir das meiste ebenso klar war, und ich immer tiefer in den Sinn eindrang.)

From 1653 on Leibniz attended the Schola Nikolaitana (Nicolai school) in Leipzig until Easter 1661. On the occasion of a school party in 1659 he presented a self-written poem in Latin consisting of 300 hexameters. All through his life Leibniz will write a flawless, perfect Latin. On admittance to school he already was enrolled at the University of Leipzig. This was a privilege of sons of professors. From 1661 on he studied there philosophy, but also attended lectures on poetry and mathematics. On the occasion of his Baccalaureate in 1663 his first scientific work *Disputatio metaphysica de principio individui* was published. It already contained elements of his later metaphysics. The summer semester 1663 he spent at the University of Jena where he attended lectures by Erhard Weigel (1625–1699).

Weigel employed mathematical arguments to reveal contradictions in scholastic philosophy. However, Leibniz seemed to have attended lectures on 'Arithmetic, elementary analysis, and combinations' (Arithmetik, die niedere Analysis und Combinationen), as Guhrauer [Guhrauer 1966, Band I, p. 26] wrote in 1846.

Back in Leipzig Leibniz started in 1663 to study both laws, canon and civil. He was an excellent student and wanted to do his doctorate, but this was not

Fig. 7.2.2. Nicolai school in Leipzig [Photo: Appaloosa, 2009]. Leibniz stayed there for eight years. Erhard Weigel; Leibniz attended his lectures in Jena

Fig. 7.2.3. The University of Altdorf in 1714

granted on threadbare grounds. Apparently elder students protested against a youngster like Leibniz becoming a doctor of both laws before them. Hence the young lawyer, whose father had died already in 1652, left his hometown, burning, as he wrote [Finster/van den Heuvel 1990, p. 14],

> *of desire to acquire greater fame in the sciences and to become familiar with the world.*
>
> (vor Begierde, größeren Ruhm in den Wissenschaften zu erwerben und die Welt kennenzulernen.)

'The world' is the University of Altdorf close to Nuremberg for the time being, where already Wallenstein had studied.

Leibniz gained the doctorate at Altdorf and was offered – being only 21 years old! – a professorship. But the young man wanted to move on 'into the world'. Yet he stayed for some time and acquired access to the secret society of the Rosicrucians where he advanced to the position of secretary, but in 1667 he was heading to new shores and aimed at Holland.

7.2.2 Leibniz in the Service of the Elector of Mainz

On his way via Frankfurt where he was to visit relatives he got caught in Mainz. A new 'Corpus juris' was to created by privy councillor Lasser in

Mainz and the young lawyer recommended himself with a text dedicated to the elector of Mainz whereupon he got a position to assist Lasser. Johann Christian von Boineburg (1622–1672), having held the position of a minister until 1664, had fallen from the grace of the elector and was just regaining reconciliation with the elector when Leibniz arrived. Boineburg had converted to Catholicism and Leibniz came into service as Boineburg's personal advisor. Immediately Leibniz started his restless activities on a large number of areas: all scientific, juridicial, political, historical, and theological questions of his days interested him and were ploughed thoroughly by him. His correspondence with scholars all over the world started in these days in Mainz and at the end of his life it will encompass about 1100 correspondents in 16 countries, China included. At the court of Mainz his hustles did not pass unnoticed: in 1670 the Protestant was moved up to a revision councillor at the High Court of Appeal (Oberappelationsgericht) in arch-Catholic Mainz as a recognition of his efforts.

About this time Louis XIV was King of France showing a strong love of expansionism. The fear was great that the French king was going to lay hands on Mainz and Leibniz received order to travel to Versailles in order to distract Louis from Mainz. Leibniz believed that a plan to conquer Egypt may do the trick and he thought he could make Louis' mouth watering when presenting such a plan. Alas, when he arrived in Paris in 1672 the invasion of Holland was imminent and Leibniz's Egyptian plan was simply ignored. Leibniz's mission failed completely but now his actual scientific career started. He is in the very centre of European sciences and very quickly made the acquaintance of major scholars. In 1672 he met Christiaan Huygens (see section 6.2.4) who pointed him to the *Arithmetica infinitorum* by John Wallis and to the *Opus geometricum* by de Saint-Vincent. First contacts to the Académie des Sciences came via meeting the Academy's secretary, Jean Gallois. However, also in 1672 his patron Boineburg died. In January 1673 Leibniz attended a legation journey to England on behalf of Mainz and found himself in London, the second European centre of the world as far as sciences were concerned. From Paris he took a model of the calculating machine with him which he had constructed. It was a sophisticated four species machine, i.e. all four elementary operations: addition, subtraction, multiplication, division, were possible, and was based on Leibniz's invention of the Leibniz wheel (Staffelwalze).

The machine was presented to the Royal Society and despite some envious remarks by Robert Hooke (1635–1703) and actual problems with the carry it made an impression. The Englishman Samuel Morland (1625–1695) had also recently developed a calculating machine but after a meeting with Morland Leibniz could safely state that his machine relied on very different principles.

Leibniz, whose supreme employer, the elector of Mainz, had meanwhile passed away, also took part in chemical experiments conducted by Robert Boyle (1627–1692). In Boyle's house Leibniz became acquainted with the

Fig. 7.2.4. Replica of Leibniz's 'four species machine' (Gottfried Wilhelm Leibniz Library – Niedersächsische Landesbibliothek Hannover, Leibniz's four species calculating machine)

mathematician John Pell (1611–1685). He proudly told Pell about his work on the representation of series by means of series of differences, but Pell pointed out that such work had in fact already been published before Leibniz even began to work on it. Leibniz thereupon felt obliged to object against any possible accusation of plagiarism in written form. But Pell also informed Leibniz of the *Logarithmotechnia* by Nicolaus Mercator, and Leibniz in fact took a copy of this book back with him to Paris. All in all Leibniz's first visit to London seems to have been a failure; at least on the English side. The calculating machine which did not function correctly, Hooke's negative remarks on it, and finally Pell's suspicion of Leibniz adopting published works of others as his own led to some distrust within the Royal Society, cp. [Hofmann 1974, Chap. 3].

The secretary of the Royal Society was the German Henry (Heinrich) Oldenburg (1618–1677), who was born in Bremen. Leibniz and his fellow countryman apparently got along very well with each other since Oldenburg assisted in writing a membership application to the Royal Society. On 19th April 1673 Leibniz was indeed received into the Royal Society. Back in Paris Leibniz faced the necessity to get a better mathematical education. Christiaan Huygens became his teacher and mentor, and Leibniz started to immerse

Fig. 7.2.5. Henry Oldenburg, John Pell [Painting: Godfrey Kneller, 17th c.], Baruch de Spinoza [unknown artist, about 1665]

himself into the mathematical literature. In 1691 he wrote in retrospective [Müller/Krönert 1969, p. 33f.]:

> *I still was an utter stranger to higher geometry when I made the acquaintance of Christiaan Huygens in Paris in the year 1672. I confess that [...] I owe the most to this man subsequent to Galilei and Descartes. When I read his book 'Horologium Oscillatorium' and Dettonville's letters*[3] *and worked through the oeuvre of Gregorius a S. Vincentio*[4]*, I began to see a light which occurreed to me, and to others knowing me being a newcomer in this area, quite unexpectedly. I soon proved this by means of examples. That way a tremendous number of theorems opened up to me which only were a markdown of a new method of which I later found parts in the works of Jac. Gregory and Isaac Barrow.*
>
> (Ich war noch ganz und gar ein Fremdling in der höheren Geometrie, als ich in Paris 1672 die Bekanntschaft Christiaan Huygens' machte. Ich bekenne, dass [...] ich ganz persönlich vor allem diesem Mann nach Galilei und Descartes das meiste verdanke. Als ich sein Buch "'Horologium Oscillatorium"' las und Dettonvilles Briefe und das Werk des Gregorius a S. Vincentio ebenfalls durcharbeitete, da ging mir plötzlich ein Licht auf, das mir und auch anderen, die mich hierin als Neuling kannten, unerwartet war. Dies stellte ich bald durch Beispiele unter Beweis. So erschloß sich mir eine ungeheure Zahl von Theoremen, die nur ein Abschlag einer neuen Methode waren, wovon ich einen Teil darauf bei Jac. Gregory und Isaac Barrow fand.)

Leibniz had detected the characteristic triangle (figure 6.1.19) in Pascal's works which the latter had applied to the quarter circle but which Leibniz now

[3] Amos Dettonville was the pseudonym of Blaise Pascal, cp. section 6.1.4
[4] Grégoire de Saint-Vincent, cp. section 5.5.5

realised as being a universal tool. If there was a particular ' annus mirabilis' for Leibniz at all then it is this year 1673 in Paris: He invented the foundations of his differential and integral calculus.

But Leibniz is in some trouble. Deprived of his employer by the death of the elector life in Paris became too expensive. Additionally the manufacture of the calculating machine swallowed enormous sums. Therefore he started to sound out the situation at some princely houses but wished secretly he would be able to commute between a position in Germany securing his living and Paris as a kind of freelancing scientist. The only offer he got was one by Duke John Frederick of Brunswick-Lüneburg, Principal of Calenberg, from Hanover who offered the position of a librarian and a privy councillor. But Hanover was not at all in the mind of a young scientist who felt comfortable in Paris and London. Only at the beginning of the 17th century had Hanover become the court of the Dukes of Calenberg. The number of inhabitants even in 1766 was not even 12000 souls and Hanover was lightyears away from what would have been an attractive town for Leibniz. But the Duke remained persistent; his offer was renewed several times and others did not offer anything. Hence Leibniz decided in 1676 to go to Hanover. How 'gladly' he went to Hanover can be seen from the fact that he did not travel directly from Paris but went to London first. He stayed there for 10 days and Oldenburg provided insights into some works by Newton and other Englishmen, cp. [Hofmann 1974, Chap. 20]. This visit will produce serious problems for Leibniz later on in the priority dispute [Sonar 2018]. The English party will argue that the German was then introduced to Newton's calculus of fluxions which he then merely dressed in a new form. From London he travelled to Holland where he visited Johannes Hudde in Amsterdam and the philosopher Baruch de Spinoza (1632–1677)

7.2.3 Leibniz in Hanover

In December 1676 Leibniz arrived in Hanover; he will stay there in service for 40 years and eventually he will die there.

In Hanover he was assigned rooms in the library in the castle. He got on well with his first Hanoverian employer – Duke John Frederick allowed him full bent and Leibniz thanked him with a firework of projects. One is the drainage of the mines in the Harz Mountains by means of horizontal windmills; the project started in 1679 and is going to fail some years later. At the beginning of January 1680 the Duke passed away and times are getting harder for Leibniz. The Duke's successor is his brother Ernest Augustus who is insensitive with regard to science and utilises his privy councillor mainly to increase the glory of his house by memoranda, surveys, coinage of mints, or birthday poems. Leibniz's library budget is cut down from 1500 Reichsthaler per annum to less than 100 Thaler; this alone says everything concerning

Fig. 7.2.6. View of Hanover from north-west about 1730, Engraving by F. B. Werner (Historisches Museum Hanover)

Ernest Augustus' attitude. However, Duchess Sophia became an important supporter of Leibniz. She is, in contrast to her husband, highly educated, witty, and she very much enjoyed intelligent conversation with Leibniz.

In the year 1684 the first work on the new differential calculus was published in the 'Acta Eruditorum': *Nova Methodus pro maximis et minimis, itemque tangentibus quae nec fractas nec irrationales quantitates moratur, & singulare pro illis calculi genus* [Leibniz/Newton 1998, p. 3–11]. After the experiments with windmills in the Harz Mountains were abandoned by the Duke in 1685 Leibniz received a new order: He has to write the history of the House of Welf (Guelf, Guelph). Ernest Augustus wanted to become elector and a glorious history of his House would have come in handy. Had the Duke expected a short work then Leibniz put a spoke in his wheel: he invented modern historical sciences! Travelling in Austria and Italy Leibniz visited archives and tried meticulously to bring to light the history of the Welfs. When he died in 1716 he will have been advanced into the year 1000.

Pointing out some urgently necessary investigations concerning the history of the House of Welfs in Wolfenbüttel's library Leibniz accepted a position as a librarian in Wolfenbüttel in 1691. This means that he now stood also in the service of the Dukes Rudolph Augustus (1627–1704) and Anthony Ulrich (1633–1714) of Brunswick-Wolfenbüttel with whom he got along much better than with his main employer in Hanover. In 1696 he received the title of a judicial privy councillor for services rendered for the House of Brunswick-Lüneburg. Even trying to list his activities would be doomed to fail; he worked on mathematical, physical, judicial, theological, philosophical,

Fig. 7.2.7. Study of Leibniz in the Leibniz house (Historisches Museum Hanover)

and historical problems and left groundbreaking works in all of these areas. *Die Theodizee von der Güte Gottes, der Freiheit des Menschen und dem Ursprung des Übels* (Theodocy) [Leibniz 1985–1992, Vols. II/1, II/2] and the *Monadologie* (Monadology) [Leibniz 2005] are arguably the most well-known of Leibniz's philosophical works, but also his *Neue Abhandlungen über den menschlichen Verstand* (Nouveaux essais sur l'entendement humain) [Leibniz 1985–1992, Vols. III/1, III/2] are works belonging still to the basic stock of philosophical literature. With the only daughter of Ernest Augustus and Sophia, Sophia Charlotte of Hanover (1668–1705) Leibniz was bound by a firm friendship. On 8th October 1684 the quick-witted Sophia Charlotte was married to electoral prince Frederick who later became Frederick I, King of Prussia, for many years. In 1698 elector George Louis, the later George I, had succeeded his deceased father Ernest Augustus as sovereign. If Leibniz's situation was already difficult under Ernest Augustus it now became even more unpleasant. Hence Leibniz turned his attention more and more to Berlin where Sophia Charlotte busily promoted the expansion of sciences. Leibniz started planning an academy in Berlin. It should have been financed by a monopoly on a calendar, by a lottery, and by breeding silkworms; a technique which Leibniz got to know from his contacts with Jesuits in China. In Hanover he even pursued an experimental garden with mulberry trees and Chinese silkworms to test whether the worms could stand the rough Central European climate. Only the first of the financing ideas turned out to be truly successful but the academy was nevertheless established and Leibniz was installed its first president on 12th July 1700. In

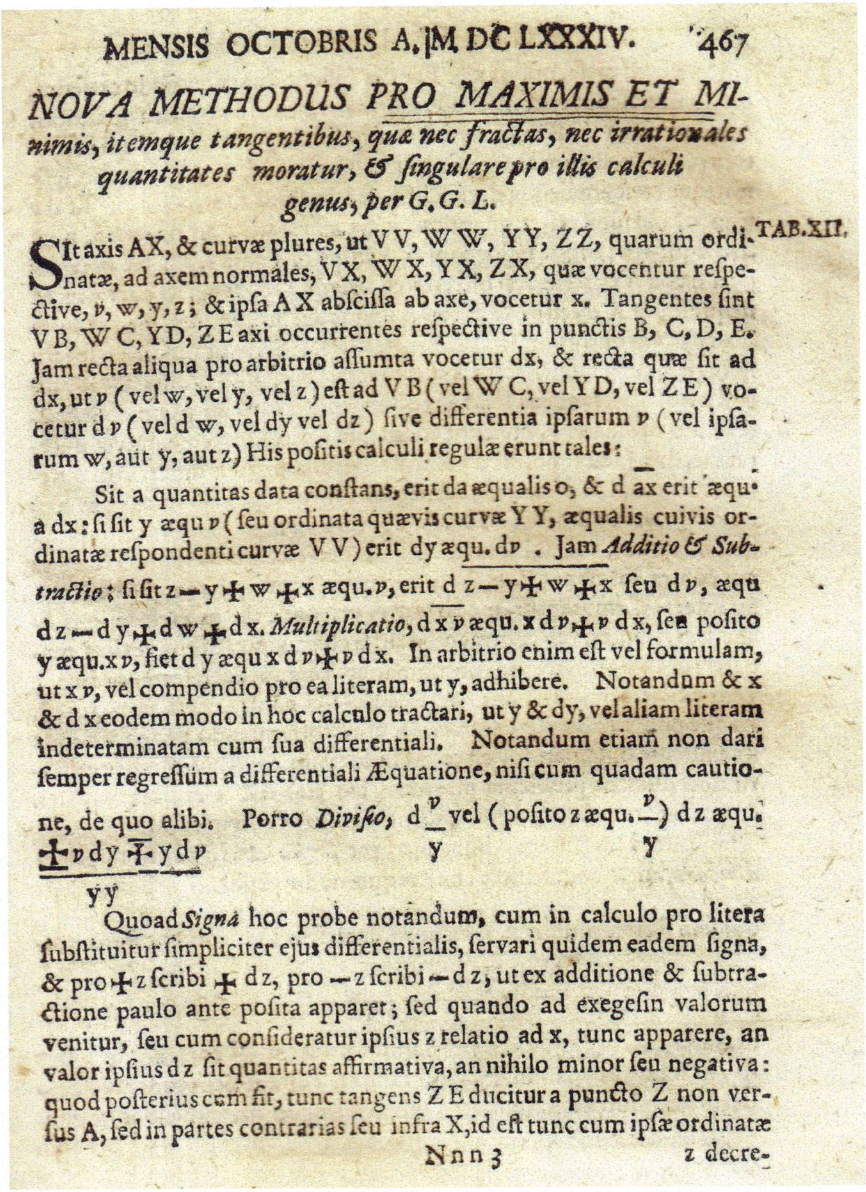

MENSIS OCTOBRIS A. M DC LXXXIV. 467

NOVA METHODUS PRO MAXIMIS ET MI-nimis, itemque tangentibus, quæ nec fractas, nec irrationales quantitates moratur, & ſingulare pro illis calculi genus, per G. G. L.

TAB. XII.

SIt axis AX, & curvæ plures, ut VV, WW, YY, ZZ, quarum ordinatæ, ad axem normales, VX, WX, YX, ZX, quæ vocentur reſpective, *v*, w, y, z; & ipſa AX abſciſſa ab axe, vocetur x. Tangentes ſint VB, WC, YD, ZE axi occurrentes reſpective in punctis B, C, D, E. Jam recta aliqua pro arbitrio aſſumta vocetur dx, & recta quæ ſit ad dx, ut *v* (vel w, vel y, vel z) eſt ad VB (vel WC, vel YD, vel ZE) vocetur d*v* (vel dw, vel dy vel dz) ſive differentia ipſarum *v* (vel ipſarum w, aut y, aut z) His poſitis calculi regulæ erunt tales:

Sit a quantitas data conſtans, erit da æqualis 0, & d $\overline{ax}$ erit æqu. a dx: ſi ſit y æqu *v* (ſeu ordinata quævis curvæ YY, æqualis cuivis ordinatæ reſpondenti curvæ VV) erit dy æqu. d*v*. Jam *Additio & Subtractio*; ſi ſit z − y + w + x æqu. *v*, erit d $\overline{z - y + w + x}$ ſeu d*v*, æqu dz − dy + dw + dx. *Multiplicatio*, d $\overline{xv}$ æqu. x d*v* + *v* dx, ſeu poſito y æqu. x*v*, fiet dy æqu x d*v* + *v* dx. In arbitrio enim eſt vel formulam, ut x*v*, vel compendio pro ea literam, ut y, adhibere. Notandum & x & dx eodem modo in hoc calculo tractari, ut y & dy, vel aliam literam indeterminatam cum ſua differentiali. Notandum etiam non dari ſemper regreſſum a differentiali Æquatione, niſi cum quadam cautione, de quo alibi. Porro *Diviſio*, d $\frac{v}{y}$ vel (poſito z æqu. $\frac{v}{y}$) dz æqu. $\frac{\pm v\,dy \mp y\,dv}{yy}$

Quoad *Signa* hoc probe notandum, cum in calculo pro litera ſubſtituitur ſimpliciter ejus differentialis, ſervari quidem eadem ſigna, & pro + z ſcribi + dz, pro − z ſcribi − dz, ut ex additione & ſubtractione paulo ante poſita apparet; ſed quando ad exegeſin valorum venitur, ſeu cum conſideratur ipſius z relatio ad x, tunc apparere, an valor ipſius dz ſit quantitas affirmativa, an nihilo minor ſeu negativa: quod poſterius cum fit, tunc tangens ZE ducitur a puncto Z non verſus A, ſed in partes contrarias ſeu infra X, id eſt tunc cum ipſæ ordinatæ

Nnn 3 z decre-

Fig. 7.2.8. Title page of the first work concerning the differential calculus of the year 1684: *Nova methodus* ... from [Acta Eruditorum 1684]

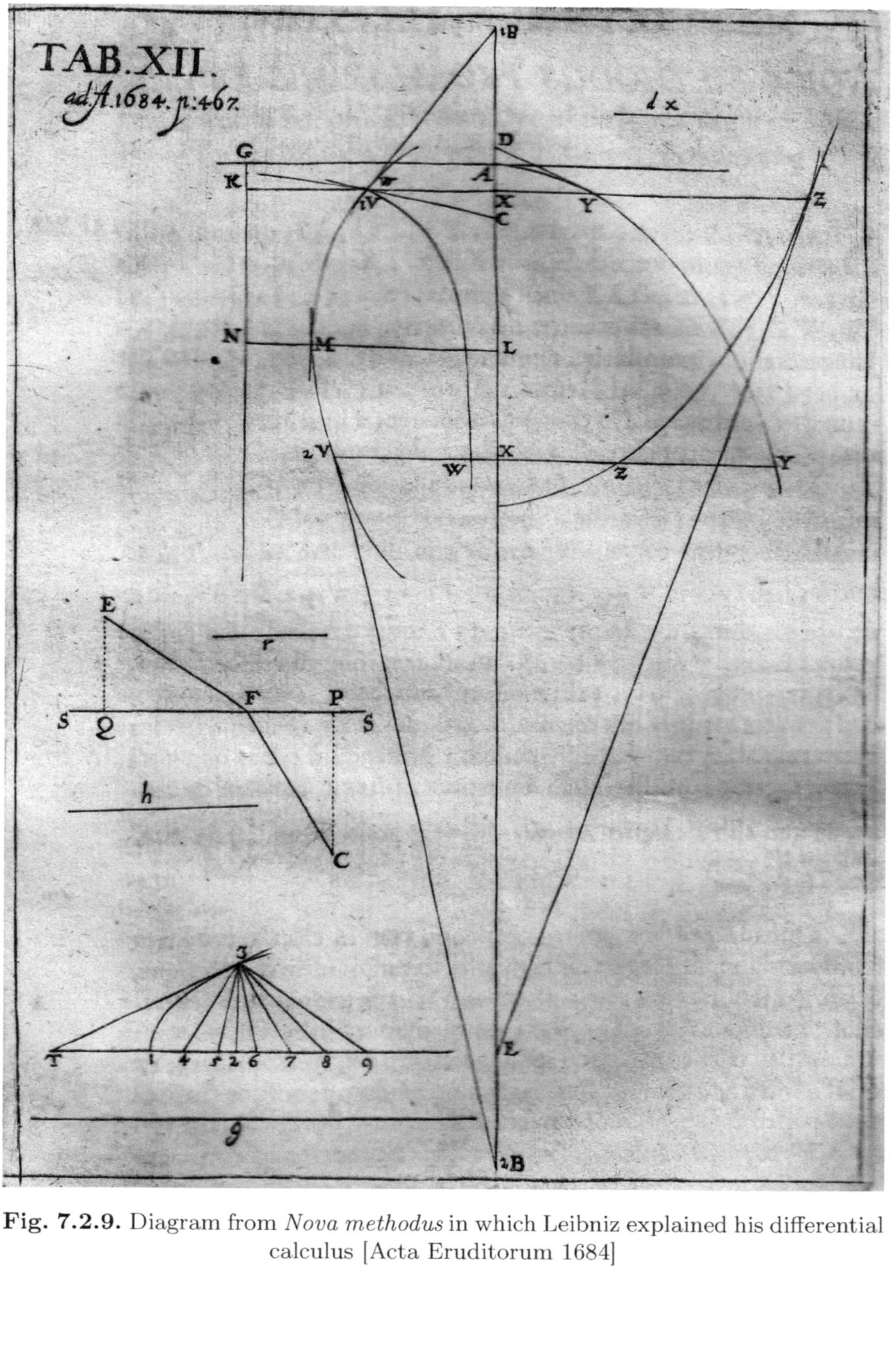

Fig. 7.2.9. Diagram from *Nova methodus* in which Leibniz explained his differential calculus [Acta Eruditorum 1684]

March 1700 he had become foreign member of the academy in Paris and now proposed the name 'Mathematisch-naturwissenschaftliche Societät' (Society for mathematics and natural sciences) to give to the Berlin academy. His frequent visits to Sophia Charlotte did by no means contribute to improve his poor relation with George Louis, who more and more insisted on a continuation of work regarding the history of the Welfs. Leibniz evaded; he made contacts with the imperial court of the Emperor in Vienna. He wanted to establish an academy in Saxony and also in Russia. When Sophia Charlotte suddenly died on 1st February 1705 Leibniz lost his most important advocate. The Brunswick Duke Anthony Ulrich provided access to Tsar Peter I who wanted to open his country to western influences and to modernise it. Though Leibniz is appointed Russian privy judicial councillor he cannot put his plans into action. Leibniz was ill and exhausted. He suffered from gout, has ulcerated legs, and is overworked and entangled in countless projects. In the summer of 1716 he took again a cure in Bad Pyrmont but a stone disease set in. On 14th November 1716, in the evening at ten o'clock, Leibniz passed away in his house in Hanover. He was buried in the court church St John in Hanover's new town without any pomp and circumstances [Sonar 2006], [Sonar 2008].

7.2.4 The Priority Dispute

Newton's first successes concerning his calculus of fluxions anteceded Leibniz's calculus by 10 years and hence Newton was effectively the first inventor of the differential and integral calculus[5]. Newton's debut work *Analysis per aequationes numero terminorum infinitas* was sent to John Collins on 10th August 1669 by Barrow. Collins can be seen as the 'English Mersenne'. He was a member of the Royal Society and its librarian, and stood in correspondence with numerous mathematicians of his days. Collins praised Newton's work, copied it, and showed it to the president of the Royal Society, William Brouncker. Henry Oldenburg had also taken note of this work as we know from a letter of Oldenburg to de Sluse of 14th September 1669 [Fleckenstein 1977, p. 20]. Although Newton had not yet published on his calculus of fluxions and fluents English mathematicians were acquainted with his new mathematics. In the contact with foreigners one remained very reticent, however. Only in a quarrel concerning methods of tangents between Newton and de Sluse did Newton send a letter to de Sluse via Collins on 20th December 1672 in which he explained his method by means of examples. During his second visit in London 1676 Leibniz got insight into Newton's papers by Collins and had also taken notes; but he had developed his own differential and integral calculus already in 1675.

[5] The complete history of the priority dispute is published in [Sonar 2018].

Fig. 7.2.10. Leibniz and his burial place in the church of the new town in Hanover, left: (Historisches Museum Hanover), right: [Photo: K. Anne Gottwald 2007]

On 26th July 1676 a letter by Newton (the 'epistola prior') was eventually sent to Leibniz via Oldenburg in which Newton reported on his mathematical techniques. The problems treated in this letter and their solutions were all well known; there was nothing on the actual method of fluxions. The letter came into Leibniz's hands only on 24th August who answered already on 27th August. He wrote openly on his 'transmutation theorem' apparently hoping that Newton would also reply more openly. One part of the letter must have interested Newton in particular [Fleckenstein 1977, p. 21]:

> [...] *If you say that most difficulties can be overcome by infinite series this does not appear to me. Many wonderful and involved things do neither depend on equations nor on quadratures. As for example the problems of the inverse tangent method of which Descartes had to confess that they were not in his command.*

([...] Wenn Ihr sagt, die meisten Schwierigkeiten liessen sich durch unendliche Reihen erledigen, so will mir das nicht recht scheinen. Vieles Wunderbare und Verwickelte hängt weder von Gleichungen noch von Quadraturen ab. So zum Beispiel die Aufgaben der umgekehrten Tangentenmethode, von welchen auch Descartes eingestand, dass er sie nicht in seiner Gewalt habe.)

The 'inverse tangent method' on which Leibniz hinted at goes back to a problem of the French mathematician Florimond de Beaune (Debeaune) (1601–1652) who was a childhood friend of Descartes'. In Debeaune's problem a curve is sought where certain properties of its tangent are prescribed. To solve this problem the integral calculus is required. Newton clearly saw the scope of Leibniz's methods and apparently feared that his own methods would stand back behind Leibniz's. In a letter to Collins of 8th November 1676 Newton at least affirmed that his own methods would not be less general and also not more cumbersome than Leibniz's methods. But had not Leibniz asked him to lay open his methods? Of what did he want to know more if he was in command of powerful methods? And anyway: why did it take so long for Leibniz to reply? In Newton's mind an answer to all of these questions brewed: Leibniz had perhaps made a chance discovery but waffled about reputed results he actually did not have, in order to break through Newton's reserve and to wrest his secrets. To make sure that his suspicion was valid another letter (the 'epistola posterior') was sent to Leibniz via Oldenburg on 2nd May 1677 which Newton had written already on 24th October 1676 – Oldenburg had simply left the letter unattended for half a year. Leibniz received this letter on 1st July 1677 and replied the same day. Apparently Newton targeted at securing his priority of the invention of the differential and integral calculus. He openly presented the proof of his binomial theorem and gave formulae for binomial integrals without proof, which Leibniz could but verify immediately as we know from notes in the margin of a letter, cp. [Fleckenstein 1977, p. 22]. Newton did not send anything on his calculus of fluxions but instead presented two anagrams to Leibniz [Fleckenstein 1977, p. 22]:

6 a c c d & 13 e f f 7 i 3 l 9 n 4 o 4 q r r 4 s 9 t 12 v x

and

5 a c c d & 10 e f f h 12 i 4 l 3 m 1 o n 6 o q q r 7 s 11 t 10 v 3 x: 11 a b 3 c d d 10 e & g 10 i l l 4 m 7 n 6 o 3 p 3 q 6 r 5 s 11 t 7 v x, 3 a c & 4 e g h 6 i 4 l 4 m 5 n 8 o q 4 r 3 s 6 t 4 v , a a d d & e e e e e i i i m m n n o o p r r r s s s s s t t u u.

Would it be possible that Leibniz could transform these riddles into Latin sentences so that he could have understood the calculus of fluxions? Surely not! Even if the solution would have been successful he would not have been able to draw conclusions from them. In the *Commercium Epistolicum* of 1712 Newton gave the solution. The first anagram meant

> *Data aequatione quotcumque fluentes quantitates involvente fluxiones invenire & viceversa*
>
> (Given the equation between arbitrary many fluent quantities to find their fluxions and vice versa)

and the second

> *Una methodus consistit in extractione fluentis quantitatis ex aequatione simul involvente fluxionem ejus: altera tantum in assumptione seriei pro quantitate qualibet incognita ex qua caetera commodo derivari possunt, & in collatione terminorum homologorum aequationis resultantis ad eruendos terminos assumptae seriei.*
>
> (One method consists in pulling out the fluents from an equation which at the same time includes their fluxions; the other method consists of assuming a series for an arbitrary unknown quantity, from which everything else can be deduced, and of the comparison of corresponding terms of the resulting equation to determine the terms of the assumed series.)

In his reply Leibniz disclosed his differential calculus but not the integral calculus. He showed his solution of Debeaune's problem by means of differential equations and not, as Newton did, by means of infinite series. By now at the latest Newton knew that the German was on a par with him and he did not reply to Leibniz's letter. Up until the second edition of his *Principia* in 1713 there was a remark concerning Leibniz in Liber II, Sect. II, Prop. VII reading very friendly and recognising [Fleckenstein 1977, p. 19]:

> *In letters which I exchanged with the very learned mathematician G. W. Leibniz about 10 years ago I showed him that I was in the possession of a method with which maxima and minima can be determined, tangents can be drawn, and similar tasks can be accomplished; namely they can be applied to irrational as to rational quantities. By interchanging the letters of the words (if an equation with arbitrary many fluents is given, to find the fluxions, and vice versa) which gave my opinion, I concealed it. The famous man replied to me thereupon that he had arrived at a method of the same kind which he communicated to me and which deviated from my own barely more as in the form of the words and symbols.*
>
> (In Briefen, welche ich vor etwa 10 Jahren mit dem sehr gelehrten Mathematiker G. W. Leibniz wechselte, zeigte ich demselben an, dass ich mich im Besitz einer Methode befände, nach welcher man

> Maxima und Minima bestimmen, Tangenten ziehen und ähnliche Aufgaben lösen könne, und zwar lassen sich dieselben ebensogut auf irrationale wie auf rationale Größen anwenden. Indem ich die Buchstaben der Worte (wenn eine Gleichung mit beliebig vielen Fluenten gegeben ist, die Fluxionen zu finden und umgekehrt), welche meine Meinung aussprachen, versetzte, verbarg ich dieselbe. Der berühmte Mann antwortete mir darauf, er sei auf eine Methode derselben Art verfallen, die er mir mitteilte und welche von meiner kaum weiter abwich als in der Form der Worte und Zeichen.)

This section disappeared in the third edition. When Leibniz's first publication concerning the differential calculus appeared in 1684 and the triumph of Leibniz's calculus started Newton could only stand back. Apparently a time bomb began to tick inside Newton's mind, probably not even recognised by Leibniz, which was going to explode only in 1699. When Jacob Bernoulli (1655–1705) in 1696 posed the problem of the brachistochrone, i.e. the curve of shortest runtime of a sphere between two points, Leibniz quickly found the solution. Together with his solution he also published a remark that only those scientists would be able to solve the problems who are adepts of the new analysis, and he mentioned explicitly Newton, Huygens, and Hudde. Thereupon Nicolas Fatio de Duillier, Newton's close fellow, felt humiliated and challenged since he had not been mentioned by Leibniz. In the year 1699 Fatio published his solution in *Lineae brevissimi descensus investigatio geometrica duplex* by means of Newton's method and used the opportunity to attack Leibniz. If one would publish the correspondence between Newton and Huygens, said Fatio on page 18 of 20 [Hess 2005, p. 65], one would recognise that

> *I am forced to accept by the evidence of the case that Newton was the first and – with many years in advance – oldest inventor of this kind of calculation. I do not want to decide whether Leibniz, the second inventor, has taken something from him, but leave that decision to those who have seen Newton's letters and his other writings.*
>
> (Ich bin durch die Evidenz der Sachlage gezwungen anzuerkennen, dass Newton der erste und – mit vielen Jahren Vorsprung – älteste Erfinder dieser Rechnungsart ist. Ob Leibniz, der zweite Erfinder, von ihm etwas übernommen hat, möchte ich weniger selbst entscheiden als dem Urteil derjenigen überlassen, die Newtons Briefe und seine anderen Handschriften gesehen haben.)

And further on [Fleckenstein 1977, p. 23]:

> *No one who has thoroughly studied what I myself have revealed will be deceived by Newton's modest silence or Leibniz's pressing bustle.*

(Niemanden, der durchstudiert, was ich selber an Dokumenten aufgerollt habe, wird das Schweigen des allzu bescheidenen Newton oder Leibnizens vordringliche Geschäftigkeit täuschen.)

In the Acta Eruditorum of 1700 Leibniz replied calmly to this attack and pointed out that Newton himself had acknowledged him in the *Principia* as being an independent inventor. He also wrote that he could not imagine Newton being aware of Fatio's attack. Fatio's reply to Leibniz's reply was not accepted for publication by the Acta Eruditorum; apparently one did not want to saddle oneself with further quarrels. When Leibniz in 1705 anonymously reviewed Newton's *Opticks: Or, a Treatise of the Reflections, Refractions, Inflections and Colours of Light* in the Acta Eruditorum he also wrote on the work *Quadratura curvarum* which was bound with the first edition of *Opticks*. He wrote that Newton's calculus of fluxions was but another notation of Leibniz's calculus. Although the review was published anonymously everyone knew that Leibniz was its author; and although the review was favourable Leibniz's remarks concerning Newton's calculus of fluxions raised Newton's wrath. But Newton remained quiet publicly; however, he pulled the strings and let John Keill (1671–1721), a young Scottish mathematician in the Royal Society, accuse Leibniz of plagiarism. In a work on the law of central forces in the Philosophical Transactions 1707/08, published in 1710, he wrote the section [Fleckenstein 1977, p. 24]:

All these things follow from the now famous method of fluxions whose first inventor doubtlessly was Sir Isaac Newton, as can everyone see who reads his letters which Wallis has published first. The same arithmetic was then later published in the Acta Eruditorum by Leibniz who only changed the name and the kind of notations.

(Alle diese Dinge folgen aus der jetzt so berühmten Methode der Fluxionen, deren erster Erfinder ohne Zweifel Sir Isaac Newton war, wie das Jeder leicht feststellen kann, der jene Briefe von ihm liest, die Wallis zuerst veröffentlicht hat. Dieselbe Arithmetik wurde dann später von Leibniz in den Acta Eruditorum veröffentlicht, der dabei nur den Namen und die Art und Weise der Bezeichnung wechselte.)

Now Leibniz made a serious mistake, probably because he was naive and believed that Newton could never condone such a behaviour: He complained to the Royal Society where Newton was president since 1703. The Royal Society quickly set up a committee to superficially deal with Leibniz's complaint. Although Newton did not appear publicly we know today that he pulled the string in the background and that no decision of the committee was taken without his intervention. The members of the committee were Edmond Halley, John Arbuthnot, William Burnett, Abraham Hill, John Machin and William Jones elected on 6th March 1712, Francis Robartes on 20th March, Louis Frederick Bonet (the representative of the Prussian

King in London) on 27th March, and Francis Aston, Brook Taylor, Abraham de Moivre on 17th April [Djerassi 2003, p. 78 f.]. The last three could have had no chance to contribute to a decision due to the late date of their appointment. Seemingly objective the committee chose manuscripts and letters to unambiguously point out that Leibniz had taken Newton's ideas. On 24th April a judgment was read to the Royal Society that Leibniz was a plagiarist. This judgment was published as *Commercium epistolicum D. Johannis Collins, et aliorum de Analysi promota: Jussu Societas Regiae in lucem editum* in 1712 in London and distributed free of charge in Europe. Later it was handed over to the public book trade and the Royal Society cared for the costs. When Leibniz died in 1716 he still did not believe that Newton was behind his condemnation. It is accepted today that Newton and Leibniz independently invented the differential and integral calculus following different paths. The priority dispute remains a thrilling episode as an unpleasant part of the history of analysis and the details can fill whole books indeed [Hall 1980], [Sonar 2018]. In his play 'Calculus ("'Newton's Whores"')' [Djerassi/Pinner 2003, p. 107-178] which takes place in the rooms of the Royal Society in London and in which the protagonists Newton and Leibniz speak for themselves, Carl Djerassi has chosen the priority dispute to reflect on the notions of 'invention' and 'priority'.

7.2.5 First Achievements with Difference Sequences

In his work *Historia et origo calculi differentialis*, written at the end of 1714, Leibniz gives an impression of the beginnings of his occupation with analysis. Shortly after he arrived in Paris he discovered an interesting property of sums of differences. If

$$a_1, a_2, a_3, \ldots, a_n$$

is a finite sequence of numbers and if

$$d_1, d_2, d_3, \ldots, d_{n-1}$$

denotes the difference sequence defined by $d_i := a_i - a_{i+1}$, then the sum of differences is

$$\begin{aligned} d_1 + d_2 + d_3 + \ldots + d_{n-1} &= (a_1 - a_2) + (a_2 - a_3) + (a_3 - a_4) + \ldots \\ &\quad \ldots + (a_{n-2} - a_{n-1}) + (a_{n-1} - a_n) \qquad (7.5) \\ &= a_1 - a_n \end{aligned}$$

and this, in nowadays language, is a 'telescoping sum'. Leibniz deduced that in case of an infinite sequence

$$a_1, a_2, a_3, \ldots$$

with $\lim_{n\to\infty} a_n = 0$ the sum of differences is just

$$d_1 + d_2 + d_3 + \ldots = a_1. \tag{7.6}$$

When Leibniz reported this result to Huygens the latter suggested to consider the series

$$\frac{1}{1} + \frac{1}{3} + \frac{1}{6} + \frac{1}{10} + \ldots + \frac{1}{n(n+1)/2} + \ldots,$$

[Edwards 1979, p. 236f.]. Of course Leibniz knew Pascal's triangle (cp. page 263)

$n=$										
0	1									
1	1	1								
2	1	2	1							
3	1	3	3	1						
4	1	4	6	4	1					
5	1	5	10	10	5	1				
6	1	6	15	20	15	6	1			
7	1	7	21	35	35	21	7	1		
8	1	8	28	56	70	56	28	8	1	
9	1	9	36	84	126	126	84	36	9	1

in which every number is the sum of the two adjacent number above. He now constructed a triangle in which every entry is the *difference* of the two adjacent numbers above and called it the 'harmonic triangle'.

$n=$							
0	$\frac{1}{1}$						
1	$\frac{1}{2}$	$\frac{1}{2}$					
2	$\frac{1}{3}$	$\frac{1}{6}$	$\frac{1}{3}$				
3	$\frac{1}{4}$	$\frac{1}{12}$	$\frac{1}{12}$	$\frac{1}{4}$			
4	$\frac{1}{5}$	$\frac{1}{20}$	$\frac{1}{30}$	$\frac{1}{20}$	$\frac{1}{5}$		
5	$\frac{1}{6}$	$\frac{1}{30}$	$\frac{1}{60}$	$\frac{1}{60}$	$\frac{1}{30}$	$\frac{1}{6}$	
6	$\frac{1}{7}$	$\frac{1}{42}$	$\frac{1}{105}$	$\frac{1}{140}$	$\frac{1}{105}$	$\frac{1}{42}$	$\frac{1}{7}$

The harmonic triangle has to be read diagonally; hence it is more lucid to represent it in rectangular shape starting with the outermost right diagonal:

$$\begin{array}{cccccccc}
\frac{1}{1} & \frac{1}{2} & \frac{1}{3} & \frac{1}{4} & \frac{1}{5} & \frac{1}{6} & \frac{1}{7} & \dots \\[2ex]
\frac{1}{2} & \frac{1}{6} & \frac{1}{12} & \frac{1}{20} & \frac{1}{30} & \frac{1}{42} & \dots & \dots \\[2ex]
\frac{1}{3} & \frac{1}{12} & \frac{1}{30} & \frac{1}{60} & \frac{1}{105} & \dots & \dots & \dots \\[2ex]
\frac{1}{4} & \frac{1}{20} & \frac{1}{60} & \frac{1}{140} & \dots & \dots & \dots & \dots \\[2ex]
\frac{1}{5} & \frac{1}{30} & \frac{1}{105} & \dots & \dots & \dots & \dots & \dots \\[2ex]
\frac{1}{6} & \frac{1}{42} & \dots & \dots & \dots & \dots & \dots & \dots \\[2ex]
\frac{1}{7} & \dots & \dots & \dots & \dots & \dots & \dots & \dots
\end{array}$$

Every row is the difference sequence of the row directly above and every row is a convergent sequence with limit 0. Following Leibniz's result (7.6) it holds for the second row

$$\frac{1}{2}+\frac{1}{6}+\frac{1}{12}+\frac{1}{20}+\ldots=1,$$

for the third

$$\frac{1}{3}+\frac{1}{12}+\frac{1}{30}+\frac{1}{60}+\ldots=\frac{1}{2},$$

for the fourth

$$\frac{1}{4}+\frac{1}{20}+\frac{1}{60}+\frac{1}{140}+\ldots=\frac{1}{3},$$

and so on. Leibniz hence succeeded at a stroke in summing infinitely many series. The series of Huygens can also be found: just multiply the second row by 2 and get

$$\frac{1}{1}+\frac{1}{3}+\frac{1}{6}+\frac{1}{10}+\ldots=2.$$

7.2.6 Leibniz's Notation

We are used to present Newton's results in Leibniz's notation simply because it turned out to be more feasible. Only in the science of mechanics time derivatives in form of Newton's dot notation $\dot{x}$, $\ddot{x}$, and so on, for dx/dt, d^2x/dt^2, and so on, have survived. The true success of the Leibnizian notation lay in his long search for a 'characteristica universalis'. This universal language or art of notation was thought to allow the decomposition of complicated notions into simpler ones which would be characterised by carefully chosen symbols. If there were such an 'index of symbols' and a corresponding 'grammar', a way to 'ars invenendi', the art of invention, should open up. By combining the symbols new knowledge should result almost automatically.

We know now that a comprehensive characteristica universalis, equally applicable to poetry, politics, theology, mathematics, and so on, can not exist. However, in mathematical analysis Leibniz succeeded in reaching a state that the higher mathematics of the differential and integral calculus can even be taught to schoolchildren, and only due to a clever choice of symbols. Indeed, Leibniz has created literally a 'calculus' so that the computation with symbols works almost on its own. Here is the motivation which will probably get sweat on the brows of maths teachers: The slope of the secant of a function $y = f(x)$ is described by

$$\frac{\Delta y}{\Delta x},$$

cp. figure 7.2.11.

If Δx becomes infinitely small (whatever that may mean!) then this happens to Δy and we get for the slope of the function at the point considered

$$\frac{dy}{dx}$$

with infinitesimal quantities dx and dy. The meaning Leibniz gave to this remains to be discussed; the d was chosen by Leibniz from the word 'difference'. Today we write rigorously

$$f'(x_0) = \frac{df}{dx}(x_0) = \lim_{h \to 0} \frac{f(x_0 + h) - f(x_0)}{h} = \lim_{x \to x_0} \frac{f(x) - f(x_0)}{x - x_0}$$

and have a clear understanding of the limit process. Leibniz deliberately chose the differential 'quotient' dy/dx since this notion then allows us to calculate with infinitesimals dx and dy like with ordinary numbers; in particular division is permitted.

If we want to compute the slope of the tangent of the composite function

$$h(g(f(x)))$$

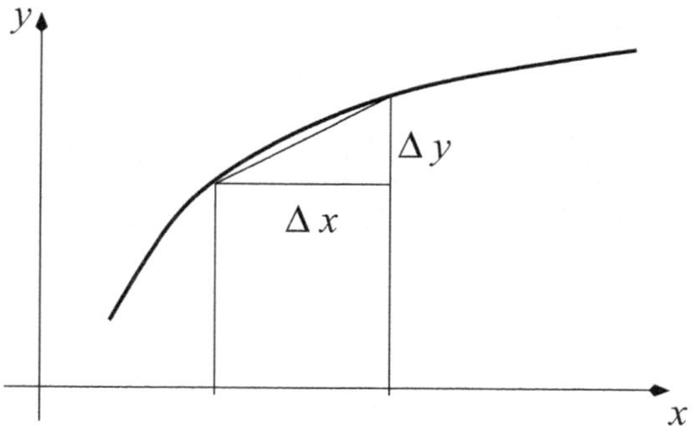

Fig. 7.2.11. Figure showing the slope of a secant

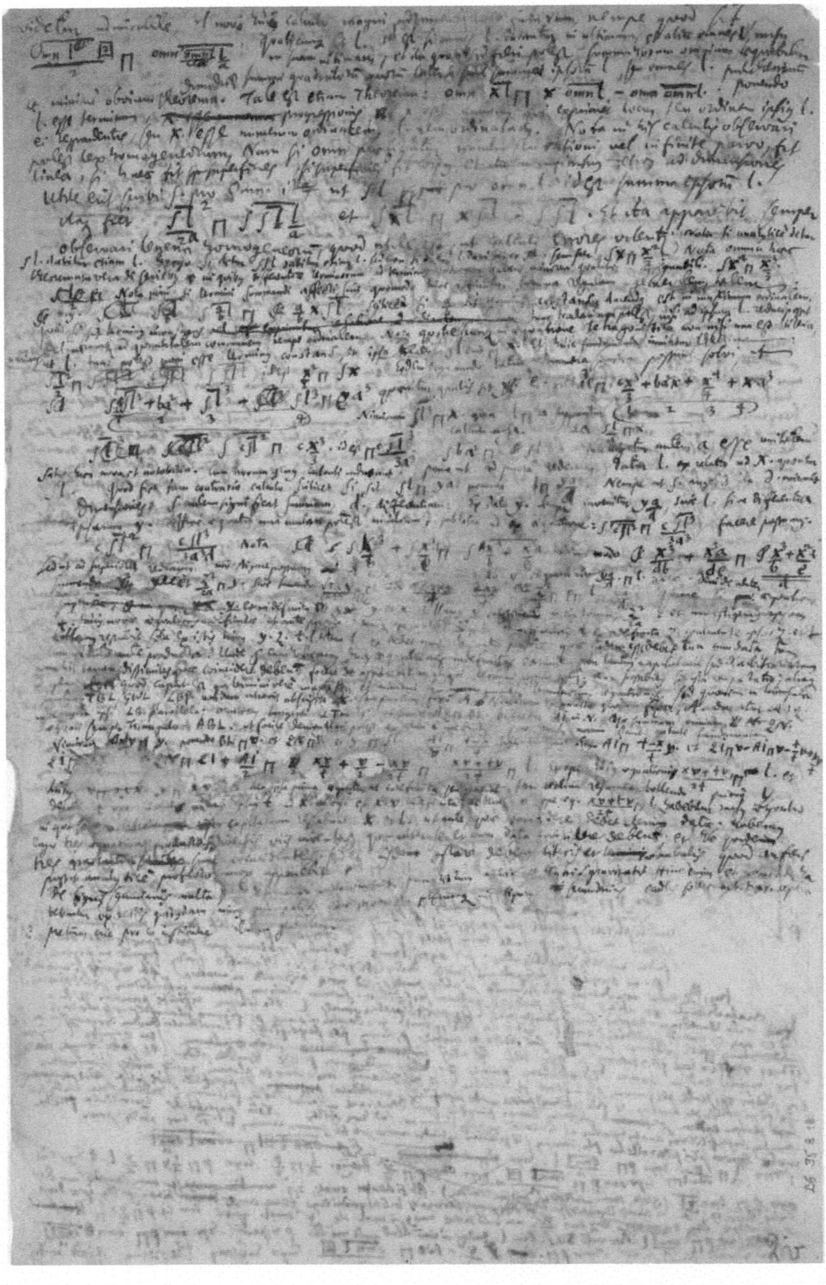

Fig. 7.2.12. Reproduction of the original handwriting by Leibniz of 29th October 1675 where he introduced the signs for integration and differentiation. Roughly in the middle we read: $\int$ .autem significat summarum, d. differentiam [Gottfried Wilhelm Leibniz Library - Niedersächsische Landesbibliothek Hanover, Signature LH XXXV, VIII, 18, Bl. 2v]

the chain rule in Leibniz's notation gives

$$\frac{dh}{dx} = \frac{dh}{dg} \cdot \frac{dg}{df} \cdot \frac{df}{dx}$$

and ordinary cancelling of dg and df shows that the equation is correct (there are the same entities on the left and the right side). Before the end of 1675 Leibniz wrote infinitesimal differences as ℓ. On the 29th October 1675 Leibniz wrote $\ell = y/d$ and on 2nd November the symbol dy was established [Edwards 1979, p. 253]. The symbol $\int$, stylising an abbreviation of 'sum', was used by Leibniz for the first time in his manuscript *Analysis tetragonistica* of 1675, where he wrote

Utile erit scribi $\int$ *pro omnia*

(It will be useful to write $\int$ instead of omnia).

'Omnia' – 'entirety' – was then the common notion for the entirety of Cavalieri's indivisibles. While before 1675 Leibniz wrote

$$\frac{\overline{\text{omn. } \ell}^2}{2} = \text{omn. } \overline{\overline{\text{omn. } \ell} \cdot \frac{\ell}{a}},$$

where the overline denotes parentheses, he turned after 1675 to (setting $a = 1$)

$$\frac{1}{2}\left(\int dy\right)^2 = \int \left(\int dy\right) dy,$$

where we have replaced the overlines by parantheses for better readability.

In Leibniz's eyes $\int f(x)\, dx$ actually *is* the sum of infinitesimal rectangles with area $f(x) \cdot dx$, cp. figure 7.2.13.

How marvellous the symbols of Leibniz work 'automatically' can now also be seen in the case of integration. If we substitute in

$$\int f(x)\, dx$$

the function $x = g(z)$, then $dx/dz = g'(z)$, hence $dx = g'(z)dz$, and it follows the substitution rule for integrals

$$\int f(x)\, dx = \int f(g(z)) \cdot g'(z)\, dz.$$

The astute symbolism – d for difference, $\int$ for sum – lets the fundamental theorem of differential and integral calculus in the form

$$\frac{d}{dx} \int f(x)\, dx = f(x)$$

appear as being quite natural: differentiation and integration are mutually inverse operations.

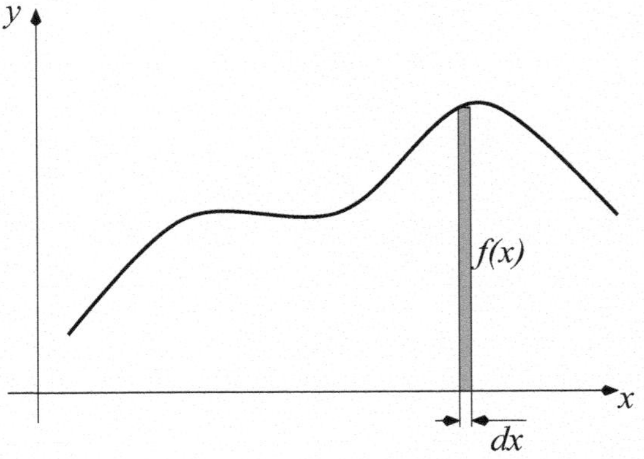

Fig. 7.2.13. An infinitesimal rectangle as area under a curve

7.2.7 The Characteristic Triangle

When Leibniz found the characteristic triangle in figure 6.1.19 in Pascal's *Lettres de A. Dettonville* in Paris it became suddenly clear to him that this tool may be not only useful for circles but for any other curve. In our presentation we follow [Edwards 1979, p. 241f.].

If as in figure 7.2.14 $n(x)$ denotes the normal to a curve in the point (x, y) and $t(x)$ the subnormal and if we draw a characteristic triangle at the point (x, y) with infinitesimal legs dx and dy, then the two triangles in the figure are similar and therefore

$$\frac{ds}{n} = \frac{dx}{y},$$

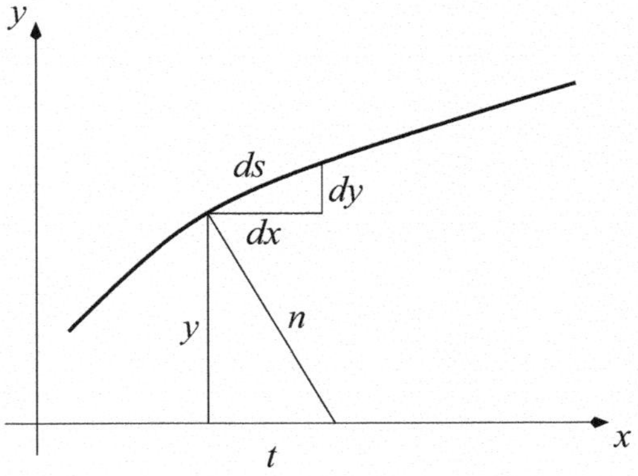

Fig. 7.2.14. The characteristic triangle at an arbitrary curve

or

$$y\,ds = n(x)\,dx.$$

The arc length element ds appears in the characteristic triangle as an infinitesimal line segment; and the Pythagorean theorem gives $ds^2 = dx^2 + dy^2$. Summing the infinitesimal quantites it follows with Leibniz's integral symbol

$$\int y\,ds = \int n(x)\,dx.$$

The integral $\int y\,ds$ is called the 'moment of the curve' and plays a major role in engineering. Multiplying by 2π the surface of the body of revolution results, i.e. the surface of the body which results from rotation of the curve about the abscissa, hence from the formula

$$O := \int 2\pi y\,ds = \int 2\pi n(x)\,dx$$

with appropriate boundaries of integration. Hence the surface of bodies of revolution can be computed by means of the normal n at the generating curve.

It also follows from figure 7.2.14 from the similarity of the triangles

$$\frac{dy}{t} = \frac{dx}{y},$$

hence $t(x)\,dx = y\,dy$. After summation of the infinitesimals it follows

$$\int t(x)\,dx = \int y\,dy. \tag{7.7}$$

To appreciate the meaning of this relation we compute the area under the curve x^n from $x = 0$ to $x = a$, hence $\int_0^a x^n\,dx$. If we could find a function $y = f(x)$ with subnormal $t(x) = x^n$ then from (7.7)

$$\int_0^a x^n\,dx = \int_{x=0}^{x=a} y\,dy = \frac{1}{2}y^2\Big|_{x=0}^{x=a} = \frac{1}{2}(f(a))^2 - \frac{1}{2}(f(0))^2$$

would follow. For the subnormal it holds

$$t = y \cdot \frac{dy}{dx}.$$

Trying $y = bx^k$ then the slope is $dy/dx = kbx^{k-1}$ and the subnormal would be

$$t = bx^k \cdot kbx^{k-1} = b^2kx^{2k-1}.$$

This should be x^n, hence we arrive at demanding

$$t = b^2kx^{2k-1} \stackrel{!}{=} x^n,$$

what we can indeed satisfy with

$$k = \frac{1}{2}(n+1), \quad b = \frac{1}{\sqrt{\frac{1}{2}(n+1)}}.$$

For $y = f(x) = bx^k$ even $f(0) = 0$ is valid, hence it follows

$$\int_0^a x^n\,dx = \frac{1}{2}f(a)^2 = \frac{1}{2}(ba^k)^2 = \frac{a^{n+1}}{n+1}.$$

As another application of the characteristic triangle we note that from the similarity of the triangles it follows from figure 7.2.15

$$\frac{ds}{r} = \frac{dy}{a},$$

where r is the length of the tangent between its point of intersection with the abscissa and a vertical straight line with a segment of fixed length a, see figure 7.2.15. Hence $a\,ds = r\,dy$ holds, or, after summation of the infinitesimals

$$\int a\,ds = \int r\,dy.$$

Putting $a = 1$ then $\int ds$ is the arc length of the curve which can easily be computed if the length r of the tangent is known as a function of y.

7.2.8 The Infinitely Small Quantities

There is no doubt that we can count Leibniz as a mathematician of infinitesimals although he started as a mathematician using indivisibles. He refused

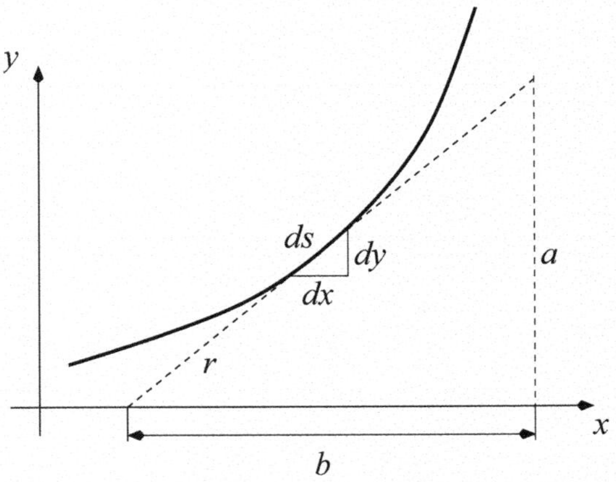

Fig. 7.2.15. The characteristic triangle again

the actual infinite and his principle of continuity which we will still have to discuss shows his acceptance of a continuum in the classical sense of the Greeks. But in which sense did Leibniz want the infinitely small quantities to be understood?

Volkert gave in [Volkert 1988, p. 98f.] three different interpretations which he substantiated with corresponding Leibniz's quotations. These three interpretations are given as follows:

1 'infinitely small = negligible'

2 'infinitely small = convergence to zero'

3 'infinitely small = 0'

Concerning the first interpretation let us read Leibniz who wrote in 1701 against a critique of his infinitesimal calculus in the 'Journal de Trévoux' [Leibniz 2004, Vol. V, p. 350] (cited after [Volkert 1988, p. 98] with corrections):

> *I add to this* [...] *that here one cannot grasp the infinite in a strong sense, but only, as one says in optics, that the rays of the sun come from a point infinitely far away and thus can be seen as parallel.*
>
> (J'ajoûterai même á ce que cet illustre Mathématicien en a dit, qu'on n'a pas besoin de prendre l'infini ici á la rigeur, mais seulement comme lorsqu'on dit dans l'optique, que les rayons du Soleil viennent d'un point infiniment éloigné, et ainsi sont estimés parallèles.)

In the work *Responsio ad nonnullas difficultates a Dn. Bernardo Niewntiit circa methodum differentialem seu infinitesimalem motas* (Riposte to some of Mister Bernard Nieuwentijt's objections brought forward against the differential or infinitesimal method), which appeared in the Acta Eruditorum in July 1695, Leibniz gave a more mathematically inclined interpretation [Leibniz 2011, p. 273f.]:

> *I namely agree with Euclid, Book 5, Definition 5*[6] *that homogeneous*[7] *quantities are comparable only if the one* [quantity], *if multiplied with a finite number, can surpass the other* [quantity]. *And what does*

[6] Meant is Definition 4 [Euclid 1956, Vol. 2, p. 114]: *'Magnitudes are said to* **have a ratio** *to one another which are capable, when multiplied, of exceeding one another.'* The background is the axiom of Archimedes: Given any two numbers $y > x > 0$ a natural number N can always be found, such that $N \cdot x$ is larger than the quantity y, i.e. that $Nx > y$ holds. The axiom of Archimedes excludes infinitesimals from the real numbers and this 'gap' was used by Leibniz in the same way as it is used today in nonstandard analysis.

[7] Quantities are homogeneous if they are 'similar in dimension', i.e. two lengths, two areas, two volumes, etc., but not an area and a volume.

not differ by such an amount I define as equal. This was done also by Archimedes and all others coming after him. And exactly that is meant if one says, that the difference [of two quantities] *is smaller than an arbitrary given* [quantity].

(Ich halte nämlich mit Euklid, [Elementa] Lib. 5, Defin. 5, homogene Größen nur dann für vergleichbar, wenn die eine [Größe], falls man sie mit einer [hier] aber endlichen Zahl multipliziert, die andere [Größe] übertreffen kann. Und was sich nicht um eine solche Größe unterscheidet, erkläre ich für gleich. Dies haben auch Archimedes und alle anderen nach ihm so gehalten. Und genau dies ist gemeint, wenn man sagt, dass die Differenz [zweier Größen] kleiner als eine beliebige gegebene [Größe] ist.)

The second interpretation of infinitely small quantities as being a kind of convergence to zero was particularly clearly expressed by Leibniz in a letter to the French mathematician and physicist Pierre de Varignon (1654–1722) of 2nd February 1702 [Leibniz 1985–1992, Vol. IV, p. 250ff.]:

At the same time one has to consider that the incomparably small quantities, taken in the usual sense, are by no means unchanged and determined; rather, since they can be assumed arbitrarily small, they play the same role as the infinitely small in the strong sense in our geometrical considerations. For if an opponent would contradict our explanations our calculus shows that the error will be smaller than any determinable error, since it is in our power to keep the incomparably little, which can be an arbitrarily small quantity, small enough for this purpose.

(Zugleich muß man jedoch bedenken, daß die unvergleichbar kleinen Größen, selbst im gebräuchlichen Sinne genommen, keineswegs unverändert und bestimmt sind, daß sie vielmehr, da man sie beliebig klein annehmen kann, in unseren geometrischen Überlegungen dieselbe Rolle spielen wie die unendlichkleinen im strengen Sinne. Denn wenn ein Gegner unserer Darlegungen widersprechen wollte, so zeigt sich durch unseren Kalkül, daß der Irrtum geringer sein wird als jeder bestimmbare Irrtum, da es in unserer Macht ist, das Unvergleichbarkleine, das man ja immer von beliebig kleiner Größe nehmen kann, für diesen Zweck klein genug zu halten.)

In *Theoria motus abstracti ...*, a work concerned with the doctrine of movement, Leibniz contributed to the third interpretation (cited after [Lasswitz 1984, Vol. 2, p. 464]):

And this is the foundation of Cavalieri's method, whereby its truth is evidently proved by thinking of certain so-called rudiments or

> *beginnings of the lines and figures, smaller than any arbitrary assignable quantity.*
>
> (Und dies ist das Fundament der cavalierischen Methode, wodurch ihre Wahrheit evident bewiesen wird, indem man gewisse sozusagen Rudimente oder Anfänge der Linien und Figuren denkt, kleiner als jede beliebige angebbare Größe.)

The third interpretation, as illogical as it seems to be, will be taken up by Leonhard Euler and he will bring the 'calculation with zeros' to perfection.

Three different interpretations of infinitesimals are also discussed by Laugwitz [Laugwitz 1990, p. 10f.] and he draws similar conlcusions. Additionally Laugwitz considered Leibniz's treatment of the infinite large. In the work *Specimen novum analyseos pro scientia infiniti circa summas et quadraturas* of 1702, concerning integration of rational functions by means of partial fraction decomposition, Leibniz included an example of the summation of series, cp. [Laugwitz 1990, p. 11f.]. He explicitly said there that the harmonic series

$$\sum_{k=1}^{\infty} \frac{1}{k}$$

has an infinitely large sum, but then he computes

$$\begin{aligned}
\frac{1}{3}+\frac{1}{8}+\frac{1}{15}+\frac{1}{24}+\ldots = \sum_{k=2}^{\infty} \frac{1}{k^2-1} &= \sum_{k=2}^{\infty} \frac{1}{2}\left(\frac{1}{k-1}-\frac{1}{k+1}\right) \\
&= \frac{1}{2}\sum_{k=2}^{\infty}\frac{1}{k-1} - \frac{1}{2}\sum_{k=2}^{\infty}\frac{1}{k+1} \\
&= \frac{1}{2}\left(1+\frac{1}{2}+\frac{1}{3}+\frac{1}{4}+\frac{1}{5}+\ldots\right) \\
&+ \frac{1}{2}\left(\qquad -\frac{1}{3}-\frac{1}{4}-\frac{1}{5}-\ldots\right) \\
&= \frac{1}{2}\left(1+\frac{1}{2}\right) = \frac{3}{4}.
\end{aligned}$$

Such a subtraction of two divergent series was accepted back then and Euler will still work with such methods.

It would be presumptuous to view the three interpretations as the sum of all that Leibniz would have had to say about his infinitesimals. Despite great efforts of the Leibniz edition in which all of the works of Leibniz will be edited[8] we now know approximately 25% of the mathematical works [Knobloch 2009]. As an example a great discovery was published only in 1993 when Eberhard

[8] Volume 5 of series seven contains important work of the time in Paris 1674–1676 [Leibniz 2008].

Knobloch transcribed and commented on Leibniz's manuscript *De quadratura arithmetica circuli ellipseos et hyperbolae cujus corollarium est trigonometria sine tabulis* [Knobloch 1993]. Leibniz had written the manuscript in the time he stayed in Paris, but he had left it there in 1676 and it was never published. Already the Leibniz scholar Joseph Ehrenfried Hofmann (1900–1973) had referred to this manuscript, cp. [Hofmann 1974], but it took until 1993 that the manuscript appeared in print. Therein Leibniz took some trouble to lay down the foundations of his analysis and there we find in Theorem VI another meaning of the infinitesimals which does not correspond to any of the three interpretations we discussed before [Knobloch 1993, p. 15]:

> *'The 'very subtle' (spinosissima) theorem 6 lays the ground of the geometry of infinitesimals by means of analytical geometry. It shows that a curvilinear bounded area can be approximated with arbitrary accuracy by a rectilinear bounded staircase-shaped area. Arbitrary accuracy means:* **the error can be made smaller than any prescribed positive number**[9]*.'*
>
> (Der 'sehr spitzfindige' (spinosissima) Satz 6 gibt eine Grundlegung der Infinitesimalgeometrie mittels der analytischen Geometrie. Er zeigt, daß eine krummlinig begrenzte Fläche durch eine geradlinig begrenzte treppenförmige Fläche beliebig genau angenähert werden kann. Beliebig genau heißt: der Fehler kann kleiner als jede vorgegebene positive Zahl gemacht werden)

Now this explanation given by Leibniz is nothing less than our modern arithmetic conception of a limit process which only crystallised in the 19th century! Note the fundamental difference between the interpretation in *Theoria motus abstracti ...* in which Leibniz speaks of 'smaller than any arbitrary assignable quantity' and the interpretation in the manuscript of 1676 in which he said 'smaller than any prescribed positive number'. A positive number being smaller than any assignable quantity is zero; one which is made smaller than any prescribed positive number is still positive but not zero! Knobloch writes in [Knobloch 2004, p. 498]:

> *The sixth theorem gives the exact foundation of the mathematics of infinitesimals by help of Riemann's sums and finite means. In nowadays terminology this means that continuous functions are integrable in the sense of Riemann.*
>
> *The proof works with the refinement of intervals of integration and a method of estimation which had satisfied even Weierstraß*
>
> (Der sechste Satz gibt die exakte Begründung der Infinitesimalmathematik mit Hilfe Riemannscher Summen und endlichen Mitteln. Er

[9] My boldface.

besagt in heutiger Terminologie, dass stetige Funktionen Riemann-integrierbar sind.

Der Beweis arbeitet mit der Verfeinerung von Integrationsintervallen und einer Abschätzungsmethode, die auch Weierstraß zufriedengestellt hätte.)

All of what we have discussed underlines that Leibniz chose different interpretations of infinitesimals to explain them depending on his correspondent (this is in particular true for the first interpretation as 'negligible quantities' as the letter [Leibniz 1985–1992, Volume IV, p. 250ff.] to Varignon clearly shows). Only for himself he seems to have develop a very clear, precise, modern idea of limits.

In [Knobloch 2004] Knobloch eventually built a bridge from Cusanus via Galilei to Leibniz. Had Galilei's work shown that 'nonquanta' like indivisibles and the actual infinite violated some underlying mathematical assumptions and were therefore no good for use in a calculus, then Leibniz's infinitesimals, 'smaller than any prescribed positive number', hence infinitely small, were still 'quanta'. Even infinitely large quantities are 'larger than any assigned quantity' and hence still 'quanta'. Knobloch [Knobloch 2004, p. 498] goes on:

> *No Cusanean coincidence of the opposite, which does not happen in mathematics, but Archimedean or Weierstsraßean rigour. Instead of Galilean nonquanta Leibniz utilised the quantified infinity which has to be differently understood than these; suitable for calculus and safe in dealing with it. Leibniz develops a corresponding 'arithmetic of the infinite'.*
>
> (Keine cusanische Koinzidenz des Entgegengesetzten, die in der Mathematik nicht stattfindet, sondern Archimedische oder Weierstraßsche Strenge. Statt der galileischen Nichtquanten verwendet Leibniz das quantifizierte Unendlich, das anders als jene verstehbar ist, das kalkülgeeignet ist und mit dem man sicher umgehen kann. Leibniz entwickelt eine entsprechende "Arithmetik des Unendlichen".)

We should remind ourselves that today we use the word 'continuum' in a completely other meaning than Leibniz; the same is true for the word 'analysis' what Breger has pointed out in [Breger 1999]. The cheap critique of Leibniz's differentials from the exclusive point of nowadays mathematics does therefore fail in many respects.

Besides the infinitesimally small differentials dx, dy analysis asks for differentials of higher order, of course; differences of differences, and so on. Differentials of second order are defined by

$$d(dx) = d\,dx = d^2x$$

and differentials of arbitrary order n are recursively defined by

$$d^n x = d(d^{n-1}x).$$

Leibniz had therefore introduced a scale in the infinitely small [Bos 1975]. The differential dx is incomparably small as compared to x and the differential of second order d^2x is incomparably small as compared to dx, and so on. It is postulated furthermore that in case of a smooth function $y = f(x)$ the n-th power $(dx)^n = dx^n$ lies in the very same scale of $d^n y$ so that the quotient

$$\frac{d^n y}{dx^n}$$

remains finite. Further investigations concerning differentials of higher order can be found in [Bos 1975].

7.2.9 The Transmutation Theorem

Leibniz has described this important theorem in his letter to Newton sent as a reply to the epistola prior, Newton's first letter.

We follow [Edwards 1979, p. 246] and imagine two infinitesimally adjacent points P and Q on the graph of a function $y = f(x)$, having coordinates (x, y) for P and $(x + dx, y + dy)$ for Q between $x = a$ and $x = b$ as shown in figure 7.2.16. The triangle PQR clearly is the characteristic triangle. Also OPQ is an infinitesimal triangle. The arc length element ds defines the tangent to f in P which intersects the ordinate in the point T with coordinates $(0, z)$. The slope of the tangent is

$$\frac{dy}{dx} = \frac{y - z}{x},$$

therefor z is given by

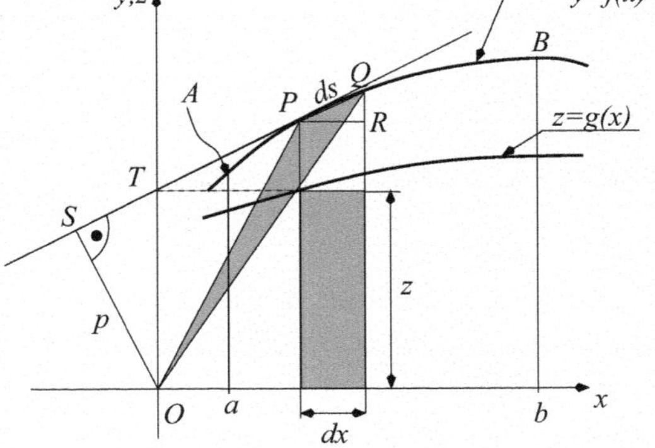

Fig. 7.2.16. Figure concerning the transmutation theorem

$$z = y - x\frac{dy}{dx}. \tag{7.8}$$

Let the segment OS be perpendicular to the tangent to f and let the length of OS be denoted by p. The triangle OST is similar to the characteristic triangle PQR, hence

$$\frac{dx}{p} = \frac{ds}{z}.$$

We can therefore compute the area of the triangle OPQ to be

$$F(OPQ) = \frac{1}{2}p \cdot ds = \frac{1}{2}z \cdot dx.$$

Summing all these infinitesimal triangles results in the area bounded above by the graph of f; hence of the curvilinearly bounded triangle OAB. This gives

$$F(OAB) = \frac{1}{2}\int_a^b z\,dx,$$

where $z = g(x)$ is defined by (7.8). We can now start to compute the area under $y = f(x)$ between $x = a$ und $x = b$. It holds

$$\int_a^b y\,dx = \frac{1}{2}b \cdot f(b) - \frac{1}{2}a \cdot f(a) + F(OAB),$$

what can easily be verified from figure 7.2.17.

It is customary to write the difference $bf(b) - af(a)$ in the form $xf(x)|_{x=a}^{x=b}$. Then we arrive at the 'transmutation theorem'

$$\int_a^b y\,dx = \frac{1}{2}\left(xy|_{x=a}^{x=b} + \int_a^b z\,dx\right). \tag{7.9}$$

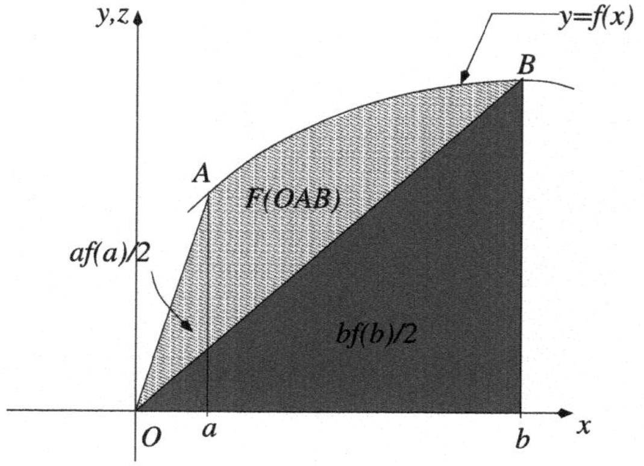

Fig. 7.2.17. Figure concerning the ratio of areas

We shall right away look at an example in which the transmutation theorem plays an important role. The reasons for Leibniz appreciating this theorem highly lay in several of its consequences. For one the theorem shows, as does the fundamental theorem, the connection between quadrature ($\int y\,dx$) and the tangent, since z is defined by the tangent. Leibniz gave the name 'quadratrix' to the function $z = g(x)$ since it enables the quadrature. It is also in this sense that we have to view the choice of the name: The quadrature problem is transmuted into another problem involving the quadratrix. Replacing z in (7.9) by (7.8) immediately gives us the important rule of integration by parts

$$\int_a^b y\,dx = xy|_{x=a}^{x=b} - \int_{f(a)}^{f(b)} x\,dy.$$

The transmutation theorem hence on the other hand is an utmost versatile tool in Leibniz's new analysis.

An important example concerning the transmutation theorem is Leibniz's 'arithmetic quadrature of the circle', i.e. the computation of the circle's area.

For this purpose Leibniz considered the half circle with radius $r = 1$ in figure 7.2.18. This half circle has the representation

$$y = \sqrt{2x - x^2}.$$

The slope of the tangent, hence the first derivative, is

$$\frac{dy}{dx} = \frac{1-x}{\sqrt{2x - x^2}} = \frac{1-x}{y}$$

and therefore it follows from (7.8) the representation of the quadratrix z to

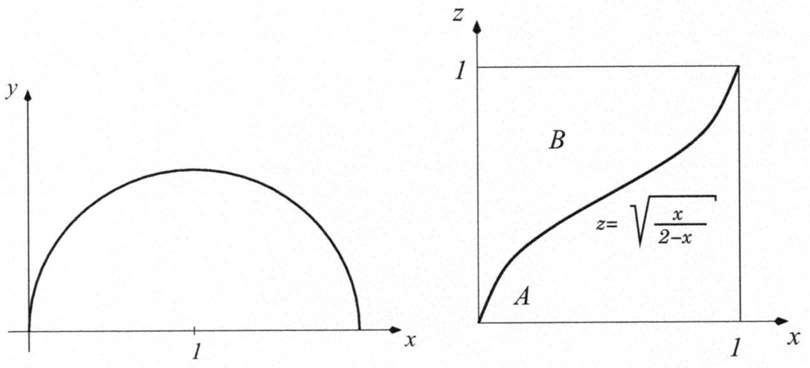

Fig. 7.2.18. Quadrature of the circle (left), decomposing the unit square by the quadratrix (right)

$$z = y - x\frac{1-x}{y} = \sqrt{2x - x^2} - x\frac{1-x}{\sqrt{2x-x^2}} = \sqrt{\frac{x}{2-x}},$$

or

$$x = \frac{2z^2}{1+z^2}.$$

We now apply the transmutation theorem (7.9) and get for the area of the quarter circle

$$I := \int_0^1 y\,dx = \frac{1}{2}\left(x\sqrt{2x-x^2}\Big|_0^1 + \int_0^1 z\,dx \right).$$

The quadratrix $z = \sqrt{\frac{x}{2-x}}$ passes through the points $(0,0)$ and $(1,1)$, and thereby divides the area of the unit square into two parts A and B as shown in the right part of figure 7.2.18.

The area A is $\int_0^1 z\,dx$, area B is $\int_0^1 x\,dz$, namely the area under $x = \frac{2z^2}{1+z^2}$ between $z = 0$ and $z = 1$. Now $B + A = 1$, hence $A = 1 - B$ or $\int_0^1 z\,dx = 1 - \int_0^1 x\,dz$, and therefore

$$I = \frac{1}{2}\left(1 + \left(1 - \int_0^1 x\,dz\right)\right) = 1 - \int_0^1 \frac{z^2}{1+z^2}\,dz.$$

Now Leibniz expanded $1/(1+z^2)$ into the geometric series $1 - z^2 + z^4 - + \ldots$, got

$$I = 1 - \int_0^1 z^2\left(1 - z^2 + z^4 - + \ldots\right)dz,$$

and integrated term by term, yielding

$$I = 1 - \left(\frac{1}{3}z^3 - \frac{1}{5}z^5 + \frac{1}{7}z^7 - + \ldots\right)\Big|_0^1 = 1 - \frac{1}{3} + \frac{1}{5} - \frac{1}{7} + - \ldots.$$

Since the area of the quarter circle with radius 1 is $\pi/4$ Leibniz found the wonderful result

$$\frac{\pi}{4} = 1 - \frac{1}{3} + \frac{1}{5} - \frac{1}{7} + \frac{1}{9} - \frac{1}{11} + - \ldots.$$

He joyfully remarked: 'Numero deus impare gaudet' (God finds delight in odd numbers) – a citation which can already be found in the writings of Virgil (8th Eclogue 76) [Vergil 2001], [Virgil 1999, p. 81]:

In an uneven number heaven delights.

7.2.10 The Principle of Continuity

Leibniz was a keen expert of ancient Greek philosophy and hence it does not wonder that Leibniz's considerations concerning continuity were based on Aristotle who in turn found the discussions on the continuum with the Eleatics. One fundamental assumption of Aristotle is 'natura non facit saltus' (nature does not jump). Hence Leibniz's principle of continuity can *not* be seen as a generalisation of ideas of his analysis. He rather let the philistophical idea flow into his analysis and this is the reason why I shun to equate 'continuity' with the mathematical notion of continuity.

In the work *A general principle which is not only useful in mathematics, but also in physics* (Ein allgemeines Prinzip, das nicht nur in der Mathematik, sondern auch in der Physik von Nutzen ist) [Leibniz 1985–1992, Volume IV, p. 230ff.] he speaks of a 'principle of general order' which has its origin in the infinite. Although the work is mostly dedicated to problems of physics we also find statements interesting for mathematics alike; e.g. when Leibniz writes [Leibniz 1985–1992, Volume IV, p. 231]:

> *If (with the given quantities) two instances come close continually so that eventually one passes into the other, then the same has necessarily to happen with the derived or depending (sought for) quantities.*
>
> (Wenn sich (bei den gegebenen Größen) zwei Fälle stetig einander nähern, so dass schließlich der eine in den anderen übergeht, muß notwendig bei den abgeleiteten bzw. abhängigen (gesuchten) Größen dasselbe geschehen.)

This depends on the following, even more general principle:

> *To an order in the prescribed corresponds an order in the sought-after.*
>
> (Einer Ordnung im Gegebenen entspricht eine Ordnung im Gesuchten.)

In a letter to Varignon of 1702 [Leibniz 1985–1992, Volume IV, p. 260ff.] Leibniz tried to explain his principle of continuity:

> *I am completely convinced by the generality and the value of this principle not only for geometry, but also for physics. Since geometry is nothing but the science of boundaries and the size of the continuum it take no wonder that this law can be observed everywhere in it: where should a sudden interruption come from in a subject where that is inadmissible by its nature? We also know that all things in this science are completely interwoven and one can not give a single example that any property suddenly cease to exist or suddenly come*

Fig. 7.2.19. George Berkeley ([John Smybert, probably 1727] National Portrait Gallery Washington, NPG.89.25), Pierre Varignon

into existence, without a transition from one to another state, which could indicate turning points and points of intersection which could explain the change so that a single algebraic equation, characterising a particular state exactly, would practically represent all others which may refer to the same state.

(Ich bin von der Allgemeingültigkeit und dem Wert dieses Prinzips nicht nur für die Geometrie, sondern auch für die Physik vollkommen überzeugt. Da die Geometrie nichts anderes als die Wissenschaft von den Grenzen und der Größe des Kontinuums ist, so ist es nicht verwunderlich, dass dieses Gesetz überall in ihr beobachtet wird: denn woher sollte eine plötzliche Unterbrechung bei einem Gegenstand kommen, der kraft seiner Natur keine zuläßt? Auch ist uns wohlbekannt, dass alles in dieser Wissenschaft vollkommen miteinander verbunden ist, und man kann hier kein einziges Beispiel dafür geben, dass irgendeine Eigenschaft plötzlich aufhörte oder entstände, ohne dass man den Übergang vom einen zum anderen Zustand, die Wende- und Schnittpunkte, welche die Veränderung erklärlich machen, angeben könnte, derart, dass eine einzige algebraische Gleichung, die einen bestimmten Zustand exakt darstellt, praktisch alle anderen darstellt, die sich auf denselben Gegenstand beziehen können.)

Thereof (and from other statements by Leibniz, cp. [Volkert 1988, p. 103]) it follows unambiguously that for Leibniz there are no functions showing discontinuities. Even more, a 'function in the sense of Leibniz' was always *smooth*, i.e. at least one time continuously differentiable. In the section 11.5

we shall find the way back to Leibniz since modern mathematics indeed has opened the way to 'smooth worlds' in analysis in which no non-smooth functions exist.

7.2.11 Differential Equations with Leibniz

As did Newton also Leibniz started to solve differential equations immediately after he invented his calculus. Differential equations appear almost natural if continually processes are modelled. During his time in Paris Leibniz observed one of his interlocutors pulling one of the then new (and very expensive!) pocket watches on its silver fob chain across the table. Immediately he asked for the curve which is described by the path of the watch.

Leibniz characterised the sought curve by having a tangent segment of constant length a. Using the notation in figure 7.2.20 yields

$$\frac{dy}{dx} = -\frac{y}{z}.$$

Pythagoras' theorem gives

$$z^2 + y^2 = a^2.$$

Hence it follows the differential equation

$$y' = \frac{dy}{dx} = -\frac{y}{\sqrt{a^2 - y^2}}.$$

Since in Leibniz's view the differential quotient is a real quotient he separates terms as follows:

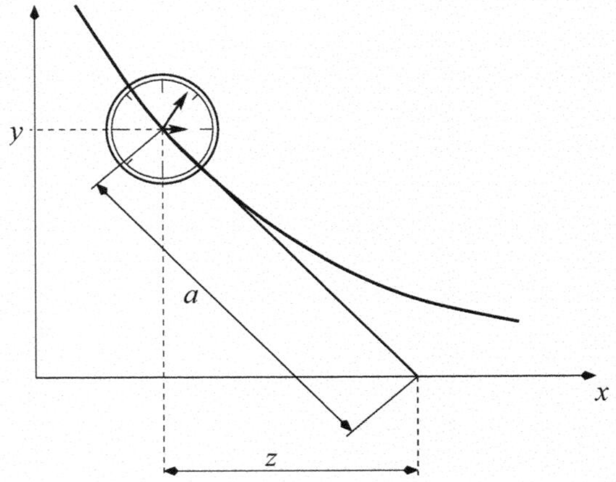

Fig. 7.2.20. The pulled pocket watch

$$-\frac{\sqrt{a^2-y^2}}{y}\,dy = dx$$

and now can solve the differential equation by quadrature:

$$-\int \frac{\sqrt{a^2-y^2}}{y}\,dy = \int dx.$$

Now $\int dx = x$ and Leibniz was already familiar with the integral on the left hand side [Hairer/Wanner 1997, p. 135]. He got the equation of the tractrix,

$$x = -\sqrt{a^2-y^2} - a\ln\frac{a-\sqrt{a^2-y^2}}{y}.$$

The formal method to solve differential equations by separating the differential quotient is nowadays called 'separation of variables'.

7.3 First Critical Voice: George Berkeley

The use of infinitesimal quantities in the differential- and integral calculus, albeit very successful in applications, was not everywhere greeted with enthusiasm. An early critique was the Irish Theologian George Berkeley (1685–1753) who became Bishop of Cloyne in 1734 [Breidert 1989]. In the quest for the infinite mathematics and theology meet and Berkeley was worried that mathematicians talking about infinity could in fact take away parts of the legitimation of theology. A first talk *On Infinity*, given on 19th November 1707 to the just founded Dublin Philosophical Society, clearly showed the inclination of young Berkeley not to leave infinity to mathematicians and philosophers. Additionally, all his life Berkeley was worried about 'freethinkers' which more and more started to show hostilities against religion publicly; among others Edmond Halley who simply saw religion as a fraud. A particularly aggressive writing by Berkeley is *The Analyst* [Berkeley 1985, p. 81–141] of 1734, carrying the full title *The Analyst; or A Discourse addressed to an Infidel Mathematician; Wherein It is examined whether the Object, Principles, and Inferences of the Modern Analysis are more distinctly conceived or more evidently deduced, than Religious Mysteries and Points of Faith*, and this writing hit the mark. We cite here §14 [Berkeley 1985, p. 97f.], [Berkeley 1992, p. 174f.]:

> *To make this Point plainer, I shall unfold the reasoning, and propose it in a fuller light to your View. It amounts therefore to this, or may in other Words be thus expressed. I suppose that the Quantity x flows, and by flowing is increased, and its Increment I call o, so that by flowing it becomes $x + o$. And as x increaseth, it follows that every Power of x is likewise increased in a due Proportion. Therefore as*

x becomes $x+o$, x^n will become $(x+o)^n$: that is, according to the Method of infinite Series,

$$x^n + nox^{n-1} + \frac{nn-n}{2}oox^{n-2} + \&c.$$

And if from the two augmented Quantities we subduct the Root and the Power respectively, we shall have remaining the two Increments, to wit,

$$o \text{ and } nox^{n-1} + \frac{nn-n}{2}oox^{n-2} + \&c.$$

which Increments, being both divided by the common Divisor o, yield the Quotients

$$1 \text{ and } nx^{n-1} + \frac{nn-n}{2}ox^{n-2}\&c.$$

which are therefore Exponents of the Ratio of the Increments. Hitherto I have supposed that x flows, that x hath a real Increment, that o is something. And I have proceeded all along on that Supposition, without which I should not have been able to have made so much as one single Step. From that Supposition it is that I get at the Increment of x^n, that I am able to compare it with the Increment of x, and that I find the Proportion between the two Increments. I now beg leave to make a new Supposition contrary to the first, i.e. I will suppose that there is no Increment of x, or that o is nothing; which second Supposition destroys my first, and is inconsistent with it, and therefore with every thing that supposeth it. I do nevertheless beg leave to retain nx^{n-1}, which is an Expression obtained in virtue of my first Supposition, which necessarily presupposeth such Supposition, and which could not be obtained without it: All which seems a most inconsistent way of arguing, and such as would not be allowed of in Divinity.

Berkeley indeed had clearly seen that the new analysis stood on very fragile ground. However, he was not only destructive in his critique but definitely acknowledged the results of the calculus of fluxions and tried to base them rigorously on arguments employing finite quantities only. This attempt failed. Inevitably mathematicians started to reply to *The Analyst.* As a rule, they did not clarify anything, got personal, and we hence do not want to discuss them and their reactions here. Berkeley could not stop the further use of infinitesimal arguments and the numerous discoveries they enabled. It is largely due to him, however, to have pointed out the weak spots in the analysis of Newton and Leibniz at a fairly early stage. It is only in the 19th century that foundational questions became so pressing that excellent mathematicians started to search for solutions.

Development of the Infinitesimal Calculus and the Priority Dispute

1664	ISAAC BARROW starts his lectures in Cambridge
1664/65	NEWTON is in possession of his binomial theorem
1665/66	NEWTON discovers the calculus of fluxions, the law of gravitation, and makes important discoveries in optics
1665	ROBERT HOOKE's book *Micrographia* published
1669	NEWTON becomes Lucasian professor
1672–1676	LEIBNIZ in Paris. The differential and integral calculus emerges
1673	LEIBNIZ presents his calculating machine to the Royal Society in London and is accepted as member
1676	The secretary of the Royal Society, HENRY OLDENBURG, allows LEIBNIZ to browse through Newton's writings. From December on LEIBNIZ is back in Hanover NEWTON sends his first letter (epistola prior) to LEIBNIZ. A correspondence starts to develop
1677	Second letter (epistola posterior) from NEWTON to LEIBNIZ
1684	The first publication of LEIBNIZ's differential calculus appears in the *Acta Eruditorum*
1687	The *Principia* are published and are NEWTON's groundbreaking work in physics
1699	The priority dispute between NEWTON and LEIBNIZ starts with an attack by FATIO
1703	NEWTON becomes president of the Royal Society for the first time
1704	NEWTON's book *Opticks* is published. In an appendix NEWTON's first publication concerning his calculus of fluxions can be found
1712	The final judgement of the Royal Society in the priority dispute is distributed as 'Commercium epistolicum'; LEIBNIZ is unjustifiably condemned as plagiarist. NEWTON has pulled the strings in the background
1716	LEIBNIZ dies in Hanover and is buried in the court church St John (Neustädter Kirche)
1727	NEWTON dies and is buried in Westminster Abbey

8 Absolutism, Enlightenment, Departure to New Shores

T. Sonar, *3000 Years of Analysis*, https://doi.org/10.1007/978-3-030-58223-4_8

1643–1715	Louis XIV King of France. Climax of absolutism
1648	The Thirty Years' War ends on 24th October with the Peace of Westphalia
1662	Establishment of the Royal Society in London
1683	The Turks before Vienna
1689–1725	Peter I (the Great) Tsar of Russia
1700	Elector Frederick III establishes the Societas Scientiarum Brandenburgica (Brandenburgische Gesellschaft der Wissenschaften) in Berlin, following a plan by Leibniz
1717	Countrywide compulsory education in Prussia
1725	Academy of Sciences founded in St Petersburg
1740–1780	Maria Theresa Archduchess of Austria, Queen of Bohemia, Croatia, and Hungary
1740–1786	Frederick II (the Great) King of Prussia
1756–1763	Seven Year War
1762–1796	Catherine II (the Great) reigns Russia as Empress
1766	Invention of the steam engine by James Watt
1768–1779	Expeditions of James Cook
about 1770	Industrial revolution starts in England
1776	Declaration of Independence of 13 colonies in the national congress of Philadelphia
1765–1790	Joseph II Holy Roman Emperor, representative of an 'enlightened absolutism'
1784	Mechanical weaving loom by E. Cartwright
1789	Start of the French Revolution. Storming the Bastille takes place on 14th July
1794	Establishment of the École Polytechnique in Paris
1799	Napoleon Bonaparte becomes first consul
1804	Napoleon I is crowned Emperor
1805	Destruction of the French fleet by the English at Trafalgar
1812	Doom of Napoleon's 'Great Army' in Russia
1813/14	Wars of liberation
1815	Napoleon finally succumbs at Waterloo
1814–1815	Congress of Vienna
1822	Charles Babbage works on the first program-controlled calculating machine. Brazil becomes an independent empire
1825	End of the Spanish colonial empire in South America. Establishment of the Polytechnic in Karlsruhe. First railway (for the transport of goods) in England (Stephenson)
1830	July Revolution in France. First railway for passenger transport from Liverpool to Manchester

8.1 Historical Introduction

The peace treaties of Münster and Osnabrück at the end of the Thirty Years' War gave rise to hope for an 'everlasting' peace but this hope was in vain. In the Holy Roman Empire absolute potentates ruled and as a rule the reign was bequeathed within a family dynasty. If such a family line became extinct a complicated inheritance trial would follow since one was connected by more or less happy intermarriages with other families. If there was more than one party entitled to inherit conflicts were inevitable and sometimes they developed into wars of succession. These wars were the true malady of the epoch after the Thirty Years' War.

Louis XIV was the prototypical absolutist monarch of his days and supported an expansive foreign policy for France. Already in 1658 the first Confederation of the Rhine had been founded in which different imperial potentates had united to support France in order to secure their sovereignty. The marriage of Louis XIV to Maria Theresa of Spain, oldest daughter of Philip IV of Spain, secured the right of inheritance on Spain. However, also the Emperor Leopold I was married to a daughter of the Spanish king so that conflicts concerning the Spanish line of succession were predetermined. When Philip IV died in 1665 Louis claimed the Spanish Netherlands and when this 'gift' was denied him by Spain Louis simply invaded these areas in the so-called War of Devolution. At the end of this war the first Confederation of the Rhine fell apart. Only when Louis's troops invaded Lorraine in 1670 and

Fig. 8.1.1. Absolutist rulers in France: Louis XIV [Painting: Pierre Mignard, before 1695, Detail]; Louis XV [Painting by Hyacinthe Rigaud, 1730, Detail] (Palace of Versailles); Louis XVI [Painting by A. F. Callet, 1788, Detail] (Musée National des Châteaux de Versailles et de Trianon)

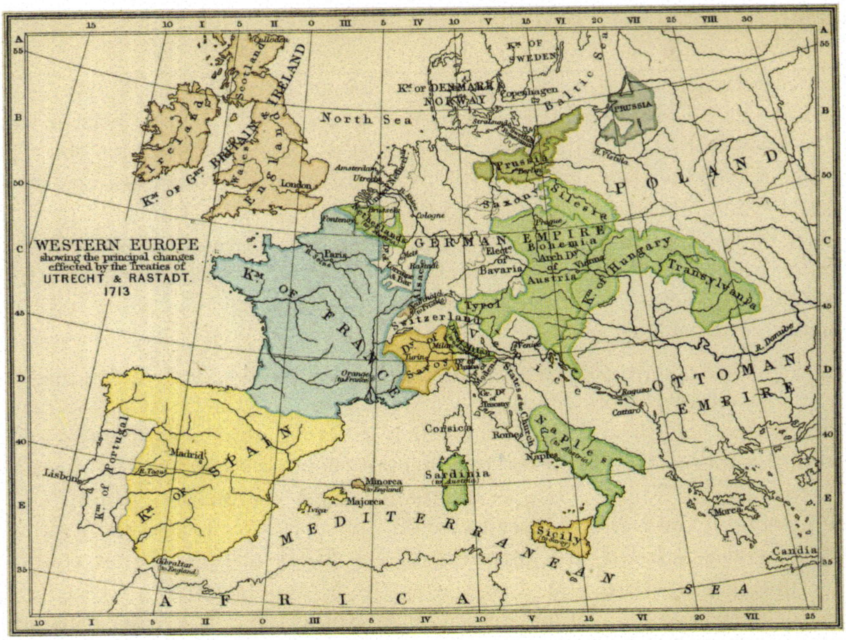

Fig. 8.1.2. Map of Europe in 1713

started an attack on the Dutch Republic in 1672 resistance formed in the Holy Roman Empire and the Empire declared war on France in 1674. The war prolonged tenaciously. Additionally the Turks stood before Vienna in 1683. In the Empire the French practised the tactics of scorched earth: Heidelberg, Mannheim, Worms, and Speyer were devastated when the French troops moved out. When piece negotiations took eventually place France got off cheaply.

In the Baltic Sea Region the nordic war raged from 1700 until 1721 which resulted from a conflict between Denmark and Sweden. Denmark had Russia and Poland on it's side and at the end of the war Russia's predominance in the Baltic Sea Region was manifest. Between 1701 and 1713/14 the War of the Spanish Succession was fought where England as well as the Netherlands were on the side of the Emperor fighting against France. When Emperor Charles VI died and was succeeded by Joseph I in 1711 who had a true claim to the rule in Spain the situation became too hot for England and the Netherlands: they withdraw from alliance with the Emperor and negotiated peace with France by a dynastic division between the House of Bourbon in Spain and its French line. To console the Emperor he was consigned the Spanish Netherlands and the Italian possessions of Spain. The Emperor felt outsmarted however and fought further on until 1714 when he eventually had to give in.

Fig. 8.1.3. 'Enlightened' monarchs: Frederick II of Prussia [handcoloured engraving, originally in black and white, 1873, New York], Emperor Joseph II ([Painting by Joseph Hickel, 1776, Detail] Heeresgeschichtliches Museum Vienna), Peter I (the Great, Russia) [Painting by Jean-Marc Nattier, after 1717, Detail] (Hermitage, St Petersburg)

The University of Halle in Germany was founded in 1694 and became a centre of early enlightenment, with which radical intellectual changes began to set in. Resignation of the people to God was replaced by a more secular view of the world. Not only could time and space be quantified by measurements but also the productivity of the economy. Formerly static views of the world changed to dynamic views in the second half of the 18th century [Dirlmeier et al. 2007, p. 220]. Astronomers like Kepler had widened the people's horizon, Newton had succeeded in explaining many processes on earth and in the heavens scientifically, and the interest in modern natural sciences was so large that the book *Le Newtonianisme pour les dames* (The Newtonianism for Ladies) by Francesco Algarotti (1712–1764) met impressive demands. In 1700 Leibniz had succeeded in the foundation of the Berlin Academy (Brandenburgische Societät der Wissenschaften) by elector Frederick III. Already in 1717 the Georgia Augusta in Göttingen was founded as a new 'reform university'. During the 18th century reading societies and salons formed in which also mathematics and natural sciences were discussed. A unique turning point in the history of philosophy occurred with Immanuel Kant (1724–1804) who in the 'Berlinische Monatsschrift' (Berlin monthly paper) in 1784 gave the *Answer to the Question: What is Enlightenment?* (Beantwortung der Frage: Was ist Aufklärung?). With his *Critique of Pure Reason* (Kritik der reinen Vernunft) he finally put the intellect into the centre of all knowledge. In *Universal Natural History and Theory of Heaven* (Allgemeine Naturgeschichte und Theorie des Himmels) [Kant 2005] Kant even speculated on the formation of sun systems and galaxies based on Newtonian physics.

Fig. 8.1.4. Philosophers of enlightenment in France: Jean-Jacques Rousseau [Pastel by Maurice Quentin de La Tour, 1753], Jean Baptiste Voltaire [Studio of Nicolas de Largillierre after 1725] and Charles-Louis de Montesquieu [unknown artist of the French School 1728]

The enlightenment spread fast across all of Europe. Influential proponents of an enlightened philosophy besides Kant were the British empiricists John Locke (1632–1704) and David Hume (1711–1776) as well as the Frenchmen Voltaire (1694–1778) and Jean-Jacques Rousseau (1712–1778). Enlightenment even entered some courts so that, among others, Frederick II of Prussia, Emperor Joseph II, and the Russian Tsar Peter I can be seen as having been enlightened monarchs.

In 1737 the first German Masonic Lodge was established in Hamburg and numerous other establishments followed all over the country in quick succession. The lodges were committed to enlightenment as was the Order of Illuminati, founded by Adam Weishaupt in 1776 and prohibited due to its radical positions in 1785. At the end of the 18th century enlightenment eventually came to the people. The book *Distress and Help Booklet for Peasants* (Noth- und Hülfs-Büchlein für Bauersleute), meant as a book to enlighten the ordinary people written by Rudolf Zacharias Becker, was published in two volumes 1778 and 1780 and in 1810 it already had sold more than a million copies [Dirlmeier et al. 2007, p. 228f.]. Educational reformers like Johann Heinrich Campe (1746–1818), private tutor to the Humboldt brothers, founded the 'Brunswick schoolbook bookstore' (Braunschweigische Schulbuchhandlung) and therewith invented the mass production of books. His translation and adaptation of *Robinson Crusoe* by Daniel Defoe was published 1779/80 in Hamburg under the title *Robinson the Younger* (Robinson der Jüngere) which became the first bestseller of youth literature.

With the reign of Louis XV (1715–1774) in 1723 began the French epoch which was later to be called the 'Ancien Régime'. While French economy flourished between 1730 and 1770 warlike quarrels continued outside the country. France got involved in the Polish war of succession 1733–1738 as

Fig. 8.1.5. Philosophers of the enlightenment in Britain and Germany: John Locke [Painting: Sir Godfrey Kneller, 1697], David Hume [Painting: Allan Ramsay, 1766], Immanuel Kant [unknown painter]

in the Austrian war of succession 1740–1748 in which France acted on the Prussian side. However, from the start of the Seven Years' War in 1756, France opposed Prussia. In the colonial war in North America France struggled with Great Britain from 1744 until 1748 until France entered the American War of Independence against England in 1778. Eventually the national finances were ruined; a reform of the state fell through. A financial politic of 'muddling through' (Durchwurstelп [Haupt et al. 2008, p. 241]) did the rest. The crisis of the Ancien Régime was additionally fuelled by scandals and affairs at the court of King Louis XVI (1754–1793, King from 1774) and his Queen Marie Antoinette (1755–1793). Since 1720 enlightenment had lead by and by to a politicisation of the people. Still the old system of estates of the realm was valid. The first two estates, clergy and nobility, were strongly privileged. In contrast, the third estate, free peasants and citizens, was a heterogenious mixture from upper-class citizens down to day labourers. This third estate made approximately 98% of the population.

All three estates were represented in the Estates General which had to approve on taxes. Due to the catastrophic financial situation of the country Louis XVI had to summon the Estates General; however it was not quite clear whether to vote by estates or by number of heads. Since the third estate had twice the number of delegates voting by the number of heads would have been advantageous, but Louis XVI did not want to accept this mode of voting. On 17th June 1798 the third estate declared themselves as being the National Assembly and took the Tennis Court Oath: One did not want to separate until a constitution for France was formulated. The king ordered troops to Paris which unsettled the population since the prices for bread were already horrendously high and the people of Paris feared a further shortage of food.

Fig. 8.1.6. A contemporary caricature: The third estate is carrying clergy and nobility

Now also members of the nobility took the side of the National Assembly. People's wrath grew further and on 14th July the storming of the Bastille took place – starting point of the French Revolution. The feudal rights were abolished as were the guilds and the clergy gave themselves a civil constitution which was condemned by the pope. In 1791 an escape attempt of the king ended in Varennes.

From outside France Emperor Leopold II in Vienna, King Frederick William II of Prussia, and the brother of the French king threatened to take military actions against France if the monarchy or their representatives should be touched. Under the impression of this provocation war was declared on Austria; Prussia took the Austrian side. The French set up a poorly trained revolutionary army which was believed could not stand the coalition army led by the Duke of Brunswick, Karl Wilhelm Ferdinand. On this military campaign the Duke of Saxony-Weimar-Eisenach was accompanied by his minister Johann Wolfgang von Goethe who described the decisive battle of Valmy in his book *Campaign in France* (Kampagne in Frankreich). Close to the village of Valmy a battle took place in which the revolutionary troops could withstand the coalition army. Since bad weather set in the battle turned into a siege in which both armies just waited opposite of each others. The rain continued and the soldiers of the coalition suffered from hunger. In the command discrepancies emerged and hence the coalition army retreated without any further exchange of fire. Goethe wrote in *Kampagne in Frankreich* that he had told a circle of officers:

Fig. 8.1.7. Napoleon Bonaparte in his study 1812 ([Painting by Jacques-Louis David 1812] National Gallery of Art, Washington DC)

> *Here and today, a new epoch in the history of the world has begun, and you can boast you were present at its birth.*
>
> (Von hier und heute, meine Herren, geht eine neue Epoche der Weltgeschichte aus, und ihr könnt sagen, ihr seid dabei gewesen.)

The revolutionary army marched on, conquered the whole left side of the Rhine and the Netherlands, and terminated this war victoriously in 1795.

A certain Napoleon Bonaparte (1769–1821) became commanding general of the French Army in Italy in 1796 and proved to be a military talent of first order and a political hope of his country. His campaign to Egypt on which he took many scientists with him was pointless, but established the science of egyptology. On 9th November 1799 a coup in France terminated the revolution and made Napoleon the first consul of the French Republic. In fact Napoleon was now the absolute ruler of France. Numerous Napoleonic reforms formed the governmental structures of France well into our time. In 1804 Napoleon crowned himself Emperor of the French. In military campaigns not seen before he succeeded to conquer wide parts of Europe. It was only the Russian winter that led to the breakdown of the French 'Grande Armée' in 1812; only a fraction of the army returned defeated. Now started the fight

Fig. 8.1.8. Battle of Waterloo 1815 [Painting by Clément-Auguste Andrieux, 1852] (Musée National du Château de Versailles et de Trianon)

against Napoleon in the wars of liberation. The Battle of the Nations close to Leipzig marks the final defeat of the French and the allied troops marched towards Paris of which they took control on 31st March 1814. Napoleon was forced to abdicate and was sent to exile on the island of Elba. In France the restoration gained pace and Louis XVIII became the new king. But very soon the French grew dissatisfied with the restoration and their new king and these news quickly spread to Elba. Napoleon left Elba and arrived in Paris on 1st March 1815 where he succeeded in taking the rule again and to form a new army. Hence the worried allies agreed in the Congress of Vienna to reunite and wage war again against Napoleon. Actually the French managed to defeat the enemy troops at Quatre-Bras and Ligny and on 18th June 1815 Napoleon attacked the English troops under Wellington, but Blücher with his Prussian troops came rushing to aid the English. The French were whitewashed and the 'rule of the 100 days' of Napoleon ended. Napoleon had to leave France and went to exile on the British isle St Helena where he died on 5th May 1821.

8.2 Jacob and John Bernoulli

With the works of Newton and Leibniz new worlds opened up for the new analysis. Newton as well as Leibniz employed their analysis immediately to solve problems in physics by differential equations. All over Europe mathematicians threw themselves into the new analysis and developed it further; on the Continent one followed Leibniz, in England one stood loyal to Newton. This had fatal consequences for the development of English analysis: Due to Newton's not very instructive notations no true 'calculus' followed which one could teach or learn, while Leibniz consciously had developed a calculus which allowed even less gifted mathematicians to apply analysis. Since 1696 the first textbook on Leibniz's analysis, *Analyse des infiniment petits pour l'intelligence des lignes courbes* [Bradley/Petrilli/Sandifer 2015] by its author Guillaume François Antoine de l'Hospital (1661–1704), was available. Leibniz's analysis started its triumphant progress. The very beginning of this stormy development was marked by a family of mathematicians which played a unique role in history – the Bernoulli family.

The family Bernoulli was a wealthy family of merchants resident in Basel. Jacob Bernoulli (1655–1705) was born in 1655 as son of Nicolaus who was a councillor in Basel. His brother Nicolaus, born in 1662, will later become a painter. Jacob was 12 years old when his brother John (1667–1748) was born in Basel. The three brothers Jacob, Nicolaus, and John became the ancestors of several generations of Bernoullis excelling in mathematics, physics, and astronomy. Since there were several Bernoullis by the names of Jacob, John, and Nicolaus (cp. figure 8.2.1) one has to assign numbers to the Bernoullis to be able to address them in a unique manner. Jacob and John are therefore often numbered as Jacob I and John I, respectively.

Jacob I studied theology and philosophy in Basel and took up studies in mathematics and astronomy against his father's wish. In 1676 Jacob completed his official studies successfully but he already was forfeited to mathematics. He travelled to Switzerland, was employed as a private tutor in different households between 1676 and 1680, and during this time period undertook several journeys to France and Holland. On a grand tour from 1681 to 1682 Jacob travelled to England, Germany, and Holland and met Robert Hooke, Robert Boyle, and Johann van Waveren (Jan) Hudde. Hence he early on came in contact with leading mathematicians of his day. Starting in 1683 he offered private lessons in experimental physics in Basel and began to study the works of John Wallis, Isaac Barrow, and René Descartes which raised his interest in the mathematics of infinitesimals. In the year 1684 when Leibniz's first publication *Pro maximis et minimis ...* was published in the Acta Eruditorum Jacob married Judith Stupanus. Two children issued from this marriage, a daughter and a son who turned to the fine arts (no family without black sheep!). In 1687 Jacob got a chair for mathematics at the

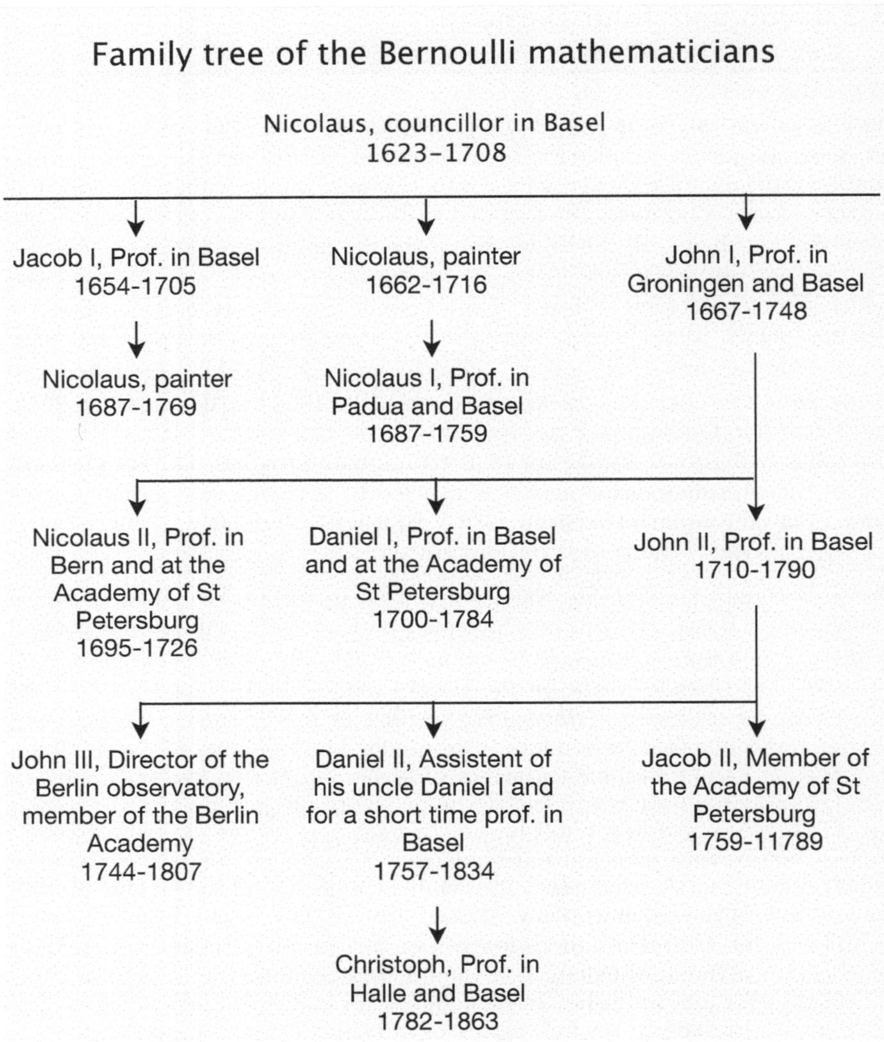

Fig. 8.2.1. Family tree of the Bernoulli family [Drawing by H. Wesemüller-Kock after Fleckenstein 1949, [Fleckenstein 1977]]. Since some first names occur repeatedly one has introduced a numbering, for example: John I, John II, John III

Fig. 8.2.2. Jacob I Bernoulli, John I Bernoulli, Daniel I Bernoulli

University of Basel which he held until his death and which was then taken over by his brother John I.

Jacob started his own scientific career with works on the theory of comets and on problems of physics. He was rather unhappy with the way in which Wallis treated induction in the *Arithmetica infinitorum* and developed the first rigorous theory of complete induction already in 1685/86. Together with his brother John I who had studied medicine and was taught mathematics by his older brother he studied Leibniz's first publication on the differential calculus. Both brothers had problems understanding this work and so Jacob wrote to Leibniz at the end of 1687 and asked for explanations. Unfortunately, Leibniz travelled Italy at this time to research the history of the House of Welfs and hence could not answer. Only three years later will Leibniz reply to this letter but the brothers then had already succeeded to understand every single detail on their own. A fruitful correspondence started between Leibniz and the Bernoullis and it is not too much to state that the Bernoullis became the most important propagandists of Leibniz's analysis.

The word 'integral' for Leibniz's '$\int$' arose from a suggestion by Jacob in the correspondence with Leibniz. Jacob used this word for the first time in his first work published in the Acta Eruditorum in which he examined his favourite curve, the spiral, by means of Leibniz's analysis. After his premature death at the age of 50 the spiral was placed on his grave in the Basel Minster together with the words 'Eadem mutata resurgo' (Although changed I return as the same), see figure 8.2.3.

Numerous mathematical results are named after Jacob Bernoulli as the Bernoulli inequality, Bernoulli's differential equation, the Bernoulli distribution in probability theory, and the Bernoulli assumptions in the theory of beams. The most important discoveries he made in a quarrel with his brother John, however. The two brothers established a whole new theory within analysis, the so-called 'calculus of variations'.

Fig. 8.2.3. Gravestone of Jacob Bernoulli showing the 'logarithmic' spiral in the Basel Minster [Photo: D. Kahle]

John I, tenth child of his parents, was destined by his father to become a merchant. Not obeying his father he studied medicine and was approbated in 1690. His older brother Jacob had introduced him to mathematics and the two brothers together had succeeded in comprehending Leibniz's analysis. John's mathematical genius was realised early on: In 1690 he solved the problem of the catenary. This is the curve an idealised chain will assume when fixed at two points. Galilei had stated that the curve looks like a parabola, cp. [Sonar 2018, p. 328ff.], but John was able to prove that it is a transcendent function, the hyperbolic cosine $\cosh x$. As John Bernoulli also Christiaan Huygens and Leibniz arrived at this result.

After his approbation as physician John travelled to Geneva and Paris where in 1691 he met the Marquis de l'Hospital (1661–1704) who was the descendent of an old noble French family. De l'Hospital was a gifted mathematician who is said to have solved Pascal's problem the cycloid at a young age. Following his family's tradition he went to the military but he was no determined soldier and so myopic that he was actually good for nothing as far as service in the army was concerned. Hence he withdrew from military service to commit himself to mathematics. De l'Hospital introduced the 24-years old John Bernoulli to the French mathematical circles so that John could lecture on Leibniz's new analysis. John Bernoulli even became the private tutor of de l'Hospital. A kind of contract was agreed upon that John Bernoulli should sell new mathematical results to de l'Hospital. The 'l'Hospital rules' concerning

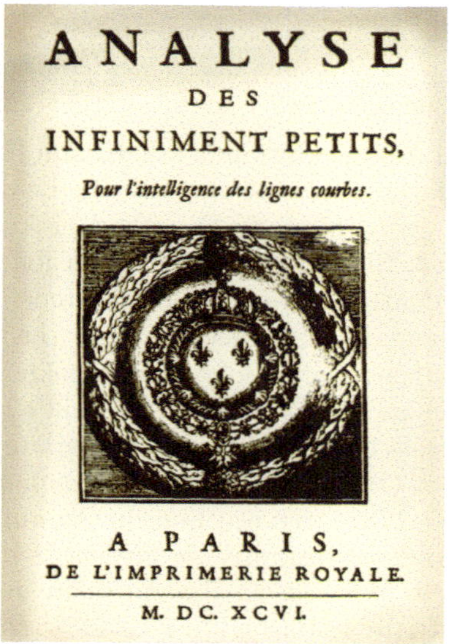

Fig. 8.2.4. Guillaume François Antoine de l'Hospital and the title page of his *Analyse des infiniment petits pour l'intelligence des lignes courbes* 1696

terms of the form '0/0' oder '∞/∞' actually are 'Bernoulli-de l'Hospital rules' because John Bernoulli sold them to de l'Hospital. The rules state that in case of a ratio $f(x)/g(x)$ where $x \to x_0$, i.e.

$$\lim_{x \to x_0} \frac{f(x)}{g(x)},$$

results in '0/0' or '∞/∞', then the ratio of the two functions may be replaced by their derivatives, if both functions are differentiable, of course. Hence

$$\lim_{x \to x_0} \frac{f(x)}{g(x)} = \lim_{x \to x_0} \frac{f'(x)}{g'(x)},$$

if $g'(x_0) \neq 0$.

A nice example is given by the so-called sinc-function ('sinus cardinalis')

$$\frac{f(x)}{g(x)} = \frac{\sin x}{x} =: \text{sinc}(x),$$

which formally leads to '0/0' in the limit $x \to 0$. The Bernoulli-de l'Hospital rule for '0/0' results in

$$\lim_{x \to 0} \frac{\sin x}{x} = \lim_{x \to 0} \frac{\cos x}{1} = 1.$$

After de l'Hospital's death in 1704 John Bernoulli claimed (correctly) these rules for himself. With help by John Bernoulli de l'Hospital succeeded in getting the first textbook on Leibniz's analysis, *Analyse des infiniment petits pour l'intelligence des lignes courbes*, published in 1696 which quickly saw further editions.

By recommendation of Christiaan Huygens John Bernoulli became professor of mathematics at the University of Groningen in the Netherlands, where John with his wife Dorothea Falkner and their common son, Nikolaus II, moved. Shortly thereafter John fell out with his older brother Jacob. After the death of Jacob John became his brother's successor as professor of mathematics in Basel. When the priority dispute between Newton and Leibniz escalated John, who was known to be very quarrelsome, clearly joined Leibniz's party and defended it even after Leibniz's death in 1716. More quarrels followed; a priority dispute with the English mathematician Brook Taylor (1685–1731) and he even quarrelled with his own son Daniel Bernoulli (1700–1782) who was born in Groningen. Daniel had presented his manuscript *Hydrodynamica, sive de viribus et motibus fluidorum commentarii*, in short: *Hydrodynamica*, to the Academy in St Petersburg in 1733. The book was eventually published in 1738 in Strasbourg and on the title page Daniel called himself 'Son of John'. But his quarrelsome father had also written a book on Hydrodynamics, the *Hydraulica, nunc primum detecta ac demonstrata directe ex fundamentis pure mechanicis*, in short: *Hydraulica*. This *Hydraulica* was begun not earlier than 1738 and was finished not later than 1740, nevertheless the book gives 1732 as its year of publication! Until today we can find the accusation that John had consciously predated his work in order to secure priority over his son. Such accusations were clearly refuted by Szabo in [Szabó 1996, p. 166ff.], however. After Leibniz's death John Bernoulli became the recognised authority in analysis. Still in his lifetime he edited and published his collected works in four volumes. In physics he could not come to terms with Newton's theory of gravitation; probably out of a stubbornness against the English. He stayed a defender of the (wrong) vortex theory of Descartes. John's pupil Leonhard Euler (1707–1783) eventually developed analysis to a unique climax of the 18th century.

8.2.1 The Calculus of Variations

The year 1696 is seen as the year of birth of a new discipline of analysis, the calculus of variations. In the June issue of the Acta Eruditorum a problem was formulated (in Latin) by John Bernoulli: *Invitation to the solution of a new problem* (Einladung zur Lösung eines neuen Problems) [Stäckel 1976, p. 3] (with notations changed to fit figure 8.2.5):

> *If two points O and A are given in a vertical plane, assign a path OPA to the moveable point P on which it, starting from O, gets to A in the shortest amount of time by virtue of his own weight.*

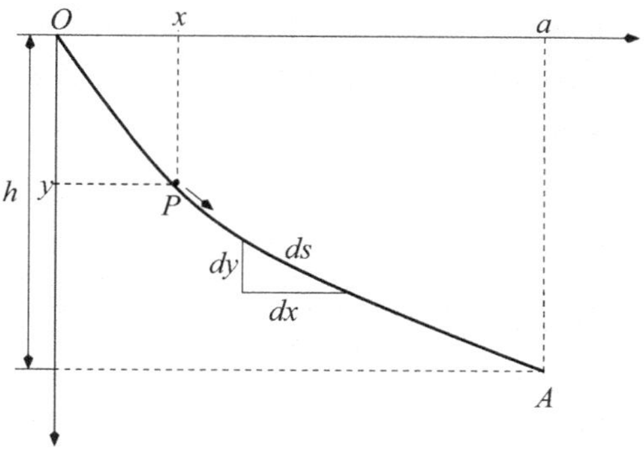

Fig. 8.2.5. Figure concerning the problem of the brachistochrone curve

(Wenn in einer verticalen Ebene zwei Punkte O und A gegeben sind, soll man dem beweglichen Punkte P eine Bahn OPA anweisen, auf welcher er von O ausgehend vermöge seiner eigenen Schwere in kürzester Zeit nach A gelangt.)

John additionally pointed out that the straight connection between O and A does *not* solve this problem which is called the 'problem of the brachistochrone curve' (*brachystos* = shortest, *chronos* = time).

What is so special with this problem? Classically maxima and minima of a function $y = f(x)$ are sought after by means of analysis. Here comes something new: Among a whole set of functions (namely among all which connect O and A and satisfy certain smoothness conditions) the one curve is sought after which it exhibits a minimal property; here it is the shortest runtime of a point moving frictionless on the curve under the influence of gravitation. Not that the problem of the brachistochrone curve was the first of its kind! In the *Principia* Newton had already asked to find the body of revolution with smallest resistance in water under all bodies of revolution. Alas, this part of the *Principia* was not received. John allowed half a year for solutions to be sent to him but except a (correct) solution by Leibniz, who called the problem 'very beautiful and unheard of until now' (sehr schön und bis jetzt unerhört) [Stäckel 1976, p. 4], no other solution arrived. Hence John published an announcement in January 1697 in Groningen that the time period for solutions was enlarged, as Leibniz had counselled. The announcement opened with the words [Stäckel 1976, p. 3]:

John Bernoulli greets the most astute mathematicians of the whole world.

(Die scharfsinnigsten Mathematiker des ganzen Erdkreises grüßt Johann Bernoulli, [...])

It is instructive to realise who the 'most astute mathematicians of the whole world' were. Huygens had died in 1695; therefore there remained brother Jacob, de l'Hospital, Newton, and Leibniz, and we have little doubt that John hoped his brother would send in a wrong solution.

In the May edition of 1697 of the Acta Eruditorum John's solution of the problem was published together with Jacob's correct solution and a note that also Leibniz had solved the problem. The sought-after curve is the cycloid; a curve which interested most of the mathematicians working with infinitesimals for quite a long time. John's solution is very elegant. He reduced the problem to the optical law of refraction and imagined the path of the point P as being a ray of light passing through a medium which continuously changed its refractive index. Jacob's solution is less elegant but, in contrast to John's solution, allows for generalisation to solve further problems! Newton also published a solution anonymously in the January issue 1669 of the *Philosophical Transactions.* It is part of the folklore that Newton wrote his solution down before going to sleep on the same day he received the problem [Goldstine 1980, p. 34], but this story was made up and Newton in fact needed a whole night, cp. [Sonar 2018, p. 346]. Leibniz saw the problem description on 9th June 1696 and John received his solution already on 16th June. Problems of this type lay outside of the mathematics taught at schools so that we can not afford to dive deeper. In all details the solutions can be found in [Goldstine 1980], [Thiele 2007], and [Stäckel 1976]. Here are only some remarks:

Looking at the geometry of the problem as shown in figure 8.2.5 and assuming an initial velocity of our moving point of zero at O, then it follows for the interrelation between velocity v and drop height h after Galilei

$$v = \sqrt{2gh},$$

where g denotes the gravitational acceleration. The ratio between the velocities at P and A is therefore

$$\frac{v_P}{v_A} = \sqrt{\frac{y}{h}},$$

that is

$$v := v_P = v_A \frac{\sqrt{y}}{\sqrt{h}} = \sqrt{2gy}.$$

Imagine an infinitesimal time intervall dt in which our mass point travels the length ds. It then follows for the velocity

$$v = \frac{ds}{dt},$$

and for the arc length element with the theorem of Pythagoras

$$ds = \sqrt{dx^2 + dy^2} = \sqrt{1 + \left(\frac{dy}{dx}\right)^2}\, dx = \sqrt{1 + (y')^2}\, dx.$$

Putting it all together yields

$$dt = \frac{1}{\sqrt{2g}} \sqrt{\frac{1+(y')^2}{y}}\, dx$$

and the 'sum' of all dt, i.e. the whole runtime, should be minimal:

$$T := \int_0^T dt = \frac{1}{\sqrt{2g}} \int_0^a \sqrt{\frac{1+(y')^2}{y}}\, dx \overset{!}{=} \text{Min.} \tag{8.1}$$

The minimisation of an *integral* is one of the typical tasks of the calculus of variations.

Now Jacob, who felt challenged by his younger brother John, could no longer hold back. In the edition of May 1697 he published *Solution of the problems of my brother, to whom I pose another* (Lösung der Aufgaben meines Bruders, dem ich dafür eine andere vorlege) in the Acta Eruditorum. Besides his solution of the problem of the brachistochrone curve Jacob posed an 'isoperimetric' problem. The most famous isoperimetric problem was already described in antiquity: The Phoenician princess Dido had to flee her brutal brother Pygmalion and arrived at the coast of Tunisia. Asking the local chief Jarbas for land where she and her faithful companions could live Jarbas promised her as much land as she could encompass with a single cowhide. Dido sliced the cowhide into very thin strips and with them encompassed a large area on which, following legend, the town of Carthago was built. It was known that among all plane figures the circle with given circumference is the one encompassing the smallest area. Jacob's problem posed to John was a generalised Dido's problem [Stäckel 1976, p. 19f.] (cp. figure 8.2.6):

> *Among all isoperimetric figures over the common basis BN the curve BFN is to be determined which does not itself have the largest area but that effects that another curve BZN has, the ordinate PZ of which is proportional to some power or root of the segment PF or of*

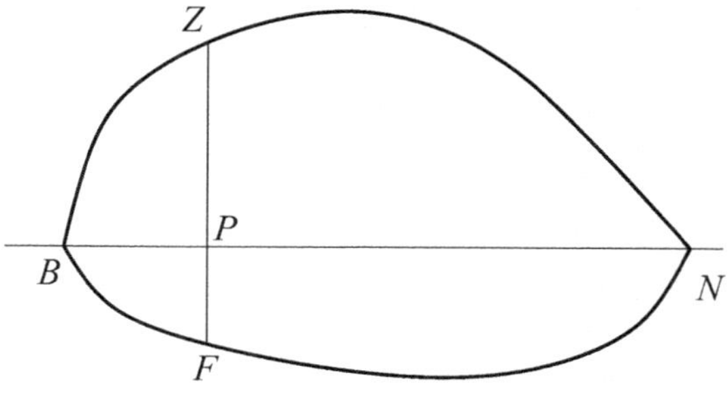

Fig. 8.2.6. Figure concerning the generalised isoperimetric problem

the arc BF. [...] And since it is unfair not to compensate someone for a work done in favour of another on the expense of his own time and to the detriment of his own matters, a man for whom I am the guarantor will pay my brother, in case he solves the problems, not only the deserved praise, but a fee of fifty ducats under the condition that he promises to try [to solve the problems] within three months after this publication, and presents the solution by means of quadratures until the end of the year, what is possible. If nobody gives it [the solutions] after this year has expired I will present mine.

(Unter allen isoperimetrischen Figuren über der gemeinsamen Basis BN soll die Kurve BFN bestimmt werden, welche zwar nicht selbst den größten Flächeninhalt hat, aber bewirkt, dass es eine andere Kurve BZN tut, deren Ordinate PZ irgend einer Potenz oder Wurzel der Strecke PF oder des Bogens BF proportional ist. [...] Und da es unbillig ist, dass jemand für eine Arbeit nicht entschädigt wird, die er zu Gunsten eines anderen mit Aufwand seiner eigenen Zeit und zum Schaden seiner eigenen Angelegenheiten unternimmt, so will ein Mann, für den ich bürge, meinem Bruder, wenn er die Aufgaben lösen sollte, ausser dem verdienten Lobe ein Honorar von fünfzig Dukaten unter der Bedingung zusichern, dass er binnen drei Monaten nach dieser Veröffentlichung verspricht es zu versuchen und bis Ende des Jahres die Lösung mittels Quadraturen, was möglich ist, vorlegt. Giebt sie Niemand nach Ablauf dieses Jahres, so werde ich die meinigen vorlegen.)

According to his own statement John needed only a few minutes for his solution, but this solution was wrong! Jacob with some pleasure inquired several times whether his brother would adhere to his solution and John affirmed. Then a terrible review of John's solution was published by Jacob including the remark that he, Jacob, never would have assumed that John would have been able to solve this problem [Wußing/Arnold 1978, p. 235].

It was only the death of Jacob that prevented further escalation in the brother's quarrel. John later took up work on isoperimetric problems again and developed further solution techniques which enabled Euler to create a general theory.

8.3 Leonhard Euler

The intellectual climate in the parsonage was inspiring: Euler's mother came from an educated family and his father had an interest in mathematics himself. Not only had he attended lectures of Jacob Bernoulli but also written a mathematical dissertation in 1688. His father was therefore Leonhard's first teacher. His first text book in mathematics was the 'Coss' (exact title: *Behend und hübsch Rechnung durch die kunstreichen regeln Algebre, so gemeinicklich die Coß genennt werden*) of Christoph Rudolff (1499–1545) in the editing by Michael Stifel (about 1487–1567) of the year 1553; a difficult book, but it was completely read, worked through, and understood by Leonhard [Thiele 1982, p. 16]. Euler grew up in a simple, devout, and scientifically open parental home and probably started to attend the Latin school in Basel in 1713. The Latin schools were poor; all mathematics had been removed from the syllabus. Hence Euler's father hired the young theologian Johann Burckhardt

Fig. 8.3.1. Leonhard Euler ([Painting by E. Handmann, 1753] Art Museum Basel)

as a private tutor to his son. Euler's friend Daniel Bernoulli will later call Burckhart the 'teacher of the great Euler in mathematics'.

On 20th October 1720, 13 years old, Leonhard entered the philosophy faculty of the University of Basel; fairly early compared with today's standards, but not unusual in those days. Euler's career must have been steep; already in 1722 he applied for a professorship of logic and law, but the application failed. In June 1724 he delivered his first public speech on the comparison of the philosophies of Descartes and Newton. Following his father's wish he matriculated at the theological faculty in 1723 but also attended introductory mathematical lectures of John Bernoulli. Leonhard strove to get private lessons with him but Bernoulli's high workload turned out to be forbidding. But Bernoulli recommended the study of some mathematical works and on Saturdays Leonhard is allowed to ask questions which came up in reading the works. What sounds like hell on earth in the ears of modern educators was later described by Euler as the *best* way to advance in mathematics. Since Euler has successfully passed the Magister exam together with Nicolaus II, one of the sons of John Bernoulli, he got in closer contact with the Bernoulli family, in particular with Nicolaus' brother Daniel. The quarrelsome and irascible John recognised the genius of his pupil and his respect even grew with the years as can be seen from initial phrases in different letters [Thiele 1982, p. 22]:

1728	To the erudite and ingenious young man (Dem hochgelehrten und ingeniosen jungen Mann)
1729	To the very famous and learned man (Dem hochberühmten und gelehrten Mann)
1737	To the very famous and most astute mathematician (Dem hochberühmten und weitaus scharfsinnigsten Mathematiker)
1745	To the uncomparable L. Euler, the prince among the mathematicians (Dem unvergleichlichen L. Euler, dem Fürsten unter den Mathematikern).

At the age of 18 the first mathematical work by Euler (in Latin), *Construction of simultaneous curves in resisting agents*, was published in the Acta Eruditorum. It contained a solution of the problem of the brachistochrone curve if the resistance of the air is taken into account. In 1726 he submitted a manuscript with a solution to a prize problem of the Academy in Paris concerning the optimal position of masts on sailing ships and received a laudatory acceptance. Overall Euler won the first prize of the academy twelve times during his life. In September 1726 a professorship for physics fell vacant in Basel and Euler applied with a dissertation on sound. The application failed and Euler started to look around for other opportunities.

Fig. 8.3.2. The Academie of Sciences in St Petersburg , established by Tsarina Cathrine I, widow of Peter the Great [Photo: H.-W. Alten]

In 1703 the Russian Tsar Peter I had built a fortress in the delta of the Neva River to secure access to the Baltic Sea and shortly after the city of St Petersburg developed there. Peter was open to Western influence and wanted to modernise his country. He saw the necessity to establish an academy following Leibniz's ideas. Only after Peter's death the academy was finally established in St Petersburg by his widow, Tsarina Catherine I. In the autumn of 1725 John Bernoulli and his sons Daniel and Nicolaus had become well-paid professors at the Academy of St Petersburg and they had promised Euler to fetch him later. With the help of Christian Goldbach (1690–1764) the Bernoullis succeeded to appoint Euler to a vacant position in physiology, hence in medicine. When his application in Basel failed Euler moved in 1727 from Basel to St Petersburg.

When Euler arrived in St Petersburg Catherine I died and the academy was thus badly influenced, but eventually Euler got a position in the mathematical class. Many academics came from German speaking countries; nevertheless Euler quickly learned Russian in writing and orally. In 1731 he became professor of physics and therefore a full-fledged member of the academy. His friend Daniel Bernoulli (1700–1782) suffered more and more from the unfavourable climate in St Petersburg so that in 1733 he returned to Basel to occupy a chair for medicine.

The political situation in Russia slowly became unstable but Euler was still well-paid and on the 7th January 1734 he married Katharina Gsell, daughter of a Swiss painter living in St Petersburg. The couple had 13 children of whom only three sons and two daughters survived. Euler's workload increased rapidly and he was a diligent worker. When the political situation became chaotic under Tsarina Anna many members of the academy left Russia and went back to their home countries; Euler stayed. He accepted the geographical department in 1735, worked in a commission concerning weights and measures, and gave lessons at the high school and at the cadet school. A general map of the Russian Empire was targeted at and Euler worked incessantly until he invented a new map projection [Hoffmann 2008, p. 455–465]. While working hard he caught an infection of his right eye and lost his right eyesight in 1738. Euler believed himself that he lost his eye due to the strenuous work on the map, but this is simply not possible, of course. Thiele pointed out in [Thiele 1982, p. 36] another well invented, but wrong, anecdote, made up to prove Euler's piety. Allegedly the French encyclopist Denis Diderot was a guest at the court of St Petersburg and spoke freely about his atheism. Euler was fetched who should have said:

> *Monsieur, it is*
> $$\frac{a+b^n}{n} = x.$$
> *Hence God exists. Answer!*

Thereupon Diderot is said to have left in silence.

Euler now dealt with problems from all areas of mathematics. In physics he introduced analysis into mechanics; he worked in hydromechanics, on ship building, and on astronomy. He solved numerous problems in number theory, established graph theory by means of the famous problem of the Seven Bridges of Königsberg [Löwe 2008a, p. 227–235], and he invented the 'Euler's polyhedron formula', stating that for every convex polyhedron with e corners, f faces, and k edges the relation

$$e + f - k = 2$$

holds. Euler's proof was not sound and many mathematicians will try to make the theorem precise and to prove it, but only Henry Poincaré (1854–1912) will establish algebraic topology and prove this mathematical treasure rigorously [Löwe 2008b, p. 207–225].

Apparently the main work of Euler's 'first Petersburg period' is the two-volume textbook *Mechanica sive motus scientia analytice exposita* (Mechanics or the science of motion, presented analytical) on mechanics of 1736 [Iro 2008, p. 237–269]. This work marks the first complete penetration of mechanics by Leibniz's analysis in which Euler was a true master [Sonar 2008c]. Further highlights of this creative period are the *Scientia Navalis* (Science of the nature of ships), [Nowaki 2008, p. 421–453] and a new theory of music [Odefey 2008, p. 467–481]. The Academy of St Petersburg became an important place for research mainly due to Euler.

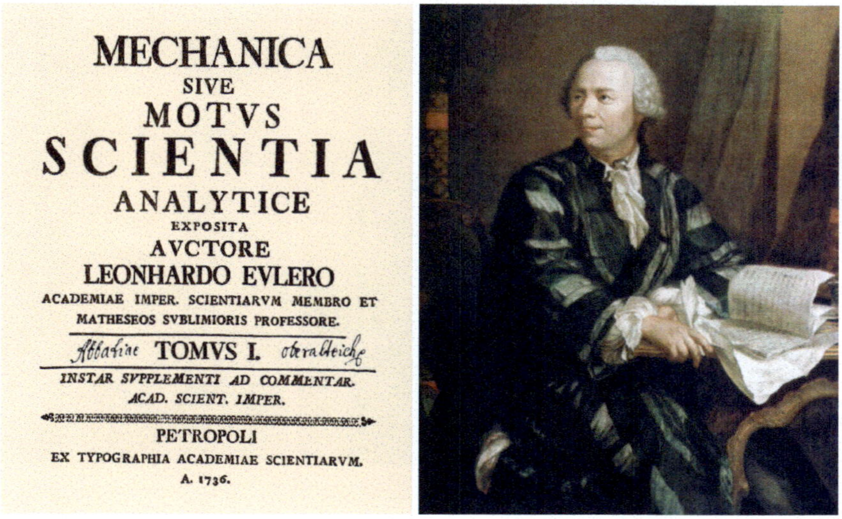

MECHANICA
SIVE
MOTVS
SCIENTIA
ANALYTICE
EXPOSITA
AVCTORE
LEONHARDO EVLERO
ACADEMIAE IMPER. SCIENTIARVM MEMBRO ET
MATHESEOS SVBLIMIORIS PROFESSORE.
TOMVS I.
INSTAR SVPPLEMENTI AD COMMENTAR.
ACAD. SCIENT. IMPER.
PETROPOLI
EX TYPOGRAPHIA ACADEMIAE SCIENTIARVM.
A. 1736.

Fig. 8.3.3. Euler ([Painting by Jakob Emanuel Handmann, about 1756] Deutsches Museum München) and the title page of his *'Mechanica'* of 1736

After the death of Tsarina Anne I in October 1740 Russia saw political turmoil and power struggles. The position of the academy became weak and hence Euler decided to accept the offer of the Prussian King Frederick II to come to Berlin.

Frederick II was philosophically educated and musically gifted. When he became King of Prussia in 1740 he followed the plan to enliven the old Brandenburg Society, designed by Leibniz, which had suffered under the soldier king. The Monarch hence invited famous scientists, among them Euler. Although the departure from St Petersburg was made difficult Euler finally moved to Berlin with his family in 1714. The Seven Years' War was still raging and the Academy in Berlin could only be prepared from 1743 on by a royal commission before it found its place in the hall of the royal palace in 1746 [Thiele 1982, p. 56]. Pierre Louis Moreau de Maupertuis (1698–1759), a French mathematician and astronomer, became the first president and Euler the director of the mathematic class. Euler and Maupertuis got along very well since they shared similar worldviews. The time at the Berlin Academy is splendidly captured by Thiele in [Thiele 2008a, p. 63–77].

In analysis Euler found the strategy for the solution of general homogeneous linear differential equations of n-th order with constant coefficients

$$y^{(n)} + a_{n-1}y^{(n-1)} + \ldots + a_2y'' + a_1y' + a_0y = 0$$

by means of the ansatz $y(x) = \mathrm{e}^{\lambda x}$; the equations

$$\mathrm{e}^x = \lim_{n\to\infty}\left(1 + \frac{x}{n}\right)^n$$

Fig. 8.3.4. Tsarina Anne I [Painting by Louis Caravaque, 1730] (Tretjakow Galerie, Moscow), Frederick II of Prussia [Painting by Anton Graff, 1781] (Schloss Charlottenburg, Berlin), Tsarina Catherine II – the monarchs and mentors of Euler in his three creative periods ([Painting by Johann-Babtist Lampi the Older about 1780] Kunsthistorisches Museum, Gemäldegalerie, Wien)

including 'Euler's number' e, and

$$\sin x = \frac{1}{2}\left(\mathrm{e}^{\mathrm{i}x} - \mathrm{e}^{-\mathrm{i}x}\right), \quad \cos x = \frac{1}{2}\left(\mathrm{e}^{\mathrm{i}x} + \mathrm{e}^{-\mathrm{i}x}\right)$$

were developed, where $\mathrm{i} := \sqrt{-1}$ denotes the 'imaginary unit'. In general Euler started with what we nowadays call 'complex analysis', that is the analysis of functions in the complex numbers $z = a + \mathrm{i}b$, cp. [Euler 1996]. We exclude the whole area of complex analysis from this book since one would need a whole book devoted to that topic, see [Bottazzini 1986] and [Bottazzini/Gray 2013]. In Euler's 'Berlin period' which lasted for 25 years he wrote approximately 380 treatises and many books. However, he always kept contacts with the Academy in St Petersburg and even drew a pension from there [Thiele 1982, p. 59].

Starting from an idea of John Bernoulli in his *Hydraulica* and by the application of a cutting principle applied to an infinitesimal volume element Euler derived the principle of linear momentum

$$dK = dm \cdot \frac{d^2x}{dt^2},$$

hence the nowadays familiar 'force=mass × acceleration' which is always assigned to Newton but which can not be found in any of Newton's writings. Euler published this result in 1750. About 25 years later he will find

Fig. 8.3.5. Pierre Louis Moreau de Maupertuis, Leonhard Euler (on the Swiss 10-Francs note): Friends, but also rivals for the presidency of the academy

the principle of angular momentum. Only then the science of mechanics is 'completed'; Newton could only consider pure translational motion of punctiform masses. Also on the field of fluid mechanics Euler ensured important breakthroughs [Sonar 2008a, p. 363–371]. Always interested in practical matters he designed the first water turbine ever [Balck 2008, p. 387–405], computed a new fountain for Sanssouci [Eckert 2008, p. 373–385], and invented cycloidal toothing for gearwheels [Gottschalk 2008, p. 311–331].

But Euler was also entangled in academic spats: The so-called 'monade dispute', see [Thiele 1982, p. 66ff.], and a dispute concerning the 'principle of least action', a kind of principle of thrift which, in the opinion of Maupertuis, ruled nature. Euler knew that this principle did not rule everywhere in nature but he took Maupertuis's side against the mathematician Johann Samuel König (1712–1757) who delivered the stumbling block. In a work of 1751 König had claimed that Leibniz had already mentioned the 'principle of least action' in 1707 in a letter and was now attacked by Maupertuis and Euler to have forged this letter of Leibniz. This quarrel drew immense circles; the Berlin Academy condemned König, but the great philosopher and scandalmonger Voltaire (1694–1778) took König's side and utilised the quarrel to mock Maupertuis.

Our main concern here is analysis, of course. Very early on Euler had addressed variational problems, now he established a whole theory which was published as *Methodus inveniendi lineas curvas maximi minimive proprietate gaudentes, sive solutio problematis isoperimetrici latissimo sensu accepti* (Method to find curves corresponding to having a property in the highest or lowest degree, or solution of the isoperimetric problem if it is understood in the broadest sense of the word) in 1744 in Lausanne. He worked on the fundamental theorem of algebra, he investigated the knight's tour, and achieved groundbreaking results in number theory. But concerning analysis Euler was a real giant. He introduced the fixed notion of 'function'

Fig. 8.3.6. 'Euler in variations' at an exhibition at the Humboldt-University Berlin during the 'year of mathematics' 2008 [Photo: H.-W. Alten]

into analysis for the first time. In his *Introductio in analysin infinitorum* (Introduction to the analysis of the infinite), cp. [Euler 1983] and [Euler 1988], published as a textbook in 1748, Euler placed the function into the centre of analysis and consequently used the notation $f(x)$ as we still do today. The significance of infinite series in analysis was already clear to Newton and Leibniz but Euler became the undefeated master of them. The power series

$$f(x) = \sum_{i=0}^{\infty} a_i x^i$$

in his work – understood as 'infinite polynomial' – became the workhorse of analysis.

Already in 1727 Euler had written to John Bernoulli and reported on his problems with the strange function

$$f(x) = (-1)^x;$$

a function considered an anomaly in the mathematics of the 18th century. Only during the second half of the 19th century could this anomaly be fully explained. Euler also worked on oscillation problems. The problem of the vibrating string had already fascinated Daniel Bernoulli but only Euler seriously worked with trigonometric series and hence initiated a development reaching far into the 19th century which finally led to the birth of set theory.

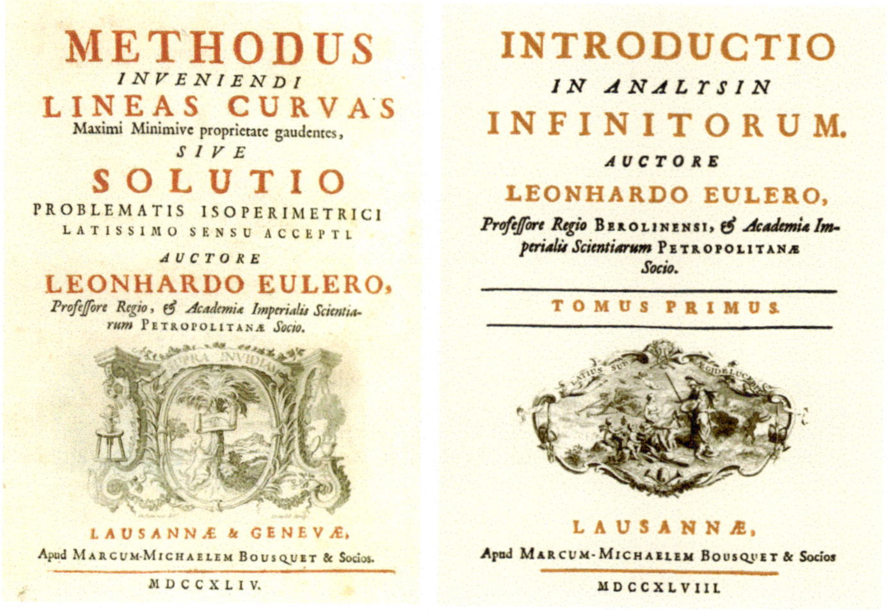

Fig. 8.3.7. *Methodus inveniendi lineas curvas* (1744) und *Introductio in Analysin Infinitorum* (1748): Two pioneering works by Leonhard Euler

The two volumes of the *Introductio* were followed by the two volumes *Institutiones calculi differentialis* concerning the differential calculus in 1755, and between 1768 and 1170 the three volumes of *Institutiones calculi integralis* treating integral calculus were published. Euler was also hands-on concerning popular sciences. Between 1760 and 1762 he wrote letters to the 16 years old Friederike of Brandenburg-Schwedt explaining sciences and philosophy because her father had asked him to do so. These letters were published in French in 1768 as *Lettres à une princesse d'Allemagne sur divers sujets de physique et de philosophie* (Letters to a German Princess on various themes of physics and philosophy). In optics he covered new ground. Newton had stated that it would not be possible to construct achromatic lenses (lenses with no colour errors) but Euler invented and investigated the achromat [Reich/Wiederkehr 2008, p. 333–347]. To please Frederick he wanted to translate the English book *New principles in gunnery* by Benjamin Robins (1707–1751), but during the work he developed ballistics based on analysis completely new and increased the size of Robins's work fivefold. He dispelled errors made by Robins, introduced new ones, but eventually *Neue Grundsätze der Artillerie* (New principles of artillery), published in 1745, became the foundation of modern ballistics, see [Sonar/Loewe 2008, p. 293–309], [Sonar 2008b].

As productive the time in Berlin was, so unworthy did it end. Frederick II loved everything French; he had a high opinion of men like Voltaire or Maupertuis who were able to parley splendidly at court. The dry Protestant Euler stayed alien to him. Euler was never poetically inclined and his attempts to treat music mathematically caused malicious remarks from Frederick. On 28th October 1746 the king's brother, Prince Augustus William, wrote to Frederick [Fellmann 1995, p. 85f.]:

> *Monseigneur Maupertuis introduced me to the mathematician Euler. I realised in him the truth of the imperfection of all things. By means of hard work he acquired a logic mind and hence made a name of himself: but his appearance and his clumsy expression darken all of these nice properties and prevent making use of it.*
>
> (Herr von Maupertuis hat mich mit dem Mathematiker Euler bekannt gemacht. Ich fand an ihm die Wahrheit von der Unvollkommenheit aller Dinge bestätigt. Durch Fleiß hat er sich logisches Denken und damit einen Namen erworben: aber seine Erscheinung und sein unbeholfener Ausdruck verdunkeln alle diese schönen Eigenschaften und verhindern, daß man sie sich zunutze macht.)

Frederick replied on 31st October:

> *Dearest brother! I already thought that your conversation with Monseigneur Euler would not uplift you. His epigrams consists of the computations of new curves, any conic sections, or astronomical measurements. Among the savants there are such enormous computers, commentators, translators, and compilaters, which are useful in the republic of science, but otherwise anything but shiny. One uses them like Doric columns in the art of building. They belong in the floor as supports of the whole building and of the Corinthian columns which form the ornament.*
>
> (Liebster Bruder! Ich dachte mir schon, daß Deine Unterhaltung mit Herrn Euler Dich nicht erbauen würde. Seine Epigramme bestehen in Berechnungen neuer Kurven, irgendwelcher Kegelschnitte oder astronomischer Messungen. Unter den Gelehrten gibt es solche gewaltige Rechner, Kommentatoren, Übersetzer und Kompilatoren, die in der Republik der Wissenschaften nützlich, aber sonst alles andere als glänzend sind. Man verwendet sie wie die dorischen Säulen in der Baukunst. Sie gehören in den Unterstock, als Träger des ganzen Bauwerkes und der korinthischen Säulen, die seine Zierde bilden.)

Even nasty remarks concerning Euler's monocular vision ('my cyclops') were handed-down by Frederick. The actual rift appeared when Maupertuis died in 1759. He had been absent from Berlin for quite some time and Euler had led the academy as a *de facto* president. It was apparent now that Euler was the only serious next president of the academy but Frederick invited the French mathematician Jean-Baptiste le Rond d'Alembert (1717–1783) and offered him the presidency.

D'Alembert certainly was an eminent mathematician but compared to Euler he was second-class. Euler felt deeply hurt but d'Alembert did not want to become the academy's president and proposed Euler instead; the king simply ignored the proposal. Although no successor of Maupertuis could be found Euler did not get the offer and submitted his petition of release in 1766. He had to repeat this submission twice before Frederick would let him go [Thiele 1982, p. 137]. Immediately after Euler had left Frederick succeeded in finding a worthy replacement of Euler by the help of d'Alembert; it was Jean-Louis Lagrange (1736–1823) with whom Euler had worked on the development of the calculus of variations.

Euler had always stayed in touch with the colleagues in St Petersburg and as an almost 60 year old man took the opportunity to return in 1766 to start his 'second Petersburg period'. He was triumphantly welcomed: Tsarina Catherine II (the Great) received him in person, he got a princely salary, accommodation free of charge including firewood, and money for the purchase of a house. When Euler had just moved to his new house he lost his remaining eye by age-related cataract. One might think that his blindness would have

Fig. 8.3.8. Jean-Baptiste le Rond d'Alembert [Painting by Maurice Qentin de la Tour, 1753] (Louvre Museum Paris)

Fig. 8.3.9. Euler's tomb on the Lazarus cemetry at the Saint Alexander Nevsky Monastery in St Petersburg [Photo: Pausanias2, 2007]

influenced his productivity negatively but approximately half of all of his works were written during the time he was completely blind! In 1771 his house burned down and many manuscripts got lost in the fire; but Euler wrote them again. In 1773 his wife died and left Euler without care. Therefore he married his former wife's half sister in 1776. On 18th September 1783 he taught one of his many grandchildren as usual and talked to his young employees. About 5 pm he sat on the sofa, smoking a pipe, when the tobacco pipe got out of his hands. Euler hailed 'my pipe!', bent down, got up without the pipe, grabbed his head and lost his conscience with the words 'I die!'. He died about 11 pm; or, as the French Marquis de Condorcet wrote: 'Euler has ceased to live and calculate.' (Euler hat aufgehört zu leben und zu rechnen.) [Thiele 1982, p. 150]. He was buried on the Lutheran Smolensky Cemetary on the Wassily island and moved to the old cemetary of Alexander Nevsky Lavra (Monastery) in 1956.

8.3.1 Euler's Notion of Function

Until now we have dodged around the notion 'function'; we cannot any longer since the notion of function became a central one in analysis due to Euler. The term 'function' can already be found in the writings of Leibniz, and John Bernoulli wrote fx to designate the value of the function f at the point

x [Thiele 1982, p. 111]. John Bernoulli had defined in 1718 [Bottazzini 1986, p. 9]:

> *I call a function of a variable magnitude a quantity composed in any manner whatsoever from this variable magnitude and from constants.*

For the time being Euler followed this understanding of functions of his teacher John Bernoulli and wrote in a manuscript composed about 1730 [Thiele 1982, p. 111]:

> *A quantity which is somehow composed of one or more quantities is called their function.*
>
> (Eine Quantität, die aus einer oder mehreren Quantitäten irgendwie zusammengesetzt ist, wird ihre Funktion genannt.)

The kind of such composition was described by Euler as: algebraic operations $(+, -, *, /)$, exponentiation, taken logarithms, and combinations thereof. In the *Introductio* of 1745 essentially the same definition of a function can be found [Euler 1988, Vol. I, p. 3]:

> *A function of a variable quantity is an analytic expression composed in any way whatsoever of the variable quantity and numbers or constant quantities.*
>
> (Eine Function einer veränderlichen Zahlgröße ist ein analytischer Ausdruck, der auf irgend eine Weise aus der veränderlichen Zahlgröße und aus eigentlichen Zahlen oder aus constanten Zahlgrößen zusammengestzt ist.[1])

However, Euler now has enlarged the permissible operation to arbitrary transcendent operations. He took for granted (but could not prove) that each function meeting his definition possesses a power series expansion

$$f(x) = \sum_{k=0}^{\infty} a_k x^k.$$

It is this class of functions which Lagrange later called the class of 'analytic functions'. Euler noticed over time though, that he had to admit also other entities as 'functions' which did not abide by his definition. This insight came solving the wave equation where functions occur which are defined by different analytic expressions on different segments of the abscissa. Heuser has described in [Heuser 2008a, p. 147–163] how uncomfortably Euler must have felt in dealing with such 'functions' since he chose adjectives like 'irregular', 'discontinuous', or 'mixed'. In 1747 he quarrelled with the French

[1] [Euler 1983, p. 4].

mathematician Jean-Baptist le Rond d'Alembert on the famous problem of the oscillating string. In 1747 d'Alembert found the differential equation of the oscillating string

$$\frac{\partial^2 u}{\partial t^2} - \frac{\partial^2 u}{\partial x^2} = 0$$

for the deflection u of the string depending on place and time, of course; hence $u = f(x,t)$, so that 'partial derivatives' have to be considered. To solve the wave equation (which is the differential equation of the oscillating string) two initial conditions have to be prescribed, the deflection $u(x,0)$ of the string at time $t = 0$, and the initial velocity of the string in every point. A 'particular solution' is therefore sought. If one does not consider initial conditions then the solution is called the 'general' solution. This general solution of d'Alembert's equation would then contain two free parameters with which it could be adjusted to given initial conditions. Following Euler's notion of function the initial conditions have to be 'analytic expressions', but already the plucked string leads to a triangular shape of the initial deflection which can only be described by two different analytic expressions (one left of the kink, another one on the right side of the kink). Finally d'Alembert gave up; he wrote [Heuser 2008a, p. 154]:

> *La nature même arrête le calcul* (nature itself stops the calculus).

Euler could not (and did not want to) come to terms with this situation! Nature could never stop the calculus because nature did function according to calculus! Hence another definition of functions had to be installed, and in 1755 (written in 1748) we find it in the *Institutiones calculi differentialis* [Euler 2000, p. 6]:

> *Those quantities that depend on others in this way, namely, those that undergo a change when others change, are called* functions *of these quantities. This definition applies rather widely and includes all ways in which one quantity can be determined by others.*
>
> (Sind nun Größen auf die Art voneinander abhängig, daß keine davon eine Veränderung erfahren kann, ohne zugleich eine Veränderung in der anderen zu bewirken, so nennt man diejenige, deren Veränderung man als die Wirkung von der Veränderung der anderen betrachtet, eine Funktion von dieser, eine Benennung, die sich so weit erstreckt, daß sie alle Arten, wie eine Größe durch eine andere bestimmt werden kann, unter sich begreift.[2])

This definition is not yet a modern one as given for the first time by Dirichlet, but it is certainly an important milestone on the way.

Incidentally Euler had already earlier used a further notion of function which corresponds to 'curves'. From a geometrical point of view Euler even

[2] [Heuser 2008a, p. 154].

considered continually hand-drawn lines as functions – a definition which is surely not compatible with his first definition.

8.3.2 The Infinitely Small in Euler's View

During our discussion of Leibniz's analysis we have seen that Leibniz was clearly aware of the nature of his infinitesimals: 'smaller than any *prescribed* quantity. It is interesting to note that Euler fell behind this definition and chose the interpretation: 'smaller than any positive number'! In the *Institutiones calculi differentialis* of 1755 Euler defines in the third chapter under number 83 [Euler 2000, p. 51]:

> *There is no doubt that any quantity can be diminished until it all but vanishes and then goes to nothing. But an infinitely small quantity is nothing but a vanishing quantity, and so it is really equal to* 0.

Hence Euler viewed the calculation with infinitely small quantities as 'computation with zeros'; and, in fact, he was a master of this computation. In the seventh chapter of his *Introductio* [Euler 1983, p. 86ff.] exponential and logarithm functions were investigated, after Euler had introduced the logarithm of base a, $\log_a x$, rigorously in chapter 6. The investigations of the seventh chapter required an infinitely small quantity ω and an infinitely large quantity i, which Euler assumed being existent without much ado. Here is the place to give a warning word: Euler did *not yet* use the symbol i denoting the imaginary unit in the *Introductio*, but still wrote $\sqrt{-1}$. Hence do not confuse the infinitely large quantity i with the imaginary unit i!

Euler started with the remark that since $a^0 = 1$ holds it also has to hold

$$a^\omega = 1 + k\omega,$$

where k is some number as a start. Now a finite number x is considered for which $x = \omega \cdot i$ is to hold. It then follows

$$a^x = a^{\omega i} = (a^\omega)^i = (1 + k\omega)^i = \left(1 + \frac{kx}{i}\right)^i. \tag{8.2}$$

Now Euler employs the binomial theorem and gets:

$$a^x = 1 + i\left(\frac{kx}{i}\right) + \frac{i(i-1)}{2!}\left(\frac{kx}{i}\right)^2 + \frac{i(i-1)(i-2)}{3!}\left(\frac{kx}{i}\right)^3 + \ldots,$$

hence

$$a^x = 1 + kx + \frac{1}{2!}\frac{i(i-1)}{i^2}k^2x^2 + \frac{1}{3!}\frac{i(i-1)(i-2)}{i^3}k^3x^3 + \ldots.$$

If you already feel dizzy in freely using ω and i then fasten your seat belts now, since Euler argues that

$$\frac{i-1}{i} = 1 - \frac{1}{i} = 1,$$

$$\frac{i-2}{i} = 1 - \frac{2}{i} = 1,$$

and so on, and therefore it follows

$$\frac{i(i-1)}{i^2} = \frac{i-1}{i} = 1,$$

$$\frac{i(i-1)(i-2)}{i^3} = \frac{i(i-1)}{i^2} \cdot \frac{i-2}{i} = 1,$$

and so on; hence we get

$$a^x = 1 + \frac{kx}{1!} + \frac{k^2x^2}{2!} + \frac{k^3x^3}{3!} + \ldots . \tag{8.3}$$

Now Euler puts $x = 1$ and gets

$$a = 1 + \frac{k}{1!} + \frac{k^2}{2!} + \frac{k^3}{3!} + \ldots,$$

and in case of $k = 1$ his famous number arises,

$$\mathsf{e} = 1 + \frac{1}{1!} + \frac{1}{2!} + \frac{1}{3!} + \ldots = \sum_{k=0}^{\infty} \frac{1}{k!},$$

which he identified as base of the natural (hyperbolic) logarithm $\ln x := \log_{\mathsf{e}} x$. Euler gives e to 23 digits:

$$\mathsf{e} = 2.718\,281\,828\,459\,045\,235\,360\,28.$$

Therewith we get from (8.2)

$$\mathsf{e}^x = \left(1 + \frac{x}{i}\right)^i,$$

which we have to interpret as $\mathsf{e}^x = \lim_{k\to\infty}\left(1 + \frac{x}{k}\right)^k$, and from (8.3) it then follows

$$\mathsf{e}^x = 1 + \frac{x}{1!} + \frac{x^2}{2!} + \frac{x^3}{3!} + \ldots = \sum_{k=0}^{\infty} \frac{x^k}{k!}.$$

Equally virtuous Euler set to work on the analysis of the logarithm function [Edwards 1979, p. 273f.] and derived Mercator's series

$$\log(1+x) = x - \frac{1}{2}x^2 + \frac{1}{3}x^3 - + \ldots . \tag{8.4}$$

Consequently Euler applied the idea of 'computing with zeros' also in case of the differentials. For $y = f(x) = x^n$ he wrote

$$dy = (x + dx)^n - x^n$$

and employed the binomial theorem again,

$$\begin{aligned} dy &= (x^n + nx^{n-1}dx + \frac{1}{2}n(n-1)x^{n-2}(dx)^2 + \ldots) - x^n \\ &= nx^{n-1}dx + \frac{1}{2}n(n-1)x^{n-2}(dx)^2 + \ldots. \end{aligned}$$

Now all differentials of higher order $(dx)^2, (dx)^3, \ldots$ are set to zero and Euler gets

$$dy = nx^{n-1}dx.$$

In case of the logarithm Euler computes

$$d(\log x) = \log(x + dx) - \log x = \log\left(\frac{x + dx}{x}\right) = \log\left(1 + \frac{dx}{x}\right)$$

and then applies Mercator's series (8.4):

$$d(\log x) = \frac{dx}{x} - \frac{dx^2}{2x^2} + \frac{dx^3}{3x^3} - + \ldots.$$

Again all differentials of higher order are set to zero and it remains

$$d(\log x) = \frac{dx}{x}.$$

Euler gives still another derivation of this formula and this lets the blood run cold even in case of inveterate friends and lovers of the handling of infinitely small and large quantities! As before Euler writes

$$a^x = a^{\omega \cdot i} = (1 + k\omega)^i$$

and denotes this as $1 + y$, hence

$$1 + y = (1 + k\omega)^i,$$

from which

$$\log_a(1 + y) = x = \omega \cdot i \tag{8.5}$$

follows. But now also $1 + k\omega = (1 + y)^{1/i}$ follows and therefore

$$\omega = \frac{(1 + y)^{1/i} - 1}{k},$$

what, if inserted into (8.5), yields the formula

$$\log_a(1+y) = \frac{i}{k}\left((1+y)^{\frac{1}{i}} - 1\right).$$

Here Euler chose $a = \mathrm{e}$ what, as we already know, entails $k = 1$. The logarithm then is the natural logarithm which we simply denote by log. Additionally we rename $1 + y$ as x and eventually arrive at

$$\log x = i\left(x^{\frac{1}{i}} - 1\right).$$

Now i is infinitely large, ω is infinitely small; what is then more obvious to put $i = 1/\omega$, yielding

$$\log x = \frac{x^\omega - 1}{\omega}.$$

With the rule of computing the derivative of the power function it follows

$$d(\log x) = \frac{1}{\omega}(d(x^\omega - 1)) = \frac{1}{\omega}\omega x^{\omega-1}dx = \frac{x^\omega dx}{x}.$$

The quantity ω is infinitely small, hence in Euler's analysis it is zero, and hence $x^\omega = x^0 = 1$ leading to

$$d(\log x) = \frac{dx}{x}.$$

8.3.3 The Trigonometric Functions

Euler is the first author to relate the trigonometric functions to a circle with radius 1, hence standardising them. This happens in the sixth chapter of the *Introductio*. In particular it follows immediately from Pythagoras's theorem

$$\sin^2 x + \cos^2 x = 1.$$

Inductively he arrived at the 'formula of Moivre'

$$(\cos z \pm \mathrm{i}\sin z)^n = \cos nz \pm \mathrm{i}\sin nz$$

using the imaginary unit $\mathrm{i} = \sqrt{-1}$. With the infinitely small quantity ω and the infinitely large quantity i Euler deduced from that for $z = i\omega$ the two equations (attention: don't mix up i and i!)

$$\cos i\omega + \mathrm{i}\sin i\omega = (\cos\omega + \mathrm{i}\sin\omega)^i$$
$$\cos i\omega - \mathrm{i}\sin i\omega = (\cos\omega - \mathrm{i}\sin\omega)^i$$

and by addition and subtraction it follows

$$\cos i\omega = \frac{1}{2}\left((\cos\omega + \mathrm{i}\sin\omega)^i + (\cos\omega - \mathrm{i}\sin\omega)^i\right)$$
$$\sin i\omega = \frac{1}{2\mathrm{i}}\left((\cos\omega + \mathrm{i}\sin\omega)^i - (\cos\omega - \mathrm{i}\sin\omega)^i\right).$$

Again Euler employs the binomial theorem, now of great importance to him, to expand the right hand sides into infinite series:

$$\begin{aligned}\cos i\omega &= \cos^i\omega - \frac{i(i-1)}{2!}\cos^{i-2}\omega\sin^2\omega \\ &+\frac{i(i-1)(i-2)(i-3)}{4!}\cos^{i-4}\omega\sin^4\omega - \ldots \\ \sin i\omega &= i\cos^{i-1}\omega\sin\omega - \frac{i(i-1)(i-2)}{3!}\cos^{i-3}\omega\sin^3\omega \\ &+\frac{i(i-1)(i-2)(i-3)(i-4)}{5!}\cos^{i-5}\omega\sin^5\omega - \ldots.\end{aligned}$$

He then sets $\cos\omega = 1$ and $\sin\omega = \omega$, since ω is an infinitely small number. Since i is infinitely large he further sets $i = i-1 = i-2 = \ldots$ and for $x := i\omega$ derives the two infinite series

$$\cos x = 1 - \frac{x^2}{2!} + \frac{x^4}{4!} - + \ldots$$
$$\sin x = x - \frac{x^3}{3!} + \frac{x^5}{5!} - + \ldots.$$

It is now fairly easy to derive the famous formula of Euler

$$\mathrm{e}^{\mathrm{i}x} = \cos x + \mathrm{i}\sin x$$

from a comparison of the series for $\mathrm{e}^{\mathrm{i}x}$ and the series of the two trigonometric functions $\sin x$ und $\cos x$. In case of $x = \pi$ it follows

$$\mathrm{e}^{\mathrm{i}\pi} = -1,$$

a beautiful formula connecting the four numbers $\mathrm{i}, 1, \pi$, and e.

8.4 Brook Taylor

English scientists after Newton stubbornly stayed attached to the calculus of fluxions and fluents of their master and hence decoupled themselves from the main currents of analysis on the Continent until well into the 19th century. Nevertheless were there ingenious British mathematicians in the 18th century, among them Brook Taylor (1685–1731) whose name is closely connected with the problem of the oscillating string. While studying law at St John's College in Cambridge Taylor came in contact with mathematics, and when he finished

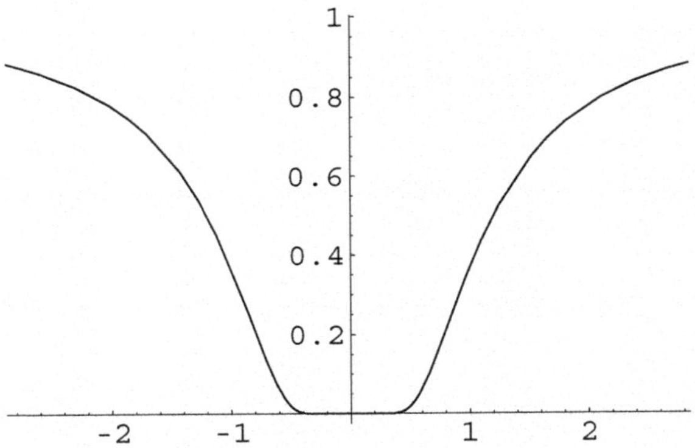

Fig. 8.4.1. Graph of an arbitrarily smooth function where Taylor's series is the zero function when expanded about $x_0 = 0$

his studies with a Bachelor's degree in 1709 he had already started his own mathematical research. In 1712 he was received into the Royal Society and was a member of the committee which had to judge the priority dispute between Newton and Leibniz. His research concerning the centre of oscillation was published only in 1714 leading to his own priority dispute with John Bernoulli.

Taylor's work belongs to the realm of mechanics and is fully based on Newton's calculus of fluxions. In 1714 he became the secretary of the Royal Society; an appointment he held until 1718. Already in 1715 two books were published which took a key position in the history of mathematics: *Methodus incrementorum directa et inversa* and *Linear Perspective*, of which the first is of importance to analysis. It contains a theory of finite differences, i.e. a theory of a kind of discrete differential calculus which is still very important today in numerical mathematics. In this book we also find the series expansion which now carries his name. If f is an arbitrarily smooth function then f can be expanded into a 'Taylor series about x_0':

$$\begin{aligned} f(x) &= f(x_0) + f'(x_0)(x - x_0) + \frac{1}{2!}f''(x_0)(x - x_0)^2 + \frac{1}{3!}f'''(x_0)(x - x_0)^3 + \ldots \\ &= \sum_{k=0}^{\infty} \frac{f^{(k)}(x_0)}{k!}(x - x_0)^k. \end{aligned} \tag{8.6}$$

In modern mathematics we have learned to be more careful. In the form given by Taylor the theorem is wrong; a counterexample is given by

$$f(x) = \begin{cases} \mathrm{e}^{-\frac{1}{x^2}} & ; x \neq 0 \\ 0 & ; x = 0 \end{cases},$$

when expanded about $x_0 = 0$, see figure 8.4.1.

This function is differentiable arbitrarily often if the value of all derivatives at $x_0 = 0$ is always continuously supplemented by $f^{(k)}(0) = 1$. However, such a function would not have occurred to Taylor (and Euler); it does not even satisfy Euler's definition of *one* analytical expression. It was finally left to the 19th century to discover such functions and to reveal the exact conditions of the expandability (= convergence) into series. The discovery of the counterexample above is owed to Augustin Louis Cauchy [Belhoste 1991, p. 79].

The Taylor series was known to Euler; Newton was aware of it and described it in the manuscript of his *De quadratura*. However, in the publication of *De quadratura* as an appendix of the *Opticks* it was left out [Edwards 1979, p. 289]. Further mathematicians knew such series for particular cases before Taylor but its role as a centrepiece of analysis was not immediately recognised. Lagrange earned the honour of clearly having seen the importance of the Taylor series in 1772. Incidentally in Taylor's writings the series took the form

$$y = y_0 + (x - x_0)\frac{\dot{y}_0}{\dot{x}_0} + \frac{(x-x_0)^2}{2!}\frac{\ddot{y}_0}{(\dot{x}_0)^2} + \frac{(x-x_0)^3}{3!}\frac{\dddot{y}_0}{(\dot{x}_0)^3} + \dots .$$

If we again translate Newton's dot notation into Leibniz's differential quotients we get

$$\dddot{y}_0 = \left.\frac{d^3y}{dt^3}\right|_{x_0}, \quad (\dot{x}_0)^3 = \left.\left(\frac{dx}{dt}\right)^3\right|_{x_0}$$

and therefore

$$\frac{\dddot{y}_0}{(\dot{x}_0)^3} = \left.\frac{d^3y \cdot dt^3}{dt^3 \cdot dx^3}\right|_{x_0} = \left.\frac{d^3y}{dx^3}\right|_{x_0} .$$

Taylor's future life was tragic as far as his private life is concerned. He married a poor girl of good descent in 1721 an thereby got into quarrels with his father. When the spouses expected their first child in 1723 mother and child died in childbed. Taylor married again in 1725 for the second time; also this time his wife died in childbed in 1730 but the baby, a daughter, survived.

8.4.1 The Taylor Series

It is very instructive to retrace the idea of the Taylor series by means of an example. We choose the function

$$f(x) = \mathrm{e}^x$$

and the expansion point as $x_0 = 0$. The actual idea of the Taylor series consists in approximating a complicated function successively by polynomials of ever higher degree. Asking for a polynomial of degree 0 approximating f close to the expansion point then only the constant polynomial

$$p_0(x) = \mathrm{e}^{x_0} = 1$$

can be our choice. What is the polynomial of degree 1 representing f about x_0 best? From the ansatz

$$p_1(x) = a_0 + a_1(x - x_0)$$

we determine a_0 and a_1 by the conditions $p_1(x_0) = f(x_0)$ and $p_1'(x_0) = f'(x_0)$, hence

$$\begin{aligned} p_1(x_0) &= a_0 = f(x_0) = 1 \\ p_1'(x_0) &= a_1 = f'(x_0) = 1. \end{aligned}$$

This gives us p_1:

$$p_1(x) = 1 + x.$$

Now we are looking for

$$p_2(x) = a_0 + a_1(x - x_0) + a_2(x - x_0)^2$$

determined by the conditions $p_2(x_0) = f(x_0)$, $p_2'(x_0) = f'(x_0)$, $p_2''(x_0) = f''(x_0)$. Since

$$\begin{aligned} p_2'(x) &= a_1 + 2a_2(x - x_0) \\ p_2''(x) &= 2a_2 \end{aligned}$$

it follows

$$\begin{aligned} p_2(x_0) &= a_0 = f(x_0) = 1 \\ p_2'(x_0) &= a_1 = f'(x_0) = 1 \\ p_2''(x_0) &= 2a_2 = f''(x_0) = 1 \end{aligned}$$

and so $p_2(x) = 1 + x + \frac{1}{2}x^2$. We proceed to $p_3(x) = a_0 + a_1(x - x_0) + a_2(x - x_0)^2 + a_3(x - x_0)^3$. An analogous computation gives

$$p_3(x) = 1 + x + \frac{x^2}{2} + \frac{x^3}{2 \cdot 3},$$

and so on. The nth Taylor polynomial hence will be

$$1 + x + \frac{x^2}{2!} + \frac{x^3}{3!} + \frac{x^4}{4!} + \ldots + \frac{x^n}{n!}$$

and this ultimately leads to the power series of the function $f(x) = \mathrm{e}^x$.

8.4.2 Remarks Concerning the Calculus of Differences

Computing with differences is fairly old and goes back to Jost Bürgi, Henry Briggs and Thomas Harriot [Goldstine 1977]. It was Taylor, however, who elevated the calculus of differences to being the 'little sister' of differential calculus. In the discrete world the operator Δ, defined by a difference, replaces the operative d. In the discrete world there is also a product rule and an integration by parts which is called 'summation by parts'. This kind of calculus of differences was expanded, among others, by the Scottish mathematician James Stirling (1692–1771) in his book *Methodus Differentialis: sive Tractatus de Summatione et Interpolatione Serierum Infinitarum* which was published in London in 1730. Due to its importance it was translated into English and published in 2003 [Tweddle 2003]. Today the calculus of differences is an indispensable part of numerical mathematics, but also of mathematical modelling and further areas within mathematics. A history of the calculus of differences has still to be written; a first attempt can be found in Goldstine's book [Goldstine 1977].

It is pivotal for the importance of the calculus of differences in Taylor's *Methodus incrementorum directa et inversa* that it holds a bridging function between Newton's and Leibniz's analysis, which was pointed out by Jahnke in [Jahnke 2003a, p. 112]. We have only to understand Leibniz's differentials as being particular, infinitely small differences to arrive at the 'correspondence'

$$dx = \dot{x} \cdot o,$$

where o denotes an infinitesimal increment of time.

Fig. 8.4.2. Brook Taylor and Colin MacLaurin

8.5 Colin Maclaurin

A particular form of the Taylor series was discovered by the Scotsman Colin Maclaurin (1698–1746). The 'Maclaurin series' is nothing but the Taylor series corresponding to the expansion point $x_0 = 0$. His name is also connected with the 'Euler-Maclaurin formula' concerning the representation of integrals. Maclaurin was a true Newtonian who received Berkeley's critique of Newton's calculus of fluxions and reasoned against it. His most famous book certainly is *A treatise of fluxions* published in two volumes in 1742. It is nothing less than the first textbook on Newton's analysis including a systematic introduction of its use. Maclaurin died when writing the book *An Account of Sir Isaac Newton's Philosophical Discoveries* [Maclaurin 1971] which had to be edited by his widow.

8.6 The Beginnings of the Algebraic Interpretation

Joseph-Louis (de) Lagrange (1736–1813) was born in Turin as Giuseppe Lodovico Lagrangia. In letters written to Euler between 1754 and 1756 he described his 'δ-method' for the solution of problems in the calculus of variations and thus became a co-founder of this field of analysis.

He could very much simplify Euler's analysis of such problems and Euler recognised the value of Lagrange's ideas, leading from a variational problem

Fig. 8.6.1. Joseph-Louis Lagrange

directly to the 'Euler-Lagrange differential equation'. Euler eventually proposed Lagrange (together with d'Alembert) as his successor at the Berlin academy, cp. page 453. In 1766 Lagrange started as the head of the mathematical class of the Berlin academy and stayed there for twenty years. Besides the calculus of variations which constitutes the foundation of mechanics his name is connected with his work *Mécanique Analytique*, written in Berlin and published in 1788 for the first time, which quickly became a standard work in the field of mechanics. Lagrange was received into the Paris Academy in 1787 and thereupon went back to Paris. He survived the French Revolution and when the École Polytechnique was established in 1794 he became the professor of analysis there. Since his students needed a textbook he published his lectures under the title *Théorie des fonctions analytiques* in 1797. Under Napoleon I he became a senator and was awarded the title count. He met Augustin Louis Cauchy's father and recognised the genius of the son which he fostered. Lagrange found his last resting place in the Panthéon; his name can be found among the 72 names of famous scientists at the Eiffel tower.

8.6.1 Lagrange's Algebraic Analysis

Lagrange felt the need to establish analysis *without* any recurrence to infinitely small quantities. He wanted an 'algebraisation' of analysis so that only finite quantities have to be considered in algebraic manipulations. As did Euler also Lagrange started with a definition of the notion of function which is very close to the one Euler gave in the *Introductio.* Like Euler also Lagrange saw the infinite series as the crux of the matter in analysis; however, Lagrange surpassed Euler. He used series for *all* functions. He wrote [Bottazzini 1986, p. 48]:

> *We therefore consider a function $f(x)$ of any variable x. If in place of x we put $x + i$, i being any indeterminate quantity whatever, it becomes $f(x + i)$ and, by the theory of series we can expand it as a series of this form*
>
> $$f(x) + pi + qi^2 + ri^3 + \dots,$$
>
> *in which the quantities $p, q, r, \dots$, the coefficients of the powers of i, will be new functions of x, derived from the primitive function x and independent of the [indeterminate] quantity i ...*

Concerning the use of infinitely small quantities in the past Lagrange remarked that *A treatise of fluxions* by Maclaurin would clearly show the difficulties in rigorously introducing them. Newton in his *Principia* therefore did not choose using fluxions but the path of classical geometry.

We have already remarked that Lagrange assumed series expansions for *all* functions. However, there is no proof that this is indeed the case. In the series (Lagrange wrote fx instead of $f(x)$, cp. [Stedall 2008, pp. 402ff.])

$$f(x+i) = fx + ip + i^2q + i^3r + \ldots$$

he substituted $x + o$ for x where o is a quantity independent of i. He then got the same result as if he had substituted $i + o$ for i, hence

$$fx + p(i+o) + q(i+o)^2 + r(i+o)^3 + \ldots.$$

This series can be written as

$$fx + pi + qi^2 + ri^3 + \ldots + po + 2qio + 3ri^2o + \ldots. \tag{8.7}$$

Replacing x by $x + o$ in the quantities $fx, p, q, \ldots$ then we get

$$\begin{aligned} f(x+o) &= fx + f'xo + \ldots, \\ p(x+o) &= p + p'o + \ldots, \\ q(x+o) &= q + q'o + \ldots \end{aligned}$$

and hence it follows

$$fx + pi + qi^2 + ri^3 + \ldots + f'xo + p'io + q'i^2o + \ldots. \tag{8.8}$$

Lagrange now compared the two series (8.7) and (8.8) and got

$$p = f'x, \quad ,2q = p', \quad 3r = q', \ldots,$$

hence

$$p = f'x, \quad q = \frac{f''x}{2}, \quad r = \frac{f'''x}{3!}, \ldots,$$

and therefore the series

$$f(x+i) = fx + if'x + \frac{f''x}{2}i^2 + \frac{f'''x}{3!}i^3 + \ldots.$$

Here now is the workhorse of the Lagrangean analysis: the Taylor series which Lagrange derived by an *algebraic* approach described above, although the problem of limits has only seemingly been overcome.

Even the notational convention f' for $\frac{dy}{dx}$ goes back to Lagrange who thereby wanted to avoid any suspicion that the derivative was actually a quotient of two infinitely small quantities.

Lagrange did not study convergence properties of his series but he introduced a representation of the remainder term of the Taylor series which has become a centrepiece of Taylor's theorem:

Taylor's theorem: *If f is a function $(n+1)$-times continuously differentiable, then*

$$f(x) = T_n(x) + R_n(x),$$

where T_n is the 'Taylor polynomial' of degree n about the expansion point x_0

$$T_n(x) = f(x_0) + f'(x_0)(x - x_0) + \frac{f''(x_0)}{2!}(x - x_0)^2 + \frac{f'''(x_0)}{3!}(x - x_0)^3 + \ldots + \frac{f^{(n)}(x_0)}{n!}(x - x_0)^n,$$

and R_n is the Lagrangean remainder term taking the form

$$R_n(x) = \frac{f^{(n+1)}(\xi)}{(n+1)!}(x - x_0)^{n+1},$$

where $x_0 < \xi < x$.

Connected with the estimate of the remainder term is the mean value theorem also going back to Lagrange.

Mean value theorem: *If $f : [a, b] \to \mathbb{R}$ is a continuous function and differentiable on $]a, b[$, then there exists a number $\xi \in]a, b[$, such that*

$$f'(\xi) = \frac{f(b) - f(a)}{b - a}$$

holds.

Lagrange can also be credited with having introduced inequalities as main technical tools of proof in analysis. Lagrange's 'algebraic' analysis had a strong influence on the development of analysis in France. However, Bottazzini has pointed out in [Bottazzini 1986, p. 54] that in *the* French textbook of analysis, *Traité du calcul différentiel et du calcul intégral* by Sylvestre Lacroix (1765–1843) of 1797–1800, see [Domingues 2008], there are large parts of Eulerian analysis besides the algebraic analysis of Lagrange. Hence we can conclude that infinitely small quantities did not vanish from teaching.

8.7 Fourier Series and Multidimensional Analysis

8.7.1 Jean Baptiste Joseph Fourier

Besides Taylor series the trigonometric and the Fourier series played a decisive role in the history of analysis; the importance of the latter surpasses the one of Taylor series by far. Some authors distinguish nicely between trigonometric series and Fourier series. Every Fourier series is a trigonometric series, but a trigonometric series is a Fourier series only if their coefficients are actually Fourier coefficients of certain functions. This kind of distinction, as important as it is, is of no concern to us here. Hence we use the terms 'trigonometric series' and 'Fourier series' synonymously.

The developments we want to describe here all started with the problem of the oscillating string. D'Alembert had derived the equation of the oscillating string in 1747,

$$\frac{\partial^2 u}{\partial t^2} - \frac{\partial^2 u}{\partial x^2} = 0, \tag{8.9}$$

(the *wave equation*) where u describes the deflection of the string, t is time, and x the position of a point on the string (cp. page 455). Brook Taylor, Johann Bernoulli, Jean le Rond d'Alembert, Daniel Bernoulli, Leonhard Euler, and Joseph-Louis Lagrange were all concerned with the derivation or the solution of the wave equation [Volkert 1988, p. 159]. D'Alembert already succeeded in a partial solution by means of a separation approach of the form $u(x,t) = f(t) \cdot g(x)$, but only Jean Baptiste Joseph Fourier (1768–1830) arrived at a satisfactory solution.

Fig. 8.7.1. Jean Baptiste Joseph Fourier
[Portrait by Julien Léopold Boilly, about 1823]

Fig. 8.7.2. Académie des Sciences 1671 [Detail, Sébastien Leclerc]. The academy was established in 1666 by Colbert and renamed Académie Royale by Louis XIV (centre left). Today their name is Académie des Sciences de l' Institut de France. Members of the academy were, among others, Maupertuis, de Roberval, Fourier, E. Cartan, A. Weil, and many others

Fourier was the son of a dressmaker. He was educated at a military college in Auxerre where he became professor when he was 18 years old. Although an adherent of the ideas of the French Revolution he almost died during the Jacobine reign of terror. As successor of Lagrange he became professor of analysis and mechanics at the École Polytechnique in 1797. He took part in the famous Egyptian campaign of Napoleon and led the secretariate of the Institut d'Égypte; he was also heavily involved in producing the famous book *Description de l'Egypte* in which the scientific yield of the campaign was described. Back in France Napoleon appointed him prefect of the Département Isère in 1802 and he received the title of a baron. When Napoleon returned from Elba in 1815 Fourier became prefect of the Département Rhône. He lived in Paris and became the lifelong secretary of the Académie des Sciences in 1815. His most famous work is *Théorie analytique de la chaleur* (Analytical theory of heat) of 1822 in which he established the Fourier analysis by means of the series which now bear his name. Today mathematics and physics can not be imagined without Fourier analysis.

8.7.2 Early Discussions of the Wave Equation

D'Alembert had succeeded to show that functions of the form

$$u(x,t) = \frac{1}{2}(y(t+x) - y(t-x))$$

are solutions of the wave equation (8.9). The function y has to satisfy the three conditions [Volkert 1988, p. 161]:

- *y is odd,*
- *y is periodic with period $2l$, if l denotes the length of the string,*
- *y equals the initial deflection of the string at $t = 0$.*

D'Alembert pointed out that $u(x, 0)$ is only defined on the length of the string $[0, l]$ while y is defined on all of $\mathbb{R}$. One could only justify to work with $y(x)$ if $u(x, 0)$ is assumed odd and periodically continued to all of $\mathbb{R}$. In d'Alembert's opinion the periodic continuation $u(x)$ should satisfy the same equation as $u(x, 0)$; an opinion which is called 'continuity principle' [Volkert 1988, p. 161]. Euler eventually was obliged to abandon this continuity principle, cp. page 456. Euler introduced the notation $f{:}(x)$ to denote a wholly arbitrary function and then gave the solution of the wave equation as

$$u = \frac{1}{2}f{:}(x+t) + \frac{1}{2}f{:}(x-t);$$

concerning the functional notation compare [Spalt 2015, S. 209 ff.]. Euler even gave a geometric construction of the solution as arithmetic mean of $f{:}(x+t)$ and $f{:}(x-t)$. But if the solution is 'continuous' then Euler proved that it has to be a trigonometric series of the form

$$u = \alpha \sin\frac{\pi y}{l} + \beta \sin\frac{2\pi y}{l} + \gamma \sin\frac{3\pi y}{l} + \ldots.$$

The question of which functions were representable as trigonometric series at all was thus born. D'Alembert argued strongly against Euler's 'arbitrary' functions; in his view the wave equation demanded the smoothness properties and therefore it made no sense to talk about functions with kinks. But under these premises the plucked string could not be treated within calculus and d'Alembert threw in the towel, cp. page 456.

A further early protagonist was Daniel Bernoulli, son of John and good friend of Leonhard Euler. He had a physician's view, cp. [Struik 1969, p. 361f.], and was of the opinion that *all* oscillations would be sinusoidal. In these early discussions of partial differential equations – and the wave equation (8.9) is such an equation – Euler was far ahead of his time. He already sensed that the classical notion of solutions à la d'Alembert which followed the smoothness requirements of the differential equation would not be sustainable in the long run. Today we consider so-called 'weak' solutions which satisfy the partial differential equations only in a weak (integral) sense, and hence Euler's idea of an 'arbitrary' function has been successfully realised.

8.7.3 Partial Differential Equations and Multidimensional Analysis

With the problem of the oscillating string and the wave equation (=equation of the oscillating string) (8.9) began the theory of partial differential equations which nowadays plays a prominent and active role within analysis and its applications. To analyse partial differential equations the analysis of functions of multiple variables or even of functions $f : \mathbb{R}^n \to \mathbb{R}^m$ is necessary. This so-called multidimensional analysis developed in the 18th century essentially in studying physical problems. In the 18th century multidimensional analysis made also large progress due to the question of the true shape of the earth. The French scientist Alexis-Claude Clairaut (Clairault) (1713–1765) succeeded after years of work to model earth as a fluid body built of strata of different densities, suffering gravitational and centrifugal forces being in equilibrium. Employing this mathematical model Clairaut was able to prove that earth necessarily had to be flattened at the poles. However, this theoretical proof was no longer necessary at this time since expeditions to measure the curvature of the earth had already confirmed this result. In the course of the work on Clairaut's main work *Théorie de la figure de la terre* of 1743 large parts of multidimensional analysis as, for example, potential theory were developed. Clairaut corresponded with Euler and stood in close contact with many excellent French mathematicians like Alexis Fontaine des Bertins (1704–1771), cp. [Greenberg 1995]. Under the strong influence of physics vector analysis was developed in the 19th century. With the integral theorems of Carl Friedrich Gauß (1777–1855), George Gabriel Stokes (1819–1903), and George Green (1793–1884) it became possible to treat moving fluids, electrical and electromagnetic fields, and general field problems [Crowe 1994].

Within this book we will neither follow the history of partial differential equations nor the development of multidimensional or even vector analysis. In each case a book on its own would be necessary.

8.7.4 A Preview: The Importance of Fourier Series for Analysis

Fourier in his book *Théorie analytique de la chaleur* of 1822 was concerned with series of the form

$$\varphi(x) = a \sin x + b \sin 2x + c \sin 3x + d \sin 4x + \dots$$

which he compared coefficientwise with the Taylor series

$$\varphi(x) = x\varphi'(0) + \frac{x^2}{2!}\varphi''(0) + \frac{x^3}{3!}\varphi'''(0) + \dots,$$

Fig. 8.7.3. Alexis Claude Clairaut [Artist: Louis-Jacques Cathelin, 18th c.] and Rudolph Lipschitz [unknown photographer]

[Stedall 2008, S. 223]. He thus got infinitely many conditional equations for the coefficients $a, b, c, d, \ldots$. Now it might happen that the Taylor series represents a completely different function than the trigonometric series or that convergence problems might occur, but all these question were not of concern to Fourier. In a *tour de force* in which Fourier had to solve various partial differential equations he finally recognised that the coefficients $a, b, c, d, \ldots$ could be given explicitly as certain integrals.

In todays notation a Fourier or trigonometric series of a 2π-periodic function f is an infinite series of the form

$$F(x) = a_0 + \sum_{k=1}^{\infty}(a_k \cos kx + b_k \sin kx),$$

where the coefficients are given by

$$a_0 = \frac{1}{2\pi}\int_{-\pi}^{\pi} f(x)\,dx$$

and

$$a_k = \frac{1}{\pi}\int_{-\pi}^{\pi} f(x)\cdot\cos kx\,dx, \quad b_k = \frac{1}{\pi}\int_{-\pi}^{\pi} f(x)\cdot\sin kx\,dx, k = 1, 2, 3, \ldots.$$

What are the conditions on f so that we really have $F = f$, i.e. when is a function represented by its Fourier series?

Augustin Louis Cauchy had tried to answer this question but the German mathematician Peter Gustav Lejeune Dirichlet found something to gripe

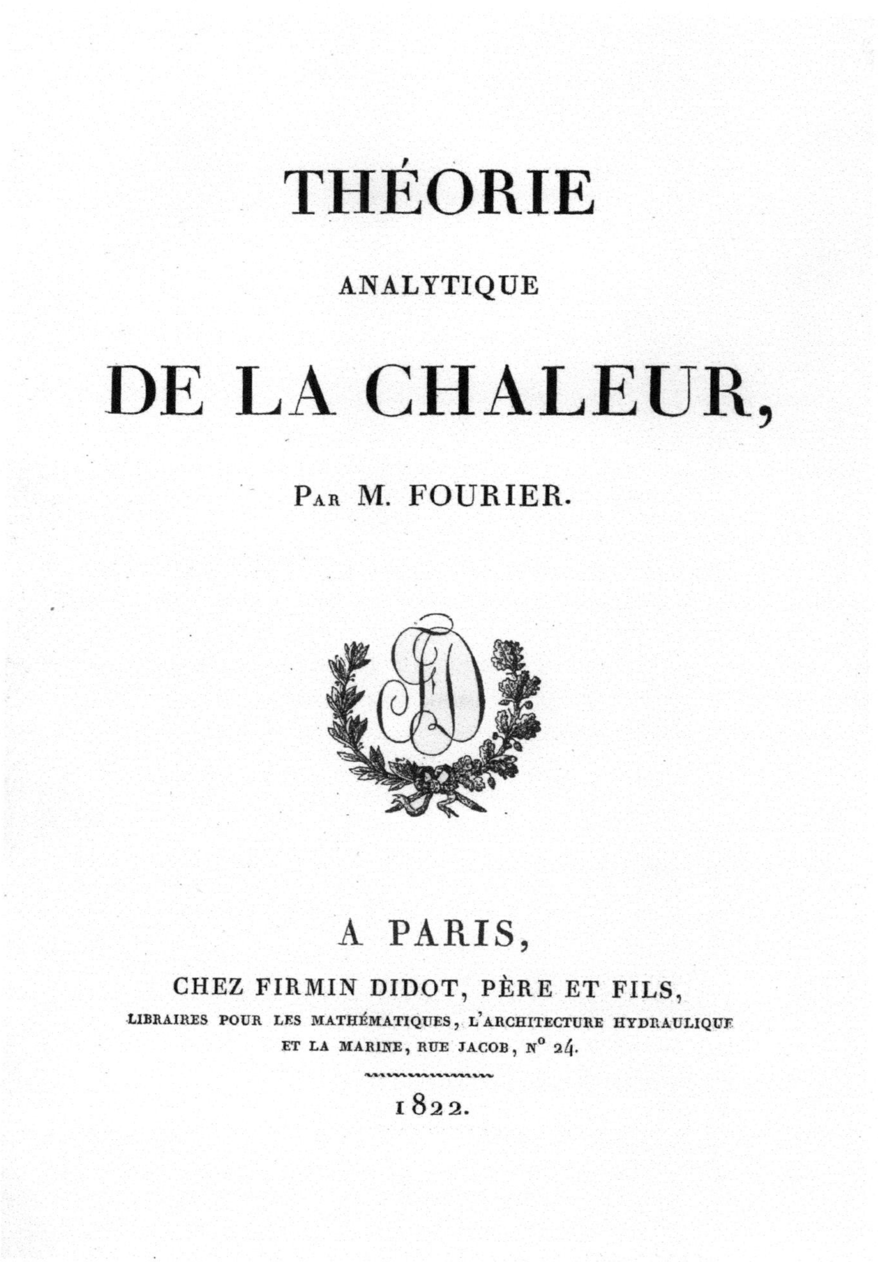

THÉORIE

ANALYTIQUE

DE LA CHALEUR,

Par M. FOURIER.

A PARIS,

CHEZ FIRMIN DIDOT, PÈRE ET FILS,

LIBRAIRES POUR LES MATHÉMATIQUES, L'ARCHITECTURE HYDRAULIQUE
ET LA MARINE, RUE JACOB, N° 24.

1822.

Fig. 8.7.4. Title page of the *Théorie analytique de la chaleur* by Fourier

about Cauchy's arguments. In a work of 1829 he formulated the following 'Dirichlet's conditions' [Dauben 1979, p. 9]:

> *The preceding considerations prove in a rigorous manner that if the function $\varphi(x)$ (for which all values are supposed finite and determined) only presents a finite number of discontinuities between the limits $-\pi$ and π, and if addition it does not have more than a determined number of maxima and minima between these same limits, then the series*
>
> $$\frac{1}{2\pi}\int \varphi(\alpha)\,d\alpha + \frac{1}{\pi}\left\{\begin{array}{l}\cos x \int \varphi(\alpha)\cos\alpha\,d\alpha + \cos 2x \int \varphi(\alpha)\cos 2\alpha\,d\alpha + \ldots \\ \sin x \int \varphi(\alpha)\sin\alpha\,d\alpha + \sin 2x \int \varphi(\alpha)\sin 2\alpha\,d\alpha + \ \ldots\end{array}\right\}$$
>
> *(where the coefficients are the definite integrals depending on the function $\varphi(x)$) is convergent and has a value expressed generally by*
>
> $$\frac{1}{2}(\varphi(x+\varepsilon) + \varphi(x-\varepsilon)),$$
>
> *where ε designates an infinitely small number.*

Dirichlet himself was not happy with the assumptions he had to put on his expandable functions and started to look for a tightened version of his result. In order to give a justification for the necessity of such assumptions Dirichlet presented in his work of 1829 a function which did *not* satisfy the assumptions and was therefore not expandable in a Fourier series. This function was the famous Dirichlet function

$$D(x) := \begin{cases} c\ ;\ x \in \mathbb{Q} \\ d\ ;\ x \in \mathbb{R}\backslash\mathbb{Q} \end{cases}, \tag{8.10}$$

taking the value c at rational numbers and the value $d \neq c$ at irrational numbers. Volkert justifiably calls those functions 'monsters' and the Dirichlet function the 'Dirichlet monster' [Volkert 1988, p. 197]. This Dirichlet monster has also decisively influenced the theory of integration and today counts as a paradigm of the superiority of the Lebesgue integral (named after Henri Léon Lebesgue (1875–1941)) over the Riemann integral.

The next person to tackle the question of the expandability of functions into Fourier series was Georg Friedrich Bernhard Riemann (1826–1886), a pupil of Dirichlet. During his work he had to observe that the integral itself was not rigorously defined and he created what we now call the Riemann integral. We will discuss this integral later. Besides Riemann also Rudolph Otto Sigismund Lipschitz (1832–1903) and Hermann Hankel (1839–1873) worked on this problem. Finally in 1873 Paul du Bois-Reymond (1831–1889) found a continuous function the Fourier series which did not converge. The groundwork done by these scientists eventually became the starting point of the first mathematical works of Georg Cantor (1845–1918), finally leading to the invention of set theory. Only a few problems have emanated that far and have influenced the fundamental questions of analysis.

Mathematicians and their Works Concerning the Analysis of the 18th Century

1687	Jacob Bernoulli gets the chair of mathematics at the University Basel
1690	John Bernoulli solves the problem of the catenary
1691	John Bernoulli gets to know Marquis de l'Hospital and signs a contract concerning the sale of mathematical results
1695	John Bernoulli gets a professorship at Groningen
1696	de l'Hospital publishes the first textbook on Leibniz's differential and integral calculus
1713	Jacob Bernoulli's *Ars conjectandi* is published posthumously In a quarrel between Jacob Bernoulli and John Bernoulli the calculus of variations started off
1715	Brook Taylor publishes a book important for analysis in which the 'Taylor series' can be found
1725	Leonhard Euler publishes his first mathematical work in the 'Acta Eruditorum'
1727	Euler moves to St Petersburg to work at the academy
1730	The book *Methodus Differentialis* by James Stirling is published in London
1731	Euler becomes professor of physics in St Petersburg
1741	Euler moves to the Berlin academy
1742	Maclaurin publishes *A treatise of fluxions*, a textbook on Newton's calculus of fluxions
1744	Euler's *Methodus inveniendi* is published as the first textbook of the calculus of variations
1747	d'Alembert derives the wave equation being the partial differential equation of the oscillating string
1748	Euler publishes *Introductio in analysin infinitorum*: Prelude to a whole string of influencial textbooks on analysis. The notion of function is generalised for analysis in the *Institutiones calculi differentialis*
1748–1766	Euler revolutionised analysis and physics. He develops the calculus of variations into a mathematical theory
1754–1756	In the correspondence between Lagrange and Euler the calculus of variations is further developed
1765	Euler's *Theoria motus corporum solidorum seu rigidorum* (mechanics) is published
1766	Euler moves back to St Petersburg. He is getting blind
1783	Euler dies and is buried in St Petersburg
1788	Lagrange's *Méchanique analytique* finally appears in print
1797	The influential textbook *Théorie des fonctions analytiques* by Lagrange is published in Paris
1822	Fourier publishes the *Théorie analytique de la chaleur* in which the Fourier analysis is described for the first time

9 On the Way to Conceptual Rigour in the 19th Century

T. Sonar, *3000 Years of Analysis*, https://doi.org/10.1007/978-3-030-58223-4_9

From the French Revolution to the German Empire

1789	Start of the French Revolution. Storming of the Bastille on 14th July
1798	British naval victory over the French fleet at Aboukir Bay (Battle of the Nile)
1799	Napoleon becomes consul for 10 years
1804	Coronation of Napoleon as Emperor Napoleon I
1805	Annihilation of the French fleet by the British under Admiral Nelson at Trafalgar
1806/07	Napoleon's military campaign against Prussia; Treaties of Tilsit
1812	Destruction of the 'Grand Armée' in Russia
1813/14	German Campaign (Wars of Liberation)
1815	Napoleon's final defeat in the Battle of Waterloo
1815-1866	German Confederation with 39 member states
1814–1815	Congress of Vienna
1821-1829	Greek War of Liberation, recognition of Greek's sovereignty
1822	Brasil becomes an independent empire
1825	End of the Spanish colonial empire in South America
1827	Destruction of the Turkey-Egypt fleet at Navarino
1830	July Revolution in France. First railway line from Liverpool to Manchester
1834	Foundation of the German Zollverein
1837	Victoria becomes Queen of Great Britain and Ireland King Ernest Augustus I of Hanover dismisses the die 'Göttingen Seven'
1845/46	Great Famine in Ireland
1848	The *Communist Manifesto* by Karl Marx and Friedrich Engels is published. Revolution in Paris. Worker's uprising. Successful revolution in many German states. Constituent National Assembly in the St Paul's Church in Frankfurt am Main
1848–1916	Franz Joseph I Emperor of Austria
1850	Establishment of the Frankfurt Parliament
1852–1870	Napoleon III Emperor of France
1854–1856	Crimean War. It ends with a defeat of Russia
1856–1871	National unification of Italy
1861–1865	American Civil War
1862	Otto von Bismarck becomes Prussian prime minister
1864	The 'Red Cross' is established Second Schleswig War
1866	Victory of Prussia over Austria and the Kingdom of Hanover
1870–1871	Franco-Prussian War
1871	King William I of Prussia is proclaimed German Emperor in Versailles

Science and Engineering in the Industrial Revolution

1765	James Watt invents the steam engine
	In Freiberg/Saxony the Mining Academy is founded
1767	The prime meridian is defined to pass through Greenwich
1772	James Cook starts his second circumnavigation of the world
	D. Rutherford discovers oxygen as a component of air
1774	J. Priestley and C.W. Scheele isolate oxygen
1779	First cast-iron bridge crossing the Severn
1781	F.W. Herschel discovers the planet Uranus
1783	J.E. and J.M. Montgolfier rise a hot air balloon in public for the first time
1784	Patent of the puddling process to produce steel
	Ch.A. Coulomb discovers his law on the attraction of electric charges
1794	Founding of the École Polytechnique in Paris
1800	A. Volta invents the Voltaic pile, the first electro-chemical source of electrical power
1802	G.F. Grotefend deciphers the cuneiform script
1803	R. Trevithick designs the first steam locomotive
1805	J.M. Jacquard designes program-controlled weaving looms
1808	É.L. Malus discovers the polarisation of light
1810	*Annales de mathématiques* by Gergonne appear in print. It is the first purely mathematics journal
1814	J.v. Fraunhofer discovers dark lines in the spectrum
	G. Stephenson builds his first locomotive
1817	K.v. Drais invents the first running machine (draisine), a precursor of the bycicle
1819	First steam ship crosses the Atlantic
1820	H.C. Oersted discovers electromagnetism
1821	T.J. Seebeck communicates his discovery of thermoelectricity
1822	J.N. Nièpce invents photography
1824	S. Carnot describes the working cycle of a thermal engine as a cycle
1825	First public railway line in England
	First German technical school is established in Karlsruhe
1826	G.S. Ohm discovers the law which is named after him
1827	F. Wöhler succeeds in the urea synthesis, the first organic syntheses ever
1829	G. Coriolis determines the inertia forces of motion in a rotating system
1831	J.v. Liebig improves the method of elementary analysis in chemistry
1832	J.-F. Champollion deciphers the Egyptian hieroglyphs

1832/33	M. Faraday explains electromagnetic induction and discovers his law of electrolysis
1833	C.F. Gauß and W. Weber design the first working telegraph line
1835	First German railway line opens 7th December connecting Nuremberg and Fürth
1837	L.J.M. Daguerre invents his method of photography
1838	First German state railway between Brunswick and Wolfenbüttel
1842	J.R. Mayer formulates the law of the conservation of energy and determines the mechanical equivalent of heat
	Ch. Darwin develops his theory of descentent
1845	G.R. Kirchhoff formulates the rule of ramified electric circuits (Kirchhoff's rules)
1848	C. Doppler discovers the effect of shift of the spectral lines in the spectra of stars
1851	Pendulum experiments of J.B.L. Foucault
	First world exhibition in London
1854	H. Goebel (Göbel) invents the electric light bulb
1855	H. Bessemer designs a method for the production of cast steel
1859	C. Darwin establishes his theory of evolution
1857	R. Clausius deduces mathematically the kinetic theory of gases
1859	Exploitation of the first oil source
1861	R. Kirchhoff and R.W. Bunsen discover spectral analysis
	Ph. Reis invents the telephone
1862	N. Riggenbach designed the first rack railway
1863	First underground railway in London
1864	É. and P.-É. Martin invent the open-hearth process and produce molten steel for the first time in the Siemens-Martin furnace
1865	J.C. Maxwell develops the electromagnetic theory of light
1866	J.G. Mendel publishes his Mendelian law
	W. v.on Siemens invents the dynamo
	P. Mitterhofer designs the first typewriter with typing lever basket and platen
1867	N.A. Otto designs the first combustion engine
	A. Nobel invents dynamite
1869	D.J. Mendelejew and L. Meyer establish the periodic table of the elements independent of each other
	Opening of the Suez Canal
1870	J.D. Rockefeller establishes the Standard Oil Company

9.1 From the Congress of Vienna to the German Empire

After the final defeat of Napoleon Europe was rearranged in the Congress of Vienna. At the expense of Poland which had to suffer a subdivision again the great powers England, Prussia, Austria, and Russia were strengthened. Together with France the so-called Concert of Europe (Pentarchie) was made permanent which was oriented towards a certain balance of powers. The blessings of the Napoleonic foreign rule – strict separation of the legislative, executive, and judiciary and the ideas of liberty, equality, and fraternity – were partly abandoned and replaced by the 'Metternich system' of political control named after the Austrian Chancellor Clemens Wenzel von Metternich (1773–1865). A restauration in all of Europe set in, in which liberal tendencies like the upcoming student movement were massively suppressed.

In the July Revolution of 1830 the Bourbons were ultimately expelled from the reign of France. When King Charles X tried to resolve the parliament the people stood up and forced Charles to resign. The July Revolution had many effects on large areas of Europe. The liberal forces got upswing and in some countries of the German Confederation as also in Italy, Poland, and the Netherlands upheavels occured. In France the 'Citizen King' Louis-Philippe I was put into place who but soon departed from his liberal roots and joined the 'Holy Alliance' which was influenced by the Metternich system.

Fig. 9.1.1. Europe after the Congress of Vienna 1815 and Clemens Wenzel von Metternich [Painting: Sir Thomas Lawrence, ca. 1820–1825]. The rearrangement of Europe was authoritatively decided by him acting as the Austrian Chancellor

Fig. 9.1.2. 'La Liberté guidant le peuple' (freedom guides the people) ([Painting by Eugene Delacroix 1830] Musée du Louvre, Paris)

This development led in 1848 to the February Revolution in France which was followed by the establishment of the French Second Republic.

In many other European states, so in the German Confederation, this revolution ignited the March Revolution 1848 which eventually finished off Metternich's restauration. But initially the March Revolution failed. From 18th May 1848 until 31st May 1849 the Frankfurt Parliament met in the Church of St Paul's; this was the first freely elected parliament of Germany. The National Assembly worked out a liberal constitution which failed when the Prussian King Frederick William IV refused to accept the title of imperator. The attempt to establish an all-German national state failed already in 1849 due to the employment of Prussian and Austrian troops. After the failure of the March Revolution the Frankfurt Parliament, existing already prior to the National Assembly, became a tool of the German rulers again. The constitutional ideas of the National Assembly could be relised only during the Weimar Republic (1918–1933).

From 1848 on the 'spectre of communism' was hunting in Europe. Karl Marx (1818–1883) and Friedrich Engels (1820–1895) had published the *Communist Manifesto* in that year; in the year 1865 the first volume of *Das Kapital* by Marx was published the two following volumes of which were published by Engels in 1885 and 1895, respectively.

Fig. 9.1.3. 'Eisenwalzwerk (Moderne Cyclopen)' (steel mill; modern cyclops) 1872–1875 by Adolph Menzel. The reality of the work of labourers in the 19th century is shown (Alte Nationalgalerie Berlin)

One of the most important reasons for the political developments in Europe certainly is the Industrial Revolution initiated by the design and use of powerful steam engines in England at the turn of the 18th and 19th century. The notion of 'Industrial Revolution' came already up during the French Revolution to characterise the upheavels connected with the industrial production showing in England. The key industry of the English Industrial Revolution was the textile sector which saw unimagined productivity increases due to modern inventions like the 'spinning Jenny' and the mechanical loom but also saw the impoverishment of the labourers. The centre of the textile industry in England was Manchester since many streams provided ideal conditions for cotton mills. Not for nothing a reckless capitalism is nowadays called 'Manchester capitalism'.

In the midst of the 19th century the Industrial Revolution had arrived on the Continent. A working class emerged and the capitalistic economic system developed. More moderate political ideas as the ones of Marx and Engels led in 1863 to the foundation of the German Workers Union (Deutscher Arbeiterverein) by Ferdinand Lasalle (1825–1864); in 1869 August Bebel (1840–1913) and Wilhelm Liebknecht (1826–1900) formed the Social Democratic Workers Party SDAP (Sozialdemokratische Arbeiterpartei) which will be transformed later into the SPD. These developments were seen as a threat by the government and the bourgeoisie and as a reaction the Anti-Socialist

Fig. 9.1.4. 'A dog's life' (Ein Hundeleben) by Gustave Doré 1872. Doré shows the situation of the life of the impoverished lower class

Laws were issued in 1878 which were repealed only in 1890. Following the motto 'carrot and stick' a health insurance was installed in 1883 followed by an accident insurance in 1884 and an old-age and disability insurance in 1889 together with Bismarck's social welfare laws as preventive measures to answer the movements of emancipation. The conservative Reich Chancellor Otto von Bismarck succeeded in asserting the Prussian dominance by an iron-fisted policies not shrinking back from wars. After the Franco-German War in 1871 the German Empire was proclaimed in Versailles. There was no future for particularism.

Analysis turned out to be a driving force for the Industrial Revolution. Now the computation of bend lines of girders or oscillations of machines became possible and whole new machines could be designed. Already in 1824 French physicist Nicolas Léonard Sadi Carnot (1796–1832) worked on the quantity

Fig. 9.1.5. Michael Faraday in his laboratory [Painting: H. J. Moore, 19th c.]. The 'Royal Institution of Great Britain' was established in 1799 for the purpose of scientific education and research. By means of lectures, experiments, and courses the new applications of sciences should be presented to the public. The Royal Institution still exists and pursues its mission

of heat of a steam engine and illustrated its work process as a cycle. A little later Benoît Pierre Émile Clapeyron (1799–1864) captured Carnot's results in mathematical form. Therewith the foundations of a new area of science within physics was laid: thermodynamics. In 1841 the German physician Julius Robert Mayer (1814–1878) formulated the law of conservation of energy and in 1854 German physicist Rudolf Clausius (1822–1888) discovered the efficiency factor of heat engines and introduced the entropy as a new state variable. Only in the 20th century thermodynamics evolved into an axiomatic theory and can not be thought of without analysis.

In physics and astronomy the beginning 19th century saw great progress. Between 1799 and 1825 the *Mécanique céleste* by Pierre Simon Laplace (1749–1827) was published and set standards in the computational treatment of celestial motion. In 1800 the Italian Alessandro Volta (1745–1827) reports his method to generate constant direct currents to the Royal Society. Electricity as a new 'force' now came under the spotlight of scientists and tycoons alike. Still in the year 1800 started extensive research works on galvanic elements in England; we would call them 'batteries' today. In Germany the high school teacher Georg Simon Ohm (1787–1854) discovered his Ohm's law $U = R \cdot I$ (voltage = resistance · current) publishing it in 1826 and

Fig. 9.1.6. George Stephenson's 'Rocket' (drawing) and a stamp in its honour (Greatbritain 1975). Stephensons 'Locomotion No. 1' pulled the first public steam rail (for the time being only for freight) 1825 from Darlington to Stockton at 15–17 km/h

1827. Gustav Kirchhoff extended Ohm's law to complicated circuits and formulated his two Kirchhoff's circuit laws (current law and voltage law). It was also Kirchhoff to point out the connection of the solution of Poisson's differential equation (being the 'potential') and the 'electroscopic' force. The connection of electricity with (electro)magnetism also became the subject of debate. Already in 1820 André Marie Ampère (1775–1836) had derived a law describing the electrodynamical interaction of currents, but it remained to an Englishmen to fully understand electrodynamics in mathematical terms.

The great experimenter Michael Faraday (1791–1867) discovered the law of induction in 1831, the law of electrolysis in 1833, and the Faraday effect in 1845. Faraday 'saw' the space around the electrical conductor filled by field lines and he began to study the effects of remote forces. These ideas were eagerly taken up by the arguably greatest theoretical physicist of the 19th century, James Clerk Maxwell (1831–1879). He developed a new theory which concluded classical physics [Simonyi 2001, p. 346]. In 1862 Maxwell's treatise *On Physical Lines of Force* was published and was followed in 1873 by the two volumes of the book *A treatise on electricity and magnetism.* To formulate ´Maxwell's equations' vector analysis is needed which also was developed in the 19th century and which we will briefly introduce in section 10.5.

Fig. 9.1.7. The 'Great Eastern'; sailing ship and steam ship alike [Painting by Charles Parsons, 1858]. In 1852 the English engineer Isambard Kingdom Brunel (1806–1859) designed and built an enormous ship of 18 916 gross register tons; the 'Great Eastern'. Much to large for a passenger ship in these days it was employed as a cable ship on the Atlantic. Among other inventions Brunel also made a name with the construction of bridges never seen before

With the industrialisation the demands of engineers increased and more mathematics became necessary, in particular analysis. One could compute streeses in steel bridges and was therefore able to design constructions which one could not think of in the 18th century. The steam engine was no longer exclusively used stationary as an engine but could be employed to drive ships and trains. Already in 1825 the first railway line was inaugurated on which George Stephenson's (1781–1848) famous 'Locomotion' operated. Already in 1829 the line Liverpool-Manchester was opened and the steam-driven railway was on its way to worldwide success.

To proudly present the new achievements to the world in 1851 the first world exhibition (todays EXPO) took place in London's Hyde Park at the suggestion of Prince Albert. A unique building, the Crystal Palace, was erected which was destroyed only in 1936 by a fire. In the world exhibition that followed 1855 in Paris besides a boat made of concrete one could also marvel at the first matches and the first espresso machine.

The school education in the 18th century was strongly denominationally oriented; at the beginning of the 19th century Europe was captured by a wave of secularisation. Church was not subordinated to the state and freedom of confession prevailed. Literature was no longer exclusively characterised by religion but fiction conquered the space in book shops. The importance of national states was emphasised. Initiated by the French polytechnics a founding wave of technical schools swept through Germany. Berlin got a technical institution already in 1821, Karlsruhe 1825, Munic 1827, Desden 1828, and Hanover 1831. They were not yet Technical Universities, however, but places of education only. There was still a long way to go until they were transformed into genuine universities.

9.2 Lines of Developments of Analysis in the 19th Century

After the death of Euler many mathematicians believed that there would not be much left in mathematics worth of study. On the other hand one felt a certain discomfort concerning the foundations of analysis which was triumphant in applications but operated still upon infinitely small quantities or even, as with Euler, upon 'zeros'. The discussions of solutions of the wave equation in particular had revealed that notions like 'function', 'continuity', and 'convergence', but also 'integral' and 'derivative', were not rigorously defined. This task was left to the mathematicians of the 19th century.

Bernhard Bolzano (1781–1848) from Prague conducted his researches as a pure leasure activity and therefore made no great impact in his days. In France Augustin Louis Cauchy worked on the foundations of analysis and had a strong effect also outside France by means of his textbooks. In Germany the high school teacher and later professor of mathematics Karl Weierstraß became the driving force behind the specification of the

Fig. 9.1.8. The *Cristall Palace* in London. Here the first world exhibition took place in 1851. The palace fell victim to a fire in 1936

foundations of analysis; the 'Weierstraß ean school' has decisively coinded modern analysis. Bernhard Riemann clarified the notion of integral. In his habilitation lecture he opened the doors towards a modern differential geometry, i.e. the application of analytical methods in geometry. Together with the wish of specification also questions concerning the construction of irrational numbers came up. One had computed with real numbers like π and $\sqrt{2}$ for centuries, but there were no clear ideas of how the structure of the number system really looked like. This question was also solved in the 19th century, namely by Richard Dedekind and Georg Cantor.

9.3 Bernhard Bolzano and the Pradoxes of the Infinite

Bernardus Placidus Johann Nepomuk Bolzano (1781–1848), in short: Bernard Bolzano, was born the son of a North-Italian art merchant and his wife from a German merchant family in Prague.

His family being very religious Bolzano was influenced so much that he studied theology at the Charles University Prague from 1801 on. After attending high school from 1791 until 1796 Bolzano had initially studied philosophy, mathematics, and physics, and all of his life he stayed a keen mathematician with a visionary talent. He studied mathematics under the

Fig. 9.3.1. Bernard Bolzano

Fig. 9.3.2. Franz Josef von Gerstner and Abraham Gotthelf Kästner

Jesuit Stanislav Vydra (1741–1804), a rather weak mathematician, and under Franz Josef Ritter von Gerstner (1756–1832), who understood how to thrill Bolzano. Gerstner also recognised Bolzano's genius although both men held different views of mathematics. In the view of Gerstner who was also a physicist and a technician mathematics was a mere ancillary science with which problems in other sciences could be solved. In contrast, Bolzano was interested mainly in mathematics as such and in problems within mathematics.

During his studies the textbooks by Abraham Gotthelf Kästner (1719–1800) had attracted him to mathematics. Kästner's textbooks had in fact a large effect in the 19th century and he even wrote a 'Geschichte der Mathematik' (history of mathematics) [Kästner 1970]. Consequently invoking reason and logic became the characteristic feature of Bolzano [Wußing/Arnold 1978, p. 321]. After having completed his studies at the philosophical faculty in 1799 Bolzano's father wanted to see his son in trading while his mother wanted to see him as a priest. After a period of reflection Bolzano decided in favour of theology. In the winter semester 1804/05 he completed his studies of theology with a doctoral thesis and was ordained priest two years later.

In 1804 two chairs became vacant at the Charles University. Stanislav Vydra had died and his chair of elementary mathematics had to be filled; a chair of theology had been newly established. Bolzano applied to both chairs and was strongly recomended by Gerstner for the mathematics chair. However, it was decided in Vienna to give the mathematics chair to a candidate who never had written an independent piece of work and Bolzano got the chair of theology. Within a short period of time Bolzano's lectures became popular with the students. He connected religious themes with social criticism and went even as far as utopian socialistic ideas and criticism of the constitution. As son of an Italian father and his German wife he tried to propagate an own Bohemian patriotism what was very well received by his Bohemian students. However, such a behaviour could not go unnoticed with the Austrian authorities. First attempts to draw him out of office could yet be parried but in the course of an investigation against him he was removed from his chair by Emperor Francis I at 24th December 1819. All further public activities were prohibited. The first volume of his *Erbauungsreden* (edification speeches) of 1813 and his *Lehrbuch der Religionswissenschaft* (textbook on the science of religion) were put on the list of forbidden books. Only after the death of Francis I in 1835 the surveillance of Bolzano was loosend and later he was even allowed to publish non-theological work in the treatises of the Royal Bohemian Society of Sciences.

Bolzano spent the years after 1819 with mathematical work and for this, and only for this, we have to be thankful to Emperor Francis. Due to his isolated position in Europe Bolzano's works had no effect in his days; from our modern point of view we have to place him in a line with Cauchy and Weierstraß . The ideas for a rigorous processing of the foundations of analysis probably might have been in mid air. Only in this way we can explain why so many mathematicians of the 19th century worked independently on the same problems.

9.3.1 Bolzano's Contributions to Analysis

In 1816 Bolzano wrote *Der binomische Lehrsatz und als Folgerung aus ihm der polynomische und die Reihen, die zur Berechnung der Logarithmen und Exponentialgrössen dienen, genauer als bisher erwiesen* (The binomial theorem and, as a conclusion of it, the polynomial [theorem] and the series which serve the computation of logarithms and exponential quantities, more accurately deduced as hitherto). In it we find Bolzano's definition of equality of two numbers [Stolz 1881, p. 257]:

> *If in the equation $A+\omega = B+\omega^1$ the quantities ω, ω^1 are allowed to become as small as one ever wants them to be while A and B remain unchanged: hence exactly $A = B$ has to hold.*
>
> (Wenn in der Gleichung $A + \omega = B + \omega^1$ die Grössen ω, ω^1 so klein werden können, als man nur immer will, während A und B unverändert bleiben: so muss genau $A = B$ sein.)

Here a characteristic feature of Bolzano's way of thinking becomes apparent: the avoidance of the notion of infinitely small quantities. Bolzano wrote [Wußing/Arnold 1978, p. 328]:

> *... as I then use instead of the so-called infinitely small quantities consistently with the same success the notion of such quantities which may become smaller than any given one, or (as I also call them for the sake of avoiding monotony ...), the quantities which may become smaller as one ever wants. Hopefully one will not misjudge the difference between quantities of this kind and those of which one otherwise thinks of under the name of the infinitely small.*
>
> (... wie ich denn auch statt der so genannten unendlich kleinen Größen mich durchgängig mit demselben Erfolge des Begriffes solcher Größen bediene, die kleiner als jede gegebene werden können, oder (wie ich sie zur Vermeidung der Eintönigkeit gleichfalls nenne, ...) der Größen, welche so klein werden können, als man nur immer will. Hoffentlich wird man den Unterschied zwischen den Größen dieser Art und dem, was man sich sonst unter dem Namen des unendlich Kleinen denkt, nicht verkennen.)

Bolzano thought that the binomial theorem was one of the most important theorems of analysis, but he is not satisfied with the previous proofs. His treatment of the binomial theorem alone puts him on the same level with Cauchy [Dieudonné 1985, p. 363].

In his work *Rein analytischer Beweis des Lehrsatzes, dass zwischen je zwei Werthen, die ein entgegengesetztes Resultat gewähren, wenigstens eine reelle Wurzel der Gleichung liege* (Purely analytical proof of the theorem that between any two values giving opposite results lies at least one real root of the equation), published in Prague in 1817, Bolzano not only rigorously proved the intermediate value theorem, but he also gave a definition of 'continuity' which is not only exact, but also came earlier than Cauchy's definition. Bolzano called a quantity assuming all possible values between two given values 'freely variable'. If a quantity takes on values in a neighbourhood of every of its assigned values so that the difference to the assigned value is arbitrarily small, then this quantity is called 'continuously variable'. Concerning the continuity of a function Bolzano wrote (quoted from [Wußing/Arnold 1978, p. 328]):

> *[...], that, if x is any such value, the difference $f(x+\omega) - f(x)$ can be made smaller as any given quantity if ω can be assumed as small as one ever wants.*
>
> ([...], daß, wenn x irgend ein solcher Werth ist, der Unterschied $f(x+\omega)-f(x)$ kleiner als jede gegebene Größe gemacht werden könne, wenn man ω so klein, als man nur immer will, annehmen kann.)

In fact this is a definition of continuity which we (in another form) even use today. Concerning convergence of infinite series Bolzano stated Cauchy's convergence criterion and rigorously employed the idea of partial sums. The *necessary* condition for convergence was described by Bolzano as follows (quoted from [Stolz 1881, p. 259]):

> *If one denotes the value of the sum of the first* n, $n+1, \ldots, n+r$ *terms of a [...] series in their order by* $F^n x, F^{n+1} x, \ldots, F^{n+r} x$, *then the quantities*
>
> $$F^1 x, F^2 x, \ldots, F^n x, \ldots, F^{n+r} x$$
>
> *now constitute a* new *series. This one has [...] the particular property that the difference between its n-th term* $F^n x$ *and any later* $F^{n+r} x$, *be it however far away from the n-th, will stay smaller than every given quantity if one has choosen* n *only large enough.*
>
> (Wenn man den Werth, welchen die Summe der ersten n, $n+1$, $\ldots$, $n+r$ Glieder einer [...] Reihe hat, der Ordnung nach durch $F^n x, F^{n+1} x, \ldots, F^{n+r} x$ bezeichnet, so stellen die Größen
>
> $$F^1 x, F^2 x, \ldots, F^n x, \ldots, F^{n+r} x$$
>
> nun eine *neue* Reihe vor. [...] Diese hat [...] die besondere Eigenschaft, dass der Unterschied, der zwischen ihrem $n$ten Gliede $F^n x$ und jedem späteren $F^{n+r} x$, es sei auch noch so weit von jenem $n$ten entfernt, kleiner als jede gegebene Grösse bleibt, wenn man erst n gross genug angenommen hat.)

And furthermore the *sufficient* convergence condition:

> *If a series of quantities*
>
> $$F^1 x, F^2 x, \ldots, F^n x, \ldots, F^{n+r} x, \ldots$$
>
> *is of the nature that the difference between its n-th term* $F^n x$ *and any later* $F^{n+r} x$, *in whatever distance it may be, stays smaller than any given quantity if one has assumed* n *large enough: hence every time there is a certain* constant *quantity, and* only one, *to which the terms of this series come ever closer and to which they can come so close as one ever wants if the series is continued far enough.*
>
> (Wenn eine Reihe von Grössen
>
> $$F^1 x, F^2 x, \ldots, F^n x, \ldots, F^{n+r} x, \ldots$$
>
> von der Beschaffenheit ist, dass der Unterschied zwischen ihrem $n$ten Gliede $F^n x$ und jedem späteren $F^{n+r} x$, sei dieses von jenem auch noch so weit entfernt, kleiner als jede gegebene Grösse verbleibt, wenn man n gross genug angenommen hat: so giebt es jedesmal eine gewisse *beständige* Grösse und zwar nur *eine*, der sich die Glieder dieser Reihe immer mehr nähern und der sie so nahe kommen können, als man nur will, wenn man die Reihe weit genug fortsetzt.)

Bolzano's 'Größenlehre' (doctrine of sizes) is particularly fascinating. It is a kind of set theory the founding paradigm of which is the 'Inbegriff' (epitome), hence the 'whole' [Spalt 1990, p. 193] which already is somehow structured or ordered. Apart from structure and order 'sets' are left in the sense of Bolzano. This is in contrast with Cantor's set theory accepted today in which 'set' is the all-embracing abstraction. We are also unaccustomed with Bolzano's treatment of infinite sets. Bolzano used the term 'Weite' (width) to compare sets. There are sets (in the sense of Bolzano, mind you) with uncomparable widths, for example the set of all cubes and the set of all spheres. The set of all triangles and the set of all right-angled triangles have comparable widths [Spalt 1990, p. 197]. 'Width' therefore has to do with the specific set of objects. Then there is the case that the widths of infinite series stand in some ratio. Following Bolzano the sets $[0,5]$ and $[0,12]$ relate as $5:12$. We may therefore conclude that Bolzano's infinite sets behave in a much more hierarchical manner than Cantor's since following Cantor the sets $[0,5]$ and $[0,12]$ share the same cardinality. However, in the view of Bolzano there are 'more' numbers in $[0,12]$ than in $[0,5]$! In fact Bolzano's notion of 'Großheit' (greatness) of an infinite set allows for a clear gradation of infinities. As Spalt has pointed out in [Spalt 1990, p. 198f.] Bolzano always took for granted:

> *The whole is always (also for an infinite whole) larger than its proper part.*
>
> (Das Ganze ist stets (auch für unendliches Ganzes) größer als sein echter Teil.)

In the light of Cantor's set theory this seems to be wrong but we have to keep in mind that in the view of Bolzano only the whole has structure and order. Taking away a part from the whole then we can exactly say where the 'empty gaps' are where certain parts of the whole are missing. We can not dive deeper here and discuss Bolzano's theory of measurable numbers but the interested reader is refered to [Spalt 1990]. In my opinion Bolzano's 'Größenlehre' is still not enough noticed and worked upon.

After Bolzano's death his book *Paradoxien des Unendlichen* (Paradoxes of the infinite) [Bolzano 2006] appeared in print in 1851, published by his friend Franz Přihonsky. It was written one year before Bolzano died and was completely ignored, until Hermann Hankel (1839–1873) pointed out that the book would (quoted from [Dieudonné 1985, p. 387])

> *contain splendid remarks on the notion of the infinite.*
>
> (treffliche Bemerkungen über den Begriff des Unendlichen enthält.)

In 70 paragraphs Bolzano listed every paradox existing in dealing with the infinite and therebye proved being the thinker of his days who not only had dealt most intensly with problems of the infinite, but who also immersed most deeply in it. We again find here the example of the infinite sets [0, 5] and [0, 12]; Bolzano wrote [Bolzano 2006, p. 28] that one set would be considerably larger than the other, but he also showed that these two sets would correspond to each other, i.e. that they are of the same cardinality as we would say today. He accomplished that by means of the equation

$$5y = 12x$$

and argues that for x between 0 and 5 the quantity y would lie between 0 and 12 and that to each x there would exist exactly one y and vice versa. Here is already the soil of the notion of cardinality of Cantor's set theory. We cannot help but accept that Bolzano was ahead of his time again.

9.4 The Arithmetisation of Analysis: Cauchy

Augustin Louis Cauchy was born in 1798 in Paris, offspring of a family of a strict Catholic royalist. He died in 1857 in Sceaux. During the French Revolution Cauchy's father had lost his position at the police while shortly after Augustin Louis was born. In fear of persecution by the terror of the revolutionaries the family fled Paris and moved to their country house in Arcueil in the Département Val-de-Marne where they lived in bitter poverty [Belhoste 1991]. Father Cauchy, a very learned and well-read man, taught his son until the family could return to Paris. With the coup of Napoleon at the 9th November 1799 Cauchy's father became General Secretary of the senate and came in contact with two eminent mathematicians: Home Secretary was Pierre-Simon Laplace (1749–1827) and Joseph-Louis Lagrange was a senator. Both men predicted a brilliant career of Cauchy's Sohn. Lagrange is said to have given the father the following advice [Kowalewski 1938, p. 274]:

> *Don't let this child touch a mathematical book before his 17th year. If you do not hurry up to give him a thorough literary education his incline will sweep him away. He will become a great mathematician but will barely be able to write in his mother's tongue.*
>
> (Lassen Sie dieses Kind vor dem siebzehnten Lebensjahr kein mathematisches Buch anrühren. Wenn Sie sich nicht beeilen, ihm eine gründliche literarische Erziehung zu geben, so wird ihn seine Neigung fortreissen. Er wird ein grosser Mathematiker werden, aber kaum seine Muttersprache schreiben können.)

Thus the son got an excellent education in classical languages before he decided to become an engineer. We have to stop here for a moment and talk about the mathematical education of French engineers in the Napoleonic era. In comparison to the education of engineers in England which traditionally was more practically oriented than mathematically, mathematics was seen crucial to engineering in France. The École Polytechnique had been established in 1794 as École Centrale des Travaux Publics to care for trainees in the ever more important engineering sciences. Only one year later followed the renaming as École Polytechnique. In 1805 this École was subordinated to the ministry of war. Now army engineers were educated. The École Polytechnique served as a blueprint in Germany to establish technical universities in the 19th century. From the start one had to pass difficult admission tests for which mathematics belonged to the necessary knowledge. Hence Cauchy took lessons in mathematics in 1804 and passed second best in the admission test in 1805. One had to decide quickly on ones specialisation since the civil service urgently needed engineers, and Cauchy chose road and bridge construction. He had superb teachers, among others Sylvestre François de Lacroix (1765–1843) who made a name with his excellent textbooks on higher mathematics [Domingues 2008], and André Marie Ampère (1775–1836) who

Fig. 9.4.1. Augustin Louis Cauchy

became immortal by his works on electricity. In these days one studied for two years and Cauchy finished as top student. He was now allowed to attend the École Nationale des Ponts et Chaussées, the Grande École of civil engineers, which he left in 1810 as an 'aspirant-ingénieur'. In contrast to many of his fellow students Cauchy was not in favour of revolutionary or even liberal thoughts; he entered the congregation of the Jesuits and stayed an outermost conservative Catholic.

Napoleon needed a new large harbour in the north if he wanted to be able to face the English. Hence a large-scale project, the Port Napoléon in Cherbourg, was launched. Cauchy was detailed to go there in 1810 but lost his love of practical engineering during his daily routine. He desired to commit himself to scientific work. In his leisure time he worked on mathematical problems; for example he improved the proof of Euler's polyhedron formula. He went on sick leave in September 1812 and travelled back to his family in Paris. Back in the French capital he found important theorems in group theory. Under no circumstances he wanted to go back to Cherbourg and hence he was glad when he got the opportunity to work in a canal project in Paris in 1813. He got married and applied to many vacant positions in Paris but nothing worked out and he went again on sick leave. The year 1814 marked the defeat of Napoleon. On 2nd April the senate declared Napoleon deposed and France had a king again. This was favourable for Cauchy since he had not to go back to the canal project and could spend more time on mathematics.

Fig. 9.4.2. École Polytechnique [Photo: Jastrow, 2004]

With the new king the conservative forces of the country gained momentum and also Cauchy profited since he became assistence professor at the École Polytechnique in November 1815. In the following year the king personally intervened in the Académie des Sciences and removed liberal members, among them Gaspard Monge (1746–1818) whose place was now taken by Cauchy.

His role as a member of the academy stayed ambivalent. He had to review submitted works of young mathematicians and his judgement was feared. However, he apparently was sloppy at times and mislayed or even lost important manuscripts; this happened to a large work of Niels Henrik Abel (1802–1829) who died from tuberculosis in the firm believe that his work were lost forever.

As a teacher Cauchy turned out to be a revolutionary. Since he considered analysis being indispensable to engineers he gave lectures on that topic. He developed a rigorous concept formation of the limit and attached much importance to utmost accuracy which discouraged his students. From this lectures a textbook evolved, the *Cours d'analyse* [Bradley/Sandifer 2009], being published in 1821 and quickly becoming a manifesto of the new rigour in analysis. In the preface Cauchy wrote [Bradley/Sandifer 2009, p. 1–2]:

> *As for the methods, I have sought to give them all the rigor which one demands from geometry, so that one need never rely on arguments drawn from the generality of algebra. Arguments of this kind, although they are commonly accepted, especially in the passage from convergent to divergent series, and from real quantities to imaginary expressions, may be considered, it seems to me, only as examples serving to introduce the truth some of the time, but which are not in harmony with the exactness so vaunted in the mathematical sciences. We must also observe that they tend to grant a limitless scope to algebraic formulas, whereas, in reality, most of this formulas are valid only under certain conditions or for certain values of the quantities involved. In determing these conditions and these values and in establishing precisely the meaning of the notation that I will be using, I will make all uncertainty disappear, so that the different formulas present nothing but relations among real quantities, relations which will always be easy to verify by substituting numbers for the quantities themselves. It is true that, in order to remain consistently faithful to these principles, I will have to accept several propositions which may appear to be a bit rigid at first. For example, I state in Chapter VI that a divergent series does not have a sum; [...]*

The last sentence may seem strange today but in Cauchy's days one still used divergent series to derive mathematical results from them. Lagrange in his *Théorie des fonctions analytiques* of 1797 had still not spent any attention to the convergence of infinite series and Cauchy of course very well knew this book. Cauchy's secret weapon to the foundation of analysis were *limits* as we shall yet see.

When in 1830 King Charles X was overthrown and replaced by a 'citizen-king' Cauchy fled Paris leaving his family behind and moved to Switzerland. He could only return to his professorship in Paris if he would have taken an oath of allegiance to the new regime and that was no option whatsoever for Cauchy. He lost all of his positions in Paris but was appointed a professorship of physics in Turin in 1831. In 1833 he went to Prague to work as a private tutor to Charles' X grandchild who was chosen to assert his claims to the French throne. Until the end of the education Cauchy had succeeded in creating a strong dislike of mathematics in this grandchild. Cauchy met with his family only infrequently during short visits in Paris but in 1834 he fetched them to join him. When the grandchild of Charles X became 18 years old the work as a private tutor was finished and Cauchy got the title of a Baron for services rendered. He and his family returned to Paris. He was still a member of the academy but without taking the oath of allegiance he could not apply for a professorship. Only in the Bureau des Longitudes, an astronomical institute founded in 1795 to solve the longitude problem, cp. [Sobel/Andrewes 1999], taking the oath was not seen seriously. Hence Cauchy became a professor there since one ignored the prohibition of recruitment for four years. The time that follows proved to be Cauchy's most productive time. Between 1839 and 1848 more than 300 mathematical works resulted which were published in the journal of the academy, the Comptes rendus. When Lacroix died in 1843 and hence a professorship became vacant at the Collège de France Cauchy applied. A politically opportune but incompetent colleague was preferred and now the government took action. Cauchy lost his position at the Bureau des

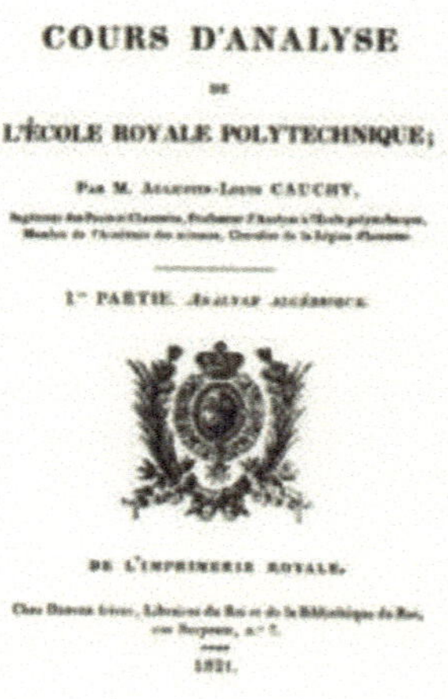
COURS D'ANALYSE

L'ÉCOLE ROYALE POLYTECHNIQUE;

PAR M. AUGUSTIN-LOUIS CAUCHY,

1821.

Fig. 9.4.3. Cauchy and his book 'Cours d'Analyse' of 1821

Longitudes and had to wait until the February Revolution of 1848 until his professional career improved. Although he was also not satisfied with the president and Emperor-to-be Napoleon III and refused to take an oath of allegiance an exception was made for him. In 1849, without taking an oath, he got a professorship. His application to a professorship at the Collège de France in 1850 failed; Joseph Liouville (1809–1882) was preferred which led to a fierce row between Cauchy and Liouville. In his last years Cauchy must have been really grumpy and argumentative but his influence on the further development of analyis was so important that we may stand back to criticise his unfortunate character traits.

9.4.1 Limit and Continuity

In the *Cours d'analyse* we find today's classical definitions of limits and continuity. Cauchy defined the limit (quoted from [Bottazzini 1986, p. 103], cp. also [Lützen 2003, p. 158]):

> *When the values successively attributed to the same variable indefinitely approach a fixed value in such a way as to end by differing from it as little as one wishes, this latter is called the* limit *of all the others.*

Also the notion of ìnfinitely small quantity' was defined by Cauchy (quoted from [Bottazzini 1986, p. 103]):

> *When the successive numerical values of the same variable decrease indefinitely in such a way as to fall below any given number, this variable becomes what one calls an* infinitesimal *or an* infinitely small *quantity. A variable of this kind has zero as a limit.*

Infinitely large quantities increasing unboundedly towards $\pm\infty$ were defined as well. It follows a definition of continuity of a function (quoted from [Bottazzini 1986, p. 104]):

> *Let $f(x)$ be a function of the variable x, and let us suppose that, for every value of x between two given limits, this function always has a unique and finite value. If, beginning from one value of x lying between these limits, we assign to the variable x an infinitely small increment α, the function itself increases by the difference*
>
> $$f(x+\alpha)-f(x),$$
>
> *which depends simultaneously on the new variable α and on the value of x. Given this, the function $f(x)$ will be a* continuous *function of this variable within the two limits assigned to the variable x if, for every value of x between these limits, the numerical value of the difference*

$$f(x+\alpha) - f(x)$$

decreases indefinitely with that of α.

Interestingly enough, Cauchy then formulated the concept of continuity by means of infinitesimal quantities (quoted from [Bottazzini 1986, p. 105]):

> *In other words, the function* $f(x)$ *will remain continuous with respect to* x *within the given limits if, within these limits, an infinitely small increase of the variable always produces an infinitely small increase of the function itself.*

The notion of function which Cauchy employed is different from the one used by Euler. Cauchy renounced completely that a function had to be given by 'one analytical term', cp. [Lützen 2003, p. 156]. Grabiner has presented an impressive study of the references and backgrounds of Cauchy's notions and ideas in [Grabiner 2005].

9.4.2 The Convergence of Sequences and Series

In the *Cours d'analyse* also the notion of convergence was rigorously defined. Cauchy was concerned with limits of certain function values for the cases $x \to \pm\infty$ and $x \to 0$. He wrote down the emerging 'undetermined forms' $0/0$, ∞/∞, $\infty - \infty$, $0 \cdot \infty$, 0^0, ∞^0, 1^∞. The following theorems have now become classics (quoted from [Bottazzini 1986, p. 109]):

> **Theorem I.** *If, for increasing values of* x*, the difference* $f(x+1) - f(x)$ *converges towards a certain limit* k*, the fraction* $f(x)/x$ *will converge towards the same limit at the same time.*
>
> **Theorem II.** *If, the function* $f(x)$ *being positive for very large values of* x*, the ratio* $f(x+1)/f(x)$ *converges towards the limit* k *as* x *increases indefinitely, the expression* $[f(x)]^{1/x}$ *will convergence towards the same limit at the same time.*

For the sake of explaining Theorem II consider the function $f(x) = \mathrm{e}^x$. Then $f(x+1) = \mathrm{e}^{x+1} = \mathrm{e} \cdot \mathrm{e}^x$ and therefore

$$\lim_{x\to\infty} \frac{f(x+1)}{f(x)} = \lim_{x\to\infty} \frac{\mathrm{e}\cdot\mathrm{e}^x}{\mathrm{e}^x} = \mathrm{e} =: k.$$

As claimed in the theorem it then also follows

$$\lim_{x\to\infty} (f(x))^{1/x} = \lim_{x\to\infty} (\mathrm{e}^x)^{1/x} = \mathrm{e} =: k.$$

Of course the theorems remain valid if f is a function of the natural numbers and hence they can be applied to real number sequences. This fact was used by Cauchy in the sixth chapter of the *Cours d'analyse* to define convergence of infinite series. As we do today he defined convergence of series as being the convergence of their partial sums. Then Cauchy proved the convergence criterion which we today call the 'Cauchy convergence criterion' (quoted after [Bottazzini 1986, p. 109]):)

> *In order for the series*
>
> $$u_0 + u_1 + u_2 + \ldots + u_n + \ldots$$
>
> *to be convergent, it is necessary and sufficient that increasing values of n make the sum*
>
> $$s_n = u_0 + u_1 + u_2 + \ldots + u_{n-1}$$
>
> *convergence indefinitely towards a fixed limit s; in other words, it is necessary and sufficient that, for infinitely large values of the number n, the sums*
>
> $$s_n, s_{n+1}, s_{n+2}, \ldots$$
>
> *differ from the limit s and consequently from each other, by infinitely small quantities.*

That s_n and s_{n+r} differ only by an infinitely small quantity if only n is large enough is the necessary condition for convergence and Cauchy could easly prove it. That this condition is also necessary depends on an important property of the real numbers, namely the 'completeness'. Cauchy was not aware of this property since a rigorous construction of $\mathbb{R}$ could be achieved only after his death. It seems that it was intuitively clear to Cauchy that a sequence had to convergence if two arbitrary numbers in a sequence s_n and s_{n+r} where n was 'far beyond' a certain large index N differed only arbitrarily little. However, as we will see later (cp. page 532), the property of completeness is something very special, since already $\mathbb{Q}$ does *not* share this property! Today we call a sequence (s_k) a 'Cauchy sequence', if for all $\varepsilon > 0$ there exists an index N, such that for all $n > N$ and all m it always holds

$$|s_{n+m} - s_n| < \varepsilon.$$

A number field is called complete if every Cauchy sequence converges.

Because of its great importance we have to consider here the notion of 'uniform convergence' going back to Christoph Gudermann (1798–1852). Beginners in mathematics have no problems with the notion of pointwise convergence: the values of a sequence of functions (f_n) converge at a point x_0 for $n \to \infty$ to a limit $\lim_{n\to\infty} f_n(x_0) = f(x_0)$. In contrast, most beginners have massive problems with uniform convergence. Let us look at an easy exampls. On $[0, 1]$ we consider the sequence of functions

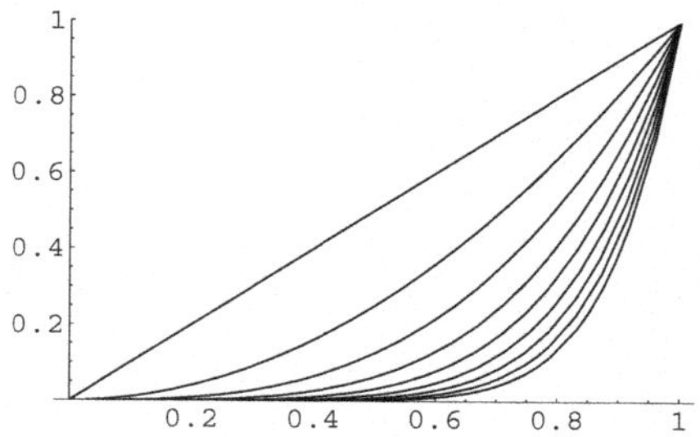

Fig. 9.4.4. Graphs of the functions $f_n(x) = x^n$ for $n = 1, \dots, 9$

$$f_n(x) := x^n.$$

The pointwise limit function f is easy to construct. For all $x \in [0, 1[$ the functions x^n will converge towards 0, but in case of $x = 1$ the limit value will be 1, hence

$$f(x) = \begin{cases} 0\ ;\ 0 \le x < 1 \\ 1\ ;\quad x = 1 \end{cases}$$

is the pointwise limit function. This easy example shows something uncomfortable: all functions f_n, and really *all*, are continuous functions, but the pointwise limit function is *not* continuous! A theorem like: 'The limit function of continuous functions is a continuous function' can simply not be expected! But if all functions f_n would stay within a hosepipe of width 2ε around the limit function f for all n beyond an N (N will depend on ε, of course), then we would say that the sequence of functions (f_n) converges uniformly. In that case we could prove the theorem: 'The limit function of a sequence of continuous functions converging uniformly is continuous.' Cauchy had proven the so-called 'summary theorem' (Summensatz): 'the limit function of a convergent sequence of continuous functions is continuous', which, as we saw, is wrong in general. Historians of mathematics have argued that this was a howler and that Cauchy did not know it better. However, Detlef Spalt has protested against such judgement in some of his works [Spalt 1996], [Spalt 2002]. Spalt has argued that we must not interpret Cauchy from our modern point of view but that we have to read his works with the eyes of his time. If we would do so we would recognise that Cauchy indeed used uniform convergence in his proof. We can not decide this controvery here but it shows that it is always valuable to put old attitudes on the test bench.

A detailed history of the notion of uniform convergence was published recently by Klaus Viertel in [Viertel 2014].

The notion of uniform convergence was not the only notion of 'uniformness' being discussed in the second half of the 19th century. After 1880 there were further modes of convergence as, for example, 'quasi-uniform convergence', which we do not want to discuss here, cp. [Grattan-Guinness 1980, p. 137].

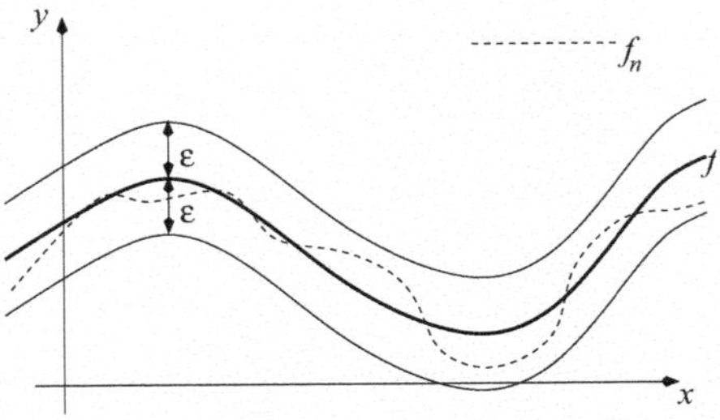

Fig. 9.4.5. Concerning the notion of uniform convergence

9.4.3 Derivative and Integral

In the *Leçons sur le calcul differentiel* of 1829 Cauchy gave the following definition (quoted from [Lützen 2003, p. 159]):

> *When the function $y = f(x)$ is continuous between two given limits of the variable x, and one assigns a value between these limits to the variable, an infinitely small increment of the value produces an infinitely small increment in the function itself. Consequently, if we then set $\Delta x = i$, the two terms of the* difference quotient
>
> $$\frac{\Delta y}{\Delta y} = \frac{f(x+i) - f(x)}{i}$$
>
> *will be infinitesimals. But while these terms tend to zero simultaneously, the ratio itself may converge to another limit, either positive or negative. This limit, when it exists, has a definite value for each particular value of x; but it varies with x. [...] The form of the new function which serves as the limit of the ratio $\frac{f(x+i)-f(x)}{i}$ will depend upon the form of the given function $y = f(x)$. In order to indicate the dependence, we give to the new function the name derivative and we designate it, using a prime, by the notation y' or $f'(x)$.*

Also in integration Cauchy took a decisive step by means of a definition of the 'Cauchy integral'. For Leibniz an integral still was the infinite sum of infinitesimal strips. By the fundamental theorem of calculus integration seemd to be the just the inversion of differentiation and hence integrals could be read off from tables of derivatives. Hence the Bernoullis and also Euler interpreted the integral as the inversion of differentiation resulting in the 'indefinite' integral. By inserting numbers for the left and right boundary points of an interval of integration then leads to a 'definite' integral. In contrast, Cauchy developed a notion of integral for the definite integral completely independent of the inverse roles of differentiation and integration. He divided an interval $[a, b]$ into n subintervals by introducing the points

$a = x_0 < x_1 < x_2 < \ldots < x_{n-1} < x_n = b$ as shown in figure 9.4.6 in the case of $n = 4$.

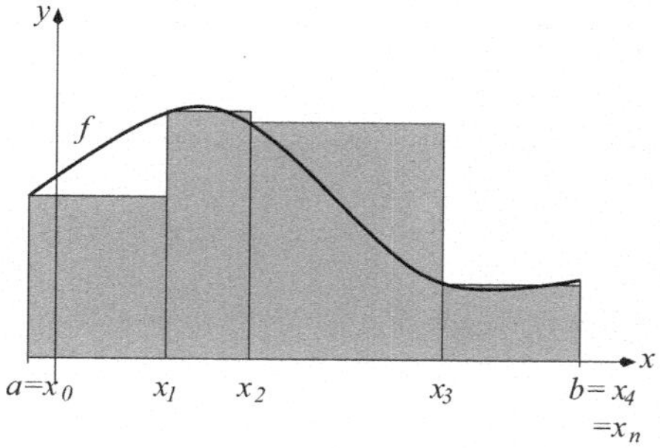

Fig. 9.4.6. Illustrating the Cauchy integral

Then Cauchy considered the sum

$$S = (x_1 - x_0) \cdot f(x_0) + (x_2 - x_1) \cdot f(x_1) + \ldots + (x_n - x_{n-1}) \cdot f(x_{n-1}).$$

This sum depends on the number n and on the kind of subdivision, of course. The dependence on the kind of subdivision, the 'mode of division', decreases with increasing n, so Cauchy said [Lützen 2003, p. 160]. If n increases without bound then the subintervals become infinitely small and hence a limit for S will result which we call 'definite' integral,

$$\int_a^b f(x)\,dx.$$

This Cauchy integral was still not sufficient to be able to answer the problems of the 19th century and it was replaced still in the 19th century by the Riemann integral. But also the latter integral was not well suited to solve subtle problems in the theory of integration. At the beginning of the 20th century the Lebesgue integral, named after the French mathematician Henri Léon Lebesgue (1875–1941), finally became *the* notion of integral used to this day [Knobloch 1983].

9.5 The Development of the Notion of Integral

We have already reported on the Cauchy integral and on the problems occuring in the treatment of general trigonometric series. With the trigonometric series came discontinuous functions into the focus of mathematicians and in the 19th century certain 'monsters' appeared; functions being continuous everywhere but nowhere differentiable, or with other disconcerting properties. The question now was: Which are the functions being integrable and what is the largest class of such functions [Wußing 2009, p. 263ff.]?

If we consider functions with jump discontinuities in a finite number of points then the Cauchy integral would not possess any problems at all. One simply uses the subdivison of the interval of integration which is defined by the jump points and then sums the integrals on the subintervals. But what happens if the number of discontinuities becomes infinite? We can not go too deep into the details here because it quickly gets very technical [Wußing 2009, p. 407–409 and p. 480]. A decisive step forward was achieved by the German mathematician Peter Gustav Lejeune Dirichlet (1805–1859) who introduced a new notion of continuous functions in 1837 (quoted after [Volkert 1988, p. 198]):

> *Imagine a and b being two fixed points and x a variable quantity taking by and by all values between a and b. If to each x there corresponds exactly one finite y in the way that while x traverses the interval from a to b continuously, $y = f(x)$ also changes gradually, then y is called a continuous or continual function of x on this interval. It is therebye not necessary that y depends on x by one law in the whole interval; one would not even need to think of a dependence expressable by mathematical operations.*
>
> (Man denke sich unter a und b zwei feste Werte, und unter x eine veränderliche Größe, welche nach und nach alle zwischen a und b liegenden Werte annehmen soll. Entspricht nun jedem x ein einziges, endliches y, und zwar so, daß während x das Intervall von a bis b stetig durchläuft, $y = f(x)$ sich ebenfalls allmählich verändert, so heißt y eine stetige oder kontinuierliche Funktion von x für dieses Intervall. Es ist dabei gar nicht nötig, daß y in diesem ganzen Intervalle nach demselben Gesetz von x abhängig sei, ja man braucht nicht einmal an eine durch mathematische Operationen ausdrückbare Abhängigkeit zu denken.)

Already in 1837 Dirichlet gave a 'modern' definition of the notion of function in a general sense: If a variable y depends on a variable x so that if x is assigned a value then, following a rule, it follows a unique value for y, then y is called a function of x. Only for the sake of completeness we cite here the nowadays accepted definition given by Bourbaki [Bourbaki 1968, p. 352]:

> *Let E and F be two sets, which may or may not be distinct. A relation between a variable element x of E and a variable element y of F is called a* functional relation in y *if,* for all $x \in E$, there exists a unique $y \in F$ which is in the given relation with x.
>
> *We give the name of* function *to the operation which in this way associates with every element $x \in E$ the element $y \in F$ which is in*

the given relation with x; y is said to be the value *of the function at the element x, and the function is said to be* determined *by the given functional relation. Two* equivalent *functional relations determine the* same *function. Such a function is said to 'take its values in F' and to be 'defined on E'. More briefly, it is also said to be a* mapping of *E* into *F*.

Bernhard Riemann (1826–1866) was student and collaborator of Dirichlet at the University of Göttingen and came in contact with the problems of trigonometric series via his teacher. In his habilitation thesis *Über die Darstellbarkeit einer Funktion durch eine trigonometrische Reihe* (On the representability of a function by a trigonometric series) of 1854 in the Abhandlungen der Königlichen Gesellschaft der Wissenschaften in Göttingen (Treatises of the Royal Society of Sciences in Göttingen), published only in

Fig. 9.5.1. Bernhard Riemann

1868, Riemann started his investigations with the question of what functions were integrable at all. To this end he introduced 'Riemann sums' as follows. The interval of integration $[a, b]$ is subdivided into n subintervalls $[x_i, x_{i+1}]$, $i = 0, 1, 2, \ldots, n-1$ where $a = x_0 < x_1 < x_2 < \ldots < x_n = b$. In every subinterval a number $0 \leq \varepsilon_i \leq 1$ is chosen and then the Riemann sum is considered,

$$S = \delta_1 \cdot f(x_0 + \varepsilon_1 \delta_1) + \delta_2 \cdot f(x_1 + \varepsilon_2 \delta_2) + \ldots + \delta_1 \cdot f(x_{n-1} + \varepsilon_n \delta_n),$$

where

$$\delta_i := x_i - x_{i-1}$$

denotes the width of the subinterval under consideration. If the Riemann sum converges to a fixed limit for arbitrary subdivisions and every choice of the ε_i if all δ_i being made infinitely small (these are Riemann's words! Cp. [Dieudonné 1985, p. 381]), then this limit is called the Riemann integral and is denoted by

$$\int_a^b f(x)\,dx.$$

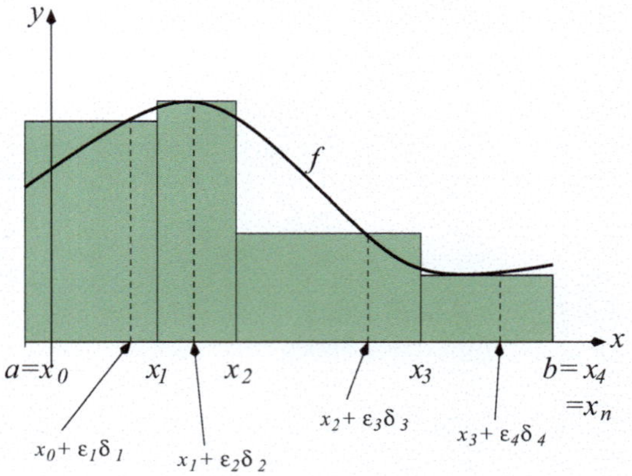

Fig. 9.5.2. The definition of the Riemann integral

Concerning the question of the convergence of the Riemann sums he noted that this would be the case if [Hawkins 1979, p. 17]:

(R_1):

$$\lim_{\|Z\| \to 0} (D_1 \delta_1 + D_2 \delta_2 + \ldots + D_n \delta_n) = 0.$$

Here Z denotes a subdivision of the set of all possible subdivisions in subintervals, $\|Z\|$ is the fineness of this subdivision (being the largest of the lengths of the subintervals), D_k is the 'oscillation' or 'variation' of the function f on the k-th subintervall, hence the differenc of the largest and smallest value of the function, and δ_k denotes again the width of the k-th subinterval. Riemann recognised a condition equivalent to (R_1) [Hawkins 1979, p. 18]:

> (R_2):
>
> *Corresponding to every pair of positive numbers ε and σ there exists a positive d such that if P is any partition of norm less than or equal to d, then $s(P, \sigma)$ is less than ε.*

Let us 'translate' this first. Obviously P is our Z; 'partition' is just another word for our subdivision. The 'norm' of P is nothing but our $\|Z\|$. But what is the strange $s(P, \sigma) = s(Z, \sigma)$? Riemann also considered functions f for which the oscillation in the k-th subinterval (width δ_k) exceeds a given threshold σ. Then $s(P, \sigma) = s(Z, \sigma)$ is the sum of all those δ_k for which the oscillation of f is larger than σ. If this s, the set on which the function is unpleasant, can be made arbitrarily small, then the function f is integrable in the sense of Riemann.

Riemann's (R_2) points already to a modern development within analysis which we can not discuss here: the development of measure theory, today underlying modern theories of integration and also stochastics.

Note that (R_2) is completely independent of all claims concerning the continuity of the function f whatsoever. Therefore Riemann pointed out rightly that even functions with an infinite number of points of discontinuities could be integrable and he gave an example. For every real number x let $m(x)$ be the integer closest to it, i.e. for which $|x - m(x)|$ is smallest. Define further

$$((x)) := \begin{cases} x - m(x) & ; x \neq \frac{n}{2}, n \text{ odd} \\ 0 & ; x = \frac{n}{2}, n \text{ even} \end{cases} .$$

The function $y = g(x) = ((x))$ is then discontinuous in all points $x = n/2$ with odd n.

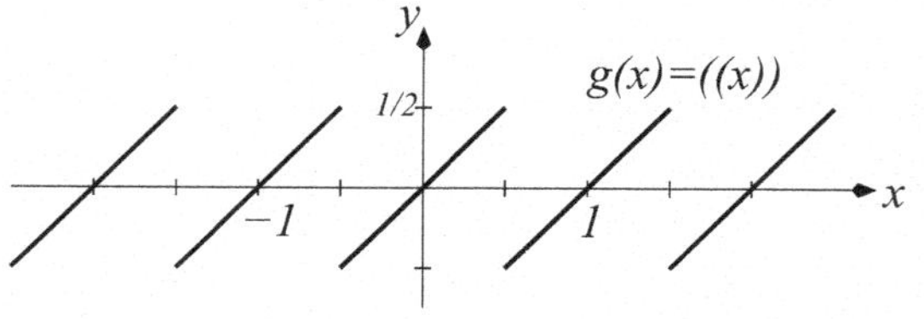

Fig. 9.5.3. The function $g(x) = ((x))$

From this function Riemann built a new one by defining

$$f(x) := ((x)) + \frac{((2x))}{2^2} + \frac{((3x))}{3^2} + \ldots + \frac{((nx))}{n^2} + \ldots .$$

This function is discontinuous in every rational point x, hence on a dense subset of the real numbers. It is not possible to plot this function but you can find the graph of a partial sum in [Bressoud 2008, p. 34]. Despite these infinitely many points of discontinuity this function is still integrable in the sense of Riemann.

A work by the Oxford professor of mathematics Henry John Stephen Smith (1826–1883), published in 1875 in the Proceedings of the London Mathematical Society [Glaisher 1965, Vol.II, p. 86–100], unfortunately was overlooked on the Continent.

Smith is actually known as a number theorist but in his work concerning integration he substantially contributed to an understanding of integration and he anticipated developments on the Continent. By means of his construction of certain not dense sets on the real axis and his investigations concerning the integration of functions showing discontinuities just on these sets he has already pointed towards modern theory of measure.

However, there were problems left with the Riemann integral, since the Dirichlet monster (8.10),

$$D(x) := \begin{cases} c \,;\, x \in \mathbb{Q} \\ d \,;\, x \in \mathbb{R}\backslash\mathbb{Q} \end{cases}$$

Fig. 9.5.4. Henri L. Lebesgue and Henry J. S. Smith [Photographers unknown]. Lebesque developed the most far reaching notion of integral which today carries his name. Smith was one of the pioneers of modern theories of integration and measure.

Fig. 9.5.5. Two early theorists of measure theory: Camille Jordan [Photographer unknown, before 1923] and Giuseppe Peano [Photographer unknown, about 1910]

is not integrable in the sense of Riemann! Additionally, the limit of a sequence of Riemann integrable functions is not necessarily Riemann integrable, which is very unpleasant. Only with the advent of measure theory at the beginning 20th century French mathematician Henri Léon Lebesgue (1875–1941) succeeded to construct an integral applicable to even a larger class of functions. His central idea in his work *Sur une généralisation de l'intégrale définie*, published in the Comptes Rendus in 1901, was not to subdivide the abscissa into subintervals, but the ordinate.

If the ordinate is subdivided between $y = y_0$ and $y = Y$ by $y = y_0 < l_1 < l_2 < \ldots < l_n = Y$ one can consider the sets

$$S_i := \{x \mid l_i \leq f(x) < l_{i+1}\}$$

and the integral can be enclosed as

$$\sum_{i=0}^{n-1} l_i \cdot |S_i| \leq \int_a^b f(x)\,dx \leq \sum_{i=0}^{n-1} l_{i+1} \cdot |S_i|,$$

where $|S_i|$ denotes the 'size' of the set S_i; the length of an interval in the most simple case. The condition on the sets S_i is 'measurability'; they have to be 'measurable' and Lebesgue in fact developed the corresponding measure theory simultaneously. Non-measurable sets are not easily found or constructed but meanwhile some of these strange sets are known and studied. Since constructions of such sets seem to be somwhat artificial (the axiom of choice is always involved) the set of all Lebesgue integrable functions is very large [Wußing 2009, p. 408f.].

The Italian Giuseppe Peano (1858–1932) and the French Camille Jordan (1838–1922) have significantly contributed to measure theory. Jordan in

Fig. 9.5.6. Emile Borel and Felix Hausdorff – two further founders of modern measure theory

particular was enormously influential by means of his textbook *Cours d'analyse* in three volumes which quickly became a standard in France and contributed to the spread of the analysis of Weierstraß.

The goal in measure theory is 'only' to associate a measured value (length, area, volume) to given sets in a way that this measured value leads to reasonable results when applied to wellknown sets like length of an interval or volume of a cube. The constructions of Peano and Jordan were *not yet* measures in nowadays sense, but so-called 'contents'. The general measure problem in the real numbers can be described as follows.

Find a function μ associating to each subset of $\mathbb{R}$ a uniquely defined number of the interval $[0, \infty]$. This number is called a 'measure' of the subset if μ satifies the four conditions

1. for all subsets A of the real numbers it holds $\mu(A) \geq 0$,
2. if A and B are two congruent sets, then $\mu(A) = \mu(B)$,
3. the measure of $[0, 1]$ has to be $\mu([0, 1]) = 1$,
4. for the union of countably (infinitely) many disjoint sets A_i it holds:

$$\mu\left(\bigcup_{i=1}^{\infty} A_i\right) = \sum_{i=1}^{\infty} \mu(A_i). \tag{9.1}$$

We actually have written down only self-evident facts: we do not want to have a negative number as the measured value, equal sets should have the same measure, a line segment of length 1 (inch, meter,...) should have the measured value of 1, and the measure of a union of disjoint sets should be the measure of the sum of the single measures. It therefore came as a shock when the

Italian Giuseppe Vitali (1875–1932) proved that the general measure problem is unsolvable! In principle there are two ways out. One could replace (9.1), the so-called σ-additivity, by an additivity of only a *finite* number of disjoint sets. But then μ is no longer called a measure but a 'content'. This was the way Felix Hausdorff (1868–1942) went. He could prove that such contents do exist in dimensions 1 and 2 (i.e. in case of $\mathbb{R}$ and $\mathbb{R}^2$), but already in three-space (and, of course, in higher dimensional spaces) such contents do no longer exist! The other way out, and the one which turned out to be successfull, is accepting the condition of σ-additivity but restricting the sets to be measured.

Hence not *all* possible subsets of the real numbers are admitted but only 'measurable' sets are permitted. In this context the 'Borel sets', named after the French mathematician Émile Borel (1871–1956), play an important role, but we can not discuss the deatils here, cp. [Wußing 2009, p. 480].

The Lebesgue integral is the one used today in analysis but the theory of integration is still an active area of analysis! The Czech mathematician Jaroslav Kurzweil (born 1926) in 1957 and the Englishman Ralph Henstock (1923–2007) about the same time invented an integral named 'Kurzweil-Henstock integral' which is closer to the Riemann integral but as powerful as the Lebesgue integral and is sometimes preferred to the latter [Bartle 1996].

9.6 The Final Arithmetisation of Analysis: Weierstraß

The final transformation of analysis into a rigorous mathematical theory took place in the 19th century by Karl Weierstraß (1815–1897). Weierstraß was born into a strict Catholic family in Ostenfelde in the Münsterland region where his father was the secretary to the mayor. When the father rose, becoming a tax inspector, the family together with eight years old Karl moved back and forth across Prussia, until the father got a permanent position in Paderborn in 1827. In that year Karl's mother died. Karl attended the Gymnasium (grammar school) Theodorianum in Paderborn and had to do work in accounting because his family was short of money. In his spare time he read Crelle's Journal of pure and applied mathematics which was the most famous mathematical journal in these days. Following his father's wish Karl studied law and finances in Bonn between 1834 and 1838; eventually the son was assumed to start a career as a Prussian civil servant. In these days as a student Karl was a member of the Corps Saxonia, one of the popular student's fraternities. He is said to have been not only one of the funniest member, but also a permanent sabre fighter on the 'Paukboden', the room in a fraternity reserved for strictly regulated fights with sabres [Klein 1926, p. 277]. Felix Klein wrote:

> *How this is compatible with the rest of his development is, as I already said, outright incomprehensible.*

> (Wie sich das mit seiner übrigen Entwicklung verträgt, ist mir, wie gesagt, gänzlich unverständlich.)

Incidentally Weierstraß read mathematical works of Laplace, Abel, and Jacobi and finally left the University of Bonn without graduation. Although his father must have been very angered and disappointed he eventually agreed to allow the study of mathematics and physics at the Academy of Münster, precursor of the modern University of Münster, in 1838.

After having successfully passed the exams Weierstraß entered the teaching profession at a Gymnasium in Münster, but in his leisure time he continued his research and developed a theory of complex functions which he did not publish. As a teacher he moved to Deutsch-Krone in West Prussia at Easter 1843 and in 1848 to Braunsberg in East Prussia (today's Braniewo in Poland). Besides teaching mathematics and physics he also had to give lectures in botany and even sports lessons for which he received an additional training 1844 in Berlin. Some of his groundbreaking research was published in a school magazine as was usual for school teachers of this time, but in 1854 *Zur Theorie der Abelschen Funktionen* (Concerning the theory of Abel's functions) was published in Crelle's Journal. Two years later a more comprehensive work concerning this theory was published also in

Fig. 9.6.1. Sofia (Sophie) Kowalewskaja and Karl Weierstraß

Crelle's. Both works made an immediate impact in the mathematical work so that the University of Königsberg granted an honorary doctorate. Leading mathematicians from Berlin, among them Dirichlet and Kummer, tried to get him to the Berline University. In 1856 he in fact moved to Berlin to teach at the Königliches Gewerbeinstitut (Royal Commercial Institute); already in the same year he became a professor at the university. Quickly a 'school' developed around Weierstraß and the 'Weierstraßean rigour' became a dictum within mathematics.

Weierstraß, the lifelong bachelor, stood up for Sofia Wassiljewna Kowalewskaja (1850–1891) who came from Moscow. Women were not allowed to study at Berlin University (and not only there!) so that Weierstraß became her private tutor from 1870 on. Due to his influence she was eventually allowed to do her doctorate at the University of Göttingen which was more forward-looking than other universities. Finally she became a private lecturer in Stockholm in 1884.

Already in his days as a school teacher Weierstraß's health was fragile. At the end of 1861 he suffered a complete breakdown but recovered. He was

Fig. 9.6.2. Karl Weierstraß in his younger years

adored by his pupils and friends so much so that they made him a moving gift on the occasion of his seventieth birthday; a photo album including all their portraits [Bölling 1994]. At his 80th birthday he was already bound to a wheelchair and he died in Berlin in 1897. When the Berlin Wall was built his gravestone on the old St Hedwig cemetry had to be transferred and the actual grave became a part in the midst of the death strip.

In mathematics Weierstraß did groundbreaking work on numerous areas, in complex analysis, the analysis of functions of one complex variable, in the calculus of variations [Thiele 2007], in differential geometry, and in the theory of elliptic functions; all of them important areas of analysis which we can not discuss in detail here. In 'classical' analysis he achieved the breakthrough resulting in a rigorous theory. Since Weierstraß the infinitely small quantities had to leave classical analysis!

9.6.1 The Real Numbers

The basis of an arithmetrisation of analysis is the (subtle) knowledge of the number field of the real numbers $\mathbb{R}$. Weierstraß had not to wait for Dedekind or Cantor but developed an own concept of irrational numbers (and therefore of $\mathbb{R}$). This theory emerged about 1863 but was published only in 1872 by Ernst Kossak in his book *Die Elemente der Arithmetik* (The elements of arithmetic). There are reasons to suspect that Weierstraß was in the possession of this theory already in 1861 [Dieudonné 1985, p. 397]. The construction of the real numbers by Weierstraß seem somehow clumsy and laborious in our modern view and we do not want to follow this trail; it has been described in [Dieudonné 1985, S. 389–391]. It is only important to us here that Weierstraß in fact could built upon a rigorous concept of real numbers. Only this made it possible to replace the notion of infintely small quantities [Dieudonné 1985, p. 397]:

> *An infinitely small quantity is a function φ of a variable h in such a way, that to a given ε one can always find a δ with the property that for all values of h, the absolute value of which is smaller than δ, $\varphi(h)$ is smaller than ε.*
>
> (Eine unendlich kleine Größe ist eine Funktion φ der Variablen h derart, daß man zu gegebenem ε immer ein δ mit der Eigenschaft finden kann, daß für alle Werte von h, deren absoluter Betrag kleiner als δ ist, $\varphi(h)$ kleiner als ε ist.)

9.6.2 Continuity, Differentiability, and Convergence

Weierstraß explained in his lectures that continuity of a function f follows from the differentiability of f, but *not* vice versa. He constructed a 'monster'; a function which is continuous everywhere but nowhere differentiable:

$$f(x) = \sum_{k=0}^{\infty} b^k \cos(a^k \cdot \pi \cdot x)$$

under the condition $a \cdot b > 1 + \frac{3}{2}\pi$. Figure 9.6.3 shows the partial sums

$$f_n(x) = \sum_{k=0}^{n} 4^k \cos(4^k \cdot \pi \cdot x)$$

for four different values of n on the interval $[-0.1, 0.1]$.

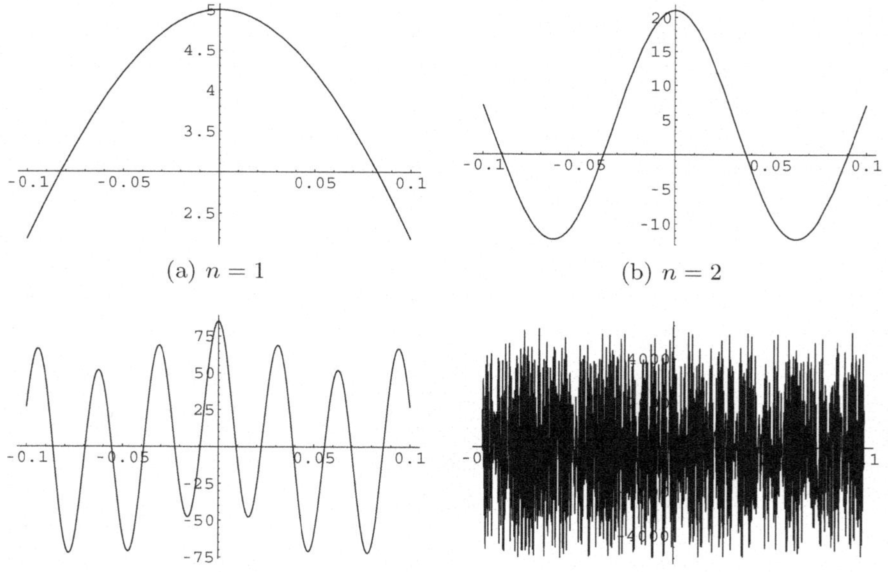

(a) $n = 1$ (b) $n = 2$ (c) $n = 3$ (d) $n = 6$

Fig. 9.6.3. Partial sums of Weierstraß's monster

An important tool of analysis, almost a powerhouse, is the theorem of Bolzano-Weierstraß:

> *Every bounded sequence of real numbers contains a convergent subsequence*

To distinguish the uniquely determined limit of a sequence of real numbers from the limits of subsequences one calls the latter limits 'accumulation points' of the sequence. The sequence

$$s_n = (-1)^n$$

is certainly bounded but not convergent; it hence has no limit. However, it contains the two convergent subsequences

$$1, 1, 1, 1, \ldots$$

and

$$-1, -1, -1, -1, \ldots.$$

The sequence therefore has two accumulation points, namely +1 and −1. The first formulation of the notion of accumulation points occured, typical for Weierstraß , in a lecture transcript by Moritz Pasch (1843–1930), [Dieudonné 1985, p. 397].

Weierstraß was the very first to distinguish pointwise convergence from uniform convergence. After the explicit construction of the real numbers by Richard Dedekind and Georg Cantor Weierstraß was able to define the notion of convergence once and for all:

> A sequence (s_n) of real numbers has a limit s, if for all $\varepsilon > 0$ there exists an index N, such that for all indices $n > N$
>
> $$|s_n - s| < \varepsilon$$
>
> holds.

The limit of functions was also precisely formulated by Weierstraß :

$$\lim_{x \to x_0} f(x) = y_0$$

> means that
>
> for all $\varepsilon > 0$ exists a $\delta > 0$, so that
>
> $$|f(x) - y_0| < \varepsilon,$$
>
> if only
>
> $$|x - x_0| < \delta$$
>
> holds.

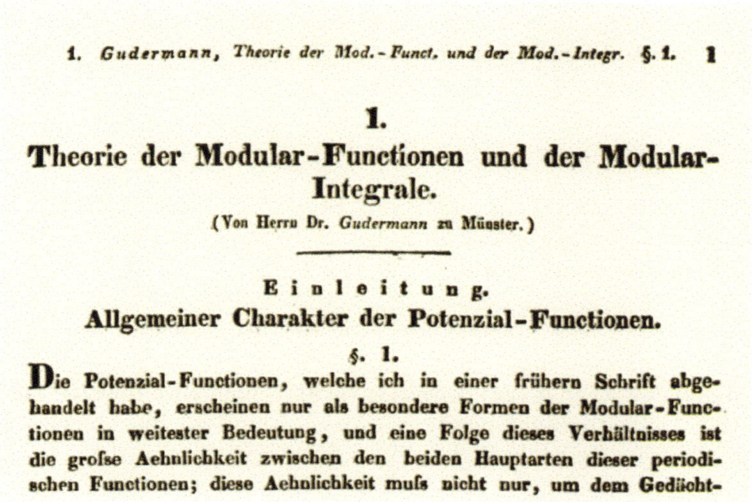

1. *Gudermann, Theorie der Mod.-Funct. und der Mod.-Integr.* §. 1. 1

1.

Theorie der Modular-Functionen und der Modular-Integrale.

(Von Herrn Dr. *Gudermann* zu Münster.)

Einleitung.

Allgemeiner Charakter der Potenzial-Functionen.

§. 1.

Die Potenzial-Functionen, welche ich in einer frühern Schrift abgehandelt habe, erscheinen nur als besondere Formen der Modular-Functionen in weitester Bedeutung, und eine Folge dieses Verhältnisses ist die große Aehnlichkeit zwischen den beiden Hauptarten dieser periodischen Functionen; diese Aehnlichkeit muſs nicht nur, um dem Gedächt-

Fig. 9.6.4. Starting page of a work by Christoph Gudermann in Crelle's Journal 1838 in which the concept of uniform convergence appeared in print for the first time

9.6.3 Uniformity

The concept of uniform convergence came from Christoph Gudermann (1798–1852), the teacher of Weierstraß [Dieudonné 1985, p. 376]. Weierstraß attended a lecture on elliptic functions by Gudermann in Münster 1839/40. This was certainly the worldwide first lecture of this kind and Weierstraß was strongly influenced by it concerning his own research. Gudermann published a work in 1838 in Crelle's Journal in which the concept of uniform convergence appeared in print for the first time. Already about 1874 Philipp Ludwig Ritter von Seidel (1821–1896) in Germany, well-known for the Gauß-Seidel method to iteratively solve linear systems, and George Gabriel Stokes (1819–1903) in England developed similar concepts which did not prevail because they were ignored.

In order to appreciate the importance of uniform convergence we have to return to the example of the sequence of functions

$$f_n(x) = x^n$$

on $[0, 1]$. We have already looked at this example in section 9.4.2 as a paradigm of pointwise convergence towards the limit function

$$f(x) = \begin{cases} 0\,;\, 0 \leq x < 1 \\ 1\,;\, x = 1 \end{cases} .$$

Now we want to exclude the point $x = 1$ which apparently causes some concern and look at the convergence on the half-open interval $[0, 1[$. The difference between pointwise and uniform convergence is as follows:

I. The sequence $(f_n(x))$ converges *pointwise* on $[0,1[$ towards a function f, if for every $\varepsilon > 0$ and for every point $x \in [0,1[$ there exists an index $N = N(\varepsilon)$ so that for all indices $n \geq N(\varepsilon)$ it holds

$$|f_n(x) - f(x)| \leq \varepsilon.$$

II. The sequence $(f_n(x))$ converges *uniformly* on $[0,1[$ towards a function f, if for every $\varepsilon > 0$ there exists an index $N = N(\varepsilon)$, so that for all indices $n \geq N(\varepsilon)$ and **for every point** $x \in [0,1[$ it holds

$$|f_n(x) - f(x)| \leq \varepsilon.$$

There are no doubts that (f_n) converges on $[0,1[$ pointwise towards $f(x) = 0$. But is this f also the limit function in case of uniform convergence? If this would be the case we could find an $N(\varepsilon)$ for every $\varepsilon > 0$ so that

$$|f_n(x) - f(x)| = |x^n - 0| = |x^n| = x^n \overset{!}{\leq} \varepsilon$$

holds for all $n > N(\varepsilon)$ and for all points $x \in [0,1[$. For $x = 0$ it holds $f_n(0) = 0 \leq \varepsilon$ for all n. For $0 < x < 1$ it follows from the inequality above by taking logarithms (the logarithm is monotonically increasing and therefore does not change the inequality) that

$$n \cdot \log x \leq \log \varepsilon.$$

If we divide by $\log x$ we have to observe that the logarithm is negative in $0 < x < 1$! Hence it follows

$$n \geq \frac{\log \varepsilon}{\log x}.$$

If we had uniform convergence than we must be able to find *one* index N which fits *all* x, but that is not possible! Let us see why. Given an $\varepsilon > 0$ (as small as you like) we can increase n arbitrarily in moving x arbitrarily close to $x = 1$! An N for all x can thus not be found and the convergence is therefore *not* uniform on $[0,1[$ although the point $x = 1$ was excluded.

Stokes talked in such a case of 'infinitely slow' convergence, since $N(\varepsilon)$ increases in case $x \to 1$ without bounds.

Weierstraß deserves credit to have fully recognised the importance of uniform convergence and to have introduced this notion into analysis. Weierstraß talked about 'convergence of the same degree' (Konvergenz in gleichem Grade). Weierstraß's theorem

> *If an infinite sequence of continuous functions on $[a,b]$ converges uniformly it can be integrated termwise to get the integral of the sum of the series.*

was published by Heinrich Eduard Heine (1821–1888) who was befriended with Weierstraß . In the year 1872 Heine introduced the notion of uniform continuity in his book *Die Elemente der Functionenlehre* (The elements of the theory of functions). He proved that on every 'compact' interval (i.e. bounded and closed intervals) $[a, b]$ continuous functions are uniformly continuous. A detailed history of the notion of uniformity was given by Viertel in [Viertel 2014].

9.7 Richard Dedekind and his Companions

Richard Dedekind (1831–1916) was born as youngest of four children in Brunswick. His father was a professor of law at the Collegium Carolinum. The Collegium Carolinum was an institute of higher education which was a necessary institution to prepare the pupils for universities. Under the chairmanship of Richard Dedekind the Collegium Carolinum will be transformed into today's Technical University of Brunswick.

The Dedekind family lived with their four children Julie, Mathilde, Adolf, and Richard in an official residence in the Collegium Carolinum. Mother Dedekind came from an influential Brunswick family; in the courtyard worked the sculptor Howaldt and the painter Heinrich Brandes, and hence the children were exposed to scientific and aesthetic influenced from the beginning. Richard Dedekind had the perfect pitch and loved music. As a gifted cellist and pianist he was making a name for himself, but music eventually lost the fight with mathematics as far as his profession was concerned. Dedekind's best friend in his youth was Hans Zincke, called 'Sommer', who also loved mathematics as well as music. In contrast to Dedekind he finally turned to music. On 2nd May 1848 Dedekind enrolled at the Collegium Carolinum to prepare for studying at the University of Göttingen where he started his studies in 1850. The former university of the Dukedom of Brunswick was in Helmstedt. It was founded in 1576 but was closed at the end of the winter semester 1809/10 under Napoleonic reign.

In Göttingen the great mathematician Carl Friedrich Gauß (1777–1855) had celebrated his 50th doctor jubilee in 1849.

It were times of revolution in Germany. Dedekind's older brother Adolf who studied law in Göttingen wrote in 1849 to his father [Dedekind 2000, p. 74]:

> *Gauß's 50th doctor jubilee went without all pomp and the students, who could not reach an agreement, were forced by the Prorector to refrain from a torchlight procession dedicated to him. By the way, it is said that he received many distinctions, for example the medal 'Heinrich des Löwen' as the honorary citizenship of Brunswick.*

Fig. 9.7.1. Richard Dedekind [unknown photographer, about 1870]

(Das Gauß'sche 50jährige Doctor-Jubiläum ist ohne allen Prunk vorbei gegangen und die Studenten, die sich nicht einigen konnten, wurden sogar vom Prorector gezwungen, einen ihm zugedachten Fackelzug zu unterlassen. Übrigens soll er viele Decorationen, z. B. den Orden "Heinrich des Löwen", sowie das Braunschweiger Ehren= Bürgerrecht erhalten.)

Richard Dedekind attended lectures by Moritz Abraham Stern, Wilhelm Weber, Johann Benedict Listing (a pioneer on the area of topology), Quintus Icilius, and Carl Friedrich Gauß! After a study of four semester Dedekind submitted his dissertation on Euler's integrals and at the age of 21 he was a 'doctoral son' of the great Gauß; in fact he was the last one. On the 30th July 1854 he habilitated successfully and became a privat lecturer. This meant that he had to make a living out of the money his students had to pay. In his first lecture there were only two students, one of them the friend Zincke who now also studied in Göttingen. Of the year 1856 we know about another lecture with only two students (one of them again Zincke). In the winter of 1856

Fig. 9.7.2. Richard Dedekind [unknown photographer, probably about 1855] and Carl Friedrich Gauß, his doctoral advisor [Detail of a painting by Gottlieb Biermann 1887, Photo: A. Wittmann]

Dedekind gave a groundbreaking lecture on modern algebra, doubtless the first of its kind but unrecognised by the rest of the world. It was meanwhile edited and published [Scharlau 1981].

In 1855 Gauß had died. It was the successor, Peter Gustav Lejeune Dirichlet (1805–1859), who became formative for Dedekind.

In contrast to Gauß, who was reclused and withdrawn, Dirichlet was an open conversation partner who always took much interest in the works of his students. Dedekind wrote [Dedekind 2000, p. 81]: *'with whom I actually start to learn'* (bei dem ich eigentlich erst recht zu lernen anfange). Dirichlet's wife was Rebecca Mendelssohn-Bartholdy, a sister of Felix Mendelssohn-Bartholdy, and quickly Dedekind's love of music ensured him being a pleasant guest for domestic music. Illustrious visitors as Brahms or Liszt also participated in the domestic music when they visited Rebecca. The Berlin mathematician Carl Gustav Jakob Jacobi (1804–1851) wrote in a letter to Alexander von Humboldt [Dieudonné 1985, p. 389]:

> *If Gauß says he had proven something I think it is very probable so, if Cauchy says it, it is as much pro as con, if Dirichlet says it, it is certain.*
>
> (Wenn Gauß sagt, er habe etwas bewiesen, ist es mir sehr wahrscheinlich, wenn Cauchy es sagt, ist eben so viel pro als contra zu wetten, wenn Dirichlet es sagt, ist es gewiß.)

Fig. 9.7.3. Peter Gustav Lejeune Dirichlet and Georg Friedrich Bernhard Riemann – two formative companions of Richard Dedekind

Another encouter became formative for Dedekind, namely the encounter with Georg Friedrich Bernhard Riemann (1826–1866).

Bernhard Riemann, born in Breselenz close to Dannenberg as son of a Lutheran pastor. He had four siblings and the children were nurtured in cramped conditions. In grammar school (Gymnasium) in Lüneburg a teacher lent him the book *Théorie des Nombres* by Adrien-Marie Legendre[1] (1752–1833); a book of about 860 pages. Riemann not only read the book within one week, but he also had understood its content. It was his father's wish that he should study theology, but love of mathematics won the day and so he changed his field of study at the University of Göttingen. In the year 1851 he finished his dissertation thesis in which he revolutionised complex analysis. In his habilitation lecture he created a theory of manifolds which has pointed the way to later generations. At times he worked as an assistent of the physicist Wilhelm Eduard Weber (1804–1891). He (almost successfully) solved an important problem of fluid mechanics: the problem we today call 'Riemann's shock tube problem'. In one sentence: Riemann was a true genius.

The two young men Dedekind and Riemann felt attracted to each other. On 3rd November 1856 Dedekind wrote to his sister Julie [Dedekind 2000, p. 18]:

[1] As Peter Duren in *Changing Faces: The Mistaken Portrait of Legendre* (Notices of the AMS, Vol.56, No.11, 2009) has proven, the well-known portrait of Adrien-Marie Legendre, reproduced in almost every book on the history of mathematics, does *not* depict our Legendre, but a politician by name Louis Legendre, who has no connection to Adrien-Marie Legendre whatsoever.

> *Furthermore I socialize very much with my excellent colleague Riemann who is, without doubt, second to or even on par with Dirichlet, the most profound mathematician and will soon be recognised as one, if his modesty allows him to publish certain things which will temporarily be intelligible by only a few.*
>
> (Außerdem verkehre ich sehr viel mit meinem vortrefflichen Kollegen Riemann, der ohne Zweifel nach oder gar mit Dirichlet der tiefsinnigste Mathematiker ist und bald als solcher anerkannt sein wird, wenn seine Bescheidenheit ihm erlaubt, gewisse Dinge zu veröffentlichen, die allerdings vorläufig nur Wenigen verständlich sein werden.)

There can be no doubts that both men mutually felt their mathematical genius. Dedekind was the calm, deep thinker, wanting to understand all inderdepencies precisely, while Riemann was rather an impetuous genius keeping in view always the final mathematical result. There can also be no doubts that a root of the 'abstract' viewpoints of mathematics, so important today, can be found in the works of Riemann and Dedekind [Ferreirós 1999, p. 31]. But Riemann did not feel well. He was overworked and suffered a breakdown in 1857 and his friend Dedekind sent him to his family to recuperate in their summer resort in Bad Harzburg in the Harz Mountains.

In January 1858 a professorship of mathematics became vacant at the Polytechnic School in Zurich which was the precursor of the modern ETH Zurich. The vacancy was tendered in all of Europe. Almost 50 applications came in, among them the ones of Dedekind and Riemann. Asked for an expert opinion Dirichlet recommended both men highly, but placed Riemann first ('den ersten Rang'). Thereupon the school governer ('Schulratspräsident') Karl Kappeler travelled to Göttingen himself to have a closer look at both men. With much insight into human nature he saw that Riemann was too much introverted to teach prospective engineers ('zu stark in sich gekehrt, um zukünftige Ingenieure zu lehren') [Sonar 2007, p. 20]. At this time the Zurich polytechnic was a genuine engineering school and the introverted, shy, and sensitive Riemann would not have been able to stand such a practical test. Hence Dedekind went to Zurich in 1858 as a professor and he started to teach future engineers. Concerning the situation he wrote to his sister Mathilde on 27th January 1859 [Dedekind 2000, p. 331]:

> *I can also not say that I am completely happy with my schoolmastery; I do not want to talk about the older students handed down to me, they are mostly spoiled; of my new pupils one third is excellent, another third is moderately good, the remaining part is weak, partly miserable. My ideas of freedom, free development of the students, have been radically destroyed; as Austria in Italy also I was too lenient for a long time; the students do not understand to appreciate it, they are children like our pupils at lower grammar schools, at least in their manners. Now I do not bother any longer to give a culprit a roasting*

in front of the others so that he collapses and learns respect. That has a very beneficial effect. But it is always annoying and makes me sick.

(Ich kann auch nicht sagen, dass ich so ganz und gar glücklich mit meiner Schulmeisterei bin; von den mir überlieferten älteren Schülern will ich gar nicht sprechen, die sind zum grossen Theil verdorben; von meinen neuen ist ein Drittel ganz vorzüglich, ein anderes Drittel mässig gut, der Rest schwach, zum Theil erbärmlich. Meine Ideen von Freiheit, freier Entwicklung der Schüler sind radical vernichtet; so wie Österreich in Italien, so bin ich auch eine Zeit lang zu milde gewesen; die Schüler verstehen das nicht zu würdigen, es sind Kinder wie unsere Progymnasiasten, wenigstens in ihrem Benehmen. Jetzt genire ich mich gar nicht mehr, einen Übelthäter vor versammelter Menge so niederzudonnern, dass er zusammensinkt und Respect kriegt. Das hat eine sehr heilsame Wirkung. Aber ärgerlich ist es immer und mir zuwider.)

It happened in Zurich in the year 1858 in the lectures to engineering students that Dedekind felt painfully the missing foundations of the real number system. In the preface of his book *Stetigkeit und Irrationale Zahlen* (Continuity and irrational numbers), published in Brunswick in 1872, he wrote [Dedekind 1963, p. 1f.]:

My attention was first directed toward the considerations which form the subject of this pamphlet in the autumn of 1858. As professor in the Polytechnic School in Zürich [sic!] *I found myself for the first time obliged to lecture upon the elements of the differential calculus and felt more keenly than ever before the lack of a really scientific foundation for arithmetic. In discussing the notion of the approach of a variable magnitude to a fixed limiting value, and especially in proving the theorem that every magnitude which grows continually, but not beyond all limits, must certainly approach a limiting value, I had recourse to geometric evidences. Even now such resort to geometric intuition in a first presentation of the differential calculus, I regard as exceedingly useful, from the didactic standpoint, and indeed indispensable, if one does not wish to lose too much time. But that this form of introduction into the differential calculus can make no claim to being scientific, no one will deny. For myself this feeling of dissatisfaction was so overpowering that I made the fixed resolve to keep meditating on the question till I should find a purely arithmetic and perfectly rigorous foundation for the principles of infinitesimal analysis.*

(Die Betrachtungen, welche den Gegenstand dieser kleinen Schrift bilden, stammen aus dem Herbst des Jahres 1858. Ich befand mich damals als Professor am eidgenössischen Polytechnikum zu Zürich zum ersten Male in der Lage, die Elemente der Differentialrechnung vortragen zu müssen, und fühlte dabei empfindlicher als jemals

Fig. 9.7.4. Main building of the 'Herzogliche Technische Hochschule Carolo-Wilhelmina' (now the Technical University of Brunswick)

> früher den Mangel einer wirklich wissenschaftlichen Begründung der Arithmetik. Bei dem Begriffe der Annäherung einer veränderlichen Größe an einen festen Grenzwert und namentlich bei dem Beweise des Satzes, daß jede Größe, welche beständig, aber nicht über alle Grenzen wächst, sich gewiß einem Grenzwert nähern muß, nahm ich meine Zuflucht zu geometrischen Evidenzen. Auch jetzt halte ich ein solches Heranziehen geometrischer Anschauung bei dem ersten Unterrichte in der Differentialrechnung vom didaktischen Standpunkte aus für außerordentlich nützlich, ja unentbehrlich, wenn man nicht gar zu viel Zeit verlieren will. Aber daß diese Art der Einführung in die Differentialrechnung keinen Anspruch auf Wissenschaftlichkeit machen kann, wird wohl niemand leugnen. Für mich war damals dies Gefühl der Unbefriedigung ein so überwältigendes, daß ich den festen Entschluß faßte, so lange nachzudenken, bis ich eine rein arithmetische und völlig strenge Begründung der Prinzipien der Infinitesimalanalysis gefunden haben würde.)[2]

Dedekind found an elegant solution. The real numbers are defined by 'cuts' in the rational numbers. The similarity of Dedekind's construction with the one by Eudoxos in the fifth book of the *Elements* of Euclid is startling.

In 1859 Dirichlet suddenly died. Dedekind will edit and publish the *Vorlesungen über Zahlentheorie* (Lectures on number theory) of his teacher Dirichlet in 1879. He will add to it his famous 'supplements' in which Dedekind published his theory of ideals, cp. [Löwe 2007]). In Brunswick the professor of mathematics at the Collegium Carolinum, August Wilhelm Julius Uhde, had died in 1861 and thus a position in Dedekind's home town became vacant. He applied for the position, but he is so frustrated of teaching engineers

[2] [Dedekind 1965, p. 3 f.].

that his application contained a condition: He never again wanted to lecture on 'inferior mathematics'! After some period of reflection the government of Brunswick accepted the condition in 1862 and Dedekind became professor in Brunswick. Already in 1863 he got the first offer of a professorship in Hanover which he declined. Further offers followed, among them two from Göttingen, but Dedekind remained faithful to Brunswick.

On a vacation trip in 1872 Dedekind accidentally met the mathematician Georg Cantor (1845–1918) from Halle (Saale). Both men went well together and it began an enduring friendship in which they will work out the first basics of set theory together. In Dedekind's book *Was sind und was sollen die Zahlen* (The nature and meaning of numbers) [Dedekind 1963] of 1888 we find the first rigorous definition of an infinite set [Dedekind 1963, p. 63]:

> *64. Definition. A system S is said to be* infinite *when it is similar to a proper part of itself.*
>
> (64. Erklärung. Ein System S heißt *unendlich*, wenn es einem echten Teile seiner selbst ähnlich ist [...]; im entgegengesetzten Falle heißt S ein *endliches* System.)[3].

Euclid's dictum, as insightful as it may seem, that a part is always smaller than the whole, is wrong in the infinite! We will have to report on this in more detail in connection with Cantor.

Meanwhile in Brunswick the Collegium Carolinum had become too small; the technical sciences played an ever increasing role. In the middle of the city a direct expansion of the Collegium was simply not possible and hence one decided building a new construction outside the city walls. Richard Dedekind led the building committee and as a working mathematician this did not fit his bill. The architect of the new building was Konstantin Uhde (1836–1905), son of the deceased professor of mathematics Uhde. Starting from 1872 Dedekind acted also as the first director of the 'Herzoglichen Technischen Hochschule Carolo-Wilhelmina' (Ducal Technical University Carolo-Wilhelmina), as the new institution of higher education was now called, and he stayed director until 1875. The new university was solemny opened on 16th October 1877.

The Dedekind family always stood steadily at the side of the Brunswick House of Welfs. When Duke Wilhelm died without heirs in 1884 the Dedekinds supported the Hanover House of Welfs what was in perfect agreement with the law. However, most of the Brunswick population had reservations against the Hanoverians and fancied a Prussian ruler so that the Hanoverians could not succeed. When membership in one of the two Welf parties became prohibited in 1897 the situation escalated. Richard Dedekind and his brother Adolf stood openly against such an act of high treason and therefore came under pressure in Brunswick. When Richard Dedekind in 1914 additionally refused to sign

[3] [Dedekind 1965, p. 13]

Fig. 9.7.5. Georg Cantor [unknown photographer, about 1894] and Richard Dedekind (Painting at the TU Brunswick, unknown painter, about 1927; right: [Photo: H. Wesemüller-Kock])

the Manifesto of the Ninety-Three in which all liabilities for the wartime atrocities of German troops in Belgium were put on the enemies he became an outsider. When he died in high age on 12th February 1916 in Brunswick the first appreciation was published in March 1816 by the Paris Academy of Sciences.

9.7.1 The Dedekind Cuts

What is at issue in Dedekind's *Stetigkeit und irrationale Zahlen*? It seems natural to most men that the natural numbers $\mathbb{N} = \{1, 2, 3, \dots\}$ are simply 'there'. Addition in $\mathbb{N}$ can be performed without problems but not subtraction, since $3 - 5$ is not a number in $\mathbb{N}$. This leads to an extension of the natural numbers in form of the integers $\mathbb{Z} = \{\dots, -3, -2, -1, 0, 1, 2, 3, \dots\}$ in which addition and subtraction are unconditionally possible. We can even multiplicate but not divide, since $2 : 5$ is not a number in $\mathbb{Z}$. This in turn leads to a further extension and one introduces the rational numbers $\mathbb{Q} = \{p/q \mid p \in \mathbb{Z}, q \in \mathbb{N}$ and p, q relatively prime$\}$. Now all four elementary operations may be performed excluding division by 0. But we are still not satisfied since we know from antiquity that there are numbers like $\sqrt{2}$, cp p. 27, not being elements of $\mathbb{Q}$. Well into the 19th century the existence of real numbers $\mathbb{R}$, consisting of $\mathbb{Q}$ *and* the irrational numbers as $\sqrt{2}, \pi$, and so on, was naively assumed and accepted. But now we have to face a serious problem! There is no other operation left which will work in $\mathbb{R}$ but not in $\mathbb{Q}$; how then do we get from $\mathbb{Q}$ to $\mathbb{R}$? The set $\mathbb{Q}$ together with its algebraic operations is already a 'field' and we can not do better!

The weakness of the set of fractions $\mathbb{Q}$ can be found in its 'holes'. Consider Hero's method to take roots. This method was known to the Babylonians already but was named after Hero of Alexandria, called 'Mechanicus' [Clagett 2001]. The length of the sides of a square is sought having area A. To start with two numbers a_0 and b_0 are chosen such that $a_0 \cdot b_0 = A$ holds. In other words, one starts with a rectangle where the area is already as it should be. In the next step one computes the side lengths of a new rectangle by means of the arithmetic mean

$$a_1 := \frac{a_0 + b_0}{2},$$

sine this lies between a_0 und b_0. The new rectangle side b_1 has to be computed from the claim $a_1 \cdot b_1 = A$, hence

$$b_1 = \frac{A}{a_1}.$$

This way one proceeds. Generally speaking we have constructed an iterative algorithm: Given numbers a_0 und b_0 in $\mathbb{Q}$ compute

$$\begin{aligned} \text{für } i = 1, 2, 3, \ldots : a_i &= \frac{a_{i-1} + b_{i-1}}{2} \\ b_i &= \frac{A}{a_i}. \end{aligned}$$

The output of this algorithm are two sequences (a_i) and (b_i) of numbers again in $\mathbb{Q}$ which hopefully converge both to the value $\sqrt{A}$. As an example consider $A = 2$ and $a_0 = 1$, $b_0 = 2$. The the algorithm gives

$$\begin{aligned} a_1 &= \frac{3}{2} = 1.5 \\ a_2 &= \frac{17}{12} = 1.41667 \\ a_3 &= \frac{577}{408} = 1.41422 \\ a_4 &= \frac{665857}{470832} = 1.41421 \\ a_5 &= \frac{886731088897}{627013566048} = 1.41421 \\ \vdots &\ldots \vdots \end{aligned}$$

and

$$
\begin{aligned}
b_1 &= \frac{4}{3} = 1.33333 \\
b_2 &= \frac{24}{17} = 1.41176 \\
b_3 &= \frac{816}{577} = 1.41421 \\
b_4 &= \frac{941664}{665857} = 1.41421 \\
b_5 &= \frac{1254027132096}{886731088897} = 1.41421 \\
&\vdots \ldots \vdots
\end{aligned}
$$

It is easy to prove that the sequence (a_i) is monotonically decreasing and bounded from below; the sequence (b_i) is monotonically increasing and bounded from above. Hence both sequences converge and the common limit is $\sqrt{2}$; not being a number in $\mathbb{Q}$!

There are several ways to rigorously define irrational numbers as we know today, cp. [Knoche/Wippermann 1986]. Dedekind decided to follow a way strikingly similar to the one taken by Eudoxos in his doctrine of proportions. The rational numbers are divided into sets U and O defining a 'Dedekind cut' as follows:

Definition (Dedekind Cut): The pair (U, O) where $U, O \subset \mathbb{Q}$ is called a Dedekind cut in $\mathbb{Q}$, if

1. $U, O \neq \emptyset, \quad U \cap O = \emptyset, \quad U \cup O = \mathbb{Q}$
2. for all numbers $u \in U$ and $o \in O$ it holds $u < o$.

In a Dedekind cut U is called lower set and O upper set. A cut is called 'gap' if U has no maximum and O no minimum.

Apparently the number $\sqrt{2}$ is a gap in a Dedekind cut. It is now easy to see that one can either work exclusively with lower sets or exclusively with upper sets; the respective other set can be seen as set theoretic complement. By means of the definition of Dedekind cuts all real numbers are identified with open intervals which are bounded from below (if working with O) or from above (if working with U).

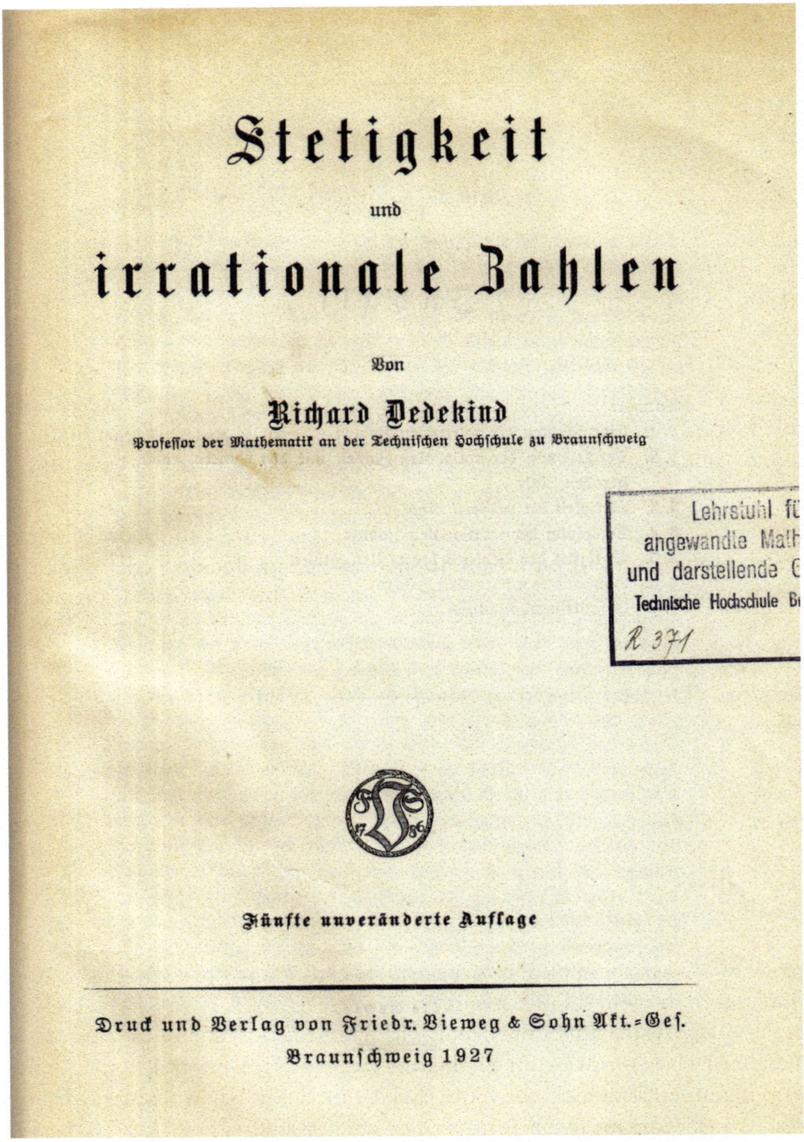

Stetigkeit

und

irrationale Zahlen

Von

Richard Dedekind

Professor der Mathematik an der Technischen Hochschule zu Braunschweig

Fünfte unveränderte Auflage

Druck und Verlag von Friedr. Vieweg & Sohn Akt.-Ges.
Braunschweig 1927

Fig. 9.7.6. *Stetigkeit und irrationale Zahlen* by Richard Dedekind. Title page of the book which saw many unrevised editions (first edt. 1872)

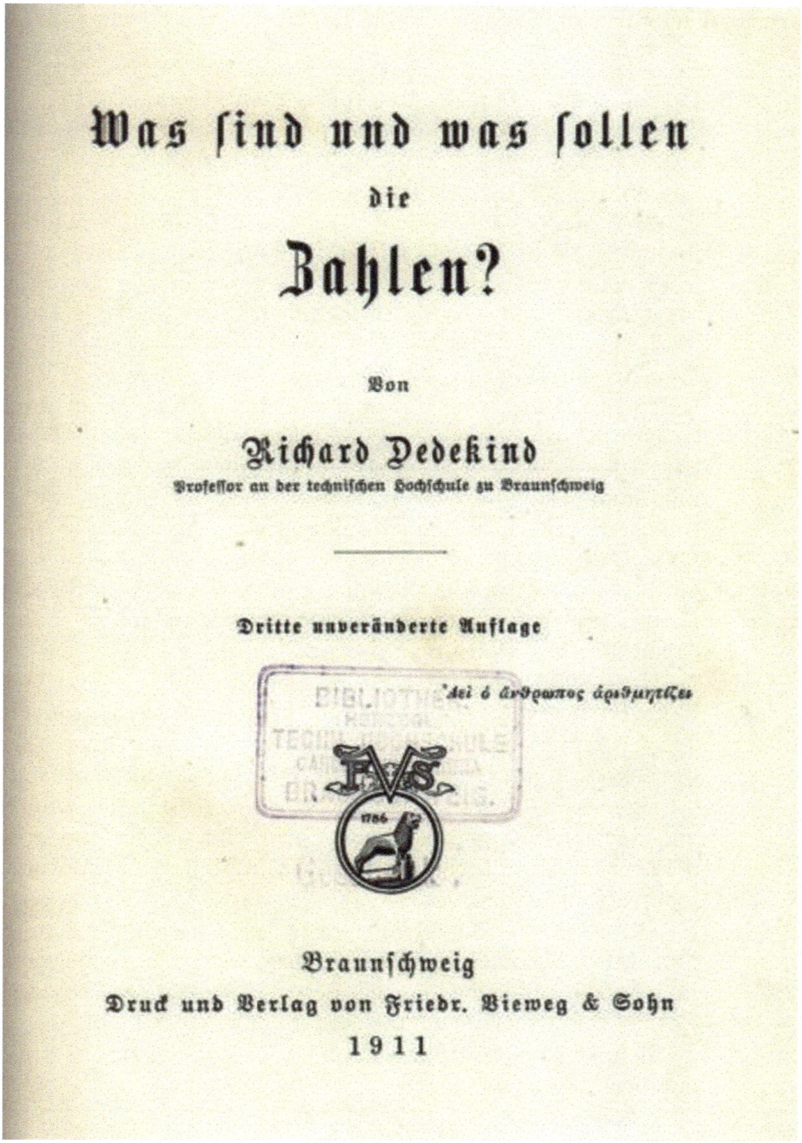

Was sind und was sollen
die
Zahlen?

Von

Richard Dedekind
Professor an der technischen Hochschule zu Braunschweig

Dritte unveränderte Auflage

Ἀεὶ ὁ ἄνθρωπος ἀριθμητίζει

Braunschweig
Druck und Verlag von Friedr. Vieweg & Sohn
1911

Fig. 9.7.7. *Was sind und was sollen die Zahlen?* by Richard Dedekind. Title page of the famous book which saw many unrevised editions

Substantial Results in Analysis 1800-1872

1816–1848	B. BOLZANO establishes rigorously many theorems of analysis
1821	CAUCHY's *Cours d'analyse* is published. In it the nowadays classical definitions of limit, continuity, and differentiability can be found
1823	CAUCHY defines in *Résumé des Leçons* the definite integral as limit of a sum, as well as improper integrals, curve and area integrals
1828	G. GREEN publishes his integral theorem and develops the method of Green's function to solve boundary value problems
1829	CAUCHY publishes his *Leçons sur le calcul differentiel*
1825–1854	G.P.L. DIRICHLET provides substantial contributions concerning the convergence of trigonometric series (Dirichlet kernel), publishes a new notion of functions, formulates the minimum principle which now carries his name, and the so-called Dirichlet-problem of potential theory
1838	CHR. GUDERMANN publishes his concept of uniform convergence
1839–1848	CAUCHY's most productive phase; more than 300 works result
about 1840	GAUSS proves his integral theorem of vector analysis, the so-called divergence theorem
1854	B. RIEMANN develops his notion of integrals
	G.G. STOKES proves his integral theorem of vector analysis
1856	The grammar school teacher KARL WEIERSTRASS gets a professorship in Berlin. He puts the whole area of analysis on rigorous grounds. He propagates an arithmetic foundation and develops a rigorous concept of real numbers and a strict definitition of limit, continuity, differentiability, and convergence by means of 'epsilontics'
1858	R. DEDEKIND becomes professor at the Zurich Polytechnic
1862	DEDEKIND becomes professor of mathematics in Brunswick
1872	DEDEKIND publishes in his book *Stetigkeit und irrationale Zahlen* the structure of the real number system by means of his definition of irrational numbers as cuts of sets of rational numbers

10 At the Turn to the 20th Century: Set Theory and the Search for the True Continuum

T. Sonar, *3000 Years of Analysis*, https://doi.org/10.1007/978-3-030-58223-4_10

General History 1871 to 1945

1871	Formation of the German Empire. Wilhelm I von Hohenzollern is proclaimed German Emperor in Versailles
1878	Congress of Berlin
1888–1918	Wilhelm II German Emperor and King of Prussia
1890	Bismarck's dismissal
1905	Revolution in Russia. End of the Russo-Japanese War
1914	World War I begins with the assassination of the Austrian successor to the throne in Sarajevo on 28th June
1917	October Revolution in Russia. Ceasefire with the Central Powers on 15th December
1918	Treaty of Brest-Litovsk on 3rd March: Collapse of the Tsarist empire, founding of the Russian Soviet Federative Socialist Republic. Kiel mutiny and November Revolution lead to the abdication of Emperor Wilhelm II and proclamation of the Republic by Scheidemann on 9th November. Dissolution of the Habsburg Monarchy and proclamation of the Republic German-Austria Ceasefire on 11th November. End of World War I
1919	Treaty of Versailles signed on 28th June The League of Nations is established. The Weimar Republic started with the Weimar constitution Treaty of Saint-Germain-en-Laye on 10th September
1919–1923	Inflation
1920	Treaty of Sèvres on 10th August: Dissolution of the Osmanian Empire
1922	Mussolini takes power in Italy
1923	Attempted coup of Hitler in Munich
1924	Death of Lenin
1926	Germany received into the League of Nations
1929	Start of the Great Depression
1932	Hindenburg wins the Reichstag election over Hitler
1933	On 30th January Hitler appointed Reich Chancellor. After the Reichstag fire during the night of 27th February the enabling act follows. Parties and unions are dissolved. The boycott of Jewish shops begins. Germany leaves the League of Nations.
1935	Nuremberg racial laws
1936–1939	Spanish civil war
1938	Crystal Night
1939	World War II begins on 1st September with the German invasion of Poland
1941–1945	Systematic mass extermination of Jews
1945	In May the German Wehrmacht surrenders unconditionally. World War II ends with the surrender of Japan on 2nd September

Technology and Natural sciences between 1871 and 1945

1875	International Metre Convention
1876	N. Otto: first four-stroke engine patented
1877	Th.A. Edison: phonograph
1882	R. Koch discovers the tubercle bacillus
1886	H. Hertz generates electromagnetic radio waves, patent vehicle by C.F. Benz
1887	Motor vehicle by G. Daimler
1889	World Exhibition in Paris; Eiffel tower by G. Eiffel
1890	First electric underground in London
1893	Invention of the Diesel engine by R. Diesel. Panama canal being build
1895	W.C. Röntgen discovers the X-rays
1896	A.H. Becquerel discovers radioactivity
1897	First International Congress of Mathematicians (ICM) in Zurich
1900	Count Zeppelin's first test flight with a rigid airship. M. Planck establishes quantum mechanics
	On the ICM in Paris Hilbert spoke about 23 problems
1901	Marconi crosses the Atlantic by means of electromagnetic waves
	First Nobel Prizes: physics (Röntgen), chemistry (van't Hoff), medicine (v. Behring)
1902	Carnegie Institution and Rockefeller Foundation established
1904	Amundsen determines the exact position of the magnetic north pole
1905	Einstein: special relativity. Explanation of the Brownian motion. Formula $e = m \cdot c^2$ discovered, photons discovered
1911	'Kaiser-Wilhelm-Gesellschaft' established
1915	Einstein: general relativity
1919	Eddington: expeditions confirms the light deflection by the sun and hence confirms relativity theory
1920	Public broadcasting in the USA
1925	Fischer-Tropsch process to synthesise liquid hydrocarbons
1927	Heisenberg: uncertainty principle
1929	Hubble discovers the expansion of the cosmos and explains the redshift
1929	N. Bohr takes the view of a dualism of wave and corpuscle
	Einstein outlines a new field theory
1931	Flight across the arctic with the airship 'Graf Zeppelin'
1932	Heisenberg: theory of the atomic nucleus
1934	Irène and Frédéric Joliot-Curie: artificial radioactivity
1938	Hahn/Strassmann: nuclear fission of uranium
1936	Public television broadcast the Olympic Games in Berlin
1939	L. Meitner/Frisch/Bohr: energetic view on the nuclear fission
	A group of French mathematician join under the name Bourbaki
1941	K. Zuse: first programmable electromechanic computer
1944/45	The construction of mainframe computers starts in the USA: ENIAC, EDVAC

10.1 From the Establishment of the German Empire to the Global Catastrophes

With the establishment of the German Empire in 1871 a German national state emerged for the first time. The Empire was significantly shaped by its chancellor Otto von Bismarck and now the political weights in Europe began to shift. When Wilhelm II, grandchild of the founding Emperor Wilhelm I, ascended to the throne in 1888 it quickly came to confrontations with Bismarck who had to leave the political stage already in 1890. In contrast to Bismarck Wilhelm II. is uncontrolled, short-tempered, and wanted to expand his Empire under all circumstances.

In these days the European world was in a fever of colonialism. Great Britain under her Queen Victoria (1819–1901) had risen to a mighty colonial power.

Fig. 10.1.1. Proclamation of the Prussian King Wilhelm as German Emperor in Versailles [Painting by Anton von Werner, about 1880, detail]. 'Dropping the Pilot', Abdication Bismarck's in a caricature of the English journal Punch in March 1890 (drawn by John Tenniel). Bismarck leaves the political stage while Emperor Wilhelm II remains satisfied

Already in 1876 Victoria (1819–1901) became Empress of India, Egypt was occupied in 1882, Kenya was incorporated 1887, Rhodesia (Zimbabwe) in 1889, British West Africa in 1891, in 1894 Uganda followed, then Sudan in 1899, and the Boer Republics in 1902. France had also early started into the era of colonialism. From 1830 to 1885 Algeria became French, Tunisia in 1881, one year later followed the lower Congo, Madagascar in 1885, Cambodia in 1887, the Ivory Coast in 1893; in 1895 French West Africa was formed, French Somaliland in 1896, French Equatorial Africa in 1919, and in 1912 Morocco came under protectorate. Although Portugal had lost it's colony Brazil in 1822 her possessions in Asia (Goa, Diu, and Damao in India, Macao in China) could be kept until after WW II; Angola, Mozambique und Portuguese Guinea in Africa could be added. Spain also had lost her colonies in South America already in the first half of the 19th century and lost Mexico, Cuba, and the Philippines in the second half; only her possessions in Africa (Spanish Morocco, Western Sahara, Río Muni) stayed Spanish colonies well into the second half of the 20th century. Italy made Eritrea and parts of Somalia a protectorate in 1889, merged it with Ethiopia which was conquered in 1936 to what became Italian East Africa, and in 1934 merged the provinces Tripolitania and Cyrenaica, entrusted to Italy by Turkey in 1911/12, resulting in the colony of Libya. The Netherlands were active as colonists in South Africa and Indonesia; Belgium held Congo under protectorate.

In this situation Wilhelm II thought that the German Reich had also to conquer a 'place in the sun' (Platz an der Sonne). Besides Togo, Cameroon, and German South West Africa as well as German East Africa, becoming 'German' in 1884 and 1885, respectively, one turned towards the east. In the year 1899 China leased Kiautschou Bay to the German Reich and Western Samoa became a German protectorate. In 1900 the Mariana Islands, Caroline Islands, Palau, Nauru, and the Solomon Islands came under German administration. Due to these developments and the excessively grown national pride in Europe the European states looked at each other in distrust. Additionally the progressing industrialisation fostered a mutual arms race.

Besides the accelerated scientific development a fast cultural development could be observed. At the beginning of the 19th century the European classicism had tried to absorb the upheavals in the civil society after the French Revolution and Napoleon's defeat at Waterloo. It also had created a universal educational ideal which was oriented at the ancient world. In philosophy and sociology progress in the cognition of the natural sciences was discovered; the positivism arose. In the German cultural area the German Idealism came to a climax with Johann Gottlieb Fichte (1762–1814), Friedrich Georg Wilhelm Hegel (1770–1831), and Friedrich Wilhelm Joseph Schelling (1775–1854), and replaced Kant's transcendental philosophy. Arthur Schopenhauer (1788–1860) and Sören Kierkegaard (1813–1855) published on the meaning of instincts and the unconsciousness in the human

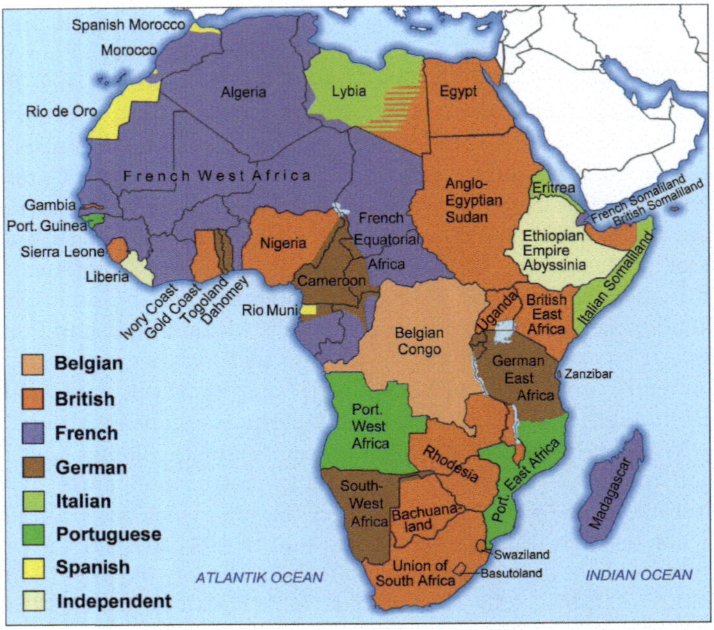

Fig. 10.1.2. Africa in the colonial era in 1914

existence in opposition to Hegel. In poetry and music Romanticism passed over to more individual forms of expression.

In research things were on the move. While the universities in the first half of the 19th century concentrated on collecting, preserving, arranging, and knowledge sharing, the technical elite universities in France and the Humboldtian model of higher education, namely the unity of research and teaching, caused a new orientation. Starting in 1880 new scientific faculty were established at German universities. Practical exercises in the form of seminars were introduced and laboratories and hospitals were affiliated to universities. Only at the end of the 19th century were women allowed to study. The famous algebraist Emmy Noether (1882–1935) could enrol 1903 in Erlangen where she finished her doctorate in 1907. In the year 1909 she was appointed to a position at the University of Göttingen by David Hilbert (1862–1943) and Felix Klein (1849–1925), but the habilitation of women at Prussian universities was still prohibited by a decree of 1908. A special permit was applied for in 1915 but that was refused in 1917. Only after WW II, in 1919, could Emmy Noether be habilitated, i.e. get her qualification as a university lecturer. Under the impression of the French polytechnics a wave of founding new technical universities went through Germany at the beginning of the 19th century which mostly developed out of former military academies. Primarily these early technical schools served only as places of education; research was still not an issue. This changed starting in 1850 when the high-tech industrialisation demanded well educated and research intensive engineers. Only in 1899 Emperor Wilhelm II granted the right to award doctoral degrees at the Prussian technical universities. The other German states followed in 1901. A notable exception was England. Here the education of engineers was traditionally practically characterised and there was no demand for technical institutions on higher levels. Only in 1907 a chair of engineering was established at Oxford University.

In Russia left wing intellectuals and communists were brutally persecuted at the beginning of the 20th century. When the economic situation in Russia decreased rapidly after the Russo-Japanese War and the problems grew due to the industrialisation, a revolution arose in 1905 which has become unforgettable due to the filming of the events about the armoured cruiser Potemkin in 1925. The revolution was terminated when Tsar Nicholas II dissolved the State Duma and introduced a new election law. However, he took care that conservative forces kept the upper hand so that the Russian revolution of 1905 can be seen as a failure.

Fig. 10.1.3. The car in which the Austrian successor to the throne was assassinated in Sarajevo in 1914. The car is on display in the Museum of Military History in Vienna. On the left in the display case the uniform jacket of the successor can be seen with the bullet hole [Photo: Pappenheim 2009]

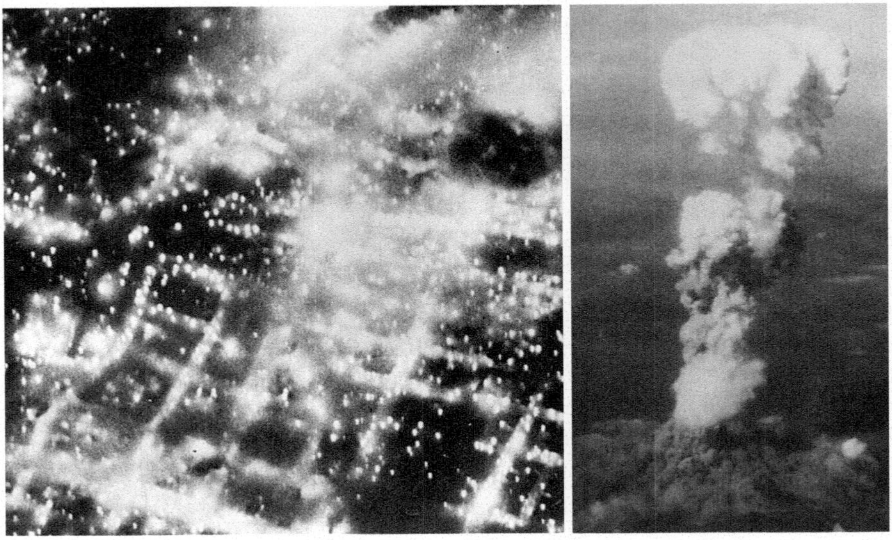

Fig. 10.1.4. The burning city of Brunswick after the firestorm on 15th October 1944 [Photo: RAF No.5 Bomber Group] Most of the German cities suffered similar catastrophes. Explosion of the atomic bomb in Hiroshima 6th August1945 [ARC ID: 542192]

With the assassination of the Austrian successor to the throne in Sarajevo the first catastrophe of the 20th century descended over the world. WW I devoured about 17 Million lives, among them about 7 Million civilians, and ended in 1918 with the capitulation of Germany and Austria-Hungary. At the end of WW I, the 'primal catastrophe of the 20th century', a whole era terminated. The German Emperor Wilhelm II went into exile, Austria-Hungary decomposed. The huge Ottoman Empire disintegrated and shrunk to become the Republic of Turkey founded by Kemal Atatürk in 1923. In Russia the bolsheviks under Wladimir Illjitsch Ulanov (1870–1924), called Lenin, came to power after the October Revolution 1917.

The catastrophe of WW I also resulted in deep cuts at the universities. In England, France, and Germany a whole generation of young men were brutally slaughtered. Only research important for the war had been pursued but after the war returnees entered the universities.

In the Treaty of Versailles of 1919 large reparation claims were put on Germany by the Allies resulting in a strong boost of national currents planting already the germ of the second catastrophe of the 20th century. In 1919 the Weimar Republic was founded. It already started in strong turbulences. The Deutsche Mark as official currency had lost one half of its value at the end of the war, but in 1919 hyperinflation set in. On 20th November 1923 the dollar exchange rate was 4 200 000 000 000 Deutsche Mark. Trust in the new republic is hence shattered in the middle classes and to make things worse the world economic crisis broke loose in 1929 and catapulted the young republic back to conditions of 1923.

With Adolf Hitler's assumption of power in 1933 Germany became fascist. The Jewish Elite was massively suppressed and threatened. Numerous excellent scientists, among them mathematicians of the first rank, left Germany [Wußing 2009, p. 363–370], [Siegmund-Schultze 2009]. Megalomania and race bigotry of the National Socialists throughout the world. In 1939 WWII began in which about 55–60 Million people found death. The attempt to extinguish the Jews costed about 6 Million victims. Besides the Holocaust (or Shoa = great catastrophe) homosexuals, Sinti and Roma, communists, social democrats, and disabled people were deliberately killed. In May 1945 the German Wehrmacht surrendered unconditionally. With the capitulation of Japan on 2nd September the second world catastrophe of the 20th century has terminated.

10.2 Saint George Hunts Down the Dragon: Cantor and Set Theory

Georg Cantor (1845–1918) was born the son of a loving, academically interested father who left a certain fortune to his son. The father was a very religious man, a devout Lutheran, and his son followed him and became an even more pious person. On the occasion of his confirmation at Pentecost 1860 the father wrote a letter to his son which the son kept for the rest of his life [Meschkowski 1967, p. 3]. Assumptions and rumours of Georg Cantor being of Jewish descent are unfounded [Purkert/Ilgauds 1987, p. 15], although this was claimed by Abraham Fraenkel in his memoirs [Fraenkel 1967, p. 152] and in his biography of Cantor [Cantor 1980, p. 481]. Georg grew up in St Petersburg with three younger siblings, Ludwig, Sophie, and Constantin. Around 1850 the metropolis St Petersburg hosted a large German community of about 40,000. Due to a lung disease of the father the family moved to Frankfurt/Main in 1856. In a letter of 4th July 1894 Georg Cantor wrote on his time in Russia [Purkert/Ilgauds 1987, p. 16]:

> *The first 11 wonderful years of my life I have spent in the delightful town at the Newa; sadly, since then I have never been returned to my homeland.*
>
> (Meine 11 ersten wundervollen Lebensjahre habe ich in der herrlichen Newastadt verlebt, seitdem bin ich leider nie wieder in meine Heimath zurückgekommen.)

After attending private schools in Frankfurt/Main and the grammar school in Wiesbaden Georg Cantor wanted to study mathematics but his father thought it economically safer to study engineering. Hence he attended the 'Höhere Gewerbeschule des Großherzogthums Hessen' (Upper industrial school of the Grand Duchy of Hesse) from 1859 on and the associated 'Realschule' in Darmstadt, from which he moved on to the Höhere Gewerbeschule which had got a polytechnic school in 1859. The Höhere Gewerbeschule had a bad reputation and lost pupils year after year. The father was additionally worried since the pupils were allowed 'student liberties' from the age of 16. Georg used these liberties to enter a student fraternity. The father spoke about the 'ridiculous, apish corps behaviour' (lächerliche, äffische Corpswesen) [Purkert/Ilgauds 1987, p. 18], but he refrained from authoritarian appearance and Georg's short intermezzo seemed to have terminated already in 1861. The mathematics teacher Jacob Külp with his profound education certainly made a strong impression on Georg Cantor and his already great love of mathematics surely was additionally fuelled. In 1862 in written form the father finally granted his son the permission to study mathematics and on 18th August 1862 Georg successfully passed the final exams.

Georg Cantor became a student at the University of Zurich in autumn 1862. While there was no excellent professor of mathematics at the university at that time Cantor could attend lectures given by professors at the Zurich polytechnic. Both institutions of higher learning, university and polytechnic, shared the same building until 1864 and the professors of both institutions gave lectures for the students of the respective other institution. However, in autumn 1862 Cantor's future friend Richard Dedekind had left Zurich for Brunswick.

In the early summer of 1863 his beloved father, who always showed much interest in the development of his son, died, and the mother moved to Berlin. Georg Cantor took a one semester break from his studies and then enrolled at the Friedrich-Wilhelms-University in Berlin. Concerning mathematics a golden age has dawned. Among others Karl Weierstraß (1815–1897), Leopold Kronecker (1823–1891), and Ernst Eduard Kummer (1810–1893) lectured in Berlin.

Kronecker lectured on the latest results in the theory of algebraic equations and number theory, Weierstraß on his current research in analysis, and Kummer on classical areas like analytical geometry and mechanics. Besides mathematics Cantor also attended lectures on physics and philosophy. In 1866 he moved to study at the University of Göttingen for one semester. A change of the place of study for one or more semester was customary in the 19th century. In Berlin Cantor submitted his dissertation *De aequationibus secundi gradus indeterminatis* (On undetermined equations of degree two) in 1867 in which Diophantine equations are considered. The first referee was Kummer, the second Weierstraß. The final result of dissertation and oral examination is 'magna cum laude', but Cantor stayed in Berlin to pass the state examination to get the teaching license for higher teaching positions

Fig. 10.2.1. Leopold Kronecker [unknown photographer, 1865] and Ernst Eduard Kummer [unknown photographer, 1870–1880]

at grammar schools. Meanwhile Hermann Amandus Schwarz (1843–1921), two years younger than Cantor, became a close friend. Both were united in their admiration of Weierstraß and both were worried concerning the influence of Kronecker who wore his heart on his sleeves and did not shy back from any confrontation whatsoever. On 17th December 1868 Cantor applied to the seminar of Karl Schellbach at the Royal Friedrich-Wilhelms-Gymnasium (grammar school) in which teachers of mathematics and physics were pedagogically educated. Cantor stayed only two month, then he got the possibility of habilitation at the University of Halle. He hence saw his future at this time clearly in a university career. To his sister Sophie he wrote on 7th February 1869 [Meschkowski 1967, p. 7]:

> *I realise more and more how close mathematics is to my heart, or rather that I am made to find happiness, satisfaction, and true pleasure in the thinking and doing in this realm.*
>
> (Ich sehe doch immer mehr ein, wie sehr mir meine Mathematik ans Herz gewachsen ist oder vielmehr, daß ich eigentlich dazu geschaffen bin, um in dem Denken und Trachten in dieser Sphäre Glück, Befriedigung und wahrhaften Genuß zu finden.)

Eduard Heine lectured in Halle, known for the Heine-Borel covering theorem, and Cantor's friend Hermann Amandus Schwarz was an associate professor there since 1867. Schwarz moved in 1869 to Zurich as a professor and certainly recommended Cantor in Halle. Cantor achieved his habilitation in the spring of 1869 with a thesis on number theory and became private lecturer in Halle. Cantor maintained a dynamic exchange of ideas with Heine. While Cantor is concerned with number theory Heine was interested in the theory

Fig. 10.2.2. Eduard Heine [unknown photographer, about 1881] and Hermann Amandus Schwarz [unknown photographer, 2nd half of the 19th c.]

of trigonometrical series, in particular Fourier series

$$a_0 + \sum_{k=1}^{\infty} (a_k \cos(kx) + b_k \sin(kx)) \, .$$

The investigations of Fourier series proved to be a true challenge and large areas of analysis have emerged from these investigations. The notion of functions became more refined by Dirichlet, and Riemann created a new notion of the integral as we have already seen in chapter 8 and section 9.5.

In his work *Beweis, daß eine für jeden reellen Wert von x durch eine trigonometrische Reihe gegebene Funktion $f(x)$ sich nur auf eine einzige Weise in dieser Form darstellen läßt* (Proof that a function $f(x)$ defined by a trigonometric series for every real x can be represented in only one way in this form) [Cantor 1980, p. 80–83] Cantor presented a uniqueness theorem on the basis of Riemann's results. According to this theorem the representation of a function by a Fourier series is unique if the Fourier series converges everywhere. Some time later Cantor published a note concerning this work. He was now able to prove that his uniqueness results stayed correct even if convergence was not required on an exception set of finitely many points. Now Cantor's curiosity was ignited. He asked for possible exception sets of *infinitely many* points so that the uniqueness theorem still holds. Hermann Hankel (1839–1873) in a work *Untersuchungen über die unendlich oft oszillierenden und unstetigen Funktionen* (Investigations on infinitely often oscillating and discontinuous functions) of 1870 had already investigated infinite point sets of real numbers and tried to characterise their properties.

Fig. 10.2.3. Georg Cantor's home in Halle from 1886 until 1918 [Photo: Thiele]

It was the friend Schwarz who pointed Cantor to Hankel's work and Cantor even wrote a review [Purkert/Ilgauds 1987, p. 36]. However, in the context of Fourier series Cantor had no precursors. Already in 1872 Cantor published the work *Über die Ausdehnung eines Satzes aus der Theorie der trigonometrischen Reihen* (On the extension of a theorem from the theory of trigonometric series) in Mathematische Annalen [Cantor 1980, p. 92–102]. In this work he had to discover that no rigorous definition of the real numbers existed at all! Compare the situation with the one of Richard Dedekind who was led to this dilemma while lecturing to students of engineering. Cantor wrote [Cantor 1980, p. 92]:

> *To this end I am compelled, if only mainly in outlines, to send ahead some discussions serving to illuminate relationships occurring always if number quantities are given in finite or infinite number; thereby I will be led to certain definitions which, for the sake of a presentation as terse as possible of the intended theorem, the proof of which is given in §3, are established.*

Fig. 10.2.4. Hermann Hankel [unknown photographer, 2nd half of the 19th c.] and Georg Cantor [unknown photographer, ca. 1900–1910]

> (Zu dem Ende bin ich aber genötigt, wenn auch zum größten Teile nur andeutungsweise, Erörterungen voraufzuschicken, welche dazu dienen mögen, Verhältnisse in ein Licht zu stellen, die stets auftreten, sobald Zahlengrößen in endlicher oder unendlicher Anzahl gegeben sind; dabei werde ich zu gewissen Definitionen hingeleitet, welche hier nur zum Behufe einer möglichst gedrängten Darstellung des beabsichtigten Satzes, dessen Beweis im §3 gegeben wird, aufgestellt werden.)

Before this §3 there are $4\frac{1}{2}$ printed pages and they contain hard material! Cantor drafted a theory of the real numbers. As Purkert and Ilgauds write in [Purkert/Ilgauds 1987, p. 36] it is a theory

> *which alone would have secured him a place in the history of mathematics.*

> (die allein hingereicht hätte, ihm einen Platz in der Geschichte der Mathematik zu sichern.)

This new theory, completely based on Cauchy sequences, was presented by Cantor already in the summer semester of 1870.

When an associate professorship (Extraordinat) became vacant in Halle Cantor and another young private lecturer, complex analyst Thomae, applied. After some while a second professorship was granted. Thomae was appointed in May 1872 with the full salary of 500 Taler; Cantor was appointed on the 16th May 1872 without any salary! When thrusts to change the financial situation failed Cantor proposed on 1st July 1873 to put his professorship to rest. This proposal finally had an impact: Cantor got a salary of 400 Taler.

Fig. 10.2.5. Georg Cantor 1880 with his wife Vally [unknown photographer, 1880]

In 1872 Cantor travelled to Switzerland for recreation where he accidentally met Richard Dedekind. From the correspondence of the two men grew the foundation of set theory which will later be further developed by Cantor so that he can hunt down the 'dragon' infinity. With Cantor the actual infinity wins full domiciliary right in modern mathematics.

In spring 1874 Cantor got engaged to Vally Guttmann, a friend of his sister. From wedlock, having taken place shortly after, arose 6 children. Cantor was now working on his theory of sets which furnished one unbelievable result after another. On 12th July 1877 Cantor submitted his work *Ein Beitrag zur Mannigfaltigkeitslehre* (A contribution to the doctrine of manifolds)[1] [Cantor 1980, p. 119–133] to Crelle's journal. We can find here statements on the same cardinality of sets of different dimensions; results so outrageous at this time that opposition grew. The printing was postponed until 1878[2] probably by Leopold Kronecker [Purkert/Ilgauds 1987, p. 51]. Kronecker developed into one of the harshest critics of Cantor and his set theory. A famous citation of Kronecker is: 'God made the integers, everything else is man's work' (Die ganzen Zahlen hat der liebe Gott gemacht, alles andere ist Menschenwerk) and that already shows his whole philosophy. Kronecker could not be persuaded with a mathematics of infinite sets and even wanted to cast off irrational numbers and 'continuous quantities' from mathematics. David Hilbert (1862–1943), arguably the greatest mathematician of the 20th century, called Kronecker 'the classic prohibition dictator' (den klassischen Verbotsdiktator) [Purkert/Ilgauds 1987, p. 53].

[1] The word 'manifold' was not used in the modern sense but to denote sets of points.

[2] It is usual today that the time between the submission of a manuscript and the time of publication takes several months. In Cantor's days this was very unusual indeed.

Already in 1876 Cantor had complained to Dedekind concerning the intellectual narrowness of Halle. He was looking for a new position at a more important university and applied for an associate professorship in Berlin. Although he (and another applicant) was put first, the ministry forbid his going away from Halle. Additionally in 1876 an already granted increase in salary of 900 Mark was withdrawn without replacement. We hence have to think that it was seen a faux pas by Cantor to have applied at another university. When in 1877 the full professor Otto August Rosenberger (1800–1890) submitted his request to be released and proposed Cantor as his successor the latter got the full professorship, again with a delay of the ministry which we can not explain, on 12th April 1879. The years between 1878 and 1884 are the most productive in Cantor's mathematical work. In this time much of the foundation of modern mathematics was created. However, in this time the group of his opponents grew. It was particularly painful that even the old friend Schwarz turned completely against Cantor.

Eventually he found a supporter for the publication of his works in the Swedish mathematician Magnus Gösta Mittag-Leffler (1846–1927) who had founded the journal 'Acta Mathematica' in 1882 where now Cantor's works were published. In 1884 Cantor wrote to Mittag-Leffler [Meschkowski 1967, p. 131],

> *that Schwarz and Kronecker for years intrigue terribly against me, [...]*
>
> (dass Schwarz und Kronecker seit Jahren fürchterlich gegen mich intriguieren, [...])

When the pressure became overwhelming Cantor wrote a reconciliation letter to Kronecker in the summer of 1884 and tried to come back to normal terms. In his reply letter Kronecker appeared conciliatory on a personal level, although in the mathematical differences there would remain a 'divergence'. For Kronecker the philosophical aspects which emerged from the treatment of the actual infinite and in which Cantor loved to dwell were simply unbearable.

Already in 1884 Cantor's health started to fail. He suffered from manic depressions which must have been minor until 1899 since he could still act as a full professor and stand his manifold duties without restraints. In 1885 Mittag-Leffler rejected a work on types of order. This work was not published in Cantor's lifetime but was rediscovered by Ivor Grattan-Guinness and published in 1970 [Grattan-Guinness 1970]. Due to this rejection Cantor turned towards other areas of interests; philosophy and theology, but also Bacon-Shakespeare theory. The latter theory became popular in Germany during the 1880s. Following this theory the English philosopher Francis Bacon (1561–1626) hid behind the pseudonym Shakespeare. Cantor was interested in Bacon per se and became an active supporter of the Bacon-Shakespeare theory where he even conducted serious studies in literary science. Somewhat

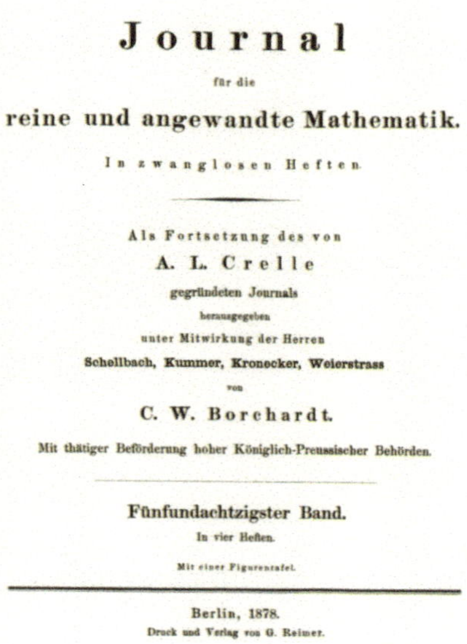

Journal

für die

reine und angewandte Mathematik.

In zwanglosen Heften.

Als Fortsetzung des von

A. L. Crelle

gegründeten Journals

herausgegeben

unter Mitwirkung der Herren

Schellbach, Kummer, Kronecker, Weierstrass

von

C. W. Borchardt.

Mit thätiger Beförderung hoher Königlich-Preussischer Behörden.

Fünfundachtzigster Band.

In vier Heften.

Mit einer Figurentafel.

Berlin, 1878.

Druck und Verlag von G. Reimer.

Fig. 10.2.6. Magnus Gösta Mittag-Leffler [unknown photographer, about 1900]; Journal für die reine und angewandte Mathematik 1878 with the contribution by Cantor on his doctrine of manifolds (*Mannigfaltigkeitslehre*) (SUB Göttingen)

more alarming were the serious attempts of Cantor to determine the 'true identity' of the philosopher Jakob Böhme from Görlitz or the 'true meaning' of the English scientist John Dee. From pertinent collections of letters as [Meschkowski/Nilson 1991] we cannot deduce that Cantor also turned more and more to the divine to discuss his transfinite numbers and his philosophical convictions concerning the infinite. Only recently has the correspondence with theologians been edited and published by Christian Tapp [Tapp 2005], cp. also [Thiele 2008]. Deep insights into Cantor's metaphysics and his relation with regard to religion can be found in [Thiele 2005].

After the death of Alfred Clebsch (1833–1872) Cantor undertook the foundation of the German Mathematical Society (Deutsche Mathematiker Vereinigung DMV). The year 1890 is said to be the founding year; in 1891 the first meeting of the DMV took place in Halle under the leadership of Cantor, who was elected the first president of the society; he stayed president of the DMV until 1893. At the first International Congress of Mathematicians 1897 in Zurich Cantor received the recognition of his set theory which was denied him for quite a long time. Admittedly he had found antinomies, logical contradictions, before, leading to serious problems within set theory. In the year 1897 Cesare Burali-Forti (1861–1931) discovered such an antinomy and published it. These antinomies emerge if sets are indefinitely formed from a

Fig. 10.2.7. Gottlob Frege [unknown photographer, before 1879] and Bertrand Russell [unknown photographer, 1907]

class of all sets having a certain property. They are best studied looking at the famous Russell's antinomy. It is likely that Bertrand Russell (1872–1970) came about this antinomy in 1901 while working on his book *The Principles of Mathematics*. In 1903 he published it.

He formed the set of all sets which do not contain themselves as element:

$$R = \{x \mid x \notin x\}.$$

If R contains itself then it does not contain itself. If R does not contain itself then it must contain itself. In 1918 Russell coined it in the form of an easy understandable paradox:

> In a village lives a barber shaving only the men who do not shave themselves.

If we now ask whether the barber shaves himself we are led to a paradoxical situation. If the barber shaves himself than he belongs to the men not shaving themselves. If he does not shave himself then he has to be one of the men who shave themselves. Russell communicated his antinomy in 1902 to the mathematician and philosopher Gottlob Frege who had published the first volume of his work *Die Grundlagen der Arithmetik* (The foundations of arithmetic) [Frege 1987] in 1884 and was working on the second volume. Frege, trying to ground arithmetic on a set theoretic system of axioms, gave up his work concerning this matter and a second volume never appeared in print.

Only the modern set theory by Zermelo and Fraenkel which we still have to discuss excluded this antinomies.

Shortly after a second residence in a sanatorium Cantor's youngest son died in 1899. From this time on the fits of depression became more frequent and Cantor withdrew from active mathematical work. In 1904, in the presence of Cantor, the Hungarian mathematician Julius (Gyula) König (1849–1913) presented a 'proof' which refuted Cantor's work on transfinite numbers. Cantor is frantic, but already the next day Ernst Zermelo (1871–1953) could show that König's proof was wrong.

The University of St Andrews in Scotland awarded an honorary doctorate to Cantor in 1912 which he could not receive in person due to his illness. He retired in 1913, was malnourished during WW I, and eventually died on 6th January 1918 in Halle in the sanatorium where he had spent the last years of his life.

10.2.1 Cantor's Construction of the Real Numbers

In Dedekind's book *Stetigkeit und irrationale Zahlen* (Continuity and irrational numbers) of 1872 the author wrote in his preface [Dedekind 1963, p. 3]:

> *While writing this preface (March 20, 1872), I am just in receipt of the interesting paper* Ueber die Ausdehnung eines Satzes aus der Theorie der trigonometrischen Reihen, *by G. Cantor (*Math. Annalen, *Vol. 5), for which I owe the ingenious author my hearty thanks. As I find on a hasty perusal, the axiom given in Section II. of that paper, aside from the form of presentation, agrees with what I designate in Section III. as the essence of continuity.*

At this point it is essential that we understand the notion of 'continuity' correctly. One has used the same word for two different things, cp. [Volkert 1988, p. 186]:

- Continuity with regard to real numbers: Here the concern is on the *completeness* of the real numbers. Today we say: Every Cauchy sequence converges.
- Continuity of functions: Small changes of the argument lead to small changes of the function value.

Here we only consider the first meaning of continuity. Dedekind had defined the real numbers by means of his cuts, cp. page 531. Cantor took to Cauchy sequences of rational numbers. A sequence of rational numbers (s_k) is a Cauchy sequence (or fundamental sequence) if to every $\varepsilon > 0$ there exists an index N, such that for all $n > N$ and all m

$$|s_{n+m} - s_n| < \varepsilon$$

holds. Contemplating a little: Cauchy sequences are those sequences in which the distances of elements of the sequence become arbitrarily small if the indices become sufficiently large. We already encountered such a sequence in the discussion of the Hero method in section 9.7.1. If such a Cauchy sequence converges towards a rational number then everything is all right. But if it does not converge towards a rational number then Cantor used this Cauchy sequence to define an irrational number. One problem occurs since many Cauchy sequences converge all to the same number. This problem can be seen already with rational limits; the Cauchy sequences $(\frac{1}{n})$, $(\frac{1}{n^2})$, $(\frac{1}{n^3})$, and so on, converge all to the limit 0. To circumvent this ambiguity Cantor aggregated all Cauchy sequences having the same limit in an equivalence class. Following Dedekind a real number is identified with an interval, following Cantor a real number is identified with an equivalence class of Cauchy sequences!

Nowadays we call the set of real numbers the 'continuum'. But this 'continuum', created by Cantor and Dedekind, has *nothing* to do with the continuum of ancient Greece! The straight line g, defined by

$$y = 2x + 3,$$

being a (linear) continuum in the Greek conception, in Cantor's mathematics takes the form

$$g = \{(x, y) \mid y = 2x + 3\},$$

and hence is a *set of points*! We may rightly say that Cantor has discretised the continuum, i.e. shattered into points. Aristotle would have been disgusted; Democritus would have gloated!

10.2.2 Cantor and Dedekind

When Cantor in his 1872 holiday met Dedekind the two men began to discuss questions which arose from Cantor's work on the exception sets concerning the convergence of Fourier series. In a fairly natural fashion the question arose 'how many' real numbers there are at all. The notion of 'quantity' or 'number' (in the sense of: how many) is only defined in the case of finite sets and fails if infinite sets are concerned. A sentence like 'The set of rational numbers contains so and so many elements' simply makes no sense. Instead, one uses Cantor's notion of 'cardinality' of a set. Galilei's investigations of square numbers had already revealed that one has to reckon with some surprises if infinite sets are concerned, cp. section 5.4.1. In the correspondence starting after their meeting in Switzerland a theory of cardinality emerged. On 29th November 1873 Cantor wrote to Dedekind [Noether/Cavaillés 1937, p. 13]:

> *Would one at first glance be not inclined to state that* (n) *can not be uniquely assigned to the epitome* $\left(\frac{p}{q}\right)$ *of all positive rational numbers*

Fig. 10.2.8. Georg Cantor [unknown photographer, probably before 1894] and Richard Dedekind [unknown painter, 1927]

> $\frac{p}{q}$? *And nevertheless is it not hard to show that* (n) *can be assigned not only to this epitome, but to still more general [...] uniquely assignable, [...]*
>
> (Wäre man nicht auch auf den ersten Anblick geneigt zu behaupten, dass sich (n) nicht eindeutig zuordnen lasse dem Inbegriffe $\left(\frac{p}{q}\right)$ aller positiven rationalen Zahlen $\frac{p}{q}$? Und dennoch ist es nicht schwer zu zeigen, dass sich (n) nicht nur diesem Inbegriffe, sondern noch dem allgemeineren [...] eindeutig zuordnen läßt, [...])

Cantor thereby has addressed a proof that the cardinality of the natural numbers does not differ from the cardinality of the rational numbers,

$$|\mathbb{Q}| = |\mathbb{N}|.$$

Hence there are 'not more' fractions than natural numbers and that seems outrageous, since even between 1 and 2 there is an infinity of fractions! Cantor presented his proof in a letter of 18th June 1886 to the grammar school teacher Goldschneider in Berlin. The letter started with a rigorous introduction and a definition [Meschkowski 1990, p. 189ff.]:

> I. *Let* M *be a given set consisting of concrete things or abstract entities which we call elements. If one abstracts from the character of the elements as well as of the order, then one gets a particular general notion which I call the* cardinality *of* M *or the* cardinal number *assigned to* M.

II. *Two particular sets M and M_1 are called equivalent, in symbols $M \sim M_1$, if it is possible to mutually assign them uniquely and completely, element for element, following a law. If $M \sim M_1$ and $M_1 \sim M_2$, so it holds $M \sim M_2$.*

⋮

III. *From I and II one deduces that equivalent sets share* always the same cardinality *and that, in reverse, sets* of the same cardinality *are equivalent.*

(I. Abstrahiert man bei einer gegebenen, bestimmten Menge M, bestehend aus konkreten Dingen oder abstrakten Begriffen, welche wir Elemente nennen, sowohl von der Beschaffenheit der Elemente, wie auch von der Ordnung ihres Gegebenseins, so erhält man einen bestimmten Allgemeinbegriff, den ich die *Mächtigkeit* von M oder die der Menge M zukommende *Cardinalität* nenne.

II. Zwei bestimmte Mengen M und M_1 heißen äquivalent, in Zeichen $M \sim M_1$, wenn es möglich ist, sie nach einem Gesetz gegenseitig eindeutig und vollständig, Element für Element einander zuzuordnen. Ist $M \sim M_1$ und $M_1 \sim M_2$, so ist auch $M \sim M_2$.

⋮

III. Aus I und II schließt man, dass äquivalente Mengen *immer dieselbe Mächtigkeit* haben und dass auch umgekehrt Mengen *von derselben Cardinalzahl* äquivalent sind.)

Hence Cantor has chosen the same method used by Galilei when it comes to comparing the 'number of elements' of infinite sets. This is the simple principle of numbering. In modern language it has to exist as a bijective mapping between the set of the natural numbers and the set of the rational numbers, and that was proven by Cantor:

VII. *Now I familiarise you with the general notion of a well-ordered set, [...] .*
Examples:

I. *(a,b,c,d,e,f,g,h,i,k) is a* well-ordered *set in contrast to*

$$\begin{array}{ccccccc} & & & a & & & \\ & & b & & c & & \\ & d & & e & & f & \\ g & & h & & i & & k; \end{array}$$

both sets consist of the same elements, hence have the same cardinality.

II. *the series of the finite cardinal numbers in their natural sequence*

$$(1, 2, 3, \ldots, \nu, \ldots)$$

the set of all positive rational numbers in the following order:

$$\left(\frac{1}{1}, \frac{1}{2}, \frac{2}{1}, \frac{1}{3}, \frac{3}{1}, \frac{1}{4}, \frac{2}{3}, \frac{3}{2}, \frac{4}{1}, \frac{1}{5}, \frac{5}{1}, \frac{1}{6}, \frac{2}{5}, \frac{3}{4}, \frac{4}{3}, \frac{5}{2}, \frac{6}{1}, \ldots\right)$$

The law of ordering here is that of two rational numbers $\frac{m}{n}$ *and* $\frac{m'}{n'}$ *in* irreducible[3] *from the first one has a lower or higher rank tha the other, depending on whether* $m+n$ *is smaller or larger than* $m'+n'$*; but if* $m+n = m'+n'$ *then the rank depends on the size of* m *und* m'.

(VII. Nun mache ich sie mit dem allgemeinen Begriff einer wohlgeordneten Menge vertraut, [...] .
Beispiele:

I. (a,b,c,d,e,f,g,h,i,k) ist eine *wohlgeordnete* Menge im Gegensatz zu

a
b c
d e f
g h i k;

beide Mengen bestehen aus denselben Elementen, haben also auch gleiche Mächtigkeit.

II. die Reihe der endlichen Cardinalzahlen in ihrer natürlichen Folge

$$(1, 2, 3, \ldots, \nu, \ldots)$$

die Menge aller positiven rationalen Zahlen in folgender Anordnung:

$$\left(\frac{1}{1}, \frac{1}{2}, \frac{2}{1}, \frac{1}{3}, \frac{3}{1}, \frac{1}{4}, \frac{2}{3}, \frac{3}{2}, \frac{4}{1}, \frac{1}{5}, \frac{5}{1}, \frac{1}{6}, \frac{2}{5}, \frac{3}{4}, \frac{4}{3}, \frac{5}{2}, \frac{6}{1}, \ldots\right)$$

Das Gesetz der Anordnung ist hier dieses, dass von zwei in der *irreduciblen* Form genommenen Rationalzahlen $\frac{m}{n}$ und $\frac{m'}{n'}$ die erstere einen niederen oder höheren Rang als die andere erhält, je nachdem $m+n$ kleiner oder größer als $m'+n'$; ist aber $m+n = m'+n'$ so richtet sich die Rangbezeichnung nach der Größe von m und m'.)

And here is already the proof! We can uniquely assign a natural number to every fraction and vice versa. The method of proof is today known as Cantor's first diagonal argument. The positive fractions can be ordered in the following tableau:

[3] I.e. they are relatively prime.

Fig. 10.2.9. One face of the Cantor cube in Halle [Photo: Richter]

$$\begin{array}{l}
\frac{1}{1}, \frac{2}{1}, \frac{3}{1}, \frac{4}{1}, \frac{5}{1}, \frac{6}{1}, \dots \\[2ex]
\frac{1}{2}, \frac{2}{2}, \frac{3}{2}, \frac{4}{2}, \frac{5}{2}, \frac{6}{2} \ \dots \\[2ex]
\frac{1}{3}, \frac{2}{3}, \frac{3}{3}, \frac{4}{3}, \frac{5}{3}, \frac{6}{3}, \dots \\[2ex]
\frac{1}{4}, \frac{2}{4}, \frac{3}{4}, \frac{4}{4}, \frac{5}{4}, \frac{6}{4}, \dots \\[2ex]
\frac{1}{5}, \frac{2}{5}, \frac{3}{5}, \frac{4}{5}, \frac{5}{5}, \frac{6}{5}, \dots \\[2ex]
\vdots \quad \vdots \quad \vdots \quad \vdots \quad \vdots \quad \vdots
\end{array}$$

The kind of ordering is clear; in the first row we have all positive fractions with denominator 1, in the second row all with denominator 2, and so on. We now travel a zigzag course through the tableau.

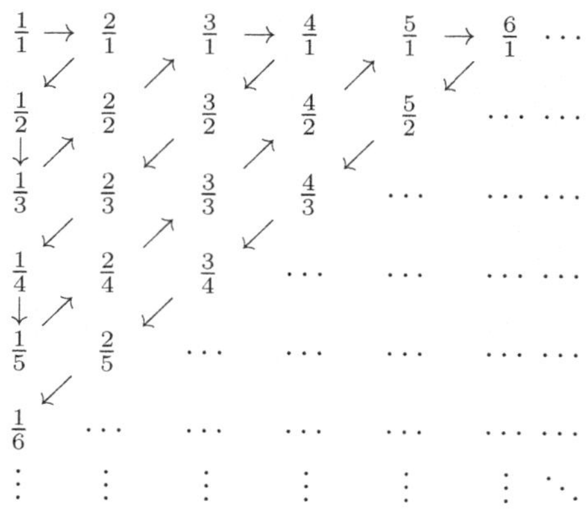

By means of this diagonal argument we have ordered all positive rational numbers like a string of pearls and hence they can be numbered. Following the zigzag course we get the list

$$\frac{1}{1}, \frac{2}{1}, \frac{1}{2}, \frac{1}{3}, \frac{3}{1}, \frac{4}{1}, \frac{3}{2}, \frac{2}{3}, \frac{1}{4}, \frac{1}{5}, \frac{5}{1}, \frac{6}{1}, \frac{5}{2}, \frac{4}{3}, \frac{3}{4}, \frac{2}{5}, \frac{1}{6}, \ldots,$$

where we have skipped the numbers $\frac{2}{2}$, $\frac{2}{4}$, $\frac{3}{3}$, $\frac{4}{2}$, since we already have $\frac{1}{1}$, $\frac{1}{2}$, and $\frac{2}{1}$. To every positive rational number in our list we can uniquely assign a natural number and vice versa. Hence the positive rational numbers can be numbered and thus $\mathbb{Q}$ is countable and has the same cardinality as $\mathbb{N}$.

The search for the cardinality of the real numbers now started off. Cantor wrote to Dedekind on 29th November 1873 saying that he could not believe that the real numbers were also countable, i.e. that they would have the same cardinality as the natural numbers. After all the real numbers form a continuum in contrast to the rational numbers. Unfortunately most of Dedekind's reply letters are lost but we can deduce from Cantor's answering letters what Dedekind has written. In a reply letter [Noether/Cavaillés 1937, p. 18ff.] Dedekind reported that he could not answer the question of the countability of the real numbers, but that he had proven the countability of the algebraic numbers. Algebraic numbers are those which are solutions of algebraic equations

$$a_0 + a_1 x + a_2 x^2 + a_3 x^3 + \ldots + a_n x^n = 0$$

with integer coefficients a_k. Cantor thanked Dedekind for this result and wrote [Noether/Cavaillés 1937, p. 13]:

> *By the way I want to add that I never seriously was concerned with it* [i.e. the question of the countability of the real numbers] *because it has no particular practical interest for me and I agree with you if you say that it does therefore not deserve too much effort.*

> (*Übrigens möchte ich hinzufügen, dass ich mich nie ernstlich mit ihr* [d. h. der Frage nach der Abzählbarkeit der reellen Zahlen] *beschäftigt habe, weil sie kein besonderes practisches Interesse für mich hat und ich trete Ihnen ganz bei, wenn Sie sagen, dass sie aus diesem Grunde nicht zu viel Mühe verdient.*)

Today Cantor's works form the basis of all higher mathematics and it seems almost grotesque to us that both men were confused concerning the importance of their reflections. Only later has Cantor appreciated the actual meaning of these early discussions. On 7th December 1873 Cantor eventually communicated to Dedekind a proof of the *un*countability of the real numbers [Noether/Cavaillés 1937, p. 15]. The proof was clumsy but already on 9th December Cantor announced a simplified proof, however, at this time Dedekind had already sent a simplified proof. With date of 7th December 1873 Dedekind noted [Noether/Cavaillés 1937, p. 19]:

> *C. sent me a rigorous proof, found on the same day, of the theorem that the epitome of all positive numbers [...] can* not *be uniquely assigned to the epitome* (n).
>
> *I answered this letter, received on 8th December, on the same day with a congratulation on the beautiful success, while at the same time I 'mirrored' the core of the proof; this account has also almost word by word passed into Cantor's paper (Crelle Bd.77); [...]*
>
> (C. theilt mir einen strengen, an demselben Tage gefundenen Beweis des Satzes mit, dass der Inbegriff aller positiven Zahlen [...] dem Inbegriff (n) *nicht* eindeutig zugeordnet werden kann.
>
> Diesen, am 8. December erhaltenen Brief beantworte ich an demselben Tage mit einem Glückwunsch zu dem schönen Erfolg, indem ich zugleich den Kern des Beweises "'wiederspiegele"'; diese Darstellung ist ebenfalls fast wörtlich in Cantor's Abhandlung (Crelle Bd.77) übergegangen; [...])

Today the proof has become known as the second Cantor diagonal argument. We show that the cardinality of the real numbers between 0 and 1 is already larger than the cardinal number $|\mathbb{N}| = |\mathbb{Q}|$. To this end we assume that all positive real numbers between 0 and 1 are countable, that is, denumerable. If our assumption would be correct we could list all real numbers in a numbered list, for example

$$\begin{aligned} 1 &: 0.000234821818.....\\ 2 &: 0.356678299100.....\\ 3 &: 0.028320477930.....\\ 4 &: 0.999998998766.....\\ 5 &: 0.005434578899.....\\ 6 &: 0.675676767676.....\\ &\vdots \qquad \vdots \end{aligned}$$

Every such list can start differently of course, but in principle *all* positive real numbers between 0 and 1 could be brought into such a list, provided our assumption would be true. We now construct a new real number $Z = 0.a_1a_2a_3a_4a_5a_6\ldots$ between 0 and 1 following the algorithm below:

The first digit (following the decimal point) a_1 is the first digit of the first number in the list to which we add 1, hence $a_1 = 1$. The second digit a_2 is the second digit of the second number in the list, to which we add 1, hence $a_2 = 6$. The third digit a_3 is the third digit of the third number in the list, to which we add 1, hence $a_3 = 9$. And so on, and so on. If one of the digits under consideration in the list is 9 and we add 1, then we want to agree upon replacing 9 by 0. Hence the fourth digit is $a_4 = 0$, the fifth $a_5 = 4$, the sixth $a_6 = 7$, and so on. Our new number

$$Z = 0.169047\ldots$$

is certainly between 0 and 1 by construction, but it is not a number in our list! It can not be the first number in the list since Z differs in the first digit. It can also not be the second number in the list, since it differs in the second digit, and so on. Hence we have constructed a real number between 0 and 1 which is not in our list of *all* of those numbers! This means that our assumption must have been false. The positive real numbers are *not* countable, they are uncountable.

Hence in 1873 it became clear that there were (at least) two different sizes of 'infinities': the cardinality of the natural and rational numbers on one hand, and the larger cardinality of the real numbers on the other hand. Certainly inspired by his theological reflections Cantor denoted the cardinalities by the first letter of the Hebrew alphabet, 'aleph':

$$\begin{aligned} \aleph_0 &:= |\mathbb{N}| = |\mathbb{Q}|,\\ \aleph &:= |\mathbb{R}| > \aleph_0. \end{aligned}$$

After this breakthrough Cantor turned towards further problems of the infinite. In a letter of 5th January 1874 [Noether/Cavaillés 1937, p. 20] Cantor asked Dedekind whether to every point on a line segment there always exists exactly one point in a square. In other words, he asked whether a line segment

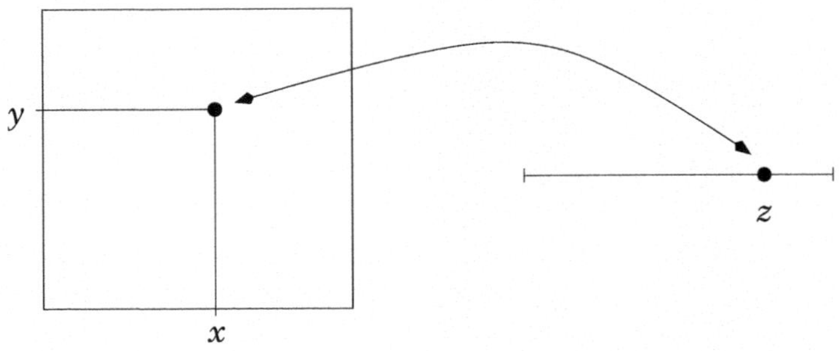

Fig. 10.2.10. Bijective mapping from a square to a line segment

and a square, viewed as point sets, share the same cardinality. Cantor was sure that this can't be, but he could not find a proof. Three years after his question, in a letter of 20th June 1877, Cantor communicated that he had found a proof for the equal cardinality of line segment and square.

The proof rests on the specification of a mapping assigning uniquely to every point of the square exactly one point on the line segment and vice versa. If the coordinates of a point in the unit square are given by

$$x = 0.a_1a_2a_3a_4\ldots$$
$$y = 0.b_1b_2b_3b_4\ldots,$$

then Cantor constructed the mapping as

$$(x, y) \leftrightarrow 0.a_1b_1a_2b_2a_3b_3a_4b_4\ldots.$$

In his reply letter [Noether/Cavaillés 1937, p. 27] Dedekind called Cantor's result an *'interesting conclusion'* (interessante Schlussfolgerung), but he found something to gripe about, having to do with the non-uniqueness of the number representation. We have $0.9999999\ldots = 1$, $0.19999999\ldots = 0.2$, and so on, and that had the effect that some points in the square had no corresponding point on the square under Cantor's mapping [Sonar 2007a, p. 95].

Cantor immediately reacted and wrote a postcard on 23rd June 1877. He accepted Dedekind's objection but he also saw that he had proven much more than a bijective mapping from the square to the line segment, namely the bijective mapping of the square to only a *subset* of the line segment. However, Cantor later modified his proof so that then the square was bijectively mapped to the whole line segment.

Cantor knew that he had proven something monstrous, but Dedekind had checked the proof and found everything as it should be. Dedekind could also not believe that a two-dimensional shape should have the same cardinality as a one-dimensional. He noted that Cantor's mapping was discontinuous and he wrote to Cantor [Meschkowski 1967, p. 41]:

> *[...] the filling of the gaps forces you to allow for a terrible, vertiginously discontinuity in the correspondence*[4]*, by which all is dissolved into atoms, so that every continuously connected piece of the one domain, however small, appears as torn, discontinuous in its image.*
>
> ([...] die Ausfüllung der Lücken zwingt sie, eine grauenhafte, Schwindel erregende Unstetigkeit in der Correspondenz eintreten zu lassen, durch welches Alles in Atome aufgelöst wird, so dass jeder noch so kleine stetig zusammenhängende Theil des einen Gebietes in seinem Bilde als durchaus zerrissen, unstetig erscheint.)

Dedekind conjectured that the reason that the dimensional difference does not matter in the light of cardinality was exactly this discontinuity. He wrote [Meschkowski/Nilson 1991, p. 44]:

> *If a mutual, unique, and complete correspondence between the points of a continuous manifold* A *of dimension* a *on one side and the points of a continuous manifold* B *of dimension* b *on the other side could be established, then this correspondence, even if* a *and* b *are different, would necessarily be a thoroughly discontinuous one.*
>
> (Gelingt es, eine gegenseitige eindeutige und vollständige Correspondenz zwischen den Puncten einer stetigen Mannigfaltigkeit A von a Dimensionen einerseits und den Punkten einer stetigen Mannigfaltigkeit B von b Dimensionen andererseits herzustellen, so ist diese Correspondenz selbst, wenn a und b ungleich sind, nothwendig eine durchweg unstetige.)

Indeed, this conjecture turned out to be true. This was proven only in 1911 by Luitzen Egbertus Jan Brouwer (1881–1961).

10.2.3 The Transfinite Numbers

The finite cardinal numbers are $1, 2, 3, 4, 5, 6, \ldots$. In case of $\mathbb{N}$ the cardinal number is $\aleph_0$. If a set M has exactly N elements we can form the set of all of their subsets, the so-called power set. The set $\{a, b, c\}$ has $N = 3$ elements. Possible subsets are $\{a\}$, $\{b\}$, $\{c\}$, $\{a, b\}$, $\{b, c\}$, $\{a, c\}$, $\{a, b, c\}$ and the empty set $\emptyset$, hence there are 8 subsets. It is easy to show that the power set $\mathcal{P}$ of a set M with N elements always has 2^N elements, hence

$$|\mathcal{P}(M)| = 2^N$$

and since $2^N > N$ no bijective mapping from M to $\mathcal{P}(M)$ can be found. Following one of Cantor's theorems the relation '$>$' can be transferred to infinite sets like $\mathbb{N}$, i.e.

[4] Meant is the mapping between square and line segment.

Beiträge zur Begründung der transfiniten Mengenlehre.

Von

Georg Cantor in Halle a./S.

(Erster Artikel.)

„Hypotheses non fingo."

„Neque enim leges intellectui aut rebus damus ad arbitrium nostrum, sed tanquam scribae fideles ab ipsius naturae voce latas et prolatas excipimus et describimus."

„Veniet tempus, quo ista quae nunc latent, in lucem dies extrahat et longioris aevi diligentia."

§ 1.

Der Mächtigkeitsbegriff oder die Cardinalzahl.

Unter einer ‚Menge' verstehen wir jede Zusammenfassung M von bestimmten wohlunterschiedenen Objecten m unsrer Anschauung oder unseres Denkens (welche die ‚Elemente' von M genannt werden) zu einem Ganzen.

In Zeichen drücken wir dies so aus:

(1) $$M = \{m\}.$$

Die Vereinigung mehrerer Mengen $M, N, P, \ldots$, die keine gemeinsamen Elemente haben, zu einer einzigen bezeichnen wir mit

(2) $$(M, N, P, \ldots).$$

Die Elemente dieser Menge sind also die Elemente von M, von N, von P etc. zusammengenommen.

‚Theil' oder ‚Theilmenge' einer Menge M nennen wir jede *andere* Menge M_1, deren Elemente zugleich Elemente von M sind.

Ist M_2 ein Theil von M_1, M_1 ein Theil von M, so ist auch M_2 ein Theil von M.

Jeder Menge M kommt eine bestimmte ‚Mächtigkeit' zu, welche wir auch ihre ‚Cardinalzahl' nennen.

‚Mächtigkeit' oder ‚Cardinalzahl' von M nennen wir den Allgemeinbegriff, welcher mit Hülfe unseres activen Denkvermögens dadurch aus der Menge M hervorgeht, dass von der Beschaffenheit ihrer verschiedenen Elemente m und von der Ordnung ihres Gegebenseins abstrahirt wird.

Mathematische Annalen. XLVI. 31

Fig. 10.2.11. First page of the *Beiträge zur Begründung der transfiniten Mengenlehre* (Contributions to the foundation of a transfinite set theory) by Georg Cantor [Math. Ann. XLVI, 1895, Heft 4, S. 481 ff.] (SUB Göttingen)

$$|\mathcal{P}(\mathbb{N})| = 2^{\aleph_0} > \aleph_0.$$

Creating further power sets we get by

$$\begin{aligned} |\mathcal{P}(\mathcal{P}(\mathbb{N}))| &> |\mathcal{P}(\mathbb{N})|, \\ |\mathcal{P}(\mathcal{P}(\mathcal{P}(\mathbb{N})))| &> |\mathcal{P}(\mathcal{P}(\mathbb{N}))|, \\ \vdots \qquad & \vdots \qquad \vdots \end{aligned}$$

ever larger cardinalities, i.e. ever growing gradation in the realm of the infinite. If we denote the cardinality of the real numbers by $\aleph$, then we already know that

$$\aleph > \aleph_0$$

holds. Georg Cantor denoted by $\aleph_1$ the smallest uncountable infinity level after the cardinality of the natural numbers. Then follow $\aleph_2$, $\aleph_3$, and so on. In case of cardinal numbers there is even an arithmetic, i.e. one can compute with them [Deiser 2009].

Cantor was particularly moved by the question whether the cardinality of the real numbers, $\aleph$, would satisfy the equation

$$\aleph = \aleph_1,$$

hence whether the cardinality of the real numbers is the cardinality of the 'second number class' (Zweite Zahlklasse). This problem is called the 'continuum problem' which can also be formulated differently: Is there a cardinality in between the cardinalities $\aleph_0$ and $\aleph$? If you think that such a cardinality 'in between' does not exist, then you advocate the 'continuum hypotheses'. Until the end of his life Cantor worked on this question but could not prove it. Only in 1938 was logician Kurt Gödel (1906–1978) able to prove the relative consistency of the continuum hypothesis with respect to the axioms of set theory which were meanwhile formulated by Zermelo and Fraenkel. The continuum hypothesis hence can not be falsified within the framework of Zermelo-Fraenkel set theory and so does not compromise mathematics.

The question was finally fully answered by the American mathematician Paul Cohen (1934–2007) only in 1963 in an unexpected manner. He proved that the assumption that the continuum hypothesis does *not* hold is also relatively consistent with respect to the axioms of set theory! Hence there are two different 'mathematics', one relying on the axioms of set theory *together with* the continuum hypothesis, and another relying on the axioms of set theory *without* the continuum hypothesis!

At the end of all the levels of infinity the devout and pious Christian Cantor had seen God. The mathematician Kowalewski wrote (quoted from [Meschkowski 1967, p. 110]):

Fig. 10.2.12. Paul Cohen [Photo: Chuck Painter] (Stanford University News Service) und Kurt Gödel [unknown photographer, about 1925]

> *These cardinalities, Cantor's alephs, were something holy for Cantor, in a sense the steps leading up to the throne of infinity, to God's throne. He was convinced that with these alephs all thinkable cardinalities were characterised.*
>
> (Diese Mächtigkeiten, die Cantorschen Alephs, waren für Cantor etwas Heiliges, gewissermaßen die Stufen, die zum Throne der Unendlichkeit, zum Throne Gottes emporführen. Seiner Überzeugung nach waren mit diesen Alephs alle überhaupt denkbaren Mächtigkeiten erschöpft.)

If we bring our inductively created power sets above into play we can even formulate a generalised continuum hypothesis If M is an infinite set then $|\mathcal{P}(M)|$ is the next cardinal number following $|M|$.

Besides the cardinal numbers Cantor has created still other interesting numbers, namely the 'ordinal numbers' or the 'ordinal'. Here the order in a set it the issue. Linguistically the cardinal number gives the answer to the question 'how many elements are there?', while the ordinal number gives the position of the elements; the first, second, third, and so on. We can use natural numbers for both purposes, of course; for a number as well as for position. The ordinal number of the set of the natural numbers is denoted by ω; this is the first 'transfinite number'. But we can count on: $\omega + 1$, $\omega + 2$, ... , 2ω, ... , ω^2, ..., ω^ω, ... There is also an arithmetic of ordinal numbers, but we have to leave the reader here and refer to the literature, for example [Deiser 2009].

10.2.4 The Reception of Set Theory

Despite the great opposition that Cantor had to face in his lifetime it did not take long until his set theory became established. Already in 1906 the textbook *The Theory of Sets of Points* of the English married couple William Henry Young (1863–1942) and Grace Chisholm Young (1868–1944) was published, [Young/Chisholm Young 1972]. The first German textbook appeared in 1914 with a dedication to Georg Cantor: *Grundzüge der Mengenlehre* by Felix Hausdorff (1868–1942), which quickly developed into the 'bible' of set theory and makes a very good read even today [Hausdorff 1978], [Hausdorff 2002]. The book was translated into English and is still available in print in its fourth corrected English edition [Hausdorff 1991].

Important for a quick reception was not only the existence of textbooks. The ghosts of antinomies had to be chased away under all circumstances since a theory including contradictions can not be successful. One had to somehow prohibit the unlimited creation of sets of sets, hence things like 'the set of all sets' had to be characterised as inadmissible. A first axiomatisation of set theory came from Ernst Zermelo in 1907 [Ebbinghaus 2007], but still was worthy of improvement. In 1921 Adolf (later Abraham) Fraenkel[5] (1891–1965) added an axiom, and in 1930 the whole system of axioms was brought

Fig. 10.2.13. Abraham Fraenkel and Ernst Zermelo [unknown photographer, probably 1907] – Creator of a system of axioms for set theory

[5] Fraenkel was a brilliant Jewish mathematician born in Munich. As a Zionist he left Germany in 1929 and stayed in what was later to become the state of Israel until the end of his life. He changed his name in 'Abraham Halevi Fraenkel'.

into its final form by Zermelo. This system of axioms is called the Zermelo-Fraenkel set theory, in short ZF. Together with the axiom of choice this system is known as ZFC. Over time different other axiomatic systems have been formulated. One of the better known is the system of Neumann-Bernays-Gödel (NBG) being equivalent to ZFC.

10.2.5 Cantor and the Infinitely Small

Cantor is the conquerer of the infinitely large; at least he has pointed out *one* possible way to overcome this 'dragon' and the majority of today's mathematicians agree to use it. What could be more obvious than, in passing so to speak, to reform the theory of the infinitely small quantities at the same time and hence to kill the two dragons with one stone? This was exactly what Cantor did *not*! Already on 29th December 1878 he wrote to Dedekind [Meschkowski/Nilson 1991, p. 50]:

> *Such a mistake was never* not even remotely *intended by me; in my work I say explicitly that every number which I denote by c can be equated with a number b. By the way is this mistake really being made from another quarter, as outrageous as this may sound; I do not know whether Thomase's outline of a theory of complex functions and of the Theta function is known to you; in the second edition pag. 9 one can find numbers which (horribile dictu) are smaller than any thinkable real number and nevertheless different from zero.*
>
> (Ein solcher Missgriff ist von mir *nie auch nur entfernt* beabsichtigt worden; ich sage ausdrücklich in meiner Arbeit, dass jede von mir mit c bezeichnete Zahl einer Zahl b gleichgesetzt werden kann. Uebrigens ist dieser Missgriff von einer andern Seite wirklich gemacht worden, so unerhört dies auch klingen mag; ich weiss nicht, ob Ihnen Thomaes Abriss einer Theorie der complexen Functionen und der Thetafunctionen bekannt ist; in der zweiten Auflage pag. 9. findet man Zahlen, welche (horribile dictu) kleiner als jede denkbare reelle Zahl und dennoch von Null verschieden sind.)

In a letter to Kerry of 4th February 1887, [Meschkowski/Nilson 1991, p. 275f.] he showed that the seemingly obvious idea to create an infinitely small quantity by taking the reciprocal of the ordinal number ω leads to contradictions. To Mittag-Leffler he made a devastating remark concerning Paul Du Bois-Reymond's book *Die allgemeine Functionentheorie* (The general theory of functions) [Meschkowski/Nilson 1991, p. 71] on 10th May 1882:

> *Therein he seems intending to restore the <u>infinitely small quantities</u> which but <u>as such</u> have no rights to exist; at least for me there are*

> *only variable quantities which become infinitely small, but no infinitely small quantities.*
>
> (Er scheint darin eine Wiederherstellung der unendlich kleinen Größen zu beabsichtigen, die doch als solche keine Existenzberechnung mehr haben; für mich wenigstens giebt es nur unendlich klein werdende, veränderliche Größen, aber keine unendlich kleinen Größen.)

One finds many such places in letters which show clearly that Cantor was a pupil of Weierstraß . He was particularly explicit to Vivanti [Meschkowski 1967, p. 117] when he called the infinitely small the 'infinitary cholera bacillus of mathematics' (infinitären Cholera Bacillus der Mathematik).

10.3 Searching for the True Continuum: Paul Du Bois-Reymond

Not later than with Cantor's set theory and Dedekind's cuts most mathematicians had accepted a 'smashed' continuum; the real numbers as a set of points. Only a few turned against such a concept of the continuum, among them the German mathematician Paul Du Bois-Reymond (1831–1889) who came from a family of Huguenots living in Berlin. Du Bois-Reymond's thoughts concerning the continuum are older than the ideas of Cantor but were unjustly displaced by them. Paul's brother Emil Heinrich Du Bois-Reymond (1818–1896) became an important physician and ranks as the father of electrophysiology. Due to public talks on science and questions of culture Emil became one of the most well-known scientists of the 19th century.

Paul wanted to emulate his older brother and began to study medicine at the University of Zurich. When he went to the University of Königsberg the physicist Franz Neumann (1798–1895) brought him to study mathematics. In 1853 he finished his studies with a dissertation thesis under Kummer. Paul began his career as a grammar school teacher in Berlin and published important contributions to the theory of partial differential equations which later served as the foundation for the works of the Norwegian Sophus Lie (1842–1899). In 1865 he became professor in Heidelberg and became full professor at the University in Freiburg im Breisgau in 1870. He went to Tübingen in 1874 and finally in 1884 to the Technical University in Berlin. He became well-known due to a work on the convergence of Fourier series published in 1873. In this work he could falsify one of Dirichlet's conjectures in that he gave a continuous function the Fourier series of which diverges at one point. We also owe him an example of a continuous function which is nowhere differentiable. His actual expertise was in the theory of differential equations, however.

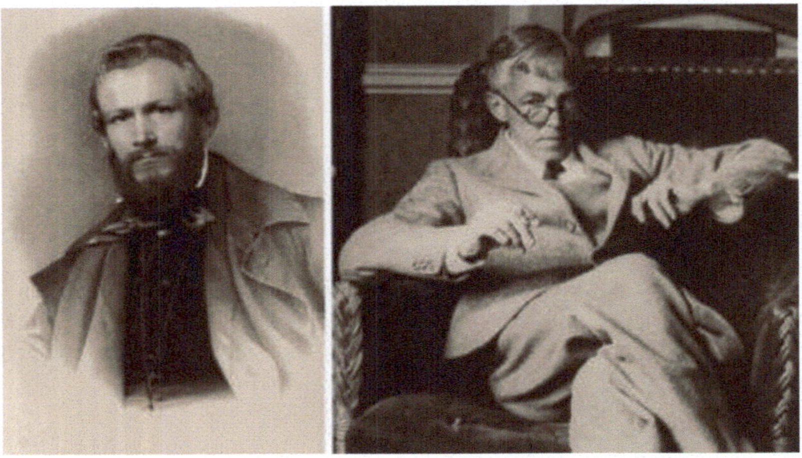

Fig. 10.3.1. Paul Du Bois-Reymond [unknown photographer, probably about 1870] and Godfrey Harold Hardy [unknown photographer, probably about 1930]

Of particular interest to us is Du Bois-Reymond's book *Die Allgemeine Functionenlehre* [Du Bois-Reymond 1968] (The general theory of functions), published in 1882. In this book Du Bois-Reymond discussed the real numbers and the continuum in a very original manner which apparently was not received well by his contemporaries. Du Bois-Reymond was not satisfied with the identification of points on a straight line with the real numbers as done by Dedekind and Cantor. The straight line is completely continuous while the real numbers are thoroughly discrete. Laugwitz noted in his epilogue [Du Bois-Reymond 1968, p. 294f.]:

> *With them* [the definitions of $\mathbb{R}$ as given by Dedekind and Cantor] *the difference between the continuous and the discrete is defined away – and it can not be gainsaid that in the last hundred years, thereby a development of analysis without scruples concerning subtle nuances of notions have been made possible. Today this belongs to the school curriculum. Du Bois-Reymond was one of the first among the mathematicians who felt uneasy by this violent chopping up of a Gordian knot.*

> (Damit [mit den Dedekindschen und Cantorschen Definitionen von $\mathbb{R}$] wird der Unterschied zwischen Kontinuierlichem und Diskretem hinwegdefiniert – und es läßt sich nicht leugnen, daß in den vergangenen hundert Jahren damit eine von Skrupeln über subtilere Feinheiten der Begriffe weitgehend unbelastete Entwicklung der Analysis ermöglicht wurde. Heute gehört das zum Schulstoff. Du Bois-Reymond war einer der ersten unter den Mathematikern, denen bei diesem gewaltsamen Zerhauen eines gordischen Knotens unbehaglich wurde.)

In his book Du Bois-Reymond gives two answers; he points out possible ways to other number systems. Two fictitious mathematicians with different philosophical inclinations appear, the 'empiricist' and the 'idealist'. In view of the empiricist there are 'too many' real numbers. He takes retreat to the argument that we can only distinguish finitely many points on a line segment due to our limitations. The idealist, in contrast, supports the hypothesis that the real numbers are 'too thin' since the difference of two such numbers is always again such a number, although there are 'infinitely small quantities' in analysis already. Then Du Bois-Reymond established an approach of infinitely small quantities which he called the 'Infinitärcalcül' (infinitary calculus).

Du Bois-Reymond's *Die Allgemeine Functionenlehre* had the potential to serve as a basis for discussions which later came up again with the intuitionists and the constructivists Brouwer, Weyl, and Lorenzen, but this fate was not granted to this important book. Only the English mathematician Godfrey Harold Hardy (1877–1947) seemed to have been the first to recognise the importance of the infinitary calculus for analysis. In 1910 his book *Orders of Infinity – The 'Infinitärcalcül' of Paul Du Bois-Reymond* [Hardy 1910] was published, which saw two further editions until 1954. In the preface Hardy wrote:

> *The ideas of Du Bois-Reymond's* Infinitärcalcül *are of great and growing importance in all branches of the theory of functions.*

10.4 Searching for the True Continuum: The Intuitionists

Prior to WW I nobody would have probably thought it possible that the Dutch mathematician Luitzen Egbertus Jan Brouwer (1881–1966) would conjure up a new foundational crisis in mathematics. Brouwer got his graduation from grammar school at the age of 16 and then went to the University of Amsterdam to study mathematics.

There he came in contact with the philosopher and mathematician Gerrit Mannoury, who introduced him to the latest developments in set theory and to the logicism of Bertrand Russell. The result of this occupation was Brouwer's dissertation *Over de grondslagen der wiskunde* (On the foundations of mathematics) of 1907. Already in 1908 Brouwer's scepticism concerning the foundations of mathematics can be seen. In his work *De onbetrouwbaarheid der logische principes* (The unreliability of the logical principles) he refuted the theorem of the excluded third, the 'tertium non datur' $P \vee \neg P$, for the first time; a statement P is either valid or not valid.

Brouwer thought that the use of the *tertium non datur* in connection with infinite sets was a true mistake which would lead to wrong results. Therewith

Brouwer aimed early at, and against, the 'Hilbert program' which the German David Hilbert (1862–1943) propagated in the 1920s, cp. [Brouwer 1924]. With finite methods the consistency of the axiomatic systems used in mathematics should be proven, but that turned out to be unenforceable. This program for Cantor's set theory was drafted by Hilbert in 1926 in the work [Hilbert 1926], see also [Taschner 2006, p. 67ff.]. Kurt Gödel essentially was instrumental to overthrow the Hilbert program already in 1931.

But Brouwer was not yet ready for the great revolution. From 1908 on Brouwer turned towards (set theoretic) topology and developed quickly into a 'star' in this area. He clarified the notion of 'dimension'; in particular could he prove Dedekind's conjecture that a continuous bijective mapping from a square to a line segment (cp. page 565) does not exist. He discovered the degree of a continuous mapping and Brouwer's fixed point theorem. His works excited David Hilbert in particular, who published many of Brouwer's papers in the important journal *Mathematische Annalen* where he acted as responsible editor. When Brouwer got a professorship at the University of Amsterdam in 1912 he returned to his mathematical-philosophical beginnings. He was against a too far going formalism and refuted the axiomatisation of set theory. He introduced the intuitionism into mathematics which saw mathematics as a purely intuitive occupation [van Stigt 1990]. A 'truth' in mathematics is only identified in the human mind by verification. Hence 'truths' are not independent of all circumstances as in the Platonic view on mathematics, but they are the product of a verificating mind. Being of this opinion Brouwer had to get in opposition to Hilbert who had raised the axiomatic method to *the* guiding principle. However, in 1914 Brouwer became yet co-editor of the Mathematische Annalen.

Fig. 10.4.1. L. E. J. Brouwer, honoured in the Netherlands by a stamp (2007)

Fig. 10.4.2. David Hilbert [unknown photographer, before 1912] (University of Hamburg, Mathematische Gesellschaft)

During WW I Brouwer developed a set theory which was independent of the *tertium non datur* and led to a constructionist foundation of analysis. Hilbert wrote in 1928 [Hilbert 1928, p. 80]:

> *To take away this tertium non datur from the mathematician would be like taking away the telescope from the astronomer or to forbid the boxer the use of his fists.*
>
> (Dieses tertium non datur dem Mathematiker zu nehmen, wäre etwa, wie wenn man dem Astronomen das Fernrohr oder dem Boxer den Gebrauch der Fäuste untersagen wollte.)

In contrast, Hilbert's student Hermann Weyl (1885–1955) was willing to sacrifice the *tertium non datur*. As an example that this theorem could not be valid Brouwer used the statement

P: In the decimal expansion of π there is the sequence 0123456789.

Fig. 10.4.3. Hermann Weyl in Göttingen ([unknown photographer, about 1930], Archives of P. Roquette, Heidelberg) and Paul Lorenzen 1967 in Erlangen [Photo: Konrad Jacobs, 1967]

How, Brouwer asked, can we verify this statement? We can not verify it, as long as we can find this sequence of digits by chance when we compute ever more digits of π. Hence we can *not* say: 'Either P is true or non-P', since we can not verify it.

Brouwer's 'intuitionistic continuum' follows from his set theory of the year 1918. Brouwer accepted points only if they could be constructed in finitely many steps, or if a construction instruction in finitely many steps can be given. Additionally so-called selection sequences are admissible which can not be described completely.

We may imagine a 'Brouwerian real number' as a sequence of overlapping intervals which get ever smaller by means of a selection process. The continuum following Brouwer hence is no fixed entity made of concrete numbers but, as Weyl wrote in [Weyl 1921, p. 50], a 'medium of free becoming' (ein Medium freien Werdens).

As a consequence of this intuitionistic continuum and the absence of the *tertium non datur* there can be no jump function

$$f(x) = \begin{cases} a \; ; \; x < c \\ b \; ; \; x \geq c \end{cases}$$

in Brouwer's analysis. Even more, no discontinuous functions exist at all. And it is still getting better! Every real function which is defined everywhere on the real line is automatically uniformly continuous; a theorem which is certainly *not* valid in classical analysis. Brouwer's intuitionism hence leads us to the statement that 'in mathematics there are no jumps' and we have here a back reference to Leibniz, who also excluded discontinuous functions in his concept of the continuum [Brouwer 1925], [Brouwer 1926], [Brouwer 1927].

In contrast to Hilbert, who was aghast at Brouwer's intuitionistic mathematics Hermann Weyl felt Brouwer's attempts to newly re-found mathematics was liberating. Already in 1917 in his treatise *Das Kontinuum* [Weyl 1917], [Weyl 1994] he had tried a similar approach to re-found the continuum. Concerning the question of the existence of real numbers he wrote in [Weyl 1921, S. 46]:

> *Due to this notional restriction* [of classical analysis] *a bunch of individual points is picked out of the flowing porridge of the continuum, so to speak. The continuum is shattered into isolated points and the intermingling of all of its parts is replaced by certain conceptional relations, based on 'larger-smaller', between these isolated elements. Therefore I speak of an* atomistic conception of the continuum
>
> (Durch diese Begriffseinschränkung [der klassischen Analysis] wird aus dem fließenden Brei des Kontinuums sozusagen ein Haufen einzelner Punkte herausgepickt. Das Kontinuum wird in isolierte Elemente zerschlagen und das Ineinanderverflossensein aller seiner Teile ersetzt durch gewisse, auf dem "'größer-kleiner"' beruhende begriffliche Relationen zwischen diesen isolierten Elementen. Ich spreche daher von einer *atomistischen Auffassung des Kontinuums*)

And further on [Weyl 1921, p. 47] he speaks of the 'continuous space sauce' (kontinuierlichen Raumsoße) gushed between two points. Weyl and his publication of 1921 have authoritatively contributed to the crisis which we today call the foundational crisis.

In the 1920's Brouwer fought for again accepting German mathematicians internationally but he failed. After the devastating WWI German colleagues had been boycotted. When eventually in 1928 an international congress with German participation was about to take place in Bologna Brouwer appealed to the German colleagues to boycott this congress for their part. Hilbert won't stand this interference from abroad and attended the Bologny congress. Still in 1928 Hilbert removed Brouwer from the editorial group of the *Mathematische Annalen* from one issue to the other. The difference in the mathematical views of the two men had grown too large.

However, the intuitionism was out of the box and found its followers in the constructivists. But Weyl pulled out, he returned to classical analysis. Others wanted to found analysis on finitary constructive arguments. Nonconstructive existence proofs are banned from constructivism. In Germany the science theorist and mathematician Paul Lorenzen (1915–1994) became a well known constructivist, [Lorenzen 1965], [Lorenzen 1986]. In the USA Erret Bishop (1928–1983) became the father of constructivistic analysis [Bishop/Bridges 1985]. An interesting example to illuminate the difference between classical and constructivistic analysis is the intermediate value theorem:

Let f be a continuous function defined on an interval $[a,b]$. If $f(a)$ has another sign than $f(b)$ (hence if $f(a) \cdot f(b) < 0$), then there exists a number c for which $f(c) = 0$ holds.

The intermediate value theorem is *not* valid in constructivistic mathematics since 'existence' has to include finite constructability. Therefore constructivistic mathematics replaces the intermediate value theorem with different alternative formulations like the following. If f is continuous on $[a,b]$ and if $f(a) \cdot f(b) < 0$, then for every $n \in \mathbb{N}$ a real number $c(n)$ can be constructively found so that $f(c(n)) < 1/n$.

Constructivistic mathematics has gained some boost again in the age of computers. We have to note however, that the overwhelming majority of mathematicians hold fast to classical analysis according to Weierstraß. A very readable introduction into constructivistic analysis has been presented recently by Rudolf Taschner [Taschner 2005].

Interestingly enough the contrast between the intuitionistic and constructivistic mathematics on one hand, and the classical conception on the other hand, can be well compared with the problem of universals of Christian scholastics. The intuitionists and constructivists, respectively, correspond to the nominalists and the classical mathematicians to the realists. That this interpretation is not too far-fetched was shown by Stegmüller and others in [Stegmüller 1978].

10.5 Vector Analysis

In the 19th century physics developed in giant strides. The Navier-Stokes equations served to describe the motion of viscid fluids, Maxwell's equation provided the foundations of electrodynamics. As different as the physical applications appeared, there always showed up a common structure. In scalar fields $\mathbb{R}^3 \ni x \mapsto f(x) \in \mathbb{R}$ (temperature, pressure, et cetera) an important role is played by the gradient

$$\nabla f(x) := \left(\frac{\partial f}{\partial x_1}(x), \frac{\partial f}{\partial x_2}(x), \frac{\partial f}{\partial x_3}(x) \right)^T$$

which is a differential operator. In vector fields (velocity, et cetera) $\mathbb{R}^3 \ni x \mapsto F(x) \in \mathbb{R}^3$ the divergence

$$\nabla \cdot F(x) := \sum_{k=1}^{3} \frac{\partial F_k}{\partial x_k}$$

and the rotation

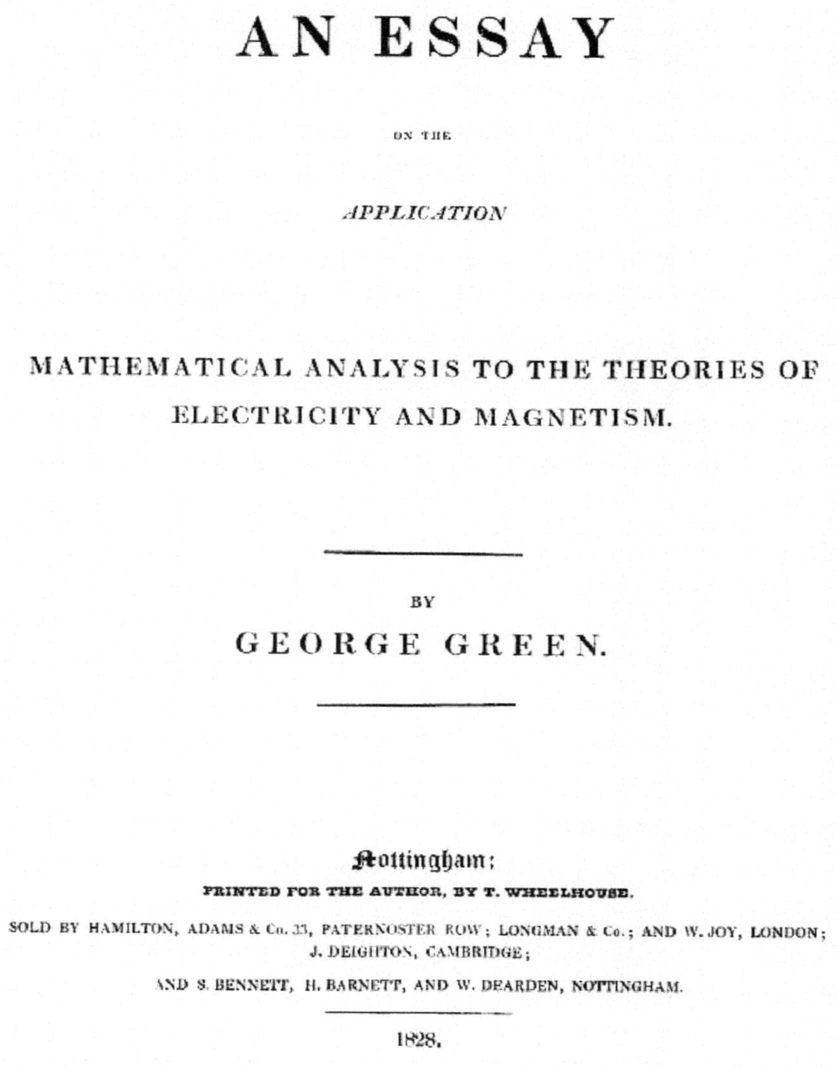

AN ESSAY

ON THE

APPLICATION

OF

MATHEMATICAL ANALYSIS TO THE THEORIES OF ELECTRICITY AND MAGNETISM.

BY

GEORGE GREEN.

Nottingham:

PRINTED FOR THE AUTHOR, BY T. WHEELHOUSE.

SOLD BY HAMILTON, ADAMS & Co. 33, PATERNOSTER ROW; LONGMAN & Co.; AND W. JOY, LONDON; J. DEIGHTON, CAMBRIDGE;

AND S. BENNETT, H. BARNETT, AND W. DEARDEN, NOTTINGHAM.

1828.

Fig. 10.5.1. George Green's famous *Essay*, published in 1828 in Nottingham

$$\nabla \times F(x) := \begin{pmatrix} \dfrac{\partial F_3}{\partial x_2} - \dfrac{\partial F_2}{\partial x_3} \\ \\ \dfrac{\partial F_1}{\partial x_3} - \dfrac{\partial F_3}{\partial x_1} \\ \\ \dfrac{\partial F_2}{\partial x_1} - \dfrac{\partial F_1}{\partial x_2} \end{pmatrix}$$

Fig. 10.5.2. Carl Friedrich Gauß shown on the 10 DM note of the German Central Bank

are the important operators.

The divergence operator gives information on the sources and sinks within a vector field and the rotation operator has in fact to do with the rotation of a field. The connections between these differential operators were clarified in the 19th century, in particular by the German Carl Friedrich Gauß (1777–1855), the Irishman George Gabriel Stokes (1819–1903), and the Englishman George Green (1793–1841), whose names are all connected today with the most important integral theorems.

The integral theorem of Green was already published in 1828 in his famous Nottingham *Essay*. Green could attend school for only two years before he had to take up work in his father's mill. In mathematics and physics Green was an autodidact who read the scientific books of the 'Nottingham Subscription Library' in his leisure time. Only at the age of 40, in 1833, Green went to Cambridge and finished his studies in 1837 with distinction. Alongside his studies he worked at the mill and made a fortune. Today he is ranked as one of the most important researchers on the areas of potential theory, electromagnetism, and hydromechanics. He died 1841 in Nottingham of the flu.

The integral theorem of Gauß reads

$$\int_G \nabla \cdot F(x)\,dx = \oint_{\partial G} \langle F(x), n(x) \rangle \, dS,$$

where G is a domain in $\mathbb{R}^3$ with boundary ∂G and n denotes the outer unit normal vector at ∂G. The Gaußian integral theorem therefore allows

Fig. 10.5.3. George Gabriel Stokes [unknown photographer, about 1860] and Elie Cartan [unknown photographer, about 1904]

to transform a volume integral into a surface integral and also allows an immediate physical interpretation. The action of all sources and sinks in the field G equals the flux $\oint_{\partial G} \langle F(x), n(x)\rangle \, dS = \oint_{\partial G}(F_1(x)n_1(x) + F_2(x)n_2(x) + F_3(x)n_3(x))\, dS$ across the boundary of G. With the integral theorems of Gauß, Green, and Stokes the most important tools of vector analysis for 19th century mathematics had been provided. It was annoying, however, that the rotation could be defined for three dimensions only (there is a definition for the two-dimensional also but in this case the rotation is a scalar quantity) while mathematics and physics of the 20th century aspired to arrive at generalisations.

In the first half of the 20th century the French mathematician Èlie Joseph Cartan (1869–1951) developed a theory of alternating differential forms which allowed an extension of vector analysis to spaces with an arbitrary number of dimensions. For example

$$\omega := f(x_1, \ldots, x_n) dx_1 \wedge dx_2 \wedge \ldots \wedge dx_n$$

is a possible differential form where $\wedge$ denotes the so-called 'outer product' and the differentials dx_k are defined as projection mappings. The integral of the differential form above then is defined as

$$\int_G \omega := \int_G f(x_1, x_2, \ldots, x_n)\, dx_1\, dx_2 \ldots dx_n$$

It is also possible to define an 'external differentiation' d and eventually the integral theorem of Stokes can be written in the form

$$\int_G d\omega = \int_{\partial G} \omega.$$

Starting from this generalised integral theorem the classical theorems of Gauß, Green, and Stokes in the three-dimensional case can be recovered.

Mathematical physics today can not be imagined without the calculus of differential forms and it plays an important role also in other parts of mathematics, for example in modern differential geometry. A readable history of vector analysis has been given by Michael Crowe in [Crowe 1994].

10.6 Differential Geometry

Differential geometry as a science starts with Carl Friedrich Gauß and his work as a land surveyor. Gauß was concerned with notions like 'curvature' of curves and surfaces and he already asked whether triangles on the surface of the earth really have an angular sum of 180°. Leonhard Euler worked on differential geometry before Gauß and already emancipated it as an independent field within analysis.

The study of curves and surfaces as elements of the surrounding two- or three-dimensional space is nowadays called 'classical' differential geometry. With Bernhard Riemann (1826–1866) and his habilitation lecture *Hypothesen, welche der Geometrie zugrunde liegen* (On the hypotheses which lie at the foundation of geometry) of 1854 'modern' differential geometry emerged.

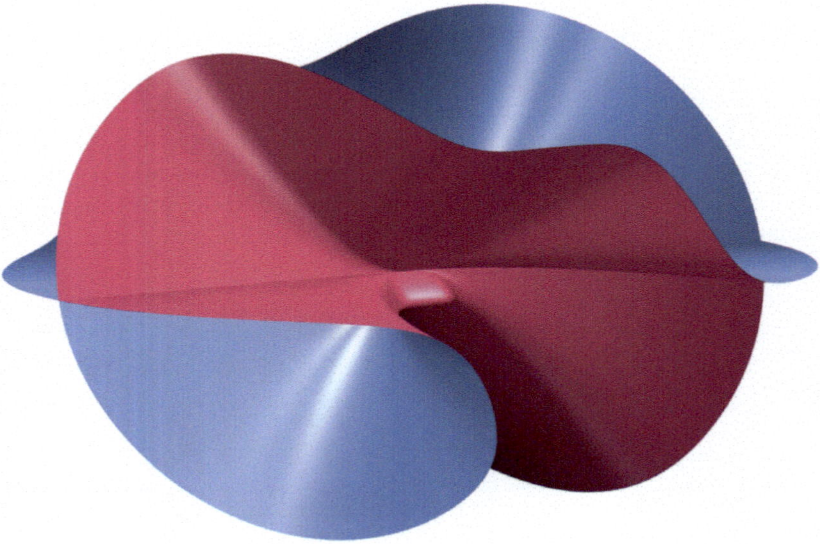

Fig. 10.6.1. Algebraic surface $x^4 + x^5 + z^5yx - y^4 = 0$

Fig. 10.6.2. Georgio Ricci-Cubastro [unknown photographer, before 1925], Tullio Levi-Civita [unknown photographer, before 1930], Jan Arnoldus Schouten [unknown photographer, 1938/1939]

Riemann liberated curves and surfaces from the space surrounding them and developed later the beginnings of an intrinsic analysis of such objects. In this way Riemann created the notion of a 'manifold'. A differentiable n-manifold is a topological space looking locally like the n-dimensional Euclidean space. Manifolds are locally described by 'charts', i.e. local parametrisations which can be carried over into each other by transition maps. Introducing a metric (i.e. a yardstick) on a manifold turns it into a so-called Riemannian manifold playing an important role in the theory of relativity, to give just one example. Manifolds can also be investigated by methods of topology. This field is then called 'differential topology'. Modern differential geometry can also not be imagined without Lie groups as developed by the Norwegian Sophus Lie (1842–1899). Lie groups are groups of continuous transformations with which symmetries of geometric structures can be analysed.

An important class of mathematical objects in differential geometry are 'tensors'. Tensors are generalisations of scalars, vectors, and matrices. The notion of tensor in today's sense appeared in a textbook of the German physicist Woldemar Voigt (1850–1919) of 1898 for the first time and modern physics without tensors and tensor fields can not be thought of. In differential geometry the works of the Italian Gregorio Ricci-Cubastro (1853–1925) and his student Tullio Levi-Civita (1873–1941) on their 'absolute differential calculus' (=tensor calculus) were decisive. Jan Arnoldus Schouten (1883–1971) of Dutch origin took the works of Ricci-Cubastro and Levi-Civita and eventually created 'tensor analysis' which he published in 1924 in his book *Der Ricci-Kalkül* (The Ricci calculus). The notation used by Schouten was so complicated that the typesetters of the Springer publishing house (remember that these were the days of lead types and if you invented a new symbol a new lead type had to be designed and casted) have set themselves a monument. The development of relativity theory and of modern differential geometry would be unfathomable without the Ricci calculus. The history of the tensor calculus has been investigated in all detail by Karin Reich in [Reich 1994].

As a fruitful mix of differential geometry, calculus of variations, and topology 'geometric measure theory' developed fairly recently. With it minimal surfaces having kinks and singularities can be studied, as is the case when several soap bubbles meet or in crystal growth when liquid metals cool off.

10.7 Ordinary Differential Equantions

Ordinary differential equations are equations for an unknown function of *one* variable in which also derivatives of this functions appear. Very naturally these equations occur in the modelling of physical, biological, chemical, or economical systems and were therefore immediately aimed at by Newton and Leibniz when they had developed their differential and integral calculus.

If a mass m hangs at the end of an ideal spring with spring constant k and is allowed to oscillate freely about its rest position, then two forces play a role. Following Newton force = mass · acceleration, hence (following Euler, in fact)

$$F = mx''(t),$$

and this force corresponds to the restoring force of the spring,

$$mx''(t) = -kx(t).$$

Division by m and $\omega^2 := k/m$ yields the differential equation of harmonic oscillations

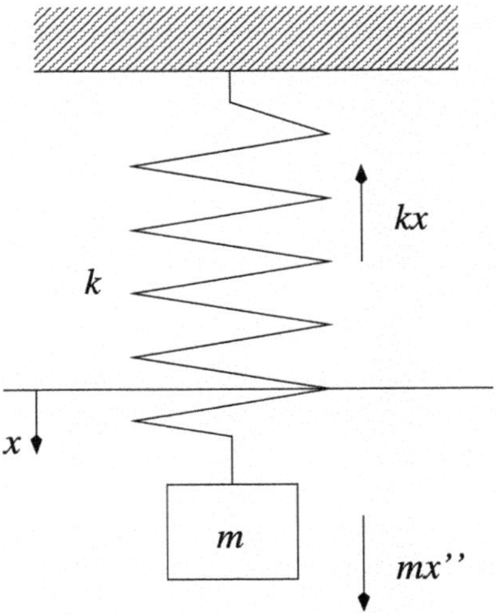

Fig. 10.7.1. The harmonic oscillator

Fig. 10.7.2. Ponte de Dom Luis I, bridge across the Douro in Porto. It was built between 1881 and 1885 by a Belgian company. The engineers of this company used the experiences made by the French engineer Gustave Eiffel who had built a railway bridge further upstream. The arched bridge with its elevated roadway connects the upper cities of Porto and Vila Nova de Gaia. The lower roadway hung at only one arch of 172 meter span connects both lower cities. [Photo: H.-W. Alten]

$$x''(t) + \omega^2 x(t) = 0.$$

Solutions of this differential equation are harmonic oscillations of the form $x(t) = c\sin(\omega t + \Phi)$. Amplitude c and phase Φ are determined by the initial conditions $x(0) = x_0$ und $x'(0) = v_0$.

In the 19th century important questions concerning the existence and uniqueness of solutions of ordinary differential equations could be answered. The Italian Giuseppe Peano (1858–1932) proved his existence theorem in 1886: If the function f in the differential equation

$$x'(t) = f(t, x(t))$$

is 'only' continuous, then a solution exists at least locally. The theorem is also valid in case of systems of differential equations of first order; linear ordinary equations of higher order (as our equation of the harmonic oscillator) can be easily transformed into such systems. The first existence *and* uniqueness theorem for initial value problems was proven by the Finnish mathematician Ernst Leonard Lindelöf (1870–1946) in 1890. To be applied the right hand side f not only has to be continuous, but 'Lipschitz continuous' in the second argument. About the same time the French Charles Èmile Picard (1856–1941) came up with a similar result so that the theorem today carries the name 'Picard-Lindelöf theorem'

In the 20th century the theory of ordinary differential equations has been further developed quite far. The theory of dynamical systems is concerned with the long time behaviour of solutions, the bifurcation theory allows for investigations of solutions of nonlinear differential equations branching out. Since only very few ordinary differential equations can be solved analytically the numerical analysis of ordinary equations gathered much momentum in the 20th century. By linearisation and discretisation the problem under consideration can be transformed into very large linear systems of equations the solutions of which can be calculated by means of modern computers. The solutions then are approximate values of the solution of the differential equation in certain points.

10.8 Partial Differential Equations

Partial differential equations are equations for an unknown function of *several* variables in which *partial derivatives* occur. We have already got to know the equation of the oscillating string, or wave equation

$$\frac{\partial^2 u}{\partial t^2} - \frac{\partial^2 u}{\partial x^2} = 0.$$

This is a partial differential equation of second order (just because second derivatives are present) for the deflection u of the string, being dependent of the point x and time t. An overwhelmingly large number of problems in the natural sciences and in engineering lead to partial differential equations. The Euler equations of gas dynamics, the Navier-Stokes equations describing the flow of viscid fluids, and Maxwell's equations of electrodynamics are classical examples of partial differential equations or systems thereof. But partial differential equations have proved important tools in other fields like in differential geometry.

It was believed for quite some time that at least in case of *linear* partial differential equations a theory of existence and uniqueness of solution could be found being as simple as in the case of ordinary differential equations. Only in 1957 Hans Lewy (1904–1988) published the work *An example of a smooth linear partial differential equation without a solution* in *Mathematische Annalen* in which he was able to prove that a certain linear partial differential equation with analytical (i.e. arbitrary smooth) coefficients had no solutions at all!

It was already Euler who had demanded the possibility to compute even with a plucked string. Although in this case the plucked string shows a kink and its initial value function $u(x, 0)$ therefore is no longer differentiable (let alone twice differentiable as the equation would require), this is certainly the real case with musical instruments. Even more disturbing, a simple one-dimensional wave equation

$$\frac{\partial u}{\partial t} + \frac{\partial u}{\partial x} = 0,$$

in which u asks for differentiablity in space and time, produces from a discontinuous initial value function

$$u(x, 0) = u_0(x) = \begin{cases} 0 \,;\, x < 0 \\ 1 \,;\, x \geq 0 \end{cases}$$

a 'solution'

$$u(x, t) = u(0, x - t) = u_0(x - t) = \begin{cases} 0 \,;\, x < t \\ 1 \,;\, x \geq t \end{cases},$$

staying discontinuous for all times! At least formally we have $\frac{\partial u}{\partial t} = -u_0'$ and $\frac{\partial u}{\partial x} = u_0'$, and therefore $\partial u/\partial t + \partial u/\partial x = -u_0' + u_0' = 0$.

Peculiarities of this kind (and others) could be rigorously explained only in the 20th century. In the Soviet Union Sergej L'vovič Sobolev (1908–1989) worked in the 1930's on a theory of function spaces which nowadays carries his name. Sobolev spaces include functions which are differentiable 'in a weak sense'. Functions showing a kink are members of certain Sobolev spaces which have weakly differentiablity. Sobolev spaces play an important role in mathematics nowadays and in particular in the theory of partial differential equations.

Fig. 10.8.1. Laurent Schwartz [Photo: Konrad Jacobs, 1970], Sergej Sobolev [Photo: Konrad Jacobs, 1970], Hans Lewy [Photo: George M. Bergman, Berkeley 1975]

In France Laurent Schwartz (1915–2002) worked on a still more radical form of weak solutions in the time after WW II, so-called 'distributions'. Distributions are no longer functions but 'linear functionals', i.e. mappings from a function space (the space of 'test functions') into the real numbers. Test functions are arbitrarily smooth functions which are different from zero only on compact intervals. Even a mapping as crazy looking as

$$\varphi \mapsto \delta(\varphi) := \varphi(0),$$

assigning to a test function φ its value at $x = 0$, is a distribution, namely the Dirac distribution, named after the English physicist Paul Adrien Maurice Dirac (1902–1984) who computed with it before the notion of distributions was even born. Eventually distributions proved the key to a theory of linear partial differential equations with constant coefficients. In the case of nonlinear partial differential equations distributions can not be used since the product of two of them can not be defined, but this is not the end of mathematics, of course. The field of partial differential equations is one of the most active in mathematics. Due to the significance of partial differential equations in science and engineering numerical analysis has developed into a large field of research since the mid-20th century. Modern computers and supercomputers with their ever increasing computer power and memory have helped, of course, and are still *the* tools in numerical analysis.

10.9 Analysis Becomes Even More Powerful: Functional Analysis

Analysis, starting from its infancy in antiquity, has developed into a mighty, independent field within mathematics which has branched out considerably and coalesced with other fields of mathematics. We could only describe the roots of this development here because analysis is still a very active research area. *One* modern development should be followed to finish this chapter, however; functional analysis.

10.9.1 Basic Notions of Functional Analysis

Here we are already in trouble since there is no uniformly accepted definition of functional analysis. Functional analysis has developed into an overarching theory in almost all fields of mathematics and no student of mathematics leaves university today without having been exposed to at least some aspects of this field. Let us attempt a definition!

We know the three-dimensional space surrounding us and we are used to measure distances and angles. The space we live in has finite dimensions, namely dimension three, or, taking time into account, dimension four. For mathematicians it is not difficult at all to abstract from the dimension of a space and to deal with five, six, or a thousand dimensions; points in those spaces simply have more coordinates. Studying mappings between such spaces takes us to classical analysis. In this book we have mainly concerned ourselves with mappings $f : \mathbb{R} \to \mathbb{R}$ but multidimensional analysis also deals with mappings as $f : \mathbb{R}^n \to \mathbb{R}^m$ for $n, m \in \mathbb{N}$. The geometry of such finite dimensional spaces belongs to the field of 'linear algebra', or, as my revered teacher Horst Tietz (1921–2012) rightfully used to call it, 'linear geometry'. But there are spaces deviating widely from finite dimensional spaces. The space of all continuous functions on the interval $[0, 1]$ is certainly not finite dimensional but we can put all such functions in a large set and call this set

$$C([0, 1])$$

(meaning *C*ontinuous). This is still not very special since we have only put all continuous functions on $[0, 1]$ into a large bag. The extra something of finite dimensional spaces was the possibility of measuring lengths and angles; additionally we can add points and multiply with a real number. In other words, the finite dimensional spaces carry a 'structure'. Is it possible to impress such a structure on the set $C([0, 1])$, i.e. can we measure distances between functions in $C([0, 1])$ or even angles between them? The answer is 'yes' and now we have arrived at the gates of functional analysis!

How can two functions $f, g \in C([0, 1])$ be added and each of them multiplied by a real number α? This turns out being trivial since it can be done pointwise,

$$(f + g)(x) := f(x) + g(x)$$
$$(\alpha \cdot f)(x) := \alpha \cdot f(x).$$

With these two operations the set $\{C([0,1]),+,\cdot\}$ is already a so-called 'vector space'; an enormously fruitful algebraic structure. Before we introduce other structures we ask ourselves what we would expect from a yardstick in three-dimensional space. We measure lengths and from a mathematical (and practical!) point of view only four requirements have to be fulllfilled,

1. The distance of x from itself should be 0,

   M1 $\qquad d(x,x)=0.$

2. If we measure the distance between x and y and the yardstick shows $d(x,y)=0$ then the two points were not really different from each other,

   M2 $\qquad d(x,y)=0 \quad \Rightarrow \quad x=y.$

3. It does not matter whether we measure the distance of x from y or of y from x,

   M3 $\qquad d(x,y)=d(y,x).$

4. In a triangle with vertices x,y,z the direct distance between x and y is never larger than the sum of the distances of x from some point z and of z from y,

   M4 $\qquad d(x,y)\leq d(x,z)+d(z,y).$

   This inequality is therefore called the 'triangle inequality'.

If a vector space carries such a yardstick, i.e. a distance function $d(x,y)$ satisfying [M1]–[M4], then the vector space is turned into a 'metric space' and d is called a 'metric'.

If we introduce as a yardstick the distance function

$$d(f,g):=\max_{x\in[0,1]}|f(x)-g(x)|$$

into the vector space $\{C([0,1]),+,\cdot\}$, then all four conditions are indeed satisfied! The vector space $\{C([0,1]),+,\cdot,d\}$, $(C([0,1],d)$ in short, is hence a metric space, i.e. we are able to measure distances between functions! But beware! This is *not* a distance which can be measured between the graphs of two functions (this would be elementary geometry in two-dimensional space) but an abstract distance. We could also have chosen the yardstick

$$d(x,y):=\sqrt{\int_0^1 |f(x)-g(x)|^2\,dx}$$

as a metric; this is also a valid distance function in $C([0,1])$.

It turns out that there is even a 'deluxe' yardstick in the vector space $\{C[0,1]),+,\cdot\}$, a mapping $C([0,1])\ni f\mapsto \|f\|\in\mathbb{R}_0^+$ which is called a 'norm'. A norm has to always satisfy three conditions, namely

N1 $\|f\| = 0 \Leftrightarrow f = 0$

N2 $\|\alpha \cdot f\| = |\alpha| \cdot \|f\|$

N3 $\|f + g\| \leq \|f\| + \|g\|$

If a vector space carrying a norm it is called a 'normed vector space'. In case of $C([0,1])$

$$\|f\|_\infty := \max_{x \in [0,1]} |f(x)| \tag{10.1}$$

is a norm, as is

$$\|f\|_2 := \sqrt{\int_0^1 |f(x)|^2 \, dx}. \tag{10.2}$$

Every normed vector space is automatically a metric space, but not every metric can be turned into a norm. Our deluxe yardstick on $C([0,1])$ in fact induces a metric

$$d(f,g) := \|f - g\|.$$

So much on lengths and distances, but what about measuring angles? In finite dimensional spaces like $\mathbb{R}^3$ we know the scalar product

$$\langle x, y \rangle := x \cdot y = |x| \cdot |y| \cdot \cos \angle(x,y),$$

with which we are able to measure the angle between the vectors x and y. Two vectors x and y are mutually orthogonal if $\langle x, y \rangle = 0$. We generalise the notion of inner products now to abstract spaces in saying that a function $\langle \cdot, \cdot \rangle : V \times V \to \mathbb{R}$ from the Cartesian product of a vector space V into the real numbers is an inner product, if the following three conditions are satisfied.

$\langle f + g, h \rangle = \langle f, h \rangle + \langle g, h \rangle$

S1 $\langle f, g + h \rangle = \langle f, g \rangle + \langle f, h \rangle$

$\langle f, \alpha g \rangle = \langle \alpha f, g \rangle = \alpha \cdot \langle f, g \rangle$

S2 $\langle f, g \rangle = \langle g, f \rangle$

S3 $\langle f, f \rangle \geq 0$ und $\langle f, f \rangle = 0 \Leftrightarrow f = 0.$

Spaces in which angles can be measured by an inner product are exceptionally important and exclusive and are named 'Pre-Hilbert spaces' after the German mathematician David Hilbert (1862–1943). It turns out that every inner product already induces a deluxe yardstick by means of

$$\|f\| := \sqrt{\langle f, f \rangle}.$$

Hence every Pre-Hilbert space automatically is a normed vector space. Every norm on a vector space defines a 'topology', i.e. a mode of convergence. If a vector space turns out being 'complete', i.e. if every Cauchy sequence in this space actually converges, then this normed vector space is called a 'Banach space' in honour of Stefan Banach (1892–1945). If a Pre-Hilbert space is complete, then it is called a 'Hilbert space'. Numerous subtleties have to be observed. Our example $C([0,1])$ is *no* Banach space with the norm (10.2), but with the norm (10.1) it turns out to be a Banach space!

We can now try to define 'functional analysis':

> Functional analysis is the analysis of mappings between infinite-dimensional spaces.

A general mapping between two spaces is called an 'operator'. A mapping from a function space to the real numbers $\mathbb{R}$ is called a 'functional' and this is where the name 'functional analysis' came from. Jacques Hadamard (1865–1963) coined the word 'fonctionelle' and Paul Pierre Lévy (1886–1971) used the notion 'analyse fonctionelle' in his book *Leçons d'analyse fonctionelle* of 1922 for the first time, [Siegmund-Schultze 1999, p. 490f.].

10.9.2 A Historical Outline of Functional Analysis

In his very instructive book [Dieudonné 1981, p. 8] Jean Dieudonné (1906–1992) presented a deterrently complex graphic concerning the development of functional analysis. Somewhat more manageable (but naturally coarser) is the graphic given in [Siegmund-Schultze 1999, p. 488]. In the opinion of Dieudonné functional analysis started already in the 18th century with the calculus of variations and the very beginnings of the theory of partial differential equations. In fact, we can identify an already large step of abstraction in the calculus of variations. Here extreme values are not sought for a single given function, but in a whole set of functions (being *points* in a function space!) we look for the one with a certain minimal property. An integral to be minimised like the one in the problem of the brachistochrone (8.1)

$$\frac{1}{\sqrt{2g}}\int_0^a \sqrt{\frac{1+(y')^2}{y}}\,dx \overset{!}{=} \min.$$

is nothing but a functional which maps functions $y = f(x)$ of a function space into the real numbers.

A further focal point for functional analysis was the Fourier series. Already Fourier himself came across infinite linear systems of equations in the computation of the coefficients of the series which he solved rather more intuitively than rigorously [Siegmund-Schultze 1999, p. 489]. When integral equations of the second kind came into focus at the end of the 19th century the problem of infinite linear systems of equations came up again. Integral equations of the second kind are equations of the form

Fig. 10.9.1. Vito Volterra [unknown photographer, about 1910] and Erik Ivar Fredholm [unknown photographer, probably before 1920]

$$\lambda u(x) + \int_a^b k(x,y)u(y)\,dy = f(x),$$

in which the 'kernel' k, the right hand side f, and the number λ are given and u is a sought-after function. The Italian mathematician Vito Volterra (1860–1940) had already in 1896 pointed out that these integral equations could be viewed as limiting cases of infinite linear systems of equations [Goodstein 2007]. Volterra had attempted already in 1887 to develop a calculus of functionals, later called 'calcolo funzionale' by Salvatore Pincherle. However, this attempt was rejected as being a 'generalisation for its own sake' (Verallgemeinerung um ihrer selbst willen) [Siegmund-Schultze 1999, p. 491]; it simply came prematurely. The Swedish mathematician Erik Ivar Fredholm (1866–1927) took up Volterra's remarks on integral equations in 1903 and saw them as the limiting case of the discretised problem

$$\lambda u_i + \sum_{j=0}^{n} \frac{1}{n} k_{ij} u_i = f_i.$$

To this end imagine the interval $[a,b]$ subdivided by $a = x_0 < x_1 < x_2 < \ldots < x_n = b$ into equidistant subintervals of width $1/n$. The kernel k as a function of two variables is a function defined on the Cartesian product $[a,b] \times [a,b]$ and k_{ij} simply means the evaluation $k(x_i, y_j)$. Fredholm developed a theory of integral equations in which time and again echoes of modern functional analytic ideas can be recognised.

Fig. 10.9.2. David Hilbert (1937) and his student Erhard Schmidt [Photo: Konrad Jacobs]

The emerging functional analysis got a creative thrust by the works of the Göttingen mathematicians, in particular by David Hilbert (1862–1943) and his student Erhard Schmidt (1876–1959).

In a sequence of communications in the *Nachrichten der Königlichen Gesellschaft der Wissenschaften zu Göttingen* from 1904 to 1910 Hilbert founded the theory of integral equations completely anew. For so-called symmetric kernels ($k(x,y) = k(y,x)$) in particular he built a bridge to the theory of eigenvalues in linear algebra. He hence treated integral operators as the linear operators (=matrices) of linear algebra. His student Schmidt got his PhD under Hilbert in 1905 with a work on integral equations and superseded his teacher in using completely different methods. He succeeded in proving Hilbert's important results only out of structural properties of a function space and he thereby achieved greater generality. The works of Hilbert and Schmidt, very important for the development of functional analysis, were reprinted recently in [Hilbert/Schmidt 1989].

On the way, which was mapped out by Volterra his student Cesare Arzelà (1847–1912) moved on. He tried to generalise Weierstraß's theorems on continuous functions to fit into Volterra's calcolo funzionale. He thereby built upon results by Giulio Ascoli (1843–1896) of the year 1884. Hence the theorem of Arzelà-Ascoli came to life providing information on the behaviour of continuous and bounded *families* of functions. In [Siegmund-Schultze 1999, p. 492] the theorem of Ascoli of 1884 is denoted as the first substantial result on infinite-dimensional function spaces.

Fig. 10.9.3. Maurice René Fréchet, Cesare Arzelà, Salvatore Pincherle

The French Maurice René Fréchet (1878–1973) immediately took up Arzelà's works and in his dissertation of 1906 he not only introduced the notion of an abstract metric space but also investigated functionals defined on them. He also succeeded in drafting the notion of compactness in an abstract sense [Siegmund-Schultze 1999, p. 492]. We have to mention also Felix Hausdorff (1868–1942) in this context since he was concerned with abstract 'topological spaces'.

The Italian Salvatore Pincherle who coined the notion 'calcolo funzionale' investigated linear operators on infinite-dimensional vector spaces already in 1901 and axiomised them. The works of Pincherle were not received though [Siegmund-Schultze 1999, p. 493], and the foundation of 'linear functional analysis' was hence left to a Polish school of mathematicians. A first victory of the Lebesgue integral was won in 1907 by the Austrian mathematician Ernst Sigismund Fischer (1875–1954) and, independently, by the Hungarian Frigyes Riesz (1880–1956) when they succeeded to prove the famous theorem of Riesz-Fischer, which eventually founded Fourier analysis in Hilbert spaces.

Fig. 10.9.4. Stefan Banach, honoured on a stamp (Poland 1982)

Fig. 10.9.5. Kawiarnia Szkocka (Schottisches Café) in Lemberg, now the "Desertniy Bar" in Lwiw (Ukraine); Hugo Steinhaus [unknown photographer, 1968]

On 30th March 1892 Stefan Banach (1892–1945) was born in Cracow into disordered family circumstances. He grew up in a foster family and attended grammar school which he finished successfully in 1910. He then worked in a bookshop in Cracow and studied mathematics in his leisure time as an autodidact. From 1911 to 1913 he studied mathematics at the University of Lemberg (Polish: Lwów, today Ukrainian: Lwiw) until the intermediate diploma. Due to his poor eyesight he was not suited to serve in WW I. Instead he became a supervisor in road construction before he returned to Cracow after the war. During wartime only very few informations on Banach are available [Kałuża 1996].

As far as we know Banach earned his living with private lessons while he privately continued his mathematical studies. In the year 1916 he met the mathematician Hugo Dionizy Steinhaus (1887–1972). Steinhaus was impressed by the young mathematician and within a short time not only a joint publication resulted but also a longstanding friendship. Steinhaus supported Banach who got an assistantship at Lemberg Polytec with Antoni Łomnicki (Łomnicki was assassinated on 4th July 1941 together with 25 further Polish professors shortly after the invasion of the German Wehrmacht) in 1920 where he stayed until 1922.

Banach published further on and took his doctoral examination in 1922 from the Jan Kazimierz University in Lemberg. The title of his dissertation thesis was *Sur les opérations dans les ensembles abstraits et leur application aux équations intégrales* (On the operations in abstract sets and their application to integral equations). This thesis can be seen as the actual birth certificate

Fig. 10.9.6. Stanisław Mazur presents the reward to Per Enflo [unknown photographer, 1972]

of functional analysis. Still in the same year Banach habilitated in Lemberg and got an extraordinary professorship. In 1927 he became full professor.

Banach was an eccentric. From 1922 until 1939 he sat in a coffee house in Lemberg to work there and not in his office at the university. In this coffee house of name 'Kawiarnia Szkocka', the 'Scottish Café', by and by the *créme de la créme* of Polish mathematicians gathered around Banach. They wrote on the table surfaces with crayons so that every day the tables of the Scottish Café had to be elaborately cleaned by the waiter. It took not long that the owner of the Café lost his patience. He bought a book with plain pages and ordered the mathematicians to write in it instead on the table tops. This book, the 'Scottish book' has come to world fame since it includes many of the most important ideas, problems, and theorems of functional analysis.

Fortunately the Scottish book has survived WW II and was partly translated into English [Mauldin 1981]. It can also be found on the internet [Szkocka]. This freely accessible internet version has been created by Stanisław Marcin Ulam (1909–1984), member of the Lemberg circle and Banach's student. The book contains all important basic theorems of functional analysis. There are problems posed and rewards for solutions are offered. Stanisław Mazur (1905–1981), one of Banach's students, offered a living goose as a reward for a proof of a theorem in the theory of Banach space in 1936. In the year 1972 the reward could finally be presented by Mazur himself to the Swedish mathematician Per Enflo (*1944).

Banach received many honours and was a member of the Polish Academy of Sciences, of the Warsaw Scientific Society, and the Scientific Society of Lemberg. Already in 1919 he became a founding member of The Polish Mathematical Society. In 1932 Banach's book *Théorie des opérations linéaires* [Banach 1987] was published and quickly became the 'bible' of functional analysis. It is in this book that Banach laid the founding stones of the theory of spaces which today carry his name.

In 1939 the Red Army reached Lemberg and incorporated the city into the Ukrainian Soviet Republic until 1941. Banach became a corresponding member of the Academy of Sciences of the Ukrainian Soviet Republic and kept his professorship in Lemberg where he was the first inhabitant of the chair of analysis and dean of the faculty of mathematics and physics. In 1941 the Nazi terror began also in Lemberg. Banach escaped the deliberate killing of 25 Lemberg professors but he was humiliated. He worked as 'living feed' for lice at the institute of the Polish bacteriologist Rudolf Weigl who did research on the lice and illnesses they could transfer to humans. This was the way Banach could unobtrusively disappear from the scene but he had to pay a price. As Lemberg mathematician Jadwiga Hallaunbrenner remembered [Kałuża 1996, p. 88f.] Banach was in a pitiful physical condition. Weigl finally succeeded in finding a means against the typhus the lice transmitted but Banach stayed 'lice feed' until the end of the Nazi occupation of Lemberg, hence until 1944. Then the Red Army took over and Banach again became professor of mathematics. At this time the chain smoker Banach was already moribundly ill with lung cancer. He passed away on 31st August 1945 and was buried in the Lytschakiwski cemetery in Lemberg.

Development of Analysis in the 19th and 20th Century

1828	George Green publishes his integral theorem and develops a method for the solution of boundary value problems by means of Green's functions
1840	C.-F. Gauss proves his divergence theorem of vector analysis
1854	George Gabriel Stokes proves his integral theorem of vector analysis
	Bernhard Riemann founds modern differential geometry with his habilitation lecture *Hypothesen, welche der Geometrie zugrunde liegen*
1870	Eduard Heine defines uniform continuity of functions and proves the theorem of termwise integration of a uniformly convergent series
1872	Georg Cantor publishes a construction of real numbers by means of Cauchy sequences. Cantor develops set theory. Heine formulates the famous Heine-Borel covering theorem which E. Borel 1895 recognises as an independent theorem and proves for points on a straight line

1882	PAUL DU BOIS-REYMOND publishes *Die Allgemeine Functionenlehre*
1883	GIULIO ASCOLI defines equicontinuity and proves a theorem, generalised by C. ARZELÀ, which is now called the theorem of Arzelà-Ascoli
	CANTOR informs DEDEKIND of his proof of the denumerability of real numbers
1887	VITO VOLTERRA develops his functional calculus
	CAMILLE JORDAN formulates the famous Jordan curve theorem (proven only 1893 in the *Cours d'Analyse*).
	GREGORIO RICCI-CUBASTRO introduces the notion of tensors (on a differentiable manifold), defines the so-called covariant derivative, and builds an analytical-algebraical tensor calculus
	LEOPOLD KRONECKER publishes his radical program to arithmetise mathematics and used it for a rigorous development of analysis
1888	*Was sind und was sollen die Zahlen* by DEDEKIND published in Brunswick
1889	GIUSEPPE PEANO provides a foundation of the natural numbers and a generic development of the whole number system in his *Arithmetices principia*
1890	ERNST LEONARD LINDELÖF and ÈMILE PICARD prove the existence of solutions of ordinary differential equations by means of successive approximations
	CANTOR founds the German Mathematical Society DMV
1891–1901	RICCI-CUBASTRO and his pupil TULLIO LEVI-CIVITA develop the absolute differential calculus, generalise it to tensors in arbitrary local coordinate systems on an n-dimensional manifold, and define the abstract tensor
1892	C. JORDAN defines the inner and outer content of bounded n-dimensional domains and publishes a general theory of measure
1895	CANTOR proofs the numerability of the rational numbers with his first diagonal method and develops transfinite arithmetic
1896	V. VOLTERRA views integral equations as limiting cases of infinite systems of equations
1899	ÈLIE JOSEPH CARTAN develops a theory of alternating differential forms and proofs a generalisation of Stokes's integral theorem
1901	HENRI LEBESGUE develops a new notion of measure and defines the Lebesgue integral
	BERTRAND RUSSELL discovers antinomies in Cantor's set theory
1904–1910	DAVID HILBERT founds the theory of integral equations

1906	The first textbook on set theory by William Henry Young and Grace Chisholm Young is published
1907	Ernst Zermelo presents the first axiomatisation of set theory
	Ernst Fischer and Frigyes Riesz prove (mutually independent) the famous theorem of Fischer-Riesz (on the completeness of the space L^2 of functions being square integrable in Lebesgue's sense)
1912	Luitzen Brouwer becomes professor of mathematics in Amsterdam. He develops intuitionistic mathematics
1914	Felix Hausdorff publishes the first German textbook on set theory
	Das Kontinuum by Hermann Weyl published
1921	Adolf (Abraham) Fraenkel starts to improve Zermelo's system of axioms
1922	Stefan Banach develops important foundations of functional analysis
1927	Banach and H. Steinhaus prove basic theorems on linear operators in Banach spaces
1929	Banach generalises a theorem by Hahn of 1927 to result in the theorem of Hahn and Banach, and proves it by means of Zorn's lemma
1930	Final form of the Zermelo-Fraenkel set theory
1932	Stefan Banach publishes *Théorie des opérations linéaires*
1938	Kurt Gödel proves the independence of the continuum hypothesis from the axioms of set theory
1945	Laurent Schwartz generalises the notion of functions to distributions
1948	Jan Mikusinski presents an abstract development of the theory of distributions
	Laurent Schwartz proceeds in his development of the theory of distributions
1950	L. Schwartz publishes his fundamental book *Théorie des distributions.*
	Sergej L'vovič Sobolew systematically builds his theory of generalised derivatives with applications in functional analysis and mathematical physics
1963	Paul Cohen proves the existence of two different mathematics; one with continuum hypothesis, the other without

11 Coming to full circle: Infinitesimals in Nonstandard Analysis

T. Sonar, *3000 Years of Analysis*, https://doi.org/10.1007/978-3-030-58223-4_11

General History From the End of WW II to Today

1945	In the Potsdam Agreement of 2nd August 1945 East Prussia, Farther Pomerania, and Silesia are put under Soviet or Polish administration, respectively. Austria announces its independence and is, like Germany, divided into 4 occupation zones of the victorious powers Establishment of the United Nations Organisation (UNO) on the 24th October
1947	On 15th August India and Pakistan arise as independent national states from dividing British Raj. This leads to a conflict in Kashmir still going on today
1948	Start of the Marshall Plan
1949	Germany divided by the establishment of the Federal Republic of Germany (FRG) and the German Democratic Republic (GDR) Countries in the sphere of influence of the Sovjet Union become People's Republics: Albania, Bulgaria, Yugoslavia, Poland, Romania, Hungary, North Korea, North Vietnam, Mongolia. Mao Zedong proclaimed in Beijing the People's Republich of China on 1st October
1950–1953	Korean War
1953–1960	Dwight D. Eisenhower president of the USA
1953	Stalin dies on 5th March Popular uprising in the GDR on 17th June, starting in East Berlin, is suppressed by Sovjet troops
1955	The FRG becomes a sovereign state and member of NATO. The GDR becomes member of the Warsaw Pact
1956	Popular uprising in Hungary on 23rd October, bloodily suppressed by Sovjet troops
1957	Sovjet Union's 'Sputnik' is the first man-made sattelite on 4th October. The 'Sputnik shock' in the USA and later in other Western countries leads to investments in the education systems. Beginnings of the European Union (EU): The European Economic Communitiy (EEC) is established
1958–1964	Nikita Sergeyevich Khrushchev head of the USSR
1959–1969	Charles de Gaulle French president
1960	'African Year': New independent African states emerge from Belgian, British, and French colonies. Bloody fights and tribal feuds follow in many places, in particular in former Belgian Congo
1961	The building of the Berlin Wall starts
1961–1963	John F. Kennedy president of the USA
1962	Cuban Missile Crisis
1964–1982	Leonid Iljitsch Breschnew party leader of the CPSU
1965–1975	Vietnam War
1965–1975	'Great Proletarian Cultural Revolution' in China

1968	Czechoslovakia invaded by troops of the Warsaw Pact. The 'Prague Spring' is terminated. May 1968 events in France: After student protests general strikes occur all over France for weeks. The student protests reach Germany
1969–1974	Willy Brandt chancellor of the FRG
1969	The Apollo 11 Mission put the first man on the moon
1976	Mao Zedong dies
1979	Theocracy of Ayatollah Khomeini terminates the reign of the Shan in Iran
1979–1989	Occupation of Afghanistan by the Sovjet Union
1979–1990	Margaret Thatcher becomes Prime Minister of Great Britain
1988–1991	Michail Sergejewitsch Gorbatschow, last head of state of the Sovjet Union
1989	Fall of the Berlin Wall on 9th November
1990	German Reunion. The GDR joins the FRG
	Second Gulf War: With an UN mandate the USA force troops of the Iraq to leave occupied Kuwait
1991	Dissolution of the Soviet Union
ab 1991	Collaps of Yugoslavia (Slovenia 1991, Croatia 1991, Macedonia 1991, Bosnia/Herzegovina 1992, Montenegro 2006)
1993	Establishment of the European Union
1996	22.6 millions infected by HIV
1999	European Ministers of Education agree on the so-called Bologna reform of university education
2001	11th September: Al-Qaeda attacks the Twin Towers in New York and further targets in the USA and provoke the 'War against terror'
	On 7th October the USA and Great Britain start the operation *Enduring Freedom* against the rule of the Taliban in Afghanistan
2001–2009	George W. Bush president of the USA
2002	Introduction of the 'Euro' as new currency in parts of Europe
2003	The USA overthrow Saddam Hussein by occupying Iraq in a 'Coalition of the willing'
2005	Election of Angela Merkel as chancellor of the FRG
2008	Insolvency of the US investment bank *Lehman Brothers* starts a global economic crisis
2009	Election of Barack Obama as president of the USA
	Intensification of the economic and financial crisis
2010	Financial crises in South Europe and Ireland lead to a crisis of the Euro
2011	Uprisings in Tunisia, Egypt, Libya, Syria, and in Yemen
2015	Large flows of migrants from Africa and Asia reach Europe. Germany deliberately violates the Dublin Regulation and welcomes migrants without checking their identity
2016	In a referendum of 23 June 2016 the majority of British voters supported leaving the EU leading to the 'Brexit'. In Turkey an attempted coup on 15th July failed
2017	20th January: Donald Trump is sworn as the 45th President of the USA

Developments in Natural Sciences and Technology

1951	Experiments of controlled nuclear fusion in the USA and Great Britain
1954	First nuclear fusion plant (Tokomak) build in the USSR
	First solar cell to gain electric energy from radiation energy
1957	USSR: First artificial satellite 'Sputnik'
1959	Lunik III takes photos from the far side of the moon
1960	Start of works on the Aswan dam in Egypt
1961	First manned spaceflight (USSR)
	First American circumvention of the earth in space
1967	First transplantation of a human heart by Christiaan Barnard
1971	First image from a computerised tomography system (CT)
1972	First commercial CT system
1974	The danger of the destruction of the earth's ozone layer is recognised
1976	Computer- aided proof of the four colour theorem
1983	Space Shuttle Columbia (USA) starts with space laboratory Spacelab
1985	Ozone hole above the Antarctic established
	Personal computers introduced in the working world
	Breakthrough in the development of magnetic resonance imaging (MRI). MRI is immediately used in clinical practice
1986	Soviet space station 'Mir'
	Nuclear power plant explodes in Chernobyl (Ukraine)
1988	Theory of the Big Bang supplemented
1990	The Hubble space telescope in orbit
1991	Neolithic circular enclosure discovered close to Goseck, Germany. The enclosure is about 6800 years old and ranks as the oldest 'sun observatory' worldwide
1993	The World Wide Web (WWW) is made available for general use
1994	Millennium event: Impact of the comet 'Shoemaker-Levy 9' on Jupiter
	Railway tunnel below the Channel opened
1995	Final proof of Fermat's Last Theorem by Andrew Wiles
	Space probe 'Galileo' (lift off 1989) reaches Jupiter
	Microsoft introduces 'Windows 95'
1997	Clone sheep 'Dolly' presented
	World chess champion Garry Kasparov competes against a chess program installed on the IBM supercomputer 'Deep Blue' and loses
1999	First balloon ride around the earth under the leadership of B. Picard.
	Discovery of the 'Nebra sky disk' from the bronze age
2000	New space station ISS manned
2004	Space probe 'Mars Express' starts charting Mars's surface (3D Photos)
2008	Smartphones replace classical cell phones and lead to far reaching changes in communication
2016	Proof of gravitation waves as a further confirmation of general relativity theory
2017	IBM announces 'IBM Q', an initiative to build a universal quantum computer for commercial use
	At the Future of Go Summit in China, Google's 'DeepMind AlphaGo' artificial intelligence program beats the world's number one Go player in the third of three matches
2018	A novel class of naturally-produced antibiotics is formally described
	Human corneas printed by an 3D printer for the first time

11.1 From the Cold War up to today

After the end of WW II in 1945 Germany was occupied by the allies. Streams of refugees pervaded a widely ruined country and amplified the social misery of the population. The allies could initially not agree on a joint politics and hence occupation zones were installed. In the Soviet occupation zone the occupying power enforced a centralistic state leadership from the beginning. The tension between the Western allies USA, Great Britain, France on one side, and the Soviet Union on the other side, noticeable since 1945, quickly led to a 'cold war' which lingered well into the 1980s. The Berlin Blockade of 1948 in which West Berlin was cut off the West by the USSR almost escalated. In China Mao Zedong proclaimed the People's Republic on 1st October 1949.

Emerging from the Western occupation zones the Federal Republic of Germany (FRG) was established in 1949 while the Soviet zone became the German Democratic Republic (GDR) in the very same year. Planned economy now stood against capitalism, communism against democracy of the West. Although the cold war smouldered mainly between the USA and the USSR and stayed below the threshold of a hot war, proxy wars were fought. The first of these proxy wars was the Korean War starting in 1950 where the USA supported South Korea while the People's Republic of China was on North Korea's side. Three years after the conflict broke out a ceasefire could be negotiated but as a consequence of the Korean War a hysteria against communism developed in the USA. Already in 1938 the House Un-American Activities Committee had been established which now became active again under the leadership of senator Joseph McCarthy. Besides communists the hunt was also opened for liberal intellectuals and McCarthy was removed only in 1954.

In the FRG the rearmament started in 1952; in the GDR the barracked People's Police (Volkspolizei) was established as a paramilitary unit. When after Stalin's death in 1953 one could hope for a slight relaxation under his successor Nikita Khrushchev the model of a 'peaceful coexistence' seemed to have come into reach. Sadly though, the Uprising of 1953 in East Germany on 17th June 1953 was overpowered by Soviet troops by means of force of arms and let the cold war back in again. The 'balance of terror' between the two superpowers was reached in 1954 when the Soviet Union caught up with the USA weapon-technologically with the H-bomb and airplanes with intercontinental range. When the FRG was received into the NATO in 1955 the Soviet Union established the Warsaw Pact as an opposing construct. In the Hungarian Revolution of 1956 the USSR used force of arms again to bloodily suppress the uprising.

The 1950s saw an 'economic miracle' in the FRG mainly fuelled by help of the USA. On 5th May 1955 the FRG became a sovereign state. The Berlin Wall between the two parts of Germany made the boundary between the capitalist West and the communist East permanent in 1961; Churchill's astute notion

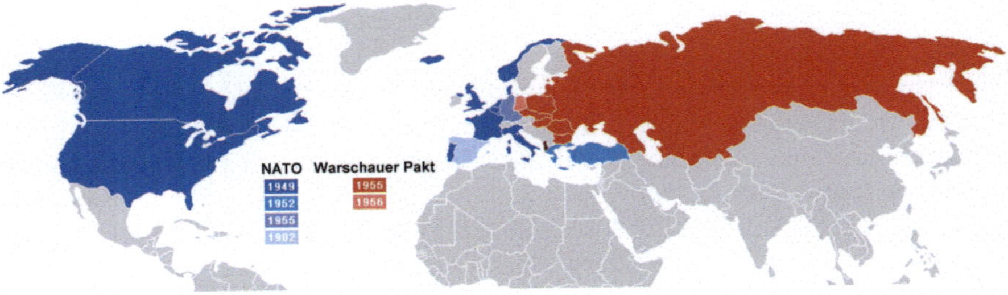

Fig. 11.1.1. Countries of the NATO and of the Warsaw Pact

of the 'iron curtain' had become manifested in concrete, barbed wire, and a mined death strip. While a social market economy developed in the FRG a system of planned economy was followed in the GDR.

In the 1950s also the aspirations establishing a European Union started which was only realised in the Maastricht Treaty in 1992. On 25th March the European Economic Community (EEC) was founded including the countries Italy, Luxembourg, the Netherlands, the FRG, Belgium, and France. The contracts concerning the establishment of the European Communities (European Coal and Steel Community ECSC 1951, European Atomic Energy Community EAEC/Euratom 1957, European Economic Community EEC 1957) were replaced with basic changes and supplements by the treaty on the European Union on 1st November 1993. This treaty was revised in the treaties of Amsterdam (effective 1st May 1999) and Nice (effective 1st February 2003) and finally in the treaty of Lisbon becoming effective on 1st December 2009.

In the second half of the 20th century the colonial times reached its end. Already in 1945 Korea and Indonesia became independent: in 1947 India and Pakistan followed. The year 1960 is called the 'African Year' since numerous African colonies were released into independence. As a consequence at many places bloody quarrels arose between long-established tribes and new regents, in particular in Belgian Congo. In North Africa Tunisia and Morocco had achieved their independence of France already in 1956. Algeria became an independent republic only after many uprisings in 1962; the Portuguese colonies Angola and Mozambique followed as late as 1975. However, it took until 1999 that the colonial era could truely be termed finished. In 1997 the British Crown Colony Hongkong including the New Territories was handed back to China, in 1999 the same happened to the Portuguese colony Macau, being in the possession of the Portuguese since 1577.

11.1.1 Computer and Sputnik Shock

After the first programmable electronic computer, built by the German civil engineer Konrad Zuse (1910–1995) in the 1940s, the computer technology experienced an enormous boost after the end of WW II, cp. [Wußing 2009, p. 517–528]. While the first large machines of IBM were reserved for large companies the introduction of the 'personal computer' at the end of the 1970s led to a wide distribution of computers even in private homes. Our modern life can not be imagined without the computer and the computer has even influenced analysis. Since the 1950s numerical analysis, dealing with the mathematics and design of computer algorithms, has made tremendous advances.

Already in the 1960s the US Department of Defence had put the order to build a computer net which went into operation in 1969 as ARPANET (Advanced Research Project Agency NET). While ARPANET connected only a few universities and research facilities in the USA to take advantage of computer resources the internet developed rapidly from it. Since 1993 the 'world wide web' www is in existence which was developed in 1989 at the research facility CERN. The www certainly has intervened in social aspects of life as strongly as the Industrial Revolution of the 19th century.

The process of globalisation was also strongly driven by computers. Today financial markets on all continents are digitally networked; consumers can order books, cloths, or jewellery from all over the world and even politics has not been able to get around globalisation. Although this globalisation has numerous advantages for men in all parts of the world the disadvantages also show clearly. If the movement of capital and goods is globalised then decisions, for example concerning production areas, can also be taken from a global point of view and hence trigger unemployments in some countries.

The Sputnik shock initiated an educational initiative in the West from which also mathematics profited. In 1957 the Soviet satellite Sputnik was the first man-made satellite orbiting the earth and caused a stir in the USA. The technological lead of the USA, thought to be secured, was suddenly lost. The budget of the National Science Foundation NSF was quadrupled, the teacher training was promoted, and new schools were built. The school subjects mathematics and physics got a high priority and in 1958 NASA was established. When Russian cosmonaut Juri Gagarin became the first man in space the space race started which was finally won by the USA with the first landing of men on the moon in 1969.

In the 1960s and 70s school education was reformed in many countries; in mathematics the reason was mainly the aftermath of the Sputnik shock. Hence the reform movement swept over from the USA where it was known as 'New Math'. Following the Bourbaki group who since 1935 had treated mathematics from a strong structural point of view and produced highly

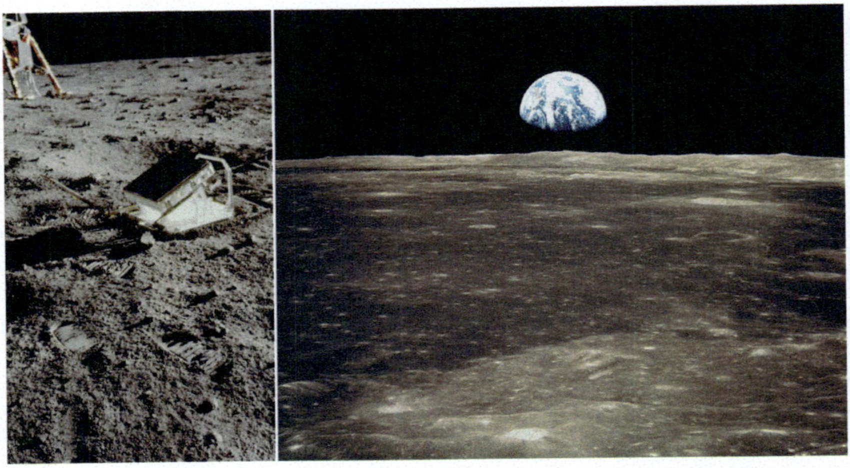

Fig. 11.1.2. Laser reflector on the moon; Rise of the earth over the horizon of the moon. With the help of laser reflectors (installed by the crew of Apollo 11) the (varying) distance earth – moon can be measured with millimeter precision. Accordingly, the distance earth – moon has increased by about 1.5 m during the last 40 years [Photos: NASA, AS11-40-5952 HR und AS11-44-6550]

influential textbooks [Mashaal 2006], one wanted to teach mathematics in schools at a more abstract level. Despite intense resistance and concerns of farer-looking university professors of mathematics the program of the New Math was installed. Even the mathematician Jean Dieudonné (1906–1992), being an important member of the Bourbaki group, dissociated clearly from New Math in schools. In Germany so-called didactics experts eagerly took up the abstraction of mathematical education under the catchword 'set theory' in the 1970s, although New Math was already considered a failure in the USA at this time. If one has understood mathematics from particular examples up to abstract theories (in this direction!) then one is proficient in it. Learning abstract notions of sets in primary schools leads to underdeveloped numeracy, however. Since also in Germany the resistance of parents and teachers against set theory was extraordinarily high and since a study proving the superiority of the new education turned out to be formally incorrect one eventually returned to classical teaching in mathematics. Today the use of powerful pocket calculators and computer algebra systems (CAS) are on the agenda of reform together with a disastrous tendency to teach only 'competences' instead of mathematics which has served to drive important content, among it analysis, out of our schools.

Fig. 11.1.3. West Berliner at the wall close to the Reichstags in December 1989 [Photo: Dr. Alexander Mayer]

11.1.2 The Cold War and its End

The Cuban Missile Crisis of 1962 comprised a serious conflict between the USA and the USSR. In 1965 the USA intervened in Vietnam militarily. While the North had become communist the South had stayed anti-communistic. When a civil war broke out in the South the USA felt their interests in the region were in danger and intervened while the North was supported by the USSR and China. War atrocities together with enormous losses of life of the population as of American soldiers led protest movements grow worldwide. Students in particular went to civil disobedience and demanded a change of social structures. The advent of protests coincided with the black civil rights movement in the USA and with a far-reaching emancipation of the youth in the hippie movement. In the FRG the positive image of the USA became heavily damaged. Only in 1973 a peace treaty was signed; in 1975 the communists marched in Saigon and all of Vietnam came under communist control.

In the FRG the student body at the end of the 60s consisted of children of a generation having lived under the Nazi regime. Disappointed by the speechlessness of the parental generation one now demanded a reappraisal of the time of the Nazi tyranny. Driven by the May revolt of the students in Paris in 1968 a complex student's movement developed, turning against the 'generation of the perpetrators'. At the same time these students protested

Fig. 11.1.4. The new map of Europe (2007) after the dissolution of the Soviet Union and Yugoslavia

against the sexual moral of chancellor Adenauer's FRG and the authoritarian tendencies of the 50s. They aimed at a far-reaching democratisation of the country. The beginnings of the student's movement could be seen already in 1961. The shooting of the student Benno Ohnesorg on 2nd July 1967 and the assassination attempt on Rudi Dutschke on 11th April 1968 led to a radicalisation of the student's movement. A radical minority went underground and formed the 'Red Army Faction' RAF the terror of which reached its climax in the 'German Autumn' September-October 1977.

A centralisticaly controlled planned economy like that of the Eastern block had no chance to win the race with the West in the long run. First attempts to open the USSR came in 1986 when Michail Sergejewitsch Gorbatschov was elected General Secretary of the CPSU. His policy of 'glasnost' (openness and transparency) brought a democratisation and a 'perestroika' (restructuring) of the Eastern bloc. The Soviet Union disintegrated in 1990/91 with the proclamation of independence of the Baltic republics Lithuania, Latvia, and Estonia. The Union was finally dissolved in 1991. In the GDR thousands of dissatisfied citizens went out on the streets and forced the regime to withdraw in 1989. On 3rd October 1990 the GDR joined the FRG on the basis of the unification treaty.

11.1.3 Bologna Reform, Crises, Terrorism

In order to create a European higher education area 29 European Ministers of Education signed the so-called Bologna Agreement in 1999. Until 2010 a unified higher education area with comparable educational certificates should have been realised. The Bologna reform met widespread criticism in university circles and is still an issue of fierce controversies. In many European countries classical and well-established university education systems were destroyed and at least in Germany, the only country I can speak of, it has destroyed even the comparability of educational certificates between universities in two different cities a few kilometers apart.

The second half of the 20th century was also characterised by great crises which may threaten our future on this planet. In 1960 about 3 billions humans populated the earth, in 1974 already 4 billions, in 1987 already 5 billions, and the 7 billion threshold was reached in 2012. This overpopulation is accompanied by controversial discussion on a climate change originating in emissions of industries, household heating, and the enormous numbers of cars.

On 11th September 2001 four regular passenger airplanes were kidnapped by members of the Islamist terror organisation Al-Quaida. Two of those airplanes were steered into the towers of the World Trade Centre in New York, one bombed the Pentagon, and another crashed in Pennsylvania after a fight of passengers with the kidnappers. President George W. Bush thereupon started a war in Afghanistan already in October 2001. The start of the war on Iraq in 2003 was also partly justified by the terror attacks. These terrorist attacks have been a historical turning point which had worldwide effects and still have.

The dangers of globalisation could clearly be seen in the collapse of the 'New Economy'. In March 2003 the so-called 'dotcom-bubble' burst. The market of companies trading via the internet having a www-ending '.com' (dotcom) was gravely overvalued at the stock exchange and led to enormous losses of investor's assets in the years from 2000 to 2003. Many of the 'dotcom'-companies went bust and disappeared from the market.

In the years 2007/08 massive losses and liquidity shortages of banks in the USA and Europa led to massive turbulences of the financial markets and a worldwide crisis of economies and financial markets was triggered by the bankruptcy of the US investment bank *Lehman Brothers* on 15th September 2008. Despite intense measures and aid programs of many states, the European Union included, this crisis reached its climax in 2009 and caused doubts on the stability of the EURO in the EU in 2010 by the impending collapse of the national budgets of Greece and Ireland.

Fig. 11.1.5. The debris of the World Trade Centre after the terrorist attack [Photo: National Oceanic and Atmospheric Administration, 2001]

11.2 The Rebirth of the Infinitely Small Numbers

Detlef Laugwitz (1932–2000), having been born in Breslau (which now is Wrocław in Poland), studied mathematics, physics, and philosophy from 1949 on at the University of Göttingen. Controlled by the British military government this was the only German university which could open its doors to students already in the winter semester 1945/46. Already in 1954, as a 22-years old, he passed his doctoral exam with the thesis *Differentialgeometrie ohne Dimensionsaxiom* (Differential geometry without an axiom of dimension) and got a research stipend of the German Research Foundation DFG. The stipend allowed him to study at the Oberwolfach Research Institute for Mathematics MFO and at the University of Erlangen. Following his habilitation at the University of Munich he became private tutor there in 1958, but he then moved to the Technische Hochschule (TH) Darmstadt (today the Technical University (TU) Darmstadt) where he became professor of mathematics in 1962. With the exception of two stays at the California Institute for Technology CALTECH as guest professor in 1976/77 and 1984/85 he kept faithful to the TU Darmstadt. He was appointed university professor by the rector of the TH at that time, Curt Schmieden, [Spalt 2001, p. 141].

Fig. 11.2.1. Detlef Laugwitz [Photo: Konrad Jacobs]

Laugwitz had met the older Curt Schmieden (1905–1991) already at the end of the 1950s. Schmieden had done his doctorate work with Richard von Mises (1883–1953) in Berlin where his doctoral thesis of 1929 was on a topic of fluid mechanics. He afterwards habilitated in Dresden in 1931 also with a thesis from fluid mechanics and became an associate professor of mathematics at the University of Rostock. In 1937 he got a professorship of mathematics at the TH Darmstadt. Schmieden was an applied mathematician in the best sense and an expert in the area of fluid mechanics. Therefore he worked for the Deutsche Versuchsanstalt für Luftfahrt (German Laboratory for Aviation) in Berlin during WW II. After the war Schmieden became rector of the TH Darmstadt in 1957/58 where he retired in 1970. Since the end of the 1940s Schmieden kept a strange book which he called 'the black book'. It is this 'black book' with which the rebirth of infinitesimals began.

11.2.1 Mathematics of Infinitesimals in the 'Black Book'

The 'black book' of Curt Schmieden has not been published up to this day. However, Detlef Spalt, one of the last of Laugwitz's students, had visited Curt Schmieden in December 1979 and was allowed to copy the manuscript *Vom Unendlichen und der Null – Versuch einer Neubegründung der Analysis* (From the infinite and the zero – attempt of a refoundation of analysis), and that *is* the 'black book'. We follow here closely the work [Spalt 2001] by Detlef Spalt.

In the 1940s Schmieden had an idea which was quite extraordinary in those days. He wanted to remove the 'paradoxes' of classical analysis. Of course, we can not speak of true paradoxes of classical analysis since the 19th century had put analysis on rigorous foundations. Let us say: Schmieden (like many students today) felt puzzled and uncomfortable by some of the results and notions of analysis and thought them too complicated or too overblown, in particular if he imagined teaching engineers and physicists. We shall discuss an example soon. To achieve his 'refoundation of analysis' Schmieden introduced a 'horizon of finiteness'. Beyond this horizon infinitely large numbers exist which can not be arrived at by the process of continual adding but only by a certain 'jump'. However, as a rule to be followed, one should be able to compute with infinitely large numbers exactly in the same way in which one computes with finite numbers. In the 'black book' Schmieden denoted an infinitely large number by 'Ω'. Hence for all n and m also

$$\Omega+1, \Omega+2, \ldots, \Omega+n$$

and

$$\Omega-1, \Omega-2, \ldots, \Omega-m$$

are infinitely large numbers, but it holds

$$\Omega-\Omega=0, \quad \Omega-(\Omega-1)=1,$$

and so on, cp. [Spalt 2001, p. 142]. Since also constructions as $\Omega \cdot \Omega$ oder Ω^{Ω} are possible *scales* of infinity open up. Schmieden chose

$$\Omega:=\lim_{n \to \infty} n$$

as infinitely large number and hence defined a scale. With this choice

$$\omega:=\frac{1}{\Omega}$$

is an infinitely small number. The different scales of infinity now quite naturally lead to different scales of the infinitely small. If in classical analysis it is true that

$$\lim_{n \to \infty} \frac{1}{n}=\lim_{n \to \infty} \frac{1}{n^2}=0,$$

the scales of Schmieden lead to

$$\omega=\lim_{n \to \infty} \frac{1}{n} \neq \lim_{n \to \infty} \frac{1}{n^2}=\omega^2,$$

and $\omega>\omega^2$. Immediately the important result of classical analysis that all subsequences of a convergent series converge to the same limit, is lost! However, introducing 'equality up to infinitely small quantities', Spalt uses the symbol '$\overset{\circ}{=}$', then

$$\lim_{n \to \infty} \frac{1}{n} \overset{\circ}{=} \lim_{n \to \infty} \frac{1}{n^2} \overset{\circ}{=} 0$$

and $\omega \overset{\circ}{=} \omega^2 \overset{\circ}{=} 0$ holds. In case of Euler's exponential functions in Schmieden's analysis

$$\mathrm{e}^x = \lim_{n\to\infty}\left(1+\frac{x}{n}\right)^n = \left(1+\frac{x}{\Omega}\right)^\Omega = (1+\omega x)^\Omega$$

is valid. Hence we can write

$$\mathrm{e}^x \overset{\circ}{=} (1+\omega x)^\Omega,$$

cp. [Spalt 2001, p. 143]. What 'paradoxes' can now be removed by Schmieden's analysis? We consider the series of $\ln 2$ as in [Spalt 2001, p. 144f.],

$$\ln 2 = 1 - \frac{1}{2} + \frac{1}{3} - \frac{1}{4} + \frac{1}{5} - \frac{1}{6} + \frac{1}{7} - \frac{1}{8} + - \ldots,$$

which can be derived from Mercator's series (6.27) by inserting $x = 1$. Division by 2 yields the series

$$\frac{1}{2}\ln 2 = \frac{1}{2} - \frac{1}{4} + \frac{1}{6} - \frac{1}{8} + - \ldots$$

and termwise summation of both series for $\ln 2$ and $\frac{1}{2}\ln 2$ eventually leads to

$$\frac{3}{2}\ln 2 = 1 + \frac{1}{3} - \frac{1}{2} + \frac{1}{5} + \frac{1}{7} - \frac{1}{4} + + - \ldots.$$

Obviously the series for $\frac{3}{2}\ln 2$ shows the same terms as the one for $\ln 2$, just in a different order, and the values of both series are obviously different. This problem is well known in classical analysis. The series for $\ln 2$ is not 'absolutely continuous' and hence must not be rearranged. Our series for $\frac{3}{2}\ln 2$ is a rearrangement of the series for $\ln 2$ and hence, from the viewpoint of classical analysis, it is no wonder that we are led to this 'paradox'. However, Schmieden solved this 'paradox' as follows. We certainly have

$$A := \ln 2 = 1 - \frac{1}{2} + \frac{1}{3} - \frac{1}{4} + - \ldots + \frac{1}{\Omega - 1} - \frac{1}{\Omega} = \sum_{n=1}^{\Omega} \frac{(-1)^{n-1}}{n}.$$

Now we are interested in equality up to infinitely small quantities only, hence we may assume that Ω is divisible by 4 without remainder. It then follows

$$\begin{aligned} B &:= \frac{3}{2}\ln 2 = 1 + \frac{1}{3} - \frac{1}{2} + \frac{1}{5} + \frac{1}{7} - \frac{1}{4} + + - \ldots + \frac{1}{\Omega - 3} + \frac{1}{\Omega - 1} - \frac{1}{\Omega/2} \\ &= \sum_{n=1}^{\Omega/4}\left(\frac{1}{4n-3} + \frac{1}{4n-1} - \frac{1}{2n}\right). \end{aligned}$$

Now there is a difference between the two series, nameley

$$C := A - B = -\frac{1}{\Omega/2+2} - \frac{1}{\Omega/2+4} - \ldots - \frac{1}{\Omega} = -\sum_{n=\frac{\Omega}{4}+1}^{\Omega/2} \frac{1}{2n}.$$

These terms are contained in A, but not in B. Schmieden expanded by ω, put $x := 2n\omega$ (and therefore $dx = 2\omega$), i.e.

$$C = -\sum_{n=\frac{\Omega}{4}+1}^{\Omega/2} \frac{1}{2n} = -\sum_{n=\frac{\Omega}{4}+1}^{\Omega/2} \frac{\omega}{2n\omega} \overset{\circ}{=} -\frac{1}{2}\int_{1/2}^{1} \frac{dx}{x} = -\frac{1}{2}\ln x\Big|_{1/2}^{1}$$
$$= \frac{1}{2}\ln\frac{1}{2} = -\frac{1}{2}\ln 2.$$

The boundaries 1/2 and 1 of the integral simply occur from inserting n. The case of $n = \Omega/4 + 1$ yields $x = 2n\omega = 2(\Omega/4+1)\omega \overset{\circ}{=} (\Omega\omega)/2$ and hence $x = 1/2$. For $n = \Omega/2$ the upper boundary value follows accordingly.

In Schmieden's analysis the series for $\frac{3}{2}\ln 2$ hence is not a rearrangement of the series for $\ln 2$, but there are infinitely many terms missing in B summing up to result in $-\frac{1}{2}\ln 2$! It follows

$$\ln 2 \overset{\circ}{=} A = B + C \overset{\circ}{=} \frac{3}{2}\ln 2 - \frac{1}{2}\ln 2.$$

In [Spalt 2001] numerous other instructive examples from the 'black book' can be found.

Schmieden also is the inventor of the 'microscope technique', i.e. the analysis of functions in infinitely small distance from an interesting point. This 'microscope' has later found its way into the textbooks of nonstandard analysis in form of a magnifying glass, see [Keisler 1986].

11.2.2 The Nonstandard Analysis of Laugwitz and Schmieden

In a manuscript [Laugwitz 1999] of 1999 Detlef Laugwitz described his acquaintance with Curt Schmieden and the 'black book'. Already in 1954 Laugwitz had received a manuscript of about fifty pages from Carl Friedrich von Weizsäcker to be presented as a talk in a seminar. Laugwitz became interested by the examples taken from analysis in which infinitely large and infinitely small numbers were worked with freely, and so von Weizsäcker revealed the author of the manuscript. It was Curt Schmieden.

After years of discussion between the two men and, as Laugwitz wrote, with Paul Lorenzen who acted as a referee, a joint work concerning the new analysis was accepted for publication in 1957 and appeared in print in 1958 [Schmieden/Laugwitz 1958]. This topic occupied Laugwitz until the end of his life. Besides numerous articles in mathematical journals also two books were published, *Infinitesimalkalkül* [Laugwitz 1978] and *Zahlen und Kontinuum* [Laugwitz 1986].

The access chosen by Laugwitz to turn Curt Schmieden's infinitesimal calculus into a waterproof mathematical theory was oriented towards Georg Cantor's construction of real numbers. Here is a short reminder: Cantor started from the rational numbers $\mathbb{Q}$, the set of all fractions, and considered Cauchy or fundamental sequences, i.e. sequences (r_n) of rational numbers so that for all $\varepsilon > 0$ there is an index N, so that for all larger indices n and m

$$|r_m - r_n| < \varepsilon$$

holds. We have already seen in examples that not every Cauchy sequence of rational numbers converges in $\mathbb{Q}$. Such a sequence would then define a new, irrational number. Since different Cauchy sequences may have the same number as their limits we even had to define a real number as being an equivalence class of Cauchy sequences.

Laugwitz did *not* restrict himself to Cauchy sequences but he permitted *all* sequences. A sequence (r_n) of numbers of $\mathbb{Q}$ is a mapping from the index space $\mathbb{N}$ to $\mathbb{Q}$; we indicate this fact by writing

$$\mathbb{Q}^{\mathbb{N}}.$$

Cantor had defined an equivalence relation on the set of all Cauchy sequences, in which two Cauchy sequences are equivalent if their difference sequence is a sequence converging to 0. Laugwitz employed a weaker equivalence relation. Two sequences (r_n), (s_n) (all are permitted!) of numbers of $\mathbb{Q}$ are equivalent if and only if for almost all (i.e. with the exception of finitely many) n

$$r_n = s_n$$

holds. The sequence (n), hence $1, 2, 3, 4, \ldots$, then defines the infinitely large number Ω. In Cantor's theory this sequence is not even admissible since it is no Cauchy sequence; the distance between two elements in the sequence is always at least 1.

Laugwitz carried the operations '+', '·', and the relation '≤' rigorously over to the 'Ω'-numbers [Laugwitz 1978, p. 27] and in that way got a set of equivalence classes out of $\mathbb{Q}$; just like Cantor had got the real numbers as equivalence classes. His new 'Rechenmenge' (number system) was denoted by Laugwitz[1] as

$$^{\Omega}\mathbb{Q}.$$

The rational numbers $\mathbb{Q}$ form an ordered field and hence also the real numbers $\mathbb{R}$. Unfortunately this is not so for Laugwitz's $^{\Omega}\mathbb{Q}$! It transpires that $^{\Omega}\mathbb{Q}$ is *not* a field but only a ring, and even worse, there are zero divisors. Zero divisors are elements a, different from zero, so that $a \cdot b = 0$ is possible although the second number b is also not zero. Zero divisors in $^{\Omega}\mathbb{Q}$ are

[1] Laugwitz in [Laugwitz 1978] actually wrote $^{\Omega}K$, because he wanted to consider an arbitrary ordered field K.

$$\alpha := (1 + (-1)^{\Omega}), \quad \beta := (1 - (-1)^{\Omega}),$$

since those Ω-numbers are defined by sequences with terms

$$a_n = 1 + (-1)^n, \quad b_n = 1 - (-1)^n,$$

and hence neither $\alpha = 0$, nor $\beta = 0$ is valid. The product

$$\alpha \cdot \beta = (1 + (-1)^{\Omega}) \cdot (1 - (-1)^{\Omega}) = 1 - (-1)^{2\Omega} = 0$$

however is zero. It is surprising what results can be achieved nevertheless in such a ring with zero divisors. To get from this ring ${}^{\Omega}\mathbb{Q}$ to a field ${}^{*}\mathbb{R}$ one has to employ heavy artillery, namely so-called ultrafilters. By using ultrafilters the construction of a number *field* gets *non*constructive, i.e. the process of becoming of this field can not be 'seen' as is possible in Cantor's or Dedekind's construction of the real numbers.

In his second book [Laugwitz 1986] followed a slightly different path. As Spalt in [Spalt 2001, p. 160] has remarked Laugwitz did not want to found his analysis on set theory. Hence he constructed a new 'theory' together with a principle that Laugwitz called 'Leibniz's principle [Laugwitz 1986, p. 88]:

> **Leibniz's principle:** *Let $A(\cdot)$ be a statement form in the language of K. If there is an $n_0 \in \mathbb{N}$ so that for all $n \geq n_0$ the statement $A(n)$ is true in the theory of K, then $A(\Omega)$ is accepted as a true statement in the new theory of ${}^{\Omega}K$.*
>
> (**Leibnizsches Prinzip:** Es sei $A(\cdot)$ eine Aussageform, formuliert in der Sprache von K. Wenn es ein $n_0 \in \mathbb{N}$ gibt so daß für alle $n \geq n_0$ die Aussage $A(n)$ in der zugrunde gelegten Theorie von K wahr ist, dann soll $A(\Omega)$ als wahrer Satz in die neue Theorie von ${}^{\Omega}K$ aufgenommen werden.)

Here K is an ordered field, for example $\mathbb{Q}$, and the 'language' consists of logical statements. This approach also led to a nonstandard analysis, but also here heavy artillery like ultrafilters has to be used to get a field in which infinitesimal quantities exist.

Already shortly after the publication of the first work [Schmieden/Laugwitz 1958] concerning nonstandard analysis Abraham Robinson published a book in the USA which became the actual bible of nonstandard analysis.

11.3 Robinson and the Nonstandard Analysis

Abraham Robinson (1918–1974) lived an eventful life [Dauben 1995]. Born in Waldenburg in Silesia as son of a traditional Zionist Jewish family he had to grow up without his father who died before his son was born.

The family left Germany in 1933 and went to Palestine where Robinson became a student of Abraham Fraenkel who had earned immortal fame by the Zermelo-Fraenkel axiomatic system of set theory. Robinson turned out to be a brilliant student and won a stipend at the Sorbonne in Paris in 1939. In these politically horrible time he travelled to Paris and took up his studies, but had to flee Paris in 1940 due to the German invasion of France. He was lucky to be able to board one of the last ships to England. In England he joined the Free France air force (forces françaises libres). Robinson, always interested most in logic and abstract algebra, was now assigned to the Royal Aircraft Establishment (RAE) in Farnborough as mathematician where he quickly developed into an expert on supersonic flows. He had turned into an 'applied' mathematician. In 1956 there even appeared the book *Wing Theory* in which Robinson impressively showed that he had become an expert on the flow about airfoils. Also in England he met his later wife Renée Rebecca Kopel from a wealthy Jewish family coming from Vienna.

Robinson's mathematical love, logic, he could not let go, despite his works on aerodynamics. In the summer of 1946 the Robinsons left England for Palestine where Abraham continued his studies of logic at the Hebrew University. He received his doctorate in 1951 with the dissertation *The metamathematics of algebraic systems*. Already in 1951 he was appointed professor of applied (!) mathematics in Toronto. During his time in Toronto he made great progress in

Fig. 11.3.1. Abraham Robinson; Flow about an airfoil. Robinson was an expert on airfoil flow and designed a nonstandard model of analysis

the field of 'model theory'. The name suggests a quite practical field but model theory is in fact a discipline of mathematical logic. To given axiomatic systems 'models' are constructed consisting of sets with certain structures. Somewhat naively this can be understood as follows. If a system of axioms is given then the mathematician working in model theory looks for mathematical structures obeying this axiomatic system.

In Palestine Abraham Fraenkel meanwhile was about to retire and his pupil Robinson became his successor at the Hebrew University in 1957. In 1962 he went to the UCLA (University of California in Los Angeles) and in 1967 he changed to Yale University. In 1973 he was diagnosed with pancreatic cancer. He underwent surgery but died a short time later.

When we discussed Greek mathematics we met the Archimedian axiom 2.1.3 prohibiting infinitely small quantities. In 1933 the Norwegian Thoralf Albert Skolem (1887–1963) had begun important work on non-Archimedean number systems and had proved that nonstandard models of arithmetic could be established by means of model theory. Robinson took up this works and tried successfully to construct a nonstandard model of analysis. His construction requires knowledge of model theory we can not discuss here and leads from the real numbers $\mathbb{R}$ to the 'hyperreal' numbers $\mathbb{R}^*$ which include the real numbers. The number system $\mathbb{R}^*$ is a field extension of $\mathbb{R}$; it is ordered but non-Archimedean. In his standard reference *Non-standard Analysis* [Robinson 1996], first published in 1961, Robinson has describe the theory and its applications to analysis. If one does not like to be concerned with the foundations of model theory one can start with an axiomatic system following Keisler [Keisler 1976], in which the existence of $\mathbb{R}^*$ is simply claimed. In $\mathbb{R}^*$ infinitesimals dx exist for which

$$|dx| < r \quad \text{for all } r \in \mathbb{R}, r > 0,$$

holds. Infinitely large numbers exist as well. Two numbers $x, y \in \mathbb{R}^*$ are called infinitesimally adjacent, $x \approx y$, if their difference $x - y$ is an infinitesimal. Every finite number x is infinitesimally adjacent to a real number r, $r \approx x$, and this uniquely defined r is called the 'standard part' of x, denoted by $r = \mathsf{std}(x)$. If $x, y \in \mathbb{R}^*$ are finite then it holds $x \approx y$ if and only if $\mathsf{std}(x) = \mathsf{std}(y)$. Hyperreal extensions f^* of real functions f exist accordingly; we write f further on. The first derivative then is nothing but

$$f'(x) = \mathsf{std}\left(\frac{f(x+dx) - f(x)}{dx}\right)$$

where dx is an infinitesimal, i.e. $dx \approx 0$. All of real analysis can now be developed on the basis of this continuum $\mathbb{R}^*$ which deserves the name 'continuum' rather more than the Cantor-Dedekind continuum $\mathbb{R}$.

While the 'bible' [Robinson 1996] is only suited to mathematically well educated readers the books [Keisler 1976] and [Keisler 1986] have served to popularise Robinson's nonstandard analysis. In this context the little book [Henle/Kleinberg 1979] by Henle and Kleinberg has to be mentioned which can even be read with profit by well educated school pupils. Today the interest in nonstandard analysis is consistently high; however, the vast majority of mathematicians still pursue classical analysis in the sense of Weierstraß.

11.4 Nonstandard Analysis by Axiomatisation: The Nelson Approach

If one is not so familiar with model theory that one can follow Robinson in his development of the hyperreal numbers then one is yearning for other paths to achieve the goal. Such a 'different path' was derived by Edward Nelson (1932 – 2014) in 1977 in his work [Nelson 1977]. Nelson had been a member of the *Institute for Advanced Study* between 1956 and 1959 and conducted research at Princeton University from 1959 on. He did not try to introduce new elements into the real numbers but enriched the Zermelo-Fraenkel axiomatic system of set theory instead. He thereby created the 'internal set theory' IST consisting of only three axioms which have to be appended to the axioms of Zermelo-Fraenkel. The three axioms themselves are also named IST; 'Idealisation', 'Standardisation', and 'Transfer'. The axiomatisation of the notion 'standard' from Robinson's nonstandard analysis is crucial since it allows to distinguish elements which in the real numbers can not be distinguished. By Nelson's IST the hyperreal numbers are axiomatically defined and the way through model theory is no longer necessary. Nelson's nonstandard analysis has been described by Robert in [Robert 2003].

11.5 Nonstandard Analysis and Smooth Worlds

Fairly recently another approach to nonstandard analysis, basically different from Robinson's and Nelson's nonstandard analysis, led to success. This approach, described in [Bell 1998], relies heavily on category theory and sheaf theory; both fairly abstract and modern mathematical theories. It was the idea of the US-American mathematician Francis William Lawvere (* 1937) to base mathematics not on set theory but on category theory. The result is the 'smooth infinitesimal analysis' the mathematical foundations of which are explained (for experts only) in [Moerdijk/Reyes 1991]. At the core of the theory is the construction of 'smooth worlds' in which an analysis of infinitesimals becomes possible. Mathematically speaking such a smooth world is a category containing all the necessary objects; Euclidean space, real numbers, and the mappings between them, whereby all mappings have

Fig. 11.4.1. Edward Nelson (June 2003, Department of Mathematics, Princeton University, USA); Francis William Lawvere [Photo: Andrej Bauer, November 2003]

to be smooth, i.e. differentiable. There is not only one smooth world but a multitude of such worlds from which we pick one which we want to denote by $\mathbb{S}$. Following the notation $\mathbb{R}$ we will denote the number line in $\mathbb{S}$ also by $\mathbb{R}$.

In $\mathbb{S}$ the *tertium non datur* is *not* valid, reminding us of Brouwer's intuitionist analysis. The absence of the *tertium non datur* of course corresponds to the nonexistence of discontinuous functions in $\mathbb{S}$. The logical theorem

$$\text{For all real numbers either } x = 0 \text{ or } \neg(x = 0) \text{ holds },$$

is *wrong* in $\mathbb{S}$ since there are no discontinuous functions as, for example,

$$f(x) = \begin{cases} -1 \,;\, x \leq 0 \\ 1 \,;\, x > 0 \end{cases} .$$

Furthermore one can easily show that logic in $\mathbb{S}$ is 'polyvalent', i.e. multi-valued. If the *tertium non datur* does not hold then the law of double negation is *not* valid, i.e. in $\mathbb{S}$ holds

$$\neg\neg P \not\Rightarrow P$$

for every statement P. What can we get from such an unaccustomed restriction? Well, it allows the existence of infinitesimals!

If two numbers a, b are called distinguishable, $a \neq b$, if they are not identical, and if we write for this

$$\neg(a = b),$$

and if we call a, b identical, if

$$\neg(a \neq b)$$

holds, then it does *not* follow in $\mathbb{S}$ that

$$\neg(a \neq b) = \neg\neg(a = b) \Rightarrow a = b,$$

i.e.: if $a \neq b$ does not hold, then this does not necessarily mean $a = b$ in $\mathbb{S}$. This now allows the existence of an infinitesimal neighbourhood I of 0, i.e. the set of all points x being not distinguishable from 0. The elements of I are called infinitesimals.

However, the infinitesimals exist in a somewhat strange sense; their existence is 'potential' only. In $\mathbb{S}$ we can not prove a theorem like: 'there exists infinitesimals $\neq 0$', since then we would have to be able to distinguish the infinitesimals from 0. But the elements of I are indistinguishable from 0 by definition.

Together with the invalidity of the *tertium non datur* the negation of the universal quantifier is no longer valid

It does not hold: for all x it holds $A(x) \Rightarrow$
there is one x, so that $A(x)$ does not hold,

or, in symbols of logic,

$$\neg(\forall x A(x)) \Rightarrow \exists x \neg A(x).$$

In contrast, the converse

$$\exists x \neg A(x) \Rightarrow \neg(\forall x A(x))$$

stays valid, as does the rule of the negation of the existential quantifier

$$\neg(\exists x A(x)) \Leftrightarrow \forall x \neg A(x).$$

Let this first look at the (discouraging as it might seem when looking at it for the first time) logic in smooth worlds suffice. We only need to remember that we are allowed to work with the set of infinitesimals

$$I := \{x \mid \neg(x \neq 0)\}$$

in a potential sense since we can neither prove its non-existence nor its existence.

Smooth worlds remind us strongly on Leibniz's conception of the continuum and of continuity, and consequently we come here to full circle which began with the Pre-Socratics philosophers. However, beware of identifing the theory of smooth worlds or any other modern theory of nonstandard analysis with the treatment of infinitesimal quantities of Leibniz or other forefathers of modern mathematics! The mathematical apparatus of the new theories have been developed only in the mid-20th century and we must not assume that Leibniz, however implicit, had a category theory in mind. The treatment and handling of infinitesimals can always only be interpreted against the background of the specific period in time under consideration.

Fairly often in this book we have encountered examples in which squares of infinitesimal quantities were equated to zero. This can be very easily accomplished in smooth worlds in that 'nilquadratic' infinitesimals are considered: If $\varepsilon \in I$, then $\varepsilon^2 = 0$. The nilquadratic infinitesimals are collected in the set

$$\Delta := \{x \in I \mid x^2 = 0\}.$$

Therewith a 'law of microaffinity' suggests itself so that in $\mathbb{S}$ the following axiom is demanded: To every function $g : \Delta \to \mathcal{R}$ with range in the smooth number line $\mathcal{R}$ there exists a $b \in \mathcal{R}$ such that for all $\varepsilon \in \Delta$

$$g(\varepsilon) = g(0) + b \cdot \varepsilon$$

holds. By shifting the law of microaffinity this holds at every point $x \in \mathcal{R}$. Thus all functions are linear in every infinitesimal neighbourhood, i.e. the graph consists of infinitesimal linear segments. This, in turn, reminds us on conceptions which could already be found in the works of John Bernoulli or Isaac Barrow ('linelets', 'timelets').

Now the path into the differential and integral calculus is clear. If at every point $x \in \mathcal{R}$ we define the function $g_x : \Delta \to \mathcal{R}$ by

$$g_x(\varepsilon) = f(x + \varepsilon),$$

then, following the law of microaffinity, there is a uniquely defined $b_x \in \mathcal{R}$ such that for all $\varepsilon \in \Delta$ it holds

$$f(x + \varepsilon) = g_x(\varepsilon) = g_x(0) + b_x \cdot \varepsilon = f(x) + b_x \cdot \varepsilon.$$

The function $f' : \mathcal{R} \to \mathcal{R}$, defined by $x \mapsto b_x$, then is the 'derivative' of f which now can be written quite naturally as

$$f'(x) = \frac{f(x + \varepsilon) - f(x)}{\varepsilon}, \quad \varepsilon \in \Delta. \tag{11.1}$$

A further important law is the 'principle of microalignment': For all $a, b \in \mathcal{R}$ it holds

$$\text{If for all } \varepsilon \in \Delta : \varepsilon a = \varepsilon b \text{ holds, then } a = b. \tag{11.2}$$

At this point we have developed enough mathematics to look at an example of analysis in smooth worlds. We want to compute the volume of a circular cone as shown in figure 11.5.1. Do you remember how Democritus tortured himself in Greek antiquity? We have cited Plutarch on page 55 who attributed the following lines to Democritus, cp. [Heath 1981, Vol. I, p. 179f.]:

> *'If a cone is cut by surfaces parallel to the base, then how are the sections, equal or unequal? If they were unequal* [(and, we might add mentally, if the slices are considered as cylinders)][2], *then the cone would have the shape of a staircase; but if they were equal, then all sections will be equal, and the cone will look like a cylinder, made up of equal circles; but this is entirely nonsensical.'*

On page 55 we have already cited Knobloch [Knobloch 2000] who has thoroughly investigated that this problem of Democritus actually lies much deeper but we do not need to care about this at this point.

In our smooth world $\mathbb{S}$ we can easily dispel Democritus's doubts; in Knobloch's words: 'the pseudo problem'. We denote the volume of the circular cylinder up to the coordinate z by $V(z)$, then we place the circular cylinder of height $\varepsilon \in \Delta$ into the circular cone. The difference in volume then is

$$V(z+\varepsilon) - V(z),$$

and since (11.1) we can rewrite this as

$$V(z+\varepsilon) - V(z) = \varepsilon V'(z).$$

If we denote the slope $BC : PC$ of the cone by b then, since $BC = \varepsilon$, the length PC is $PC = \varepsilon b$ and the area of the triangle PBC in figure 11.5.1 is given by

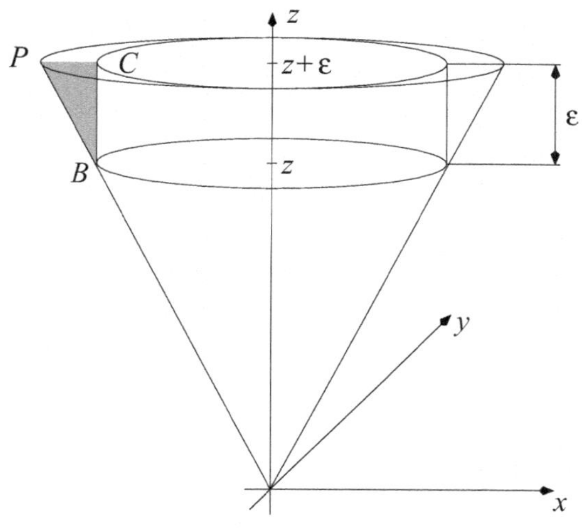

Fig. 11.5.1. Computation of the volume of a circular cone by means of infinitesimal cylindrical slices

[2] Remark by van der Waerden.

$$\frac{1}{2} \cdot \varepsilon \cdot \varepsilon b = \frac{1}{2}\varepsilon^2 b,$$

but ε is nilquadratic and therefore the area of the triangle is zero! Thus also the volume of the ring element created by the triangle is zero.

Our cylindrical slice has a radius of $r = bz$ and thus a volume of

$$\pi b^2 z^2 \cdot \varepsilon,$$

hence

$$\varepsilon V'(z) = V(z+\varepsilon) - V(z) = \varepsilon \pi b^2 z^2.$$

Now the principle of microalignment comes into effect and we deduce from (11.2):

$$V'(z) = \pi b^2 z^2.$$

The principle of microalignment has thus made it possible to leave the infinitesimal cylindrical slice. Integration is also possible in smooth worlds, of course, and we finally get

$$V(z) = \frac{1}{3}\pi b^2 z^3.$$

Development of Nonstandard Analysis

1933	Thoralf Albert Skolem starts his works on non-Archimedean number systems
End of 1940s	Curt Otto Schmieden begins writing his 'black book'
1958	Detlef Laugwitz and C. Schmieden create a model of nonstandard analysis in the space of infinite sequences and apply their formalised principles to many topological structures
1961	Abraham Robinson publishes his book on nonstandard analysis
1962	W.A.J. Luxemburg constructs (tying in ideas of Laugwitz, Robinson, and Schmieden) a totally ordered, non-Archimedean, zero divisior free extension of the real number field, and thus a model of nonstandard analysis, by the use of ultrafilters in the space of infinite sequences of real numbers
1966	A. Robinson develops and publishes a general method to create a nonstandard analysis in which infinitesimal as well as infinitely large quantities are contained. The method relies on a non-Archimedean extension of the field of real numbers; the so-called 'hyperreal numbers'
1977	Edward Nelson publishes an axiomatic definition of hyperreal numbers based on his 'internal set theory' IST
1977	C. Schmieden. The 'black book' is in the form of a manuscript but is not being published
1978	D. Laugwitz: *Infinitesimalkalkül – eine elementare Einführung in die Nichtstandard-Analysis*
> 1980	Francis William Lawvere develops the 'smooth infinitesimal analysis' based on category theory. All functions are differentiable but the *tertium non datur* is not valid
1986	D. Laugwitz: *Zahlen und Kontinuum – Eine Einführung in die Infinitesimalmathematik*
1988	A.M. Robert: *Nonstandard Analysis*
1996	To date the last reprint of Robinson's *Non-standard Analysis* is published

12 Analysis at Every Turn

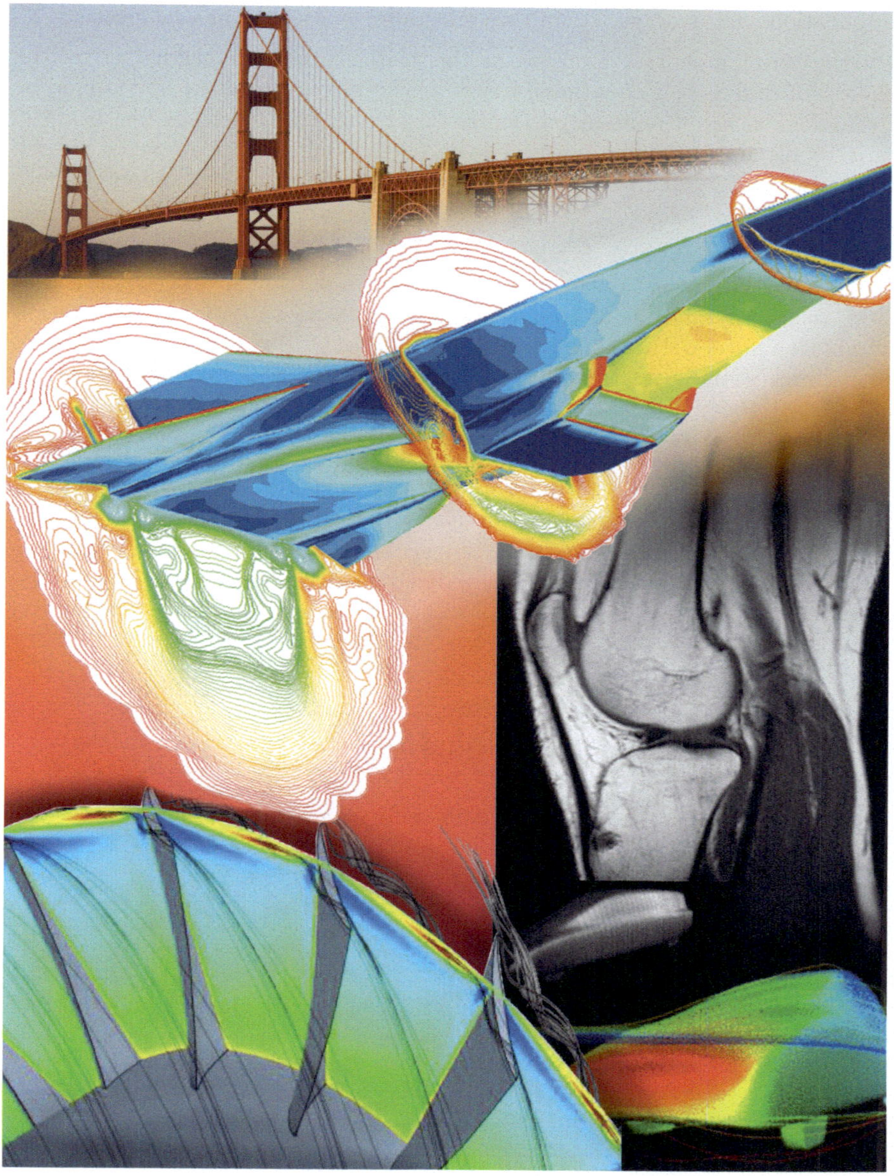

T. Sonar, *3000 Years of Analysis*, https://doi.org/10.1007/978-3-030-58223-4_12

In 'The Nature and Meaning of Numbers' the Brunswick mathematician Richard Dedekind asked in 1888 what numbers really are and what's the point of them? 'What is analysis and what's the point of it?' is the corresponding question of the mathematician working on analysis. We have spent many words in this book to answer the question of what analysis really is. The point of it, what to do with it, where it is of use, and where we can meet it in daily life is worth thinking about. The answer is easy: Analysis and its applications are everywhere and we encounter it literally at every turn!

By the development of analysis in the 18th century, in particular by the theory of differential equations, many possibilities opened up for numerous technological inventions, in particular large steel constructions like bridges, large ships, and the Eiffel Tower. The excellent book by Kurrer on the history of structures in civil engineering [Kurrer 2008] also contains a unique tribute to analysis.

Two different traditions developed within engineering science in the 19th century. While on the Continent and in France in particular 'higher mathematics', including analysis of course, was seen indispensable for the education of able engineers, British engineering traditionally was not very close to mathematics. In Great Britain 'engineers' had been craftsmen and the most capable of them joined the 'Institution of Civil Engineers' to distinguish themselves from mere artisans [Kaiser/König 2006]. The greatest English

Fig. 12.0.1. The Clifton Suspension Bridge was designed by the English engineer Isambard Kingdom Brunel (1806–1859). This bridge across the River Avon was opened in 1864 only after Brunel's death and stayed in use ever since. It ranks as one of the greatest engineering achievements of its day

engineer of the 19th century, and arguably the greatest engineer of all times, Isambard Kingdom Brunel (1806–1859), was an exception from the rule. His family came from France where Brunel attended the Lycée Henri IV, one of the most famous higher schools in France. Hence Brunel was in very good command of higher mathematics.

In the year 2000 the 'Millennium Bridge' across the River Thames was opened to the public in London. The bridge had cost 18.2 Million £ and was a purely pedestrian bridge in a breathtakingly beautiful modern design. However, the bridge proved unstable at the very day of the opening and began to swing horizontally when many pedestrians at one time crossed the bridge. Obviously, and due to the new design, one had not taken the phenomenon of resonance seriously. Every building has an inherent resonance frequency coming from the differential equations describing the oscillations of the building under consideration. If a bridge induced by vibrations of this frequency, for example by pedestrians moving in a periodic fashion or by wind, then the oscillations of the bridge increase without bound and the bridge may even be destroyed completely. Was it really the new design which forbid a mathematical analysis or was it an oversight on the side of the civil engineers which only relied on their Finite Element (FE) computer programs without taking the mathematics seriously? While the Millennium Bridge was retrofitted with a number of dampers and can be crossed today safely even by masses of pedestrians, the Tacoma Narrows bridge, a suspension bridge in the state of Washington, was completely destroyed by resonance on 7th November 1940. Shortly after its opening on 1st July 1940 the bridge got its nickname 'Galloping Gertie' due to its oscillating behaviour. On 7th November the same year strong cross winds moved across the bridge. Periodic vortices

Fig. 12.0.2. The Millennium Bridge in London [Photo: Adrian Pingstone, June 2005] and the Tacoma Narrows Bridge in the USA while suffering a catastrophe triggered by resonance phenomena. The bridge was completely destroyed on the 7th November 1940

occurred, the frequency of which corresponded with the resonance frequency of the bridge and finally brought the bridge down. In the former English colonies in the USA the English type of the educated practitioner prevailed as the engineer; however, in the 19th century special engineering schools developed from which finally some elite universities developed, for example the Massachusetts Institute of Technology MIT in Cambridge, Mass., or the California Institute of Technology CALTECH in Pasadena.

With the ever expanding ideas of engineers the demands for analysis increased. Stationary steam engines, having fuelled the industrial revolution, were usually controlled by centrifugal governors. When the supply of steam was too large the rotation speed of the machine increased and two metal spheres moved outwards by centrifugal force and thus reduced the steam supply by means of a slider. In the meantime the speed of the engine was kept almost constant. The control by centrifugal governors could be modelled by coupled systems of differential equations. To compute a solution, analysis was demanded. Today control technology relies completely on analytical methods and even has triggered some developments within analysis.

The possibilities of engineering have been multiplied by the development of the computer and hence of numerical analysis in the 20th century. The worldwide first program controlled computer was the *Analytical Engine* of the English mathematician and inventor Charles Babbage (1791–1871). Babbage had seen the program controlled weaving looms in the industrial centres in England and had the idea to control his fully mechanical computer also by means of punched cards. The machine sadly was never built; however, the *Difference Engine*, an earlier project of Babbage, was completed, but not by Babbage. It already had a CPU which Babbage called ‘the mill’ and was able

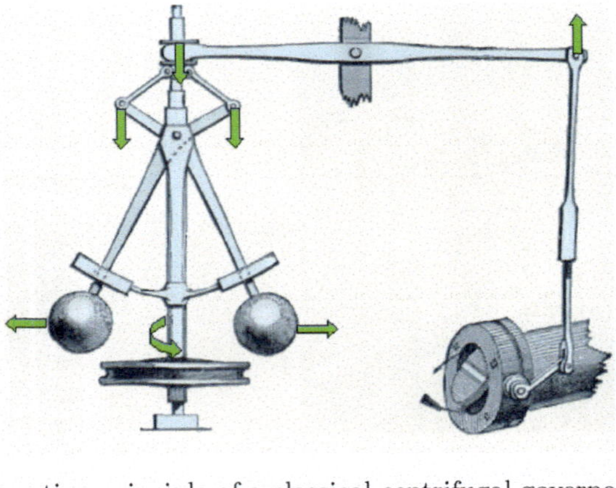

Fig. 12.0.3. Operating principle of a classical centrifugal governor at a stationary steam engine. If the engine’s speed becomes too high the spheres are forced outward by centrifugal force and the steam supply is throttled by means of a slider. If the speed becomes too low the slider opens the steam supply accordingly.

Fig. 12.0.4. The CPU, called 'the mill', of the *Difference Engine* by Charles Babbage of 1833 (Science Museum / Science & Society Picture Libary)

to produce error-free tables of logarithms or insurance tables for the first time by means of a mechanical printing unit. A modern and fully functional replica of the *Difference Engine* can be seen today in the London *Science Museum*, cp. [Swade 2000].

The next big step towards the development of powerful computers was achieved in the 1940s by a German civil engineer. Konrad Zuse (1910–1995) loathed the extensive computations which had to be performed b engineers of the early 20th century. In the living room of his parents he built his first programmable computer 'Z1' which still was a mechanical machine. The Z1 then served Zuse as a blueprint for his famous Z3 which was based on telephone relays. The Z3 came into existence in the year 1941 which was not a good year concerning groundbreaking inventions in Germany which had to be further developed advisedly and in tranquility. The original first Z3 was destroyed in the war; a replica can be seen in the German Museum in Munich. Although Zuse established his own company to build computers after the war the spirit of the time in German banks was against him. The company quickly expanded. The worldwide first plotter was built there, but the banks refused to financially assist the unknown computer business so that the 'Zuse KG' finally became a part of the Siemens company in 1969.

Meanwhile the English and the Americans had taken over the construction of powerful computers, cp. [Goldstine 1972], [Ifrah 2001], [Williams 1997]. With the advent of the transistor the electronic valves were displaced by and by in the 1950s, but only with the advent of the technology of integrated circuits at the beginning of the 1980s components of VLSI (Very Large Scale Integration) entered the computer business which contained hundreds of thousands of transistors on one chip. Incidentally, recent graphic processors contain 2.1 thousand million transistors. Through the big mainframe machines of IBM, but also through the specialised supercomputers of the Cray Research company, the numerical treatment of really large and complex problems in engineering and physics became possible. Since the 1950s 'numerical analysis' developed as a branch of analysis and linear algebra. Sadly, this designation is often misused and one frequently misunderstands numerical analysis as the implementation of a problem on a computer and the production of 'numerical solutions' by means of a coloured computer plot. But numerical analysis has nothing to do with that but is concerned with the analysis (i.e. mathematics) applied to numerical methods.

One such method is the finite element method FEM. Their foundations were laid in about 1940 by the German mathematician Richard Courant (1888–1972) who had to flee Germany from the Nazis and went to the USA where he started a very successful second mathematical career. The FEM reached its climax only when computers became widely available. Domains under consideration are subdivided into small contiguous pieces; e.g. the chassis of a car will be subdivided into small contiguous triangles or rectangles. To know the deformation of the car in a crash complicated systems of partial differential equations would have to be solved which describe the deformation of the chassis. The exact analytical solution of such systems is not possible concerning the state of the art of mathematics and hence these systems are restricted to the subdivisions of the chassis and their discrete analogues are solved on a computer. These discrete equations have only polynomials as solutions and numerical analysis is needed to prove that the numerical solutions (hopefully) are good approximations of the original systems.

In case of other types of partial differential equations numerical analysis provides further methods like the finite difference method FDM or the finite volume method FVM which is of use particularly in fluid mechanical problems.

Beside the possibility to compute unprecedented buildings and to realise new machinery analysis, and in particular numerical analysis, opened further

Fig. 12.0.5. Result of a FEM computation when a vehicle hits a fixed obstacle asymmetrically. Different colours correspond to different stresses in the material.

opportunities in the engineering sciences. Already in the 18th century Leonhard Euler had formulated the basic equations of inviscid fluid dynamics in the form of a nonlinear system of partial differential equations. The French mathematician and physicist Claude Louis Marie Henri Navier (1785–1836) and the Irish mathematician George Gabriel Stokes (1819–1903) succeeded in the 19th century to include the influence of fluid friction. The Navier-Stokes equations are a system of partial differential equations the solutions of which describe the flow of liquids and gases in a (hopefully) realistic manner. Up to this day no theorem concerning the existence of solutions in realistic three-dimensional cases is known and such a theorem would yield its discoverer the sum of one million US-$ [Sonar 2009]. With many restrictions on the flow conditions mathematicians at the end of the 19th century were already able to compute the flow about two-dimensional airfoils. In the case of incompressible, inviscid, and rotation-free flow Euler's system of the differential equations of gas dynamics actually reduce to a simple linear partial differential equation of second order which can be solved by means of complex analysis in the Gaussian plane. In airfoil theory the German physician Albert Betz (1885–1968) became well-known for the development and use of such methods. Betz was working in Göttingen with the 'father of German fluid mechanics', Ludwig Prandtl (1875–1953). This theory of airfoils is by no means sufficient to assess the flow about a complete aircraft. This in turn became possible only with the advent of powerful computers and, of course, with numerical analysis. Today all around the globe an enormous number of hours of wind tunnel testing can be spared since the flow about aircrafts can be computed numerically in a number of simulations on high performance computers.

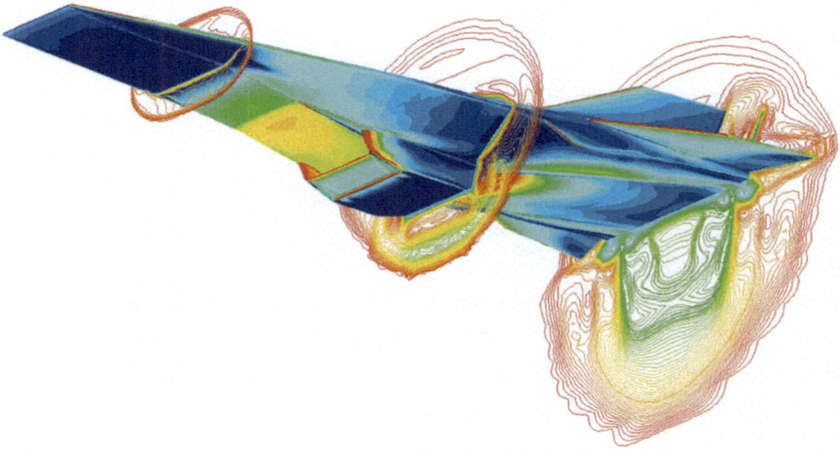

Fig. 12.0.6. Result of a computer simulation performed at NASA. The experimental airplane 'Hyper-X' (X43A) can be seen in flight at Mach seven (sevenfold speed of sound) with engine still running. The colours on the surface of the futuristic flying apparatus correspond to the heat flow while the coloured lines shown at three different positions in the field give an idea of the local velocities of the flow. In the colour coding red corresponds to high, and green to low values. The art to compute solutions of complex systems of differential equations modelling fluid mechanics by means of numerical analysis and computers is called *Computational Fluid Dynamics* CFD.

In the case of flow about cars one also has turned by and by towards computer simulations and methods of 'Computational Fluid Dynamics' CFD. On the one hand a large resistance of the vehicle leads to a high consumption of fuel so that from the point of view of sparing resources alone a streamlined design of cars can be wished for. On the other hand modern engines have become so quiet that noise coming from the flow of air around the car may be disturbing. Even tiny details like the design of the exterior mirrors or small rain gutters on the top of cars can be truelly on the noisy side. The field of 'aeroacustics' therefore connects fluid mechanics with accustics and has become more and more important in automobile manufacture over the past years. Needless to say, here also nothing could have been achieved without analysis.

Younger people in our times can often be seen on the streets or in trains or coaches with small earphones or even very large headphones with which they listen to music coming from miniaturised 'MP3' or 'MP4-players'. Only a very few of these young people anticipate that they would not hear a single note without analysis! Mathematically a piece of music is nothing but a (sometimes complicated) sequence of oscillations which, transferred to a loudspeaker, generates pressure variations in the air which can be

Fig. 12.0.7. Among the most popular MP3 players of our time is the iPod family of Apple. One can store whole music libraries and even videos which are compressed at storing and decompressed at playing or viewing. Without analysis such small devices to be put in trouser pockets would not exist [Photo: Matthieu Riegler]

sensed by our eardrums. By means of analysis every piece of music can be expanded frequence-wise in Fourier series. Every coefficient of the Fourier series corresponds to a certain frequency. If one combines closely neighbouring frequencies a certain compression effect can be achieved since our ear does not really need the complete signal. Eventually a much more sophisticated method of compression is the MP3-compression; actually named 'MPEG-1 Audio Layer 3'. Using it extensive pieces of music and whole music libraries may be compressed to occupy a small amount of computer memory only. Computer users like to use the 'jpg'-format for images, photos, and graphics. This format also relies on a compression algorithm from analysis and every device for playing DVDs reconstructs the movie from a compressed data file.

The JPEG-2000 standard for the compression of images uses a relatively recent technique based on so-called 'wavelets'. The disadvantage of Fourier representations of signals is the large number of coefficients which have to be stored, since the information on frequencies can not be localised satisfactorily. With the advent of wavelets (= small waves) one has command of functions being non-zero only on small, adjustable domains. These small domains can not only be adjusted but also shifted. Sampling a signal with such a wavelet local information on frequencies can be gained and the rate of compression in general is very high. In the past the FBI stored fingerprints by means of a technique based on Fourier series. Recently this has changed and wavelets are used instead. Some of these techniques appear to be completely lossless,

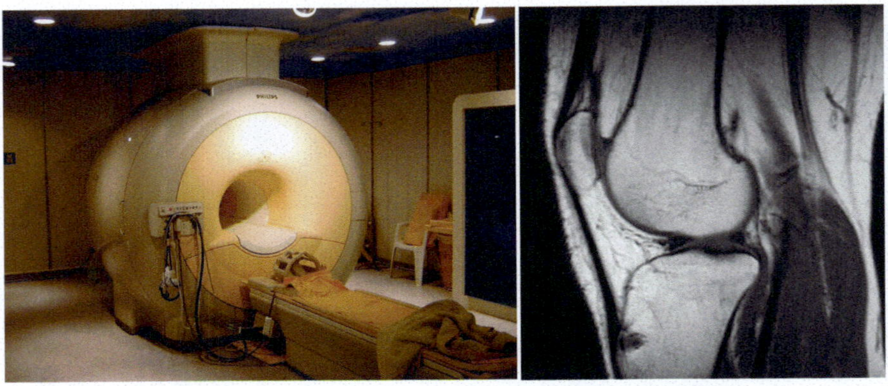

Fig. 12.0.8. A modern MRT machine [Photo: Kasuga Huang, 2006] beside the MRT scan of a human knee. With the help of tomography organs may be viewed in real time and three dimensional images of the entrails of patients may be created

i.e. the initial signal can be reconstructed perfectly. But also lossy methods are employed when the accuracy of the initial signal does not play a role and this is often the case.

Modern compression techniques of this kind can also be found in the numerical treatment of partial differential equations of physics and engineering, in particular when problems have to be solved involving a large variety of different scales. Standard example here is the turbulent flow about an aircraft. While the flow field far away from the aircraft shows large scales the turbulent flow close to the surface of the aircraft can only be resolved on small scales. In such cases one often uses 'multiple scale methods' in which different numerical methods are used for different scales.

In medicine analysis also appears at every turn. Today medical screenings in computer tomographs CTS or magnetic resonance scanners MRSs belong to the daily routine. In 2003 the inventors of the MRS even won the Nobel Prize for Medicine. In tomography the patient is horizontally driven through a ring-shaped apparatus in which a detector rotates around the patient by 360°. In this way the intensity of signals leaving the body is measured and compared to the signals sent. One such measurement gives no information but if many measurements under different angles are made then a three dimensional cross section of the patient can be created. Taking all such cross sections together results in a three dimensional image of the patient's entrails. This has only become possible by a development of the Austrian mathematician Johann Radon (1887–1956) which he published in 1917 (!). A more beautiful example for the importance of abstract mathematical research can hardly be given. The Radon transform allows for the computation of an integral along a straight line. The unknown function being integrated is the density of the tissue along the line. The reconstruction of the values of this function at any point on the line is difficult to achieve without more ado; it is an ill-conditioned so-called inverse problem. Only with the help of regularising techniques (coming from analysis, of course) the computation becomes feasible. We owe the mathematical theory behind computer tomography significantly to the German mathematician Frank Natterer (b. 1941) of the University of Münster who started in the 1970s to consider the Radon transform with respect to tomography [Natterer 2001].

Whether cars, airplanes, ships, bridges, transmission of signals, playing of music, wireless telephony, or medical technology; there is hardly one field in our lives which is not characterised by the achievements of analysis. May the readers of this book walk the earth with open eyes and sharp mind and experience the immense importance of analysis at every turn!

References

[Aaboe 1998] Aaboe, A.: Episodes from the Early History of Mathematics. 12th printing, Mathematical Association of America, 1998.

[Alten et al. 2005] Alten, H.-W.; Djafari Naini, A.; Folkerts, M.; Schlosser, H.; Schlote, K.-H.; Wußing, H.: 4000 Jahre Algebra. Hrsg. von: Alten, Djafari Naini, Wesemüller-Kock. Berlin, Heidelberg, New York 2003.

[Alten et al. 2014] Alten, H.-W.; Djafari Naini, A.; Eick, B.; Folkerts, M.; Schlosser, H.; Schlote, K.-H.; Wesemüller-Kock, H.; Wußing, H.: 4000 Jahre Algebra. 2. Auflage. Hrsg. von: Alten, Sander, Sonar, Djafari Naini, Schmidt-Thieme, Wagner, Förster, Schlote, Wesemüller-Kock. Berlin, Heidelberg 2014.

[Altgeld 2001] Altgeld, W.: Kleine italienische Geschichte. Ditzingen, 2001.

[Andersen 1985] Andersen, K.: Cavalieri's Method of Indivisibles. Archive for History of Exact Sciences, Vol.31, No.4, S. 291–367, 1985.

[Anselm 1996] Anselm: Monologion and Proslogion with the replies of Guanilo and Anselm. Translated, with introduction by Thomas Williams. Indianapolis and Cambridge, 1996.

[Archimedes 2002] The Works of Archimedes. Edited by T.L. Heath. Mineola, N.Y., 2002. Unabridged republication of the edition published by Cambridge University Press in 1897 under the title *The Works of Archimedes, Edited in Modern Notation with Introductory Chapters*, and of the supplement published by Cambridge University Press in 1912 under the title *The Method of Archimedes, Recently Discovered by Heiberg. A Supplement to The Works of Archimedes.*

[Aristoteles 1876] Aristoteles: Kategorien oder Lehre von den Grundbegriffen. Übersetzung von Heinrich von Kirchmann 1876,
http://www.zeno.org/Philosophie/M/Aristoteles/Organon/
Kategorien+oder+Lehre+von+den+Grundbegriffen/7.+Kapitel.

[Aristotle 1995] Aristoteles: Physics. in: [Barnes 1995, Vol. I, p. 315–446].

[Aubrey 1982] Aubrey, J.: Brief Lives. Edited by Richard Barber, Woodbridge 1982.

[Balck 2008] Balck, F.: Eulers Aufsatz zur Physik der Reaktionsturbine – ein wichtiger Baustein zur Technikgeschichte der Wasserräder, Turbinen und anderer Energiewandlungs-Maschinen. in: [Biegel/Klein/Sonar 2008].

[Banach 1987] Banach, S.: Theory of Linear Operations. Amsterdam, New York, etc., 1987.

[Barner 2001] Barner, K.: Das Leben Fermats. Mitt. DMV Heft 3/2001, S. 12–26, 2001.

[Barner 2001a] Barner, K.: How old did Fermat become? N. T. M. 9, S. 209–228, 2001.

[Barnes 1995] Barnes, J. (edt.): The Complete Works of Aristotle. The Revised Oxford Translation Vol. I and II. Bollingen Series LXXI.2, Princeton, 1995.

[Baron 1987] Baron, M. E.: The Origins of the Infinitesimal Calculus. Nachdruck der Ausgabe Oxford 1969, New York, 1987.

[Barrow 1973] Barrow, I.: The Mathematical Works. Edited by W. Whewell, 2 Bände in einem Band, Hildesheim, New York, 1973.

[Bartle 1996] Bartle, R. G.: Return of the Riemann Integral. The American Mathematical Monthly 103, S. 625–632, 1996.

[Batho 2000] Batho, G. R.: The possible portraits of Thomas Harriot. in: [Fox 2000].

T. Sonar, *3000 Years of Analysis*, https://doi.org/10.1007/978-3-030-58223-4

[Baumann 1991] Baumann, U.: Heinrich VIII. Reinbek, 1991.
[Becker 1964] Becker, O.: Grundlagen der Mathematik in geschichtlicher Entwicklung. 2. Aufl., Freiburg, 1964.
[Becker 1998] Becker, O.: Das mathematische Denken der Antike. 2. Auflage, Göttingen, 1998.
[Beckmann 1971] Beckmann, P.: A History of π. New York, 1971.
[Beery/Stedall 2009] Beery, J., Stedall J. (eds.): Thomas Harriot's Doctrine of Triangular Numbers: the 'Magisteria Magna'. Zürich, 2009.
[Beguin 1998] Béguin, A.: Pascal. 13. Auflage, Hamburg, 1998.
[Behrends/Colognesi 1992] Behrends, O., Colognesi, L.C. (Hrsg.): Die römische Feldmeßkunst. Interdisziplinäre Beiträge zu ihrer Bedeutung für die Zivilisationsgeschichte Roms. Göttingen, 1992.
[Belhoste 1991] Belhoste, B.: Augustin-Louis Cauchy – A Biography. New York, Berlin, Heidelberg, 1991.
[Bell 1998] Bell, J. L.: A Primer of Infinitesimal Analysis. Cambridge, 1998.
[Berkeley 1985] Berkeley, G.: Schriften über die Grundlagen der Mathematik und Physik. Frankfurt am Main, 1985.
[Berkeley 1992] Berkeley, G.: De Motu and The Analyst. A Modern Edition, with Introductions and Commentary. Edited and translated by Douglas M. Jesseph. The New Synthese Historical Library, Volume 41. Dordrecht 1992.
[Biegel/Klein/Sonar 2008] Biegel, G.; Klein, A.; Sonar, Th. (Hrsg.): Leonhard Euler – Mathematiker, Mechaniker, Physiker. Braunschweig, 2008.
[Bishop/Bridges 1985] Bishop, E.; Bridges, D.: Constructive Analysis. New York, Heidelberg, etc., 1985.
[Bölling 1994] Bölling, R. (Hrsg.): Das Photoalbum für Weierstraß. Braunschweig, 1994.
[Boethius 1999] Boethius: The Consolation of Philosophy. London, New York, etc., 1999.
[Bolzano 2006] Bolzano, B.: Paradoxien des Unendlichen. Photokopiertes Taschenbuch nach der Erstausgabe 1851, Saarbrücken, 2006.
[Bos 1975] Bos, H. J. M.: Differentials, higher-order differentials and the derivative in the Leibnizian calculus. Arch. Hist. Exact Sci. 14, S. 1–90, 1975.
[Bottazzini 1986] Bottazzini, U.: The Higher Calculus – A History of Real and Complex Analysis from Euler to Weierstrass. New York, Berlin, Heidelberg, etc., 1986.
[Bottazzini/Gray 2013] Bottazzini, U.; Gray J.: Hidden Harmony – Geometric Fantasies. The Rise of Complex Function Theory. New York, Heidelberg, Dordrecht, etc., 2013.
[Bourbaki 1968] Bourbaki, N.: Theory of Sets. Berlin, Heidelberg, 1968.
[Bourbaki 1971] Bourbaki, N.: Elemente der Mathematikgeschichte. Göttingen, 1971.
[Boyer 1959] Boyer, C. B.: The History of the Calculus and its Conceptual Development. Nachdruck der Erstausgabe 1949, New York, 1959.
[Bradley/Sandifer 2009] Bradley, R.E.; Sandifer, C.E.: Cauchy's Cours d'analysis. An Annotated Translation. Dordrecht, Heidelberg, London, New York, 2009.
[Bradley/Petrilli/Sandifer 2015] Bradley, R.E.; Petrilli, S.J.; Sandifer, C.E.: L'Hopital's Analyse des infiniments petits. An Annotated Translation with Source Material by Johann Bernoulli. Cham, Heidelberg, New York, etc., 2015.
[Braunmühl 1900–1903] Braunmühl, A. von: Vorlesungen über die Geschichte der Trigonometrie. Two volumes. Leipzig 1900–1903.

[Breger 1999] Breger, H.: Analysis und Beweis. Internationale Zeitschrift für Philosophie, 1, S. 95–106, 1999.

[Breidert 1979] Breidert, W.: Das aristotelische Kontinuum in der Scholastik. 2. Auflage, Münster, 1979.

[Breidert 1989] Breidert, W.: George Berkeley 1685–1753. Basel, Boston, Berlin, 1989.

[Bressoud 2008] Bressoud, D. M.: A Radical Approach to Lebesgue's Theory of Integration. Cambridge, New York, etc., 2008.

[Brewster 1855] Brewster, D.: Memoirs of the Life, Writings, and Discoveries of Sir Isaac Newton. 2 Vols. Edinburgh 1855.

[Briggs 1976] Briggs, H.: Arithmetica Logarithmica. Nachdruck der Ausgabe London 1628, Hildesheim, New York, 1976.

[Brouwer 1924] Brouwer, L. E. J.: Über die Bedeutung des Satzes vom ausgeschlossenen Dritten in der Mathematik, insbesondere in der Funktionentheorie. Journal für reine und ang. Math. 154, S. 1–7, 1924.

[Brouwer 1925] Brouwer, L. E. J.: Zur Begründung der intuitionistischen Mathematik I. Math. Annalen 93, S. 244–257, 1925.

[Brouwer 1926] Brouwer, L. E. J.: Zur Begründung der intuitionistischen Mathematik II. Math. Annalen 95, S. 453–472, 1926.

[Brouwer 1927] Brouwer, L. E. J.: Zur Begründung der intuitionistischen Mathematik III. Math. Annalen 96, S. 451–488, 1927.

[Brown 2010] Brown, N.M.: The Abacus and the Cross. The Story of the Pope Who Brought the Light of Science to the Dark Ages. New York, 2010.

[Burn 2001] Burn, R. P.: Alphonse Antonio de Sarasa and Logarithms. Historia Mathematica, No.28, S. 1–17, 2001.

[Cajori 1918] Cajori, F.: Pierre Laurent Wantzel. Bull. Amer. Math. Soc. 24(7), 339–347, 1918.

[Cajori 2000] Cajori, F.: History of Mathematics. Reprint of the 5th edition 1991, Rhode Island, 2000.

[Cantor 1875] Cantor, M.: Die römischen Agrimensoren und ihre Stellung in der Geschichte der Feldmesskunst. Eine historisch-mathematische Untersuchung. Leipzig, 1875.

[Cantor 1980] Cantor, G.: Gesammelte Abhandlungen mathematischen und philosophischen Inhalts. Nachdruck der Erstauflage 1932, Berlin, Heidelberg, New York, 1980.

[Cantor 2002] Cantor, N.: In the Wake of the Plague. New York, 2002.

[Capelle 2008] Capelle, W. (Hrsg.): Die Vorsokratiker. 6. Auflage, Stuttgart, 2008.

[Caspar 1993] Caspar, M.: Kepler. New York, 1993.

[Cauchy 1885] Cauchy, A. L.: Algebraische Analysis. Deutsch von Carl Itzigsohn. Berlin, 1885.

[Chaucer 2003] Chaucer, G.: The Canterbury Tales. Translated into modern English by Nevill Coghill. First publ. 1951, revised 1958, multiple reprints. London, New York, etc., 2003.

[Chaucer 2005] Chaucer, G.: The Canterbury Tales. Edited with an Introduction and Notes by Jill Mann. London, New York, etc., 2005.

[Child 1916] Child, J. M.: Geometrical Lectures of Isaac Barrow. Chicago, London, 1916.

[Child 2005] Child, J. M. (Übers.): The Early Mathematical Manuscripts of Leibniz. Nachdruck der Originalausgabe 1920, Mineola, New York, 2005.

[Cicero 1877] Cicero: Cicero's Tusculan Disputations. Translated chiefly by C.D. Yonge. New York, 1877.

[Cicero 1931] Cicero: On Ends. Loeb Classical Library 40, Cambridge, Mass., London, 2nd edition 1931.

[Clagett 1961] Clagett, M.: The Science of Mechanics in the Middle Ages. Wisconsin, 1961.

[Clagett 1968] Clagett, M.: Nicole Oresme and the Medieval Geometry of Qualities and Motions. Madison, Milwaukee, London, 1968.

[Clagett 2001] Clagett, M.: Greek Science in Antiquity. Nachdruck der Erstauflage 1955, Mineola, New York, 2001.

[Clark 2015] Clark, K.: Jost Bürgi's Aritmetische und Geometrische Progreß Tabulen (1620). Edition and Commentary. New York, Heidelberg, Dordrecht, London, 2015.

[Copernicus 1992] Copernicus, N.: On the Revolutions. Baltimore, London, 1992.

[Crombie 1953] Crombie, A. C.: Robert Grosseteste and the Origins of Experimental Science 1100–1700. Oxford, 1953.

[Crombie 1995] Crombie, A. C.: The History of Science. From Augustine to Galileo. 2 vols. bound as one, New York, 1995.

[Crowe 1994] Crowe, M. J.: A History of Vector Analysis. Korrigierter Nachdruck der Erstauflage 1967, New York, 1994.

[Cuomo 2001] Cuomo, S.: Ancient Mathematics. London and New York, 2001.

[Dante 1920] Dante Alighieri: The Divine Comedy of Dante Alighieri. The Italian Text with a Translation in English Blank Verse and a Commentary by Courtney Langdon, vol. 2 (Purgatorio), Cambridge, Massachusetts, 1920.

[Dauben 1979] Dauben, J. W.: Georg Cantor – His Mathematics and Philosophy of the Infinite. Princeton, 1979.

[Dauben 1995] Dauben, J. W.: Abraham Robinson – The Creation of Nonstandard Analysis, a Personal and Mathematical Odyssey. Princeton, 1995.

[Davidenko 1993] Davidenko, D.: Ich denke, also bin ich – Descartes' ausschweifendes Leben. Frankfurt/Main, 1993.

[De Crescenzo 1990] de Crescenzo, L.: Geschichte der griechischen Philosophie. 2 Bände, Zürich, 1990.

[Dedekind 1963] Dedekind, R.: Essays on the Theory of Numbers. I. Continuity and Irrational Numbers. II. The Nature and Meaning of Numbers. Authorized Translation by Wooster Woodruff Beman. Mineola, New York, 1963.

[Dedekind 1965] Dedekind, R.: Was sind und was sollen die Zahlen?, Zehnte Auflage, Stetigkeit und Irrationale Zahlen, Siebente Auflage. Berlin, 1965.

[Dedekind 2000] Dedekind, I.: Unter Glas und Rahmen – Briefe und Aufzeichnungen 1850–1950. Braunschweig, 2000.

[Dee 1570] Dee, Dr. J.: Mathematicall Praeface to the Elements of Geometrie of Euclid of Megara (1570). Kessinger Publishing, USA, ohne Druckdatum.

[Deiser 2009] Deiser, O.: Einführung in die Mengenlehre. Dritte Auflage, Berlin, Heidelberg, 2009.

[de Santillana 1955] de Santillana, G.: The Crime of Galileo. Chicago, London, 1955.

[Descartes 1954] Descartes, R.: The Geometry of Rene Descartes with a facsimile of the first edition. Unabrigded and unaltered republication of the English translation by David Eugene Smith and Marcia L. Latham originally published by the Open Court Publishing Co. in 1925. Mineola, N.Y., 1954.

[Descartes 2001] Descartes, R.: Discours on Method, Optics, Geometry, and Meteorology. Translated, with an Introduction, by Paul J. Olscamp. Revised edition. Indianapolis, Cambridge, 2017.

[Descartes 2017] Descartes, R.: Meditations on First Philosophy with Selections from the Objections and Replies. Translated and edited by John Cottingham. Second edition with an introductory essay by Bernard Williams and a general introduction by John Cottingham. Cambridge, 2017.

[Dieudonné 1981] Dieudonné, J.: History of Functional Analysis. Amsterdam, New York, Oxford, 1981.

[Dieudonné 1985] Dieudonné, J.: Geschichte der Mathematik 1700–1900 – Ein Abriß. Braunschweig, Wiesbaden, 1985.

[Dijksterhuis 1987] Dijksterhuis, E.J.: Archimedes. Translation of the first Dutch edition 1938. Princeton, New Jersey, 1987.

[Diogenes Laertius 1931] Diogenes Laertius: Lives of Eminent Philosophers. Volume I, Books 6-10. Translated by R.D. Hicks. Cambridge, Mass., London 1925. Revised and reprinted 1931.

[Diogenes Laertius 1972] Diogenes Laertius: Lives of Eminent Philosophers. Volume I, Books 1-5. Translated by R.D. Hicks. Cambridge, Mass., London 1925. Reprinted with new introductory material 1972.

[Dirlmeier et al. 2007] Dirlmeier, U.; Gestrich, A.; Herrmann, U.; Hinrichs, E.; Jarausch, K. H.; Kleßmann, Ch.; Reulecke, J.: Kleine deutsche Geschichte. Ditzingen, 2007.

[Djerassi 2003] Djerassi, C.: Kalkül/Unbefleckt - Zwei Theaterstücke aus dem Reich der Wissenschaft. Innsbruck, 2003.

[Djerassi/Pinner 2003] Djerassi, C., Pinner, D.: Newton's Darkness. Two Dramatic Views. London, 2003.

[Dobbs 1991] Dobbs, B.J.T.: The Janus Face of Genius – The Role of Alchemy in Newton's Thought. Cambridge, 1991.

[Domingues 2008] Domingues, J. C.: Lacroix and the Calculus. Basel, Boston, Berlin, 2008.

[Drake 2003] Drake, S.: Galileo at Work – His Scientific Biography. Nachdruck der Erstausgabe 1978, Mineola, New York, 2003.

[Du Bois-Reymond 1968] Du Bois-Reymond, P.: Die Allgemeine Functionenlehre – Erster Teil. Darmstadt, 1968.

[Du Sautoy 2008] du Sautoy, M.: The Story of Maths. DVD, Open University International / BBC, 2008.

[Dudley 1987] Dudley, U.: The Trisectors.Washington, 1987.

[Ebbinghaus 2007] Ebbinghaus, H.-D.: Ernst Zermelo – An Approach to His Life and Work. Berlin, Heidelberg, 2007.

[Early Greek Philosophy 2016] Early Greek Philosophy. Vol. I – Vol. VII, Loeb Classical Library, Cambridge, Mass., London, 2016.

[Ebert 2009] Ebert, Th.: Der rätselhafte Tod des René Descartes. Aschaffenburg, 2009.

[Eckert 2008] Eckert, M.: Hydraulik im Schloßpark – War Euler schuld am Versagen der Wasserkunst in Sanssouci? in: [Biegel/Klein/Sonar 2008].

[Edwards 1979] Edwards, C. H., Jr.: The Historical Development of the Calculus. New York, Heidelberg, Berlin, 1979.

[Euclid 1956] Euclid: The Thirteen Books of The Elements. Translated with introduction and commentary by Sir Thomas L. Heath. Unabridged and unaltered republication of the second edition. Three vols., New York 1956.

[Euler 1983] Euler, L.: Einleitung in die Analysis des Unendlichen. Erster Teil. Berlin, Heidelberg, New York, 1983.

[Euler 1988] Euler, L.: Introduction to Analysis of the Infinite. 2 vols., New York, Berlin, Heidelberg, etc.,1988.

[Euler 1996] Euler, L.: Zur Theorie komplexer Funktionen. Frankfurt am Main, 1996.

[Euler 2000] Euler, L.: Foundations of Differential Calculus. New York, Berlin, Heidelberg, etc., 2000.

[Fauvel/Flood/Wilson 2000] Fauvel, J.; Flood, R.; Wilson, R.: Oxford Figures. 800 Years of the Mathematical Sciences. Oxford, New York, 2000.

[Feingold 1990] Feingold, M. (edt.): Before Newton – The Life and Times of Isaac Barrow. Cambridge, New York, etc., 1990.

[Feingold 1990a] Feingold, M.: Isaac Barrow – Divine, Scholar, Mathematician. in: [Feingold 1990].

[Fellmann 1995] Fellmann, E. A.: Leonhard Euler. Reinbek, 1995.

[Ferreirós 1999] Ferreirós, J.: Labyrinth of Thought – A History of Set Theory and its Role in Modern Mathematics. Basel, Boston, Berlin, 1999.

[Finster/van den Heuvel 1990] Finster, R.; van den Heuvel, G.: Gottfried Wilhelm Leibniz. Reinbek, 1990.

[Flasch 2004] Flasch, K.: Nikolaus von Kues in seiner Zeit – Ein Essay. Stuttgart, 2004.

[Flasch 2005] Flasch, K.: Nicolaus Cusanus. Zweite Auflage, München, 2005.

[Flasch 2008] Flasch, K.: Nikolaus von Kues: Geschichte einer Entwicklung. Dritte Auflage, Frankfurt/Main, 2008.

[Fleckenstein 1977] Fleckenstein, J. O.: Der Prioritätsstreit zwischen Leibniz und Newton – Isaac Newton. 2. Auflage, Basel, 1977.

[Folkerts 2014] Folkerts, M.: Eine bisher unbekannte Schrift von Jost Bürgi zur Trigonometrie. In: Gebhardt, R. (edt.): Arithmetik, Geometrie und Algebra in der frühen Neuzeit. Adam-Ries-Bund, Annaberg-Bucholz, pp. 107-114, 2014.

[Folkerts/Launert/Thom 2016] Folkerts, M.; Launert, D.; Thom, A.: Jost Bürgi's Method for Calculating Sines. (Historia Mathematica 43, 133-147, 2016).

[Fowler 1999] Fowler, D.: The Mathematics of Plato's Academy. A New Reconstruction. Second Edition, Oxford, 1999.

[Fox 2000] Fox, R. (edt.): Thomas Harriot: An Elizabethan Man of Science. Aldershot, Burlington, Singapore, Sydney, 2000.

[Fraenkel 1967] Fraenkel, A. A.: Lebenskreise – Aus den Erinnerungen eines jüdischen Mathematikers. Stuttgart, 1967.

[Frege 1987] Frege, G.: Die Grundlagen der Arithmetik. Stuttgart, 1987.

[French 1972] French, P.: John Dee: The World of an Elizabethan Magus. New York, 1972.

[Galilei 1967] Galilei, G.: Dialogue Concerning the Two Chief World Systems. Translated, with revised notes, by Stillman Drake. Forword by Albert Einstein. 2nd edt., Berkeley, Los Angeles, London, 1967.

[Galilei 1982] Galilei, G.: Dialog über die beiden hauptsächlichsten Weltsysteme. Aus dem Italienischen übersetzt und erläutert von Emil Strauss. Mit einem Beitrag von Albert Einstein sowie einem Vorwort zur Neuasugabe und weiteren Erläuterungen von Stillman Drake. Herausgegeben von Roman Sexl und Karl von Meyenn. Darmstadt, 1982.

[Galilei 2014] Galilei, G.: Two New Sciences Including Centers of Gravity and Force of Percussion. Second edition. Translated, with a New Introduction and Notes, by Stillman Drake. Third printing 2014, Toronto, 2014.

[Galilei 2015] Galilei, G.: Siderius Nuncius or The Siderial Messenger. Translated and with Commentary by Albert van Helden. 2nd edt., Chicago, London, 2015.

[Gericke 2003] Gericke, H.: Mathematik in Antike, Orient und Abendland. Sonderausgabe in einem Band, 6. Auflage, Wiesbaden, 2003.

[Gerlach/List 1987] Gerlach, W.; List, M.: Johannes Kepler – Der Begründer der modernen Astronomie. München, Zürich, 1989.

[Gibson 1914] Gibson, G. A.: Napier's Life and Works. in: [Horsburgh 1982].

[Gilbert 1958] Gilbert, W.: De Magnete. Translated by P. Fleury Mottelay in 1893, New York, 1958.

[Gilder 2005] Gilder, J.; Gilder, A.-L.: Der Fall Kepler – Mord im Namen der Wissenschaft. Berlin, 2005.

[Gillings 1982] Gillings, R. J.: Mathematics in the Time of the Pharaohs. Nachdruck der Ausgabe 1972. New York, 1982.

[Gilson 1989] Gilson, E.: History of Christian Philosophy in the Middle Ages. 3. Auflage, London, 1989.

[Glaisher 1965] Glaisher, J. W. L. (edt.): The Collected Mathematical Papers of Henry John Stephen Smith. 2 Volumes. Nachdruck der Originalausgabe 1894, New York, 1965.

[Goldstine 1972] Goldstine, H.H.: The Computer from Pascal to von Neumann. Princeton, 1972.

[Goldstine 1977] Goldstine, H. H.: A History of Numerical Analysis from the 16th to the 19th Century. New York, Heidelberg, Berlin, 1977.

[Goldstine 1980] Goldstine, H. H.: A History of the Calculus of Variations from the 17th through the 19th Century. New York, Heidelberg, Berlin, 1980.

[Goodstein 2007] Goodstein, J. R.: The Volterra Chronicles – The Life and Times of an Extraordinary Mathematician 1860–1940. Providence, Rhode Island, 2007.

[Gottschalk 2008] Gottschalk, J.: Kurze geschichtliche Entwicklung des Zahnrades und Eulers Einfluss auf die Entwicklung der Verzahnung. in: [Biegel/Klein/Sonar 2008].

[Grabiner 2005] Grabiner, J. V.: The Origins of Cauchy's Rigorous Calculus. Unveränderter Nachdruck der Erstauflage 1981, Mineola, New York, 2005.

[Grattan-Guinness 1970] Grattan-Guinness, I.: An unpublished paper by Georg Cantor: Principien einer Theorie der Ordnungstypen. Erste Mitteilung. Acta math. 124, S. 65–107, 1970.

[Grattan-Guinness 1980] Grattan-Guinness, I.: The emergence of mathematical analysis and its foundational progress, 1780-1880. in: [Grattan-Guinness 1980a, S. 94-148].

[Grattan-Guinness 1980a] Grattan-Guinness, I. (edt.): From Calculus to Set Theory 1630–1910. An introductory History. Princeton and Oxford, 1980.

[Grattan-Guinness 2005] Grattan-Guinness, I.: Landmark Writings in Western Mathematics 1640–1940. Elsevier, Amsterdam, Boston, etc., 2005.

[Greenberg 1995] Greenberg, J. L.: The Problem of the Earth's Shape from Newton to Clairaut – The Rise of Mathematical Science in Eighteenth-Century Paris and the Fall of 'Normal' Science. Cambridge, 1995.

[Gronau 2009] Gronau, D.: Paulus Guldin, 1577-1643, Jesuit und Mathematiker. in: [Pichler/von Renteln 2009], S. 101-120, 2009.

[Guhrauer 1966] Guhrauer, G. E.: Gottfried Wilhelm Freiherr von Leibniz – Eine Biographie. 2 Bände, Nachdruck der Originalausgabe von 1846, Hildesheim, 1966.

[Gutas 1988] Gutas, D.: Avicenna in the Aristotelian Tradition, 4 vols. Leiden, 1988.

[Guthrie 1987] Guthrie, K. S.: The Pythagorean Sourcebook and Library. Grand Rapids, Michigan, 1987.

[Haan/Niedhart 2002] Haan, H.; Niedhart, G.: Geschichte Englands vom 16. bis zum 18. Jahrhundert. München 2002.

[Hairer/Nørsett/Wanner1987] Hairer, E.; Nørsett, S. P.; Wanner, G.: Solving Ordinary Differential Equations I. Berlin, Heidelberg, New York, etc., 1987.

[Hairer/Wanner 1997] Hairer, E.; Wanner, G.: Analysis by its History. New York, Heidelberg, Berlin, 2nd edt., 1997.

[Hall 1980] Hall, A. R.: Philosophers at war – The quarrel between Newton and Leibniz. Cambridge, London, New York, etc., 1980.

[Harborth/Heuer/Löwe/Löwen/Sonar 2007] Harborth, H.; Heuer, M.; Löwe, H.; Löwen, R.: Gedenkschrift für Richard Dedekind. Braunschweig, 2007.

[Hardy 1910] Hardy, G.H.: Orders of Infinity. The 'Infinitärcalcül' of Paul Du Bois-Reymond. Cambridge, 1910.

[Harrison 1978] Harrison, J.: The Library of Isaac Newton. Cambridge, New York, etc., 1978.

[Hasse/Scholz 1928] Hasse, H.; Scholz, H.: Die Grundlagenkrise der griechischen Mathematik. (Kant-Studien 33, 4–34, 1928).

[Haupt et al. 2008] Haupt, H. G.; Hinrichs, E.; Martens, S.; Müller, H.; Schneidmüller, B.; Tacke, Ch.: Kleine Geschichte Frankreichs. Ditzingen, 2008.

[Hausdorff 1978] Hausdorff, F.: Grundzüge der Mengenlehre. Nachdruck der Erstauflage 1914, New York, 1978.

[Hausdorff 1991] Hausdorff, F.: Set Theory. Translated from the German by John R. Aumann, et al. Fourth corrected edition, New York, 1991.

[Hausdorff 2002] Hausdorff, F.: Gesammelte Werke, Band II: Grundzüge der Mengenlehre. Berlin, Heidelberg, etc., 2002.

[Havil 2014] Havil, J.: John Napier. Life, Logarithms, and Legacy. Princeton and Oxford, 2014.

[Hawking 1988] Hawking, S. W.: A Brief History of Time. Toronto, New York, London, Sidney, Auckland, 1988.

[Hawkins 1979] Hawkins, T.: Lebesgue's Theory of Integration – Its Origins and Development. Korrigierter Nachdruck der Erstauflage 1970, Providence, Rhode Island, 1979.

[Heath 1981] Sir Heath, T.: A History of Greek Mathematics, 2 Vols. New York, 1981.

[Heath 2002] Sir Heath, T. L.: The Works of Archimedes. Nachdruck der Erstauflage 1897, Mineola, New York, 2002.

[Heath 2004] Sir Heath, T.: A Manual of Greek Mathematics. New York, 2004.

[Hein 2010] Hein, W.: Die Mathematik im Mittelalter – Von Abakus bis Zahlenspiel. Darmstadt, 2010.

[Hemleben 2006] Hemleben, J.: Galilei. 19. Auflage, Reinbek, 2006.

[Henle/Kleinberg 1979] Henle, J. M.; Kleinberg, E. M.: Infinitesimal Calculus. Cambridge, Mass., 1979.

[Hess 2005] Hess, H.-J.: Leibniz auf dem Höhepunkt seines mathematischen Ruhms. (Studia Leibnitiana, Band 37, 2005, Heft 1, S. 48-67).

[Heuser 2008] Heuser, H.: Unendlichkeiten – Nachrichten aus dem Grand Canyon des Geistes. Wiesbaden, 2008.

[Heuser 2008a] Heuser, H.: Eulers Analysis. in: [Biegel/Klein/Sonar 2008].

[Hilbert 1926] Hilbert, D.: Über das Unendliche. Math. Annalen 95, S. 161–190, 1926.

[Hilbert 1928] Hilbert, D.: Die Grundlagen der Mathematik. Abhandlungen aus dem mathematischen Seminar der Hamburgischen Universität, Band VI, 1928.

[Hilbert/Schmidt 1989] Hilbert, D.; Schmidt, E.: Integralgleichungen und Gleichungen mit unendlich vielen Unbekannten. Leipzig, 1989.

[Hill 1997] Hill, Ch.: Intellectual Origins of the English Revolution Revisited. Oxford, New York, 1997.

[Hinrichs 1992] Hinrichs, F. T.: Die 'agri per extremitatem mensura comprehensi'. Diskussion eines Frontintextes und der Geschichte seines Verständnisses. in: [Behrends/Colognesi 1992], S. 348–374, 1992.

[Höppner 1992] Höppner, H.-J.: Die Rache der Schildkröte – über akataleptische Zenon-Bahnen. in: Das Wilhelm-Gymnasium. Mitteilungsblatt des Vereins "'Ehemalige Wilhelm-Gymnasium e. V."', Heft 59, S. 59–69, Hamburg, 1992.

[Hoffmann 2008] Hoffmann, P.: Die Entwicklung der Geographie in Russland und Leonhard Euler. in: [Biegel/Klein/Sonar 2008].

[Hofmann 1939] Hofmann, J. E.: On the Discovery of the Logarithmic Series and Its Development in England up to Cotes. National Mathematics Magazine, Vol.14, No.1., 37–45, 1939.

[Hofmann 1951] Hofmann, J. E.: Zum Gedenken an Thomas Bradwardine. Centaurus, 1, 293–308, 1951.

[Hofmann 1968] Hofmann, J. E.: Michael Stifel (1487?–1567): Leben, Wirken und Bedeutung für die Mathematik seiner Zeit. Sudhoffs Archiv, Beiheft 9, Wiesbaden, 1968.

[Hofmann 1974] Hofmann, J. E.: Leibniz in Paris 1672–1676. His Growth to Mathematical Maturity. Cambridge, New York, Melbourne, etc., 1974.

[Hogrebe/Broman 2004] Hogrebe, W.; Bromand, J. (Hrsg.): Grenzen und Grenzüberschreitungen – XIX. Deutscher Kongress für Philosophie. Berlin, 2004.

[Honnefelder/Wood/Dreyer/Aris 2005] Honnefelder, L., Wood, R., Dreyer, M., Aris, M.-A. (Hrsg.): Albertus Magnus und die Anfänge der Aristoteles-Rezeption im lateinischen Mittelalter. Von Richardus Rufus bis zu Franciscus de Mayronis. Münster 2005.

[Hopkins 1985] Hopkins, J.: On Learned Ignorance (De docta ignorantia) by Nicholas of Cusa. Translation of Book I from [Nikolaus 2002a] by Paul Wilpert, revised by Hans G. Senger. http://jasper-hopkins.info/DI-I-12-2000.pdf, 2nd edt., 1985

[Horsburgh 1982] Horsburgh, E. M. (edt.): Handbook of the Napier Tercentenary Celebration or Modern Instruments and Methods of Calculation. Nachdruck der Originalausgabe 1914, Los Angeles, San Francisco, 1982.

[Ifrah 2001] Ifrah, G.: The Universal History of Computing – From the Abacus to the Quantum Computer. New York, Chichester, Weinheim, etc., 2001.

[Iliffe 2017] Iliffe, R.: Priest of Nature. The Religious Worlds of Isaac Newton. New York 2017.

[Iro 2008] Iro, H.: Eulers analytische Mechanik. in: [Biegel/Klein/Sonar 2008].

[Jahnke 2003a] Jahnke, H. N. (Edt): A History of Analysis. American Mathematical Society 2003.

[Jahnke 2003b] Jahnke, H. N.: Algebraic Analysis in the 18th Century. in: [Jahnke 2003a], p. 105–136.

[Jesseph 1999] Jesseph, D. M.: Squaring the Circle – The War between Hobbes and Wallis. Chicago, London, 1999.

[Johnson/Wolbarsht 1979] Johnson, L. W.; Wolbarsht, M. L.: Mercury Poisoning: A Probable Cause of Isaac Newton's Physical and Mental Ills. Notes and Records of the Royal Society of London, Vol.34, No.1, 1–9, 1979.

[Juschkewitsch 1964] Juschkewitsch, A. P.: Geschichte der Mathematik im Mittelalter. Leipzig, 1964.

[Kästner 1970] Kästner, A. G.: Geschichte der Mathematik. 4 Bände, Hildesheim, 1970.

[Kaiser/König 2006] Kaiser, W., König, W. (Hrsg.): Geschichte des Ingenieurs – Ein Beruf in sechs Jahrtausenden. München, Wien, 2006.

[Kałuża 1996] Kałuża, R.: Through a Reporter's Eyes – The Life of Stefan Banach. Boston, Basel, Berlin, 1996.

[Kant 2005] Kant, I.: Allgemeine Naturgeschichte und Theorie des Himmels oder Versuch von der Verfassung und dem mechanischen Ursprunge des ganzen Weltgebäudes nach Newtonischen Grundsätzen abgehandelt. Frankfurt am Main, 2005.

[Katz 2007] Katz, V. J. (edt.): The Mathematics of Egypt, Mesopotamia, China, India, and Islam – A Sourcebook. Princeton and Oxford, 2007.

[Keisler 1976] Keisler, H. J.: Foundations of Infinitesimal Calculus. Boston, 1976.

[Keisler 1986] Keisler, H. J.: Elementary Calculus: An Infinitesimal Approach. 2nd edt., Boston, 1986.

[Kelly 2005] Kelly, J.: The Great Mortality. An Intimate History of the Black Death. London, New York, Toronto, Sidney, 2005.

[Kepler 1908] Kepler, J.: Neue Stereometrie der Fässer. Besonders der in der Form am meisten geeigneten österreichischen, und Gebrauch der kubischen Visierrute; Mit einer Ergänzung zur Stereometrie des Archimedes. Leipzig, 1908.

[Kepler 2003] Kepler, J.: Dioptrik. Frankfurt/Main, 2003.

[Kepler 2005] Kepler, J.: Was die Welt im Innersten zusammenhält: Antworten aus Keplers Schriften. Wiesbaden, 2005.

[Kepler 2005a] Kepler, J.: Mysterium Cosmographicum. in: [Kepler 2005].

[Kepler 2005b] Kepler, J.: Astronomia Nova. Wiesbaden, 2005.

[Kepler 2006] Kepler, J.: Weltharmonik. München, 2006.

[Keynes 1980] Keynes, M.: Sir Isaac Newton and his Madness of 1692–93. The Lancet, March 8, 1980.

[Klein 1926] Klein, F.: Vorlesungen über die Entwicklung der Mathematik im 19. Jahrhundert. Teil I. Berlin, 1926.

[Knobloch 1983] Knobloch, E.: Von Riemann zu Lebesgue – Zur Entwicklung der Integrationstheorie. Historia Mathematica 10, 318-343, 1983.

[Knobloch 1988] Knobloch, E.: La vie et l'oeuvre de Christoph Clavius. Revue d'Histoire des Sciences 41, 331-356, 1988.

[Knobloch 1993] Knobloch, E. (Hrsg.): Gottfried Wilhelm Leibniz – De quadratura arithmetica circuli ellipseos et hyperbolae cujus corollarium est trigonometria sine tabulis. Göttingen, 1993.

[Knobloch 1999] Knobloch, E.: Galileo und Leibniz: Different approaches to infinity. Archive for History of Exact Sciences 54, 87-99, 1999.

[Knobloch 2000] Knobloch, E.: Archimedes, Kepler, and Guldin: the role of proof and analogy. in: [Thiele 2000], 82-100, 2000.

[Knobloch 2004] Knobloch, E.: Von Nicolaus von Kues über Galilei zu Leibniz – Vom mathematischen Umgang mit dem Unendlichen. in: [Hogrebe/Broman 2004], S. 490–503.

[Knobloch 2005] Knobloch, E.: Archimedes, Kepler und Guldin - Zur Rolle von Beweis und Analogie. in: [Peckhaus 2005], 15-34, 2005

[Knobloch 2009] Knobloch, E.: persönliche Mitteilung, 2009.

[Knobloch 2010] Knobloch, E.: persönliche Mitteilung, 2010.

[Knoche/Wippermann 1986] Knoche, N.; Wippermann, H.: Vorlesungen zur Methodik und Didaktik der Analysis. Mannheim, Wien, Zürich, 1986.

[König 1990] König, G. (Hrsg.): Konzepte des mathematisch Unendlichen im 19. Jahrhundert. Göttingen, 1990.

[Körle 2009] Körle, H.-H.: Die phantastische Geschichte der Analysis. München, 2009.

[Koestler 1989] Koestler, A.: The Sleepwalkers. London, 1989.

[Koetsier/Bergmans 2005] Koetsier, T.; Bergmans, L. (edt.): Mathematics and the Divine – A historical study. Amsterdam, 2005.

[Kowalewski 1938] Kowalewski, G.: Große Mathematiker – Eine Wanderung durch die Geschichte der Mathematik vom Altertum bis zur Neuzeit. München, Berlin, 1938.

[Kühn 2009] Kühn, E.: Persönliche Korrespondenz, 2009.

[Kuhn 1957] Kuhn, T. S.: The Copernican Revolution. Planetary Astronomy in the Development of Western Thought. Harvard, 1957.

[Kuhn 1962] Kuhn, T. S.: The Structure of Scientific Revolutions. Chicago, 1962.

[Kurrer 2008] Kurrer, K.-E.: The History of the Theory of Structures. Berlin, 2008.

[Lasswitz 1984] Lasswitz, K.: Geschichte der Atomistik vom Mittelalter bis Newton, 2 Bde. 2. Nachdruck von 1890, Hildesheim, 1984.

[Lattis 1994] Lattis, J.M.: Between Copernicus and Galileo. Christoph Clavius and the Collapse of Ptolemaic Cosmology. Chicago, London, 1994.

[Laugwitz 1978] Laugwitz, D.: Infinitesimalkalkül – Eine elementare Einführung in die Nichtstandard-Analyse. Mannheim, Wien, Zürich, 1978.

[Laugwitz 1986] Laugwitz, D.: Zahlen und Kontinuum – Eine Einführung in die Infinitesimalmathematik. Mannheim, Wien, Zürich, 1986.

[Laugwitz 1990] Laugwitz, D.: Das mathematisch Unendliche bei Euler und Cauchy. in: [König 1990].

[Laugwitz 1999] Laugwitz, D.: Curt Schmieden's approach to infinitesimals – an eye-opener to the historiography of analysis. Preprint Nr. 2053, Fachbereich Mathematik, Technische Universität Darmstadt, August 1999.

[Launert 2015] Launert, D.: Nova Kepleriana. Bürgis Kunstweg im *Fundamentum Astronomiae*. Entschlüsselung seines Rätsels. Bayerische Akademie der Wissenschaften, München, 2015.

[Leibniz 1985–1992] Leibniz, G. W.: Philosophische Schriften. Darmstadt, 1985–1992.

[Leibniz/Newton 1998] Leibniz, G. W.; Newton, I.: Über die Analysis des Unendlichen / Abhandlung über die Quadratur der Kurven. Frankfurt am Main, 1998.

[Leibniz 2004] Leibniz, G. W.: Mathematische Schriften. Hrsg. v. C. I. Gerhardt. Nachdruck. Faks.-Ausg. 1971-2004. 7 Bde. Hildesheim, Zürich, New York 2004.

[Leibniz 2005] Leibniz, G. W.: Monadologie. Stuttgart, 2005.

[Leibniz 2008] Leibniz, G. W.: Mathematische Schriften, Fünfter Band. Berlin, 2008.

[Leibniz 2011] Leibniz, G. W.: Die mathematischen Zeitschriftenartikel. Mit einer CD: Die originalsprachlichen Fassungen. Übersetzt und kommentiert von Heinz-Jürgen Hess und Malte-Ludolf Babin. Hildesheim, Zürich, New York 2011.

[Lelgemann 2010] Lelgemann, D.: Die Erfindung der Messkunst – Angewandte Mathematik im antiken Griechenland. Darmstadt, 2010.

[Lemcke 1995] Lemcke, M.: Johannes Kepler. Reinbek, 1995.

[Lewis 2005] Lewis, N.: Robert Grosseteste and the Continuum. in: [Honnefelder/Wood/Dreyer/Aris 2005, S. 159–187].

[Lieb/Hershman 1983] Lieb, J.; Hershman,D.: Isaac Newton: Mercury Poisoning or Manic Depression? The Lancet, December 24/31, 1479–1480, 1983.

[Livy 1940] Livy: History of Rome. Books XXIII–XXV. With an English Translation by Frank Gardner Moore. Loeb Classical Library LCL 355, Livy VI. Cambridge, Mass., London, 1940.

[Loeffel 1987] Loeffel, H.: Blaise Pascal 1623–1662. Basel, Boston, 1987.

[Löwe 2007] Löwe, H.: Dedekinds Theorie der Ideale. In: [Harborth/Heuer/Löwe/Löwen/Sonar 2007], S. 51–82.

[Löwe 2008a] Löwe, H.: Das Königsberger Brückenproblem – Der Beginn der modernen Graphentheorie. in: [Biegel/Klein/Sonar 2008].

[Löwe 2008b] Löwe, H.: Eulers Polyederformel. in: [Biegel/Klein/Sonar 2008].

[Lorenzen 1965] Lorenzen, P.: Differential und Integral – Eine konstruktive Einführung in die klassische Analysis. Frankfurt am Main, 1965.

[Lorenzen 1986] Lorenzen, P.: Die Theoriefähigkeit des Kontinuums. in: Jahrbuch Überblicke Mathematik 1986, S. 147–153, Mannheim, 1986.

[Loyson 2012] Loysen, P.: Galilean Thermometer Not So Galilean. Journal of Chemical Education 89(9), 1095–1096, 2012.

[Lüneburg 2008] Lüneburg, H.: Von Zahlen und Größen – Dritthalbtausend Jahre Theorie und Praxis, Band 2. Basel, Boston, Berlin, 2008.

[Lützen 2003] Lützen, J.: The Foundation of Analysis in the 19th Century. in: [Jahnke 2003a], p. 155–195.

[Lutsdorf/Walter 1992] Lutsdorf, H.; Walter, M.: Jost Bürgis "‘Progress Tabulen"’ (Logarithmen). Schriftenreihe der ETH-Bibliothek Nr.28, Zürich, 1992.

[Maclaurin 1971] Maclaurin, C.: An Account of Sir Isaac Newton's Philosophical Discoveries. Nachdruck der Erstausgabe London 1748, Hildesheim, 1971.

[Mahoney 1990] Mahoney, M. S.: Barrow's mathematics – between ancients and moderns. in: [Feingold 1990].

[Mahoney 1994] Mahoney, M. S.: The Mathematical Career of Pierre de Fermat 1601–1665. Überarbeitung der Erstauflage 1973, Princeton, 1994.

[Maier 1949] Maier, A.: Die Vorläufer Galileis – Studien zur Naturphilosophie der Spätscholastik. Rom, 1949.

[Maier 1964] Maier, A.: Ausgehendes Mittelalter – Gesammelte Aufsätze zur Geistesgeschichte des 14. Jahrhunderts. Rom, 1964.

[Mancosu 1996] Mancosu, P.: Philosophy of Mathematics & Mathematical Practice in the Seventeenth Century. New York, Oxford, 1996.

[Mancosu/Vailati 1991] Mancosu, P.; Vailati, E.: Torricelli's Infinitely Long Solid and Its Philosophical Reception in the Seventeenth Century. ISIS No. 82, S. 50–70, 1996.

[Mansfeld 1999] Mansfeld, J. (Übers.): Die Vorsokratiker. 2 Bände, Stuttgart, 1999.

[Manuel 1968] Manuel, F. E.: A Portrait of Isaac Newton. Cambridge, Mass., 1968.

[Martzloff 2006] Martzloff, J.-C.: A History of Chinese Mathematics. Berlin, Heidelberg, 2006.

[Mashaal 2006] Mashaal, M.: Bourbaki – A Secret Society of Mathematicians. Rhode Island, 2006.
[Mauldin 1981] Mauldin, R. D.: The Scottish Book – Mathematics from the Scottish Cafe. Basel, 1981.
[Maurer 2002] Maurer, M.: Kleine Geschichte Englands. Stuttgart, 2002.
[Mercator 1975] Mercator, N.: Logarithmotechnia. Reprint of the edition 1666, Hildesheim, New York, 1975.
[Merton 1985] Merton, R. K.: The Vicennial Edition: On the Shoulders of Giants. A Shandean Postscript. San Diego, New York, London, 1985.
[Meschkowski 1967] Meschkowski, H.: Probleme des Unendlichen – Werk und Leben Georg Cantors. Braunschweig, 1967.
[Meschkowski 1990] Meschkowski, H.: Denkweisen großer Mathematiker – Ein Weg zur Geschichte der Mathematik. Braunschweig, 1990.
[Meschkowski/Nilson 1991] Meschkowski, H.; Nilson, W. (Hrsg.): Georg Cantor – Briefe. Berlin, Heidelberg, New York, etc., 1991.
[Moerdijk/Reyes 1991] Moerdijke, I.; Reyes, G. E.: Models for Smooth Infinitesimal Analysis. New York, Berlin, Heidelberg, etc., 1991.
[Montaigne 1998] de Montaigne, M.: Essais. Frankfurt/Main, 1998.
[Müller/Krönert 1969] Müller, K.; Krönert, G.: Leben und Werk von Gottfried Wilhelm Leibniz – Eine Chronik. Frankfurt am Main, 1969...
[Natterer 2001] Natterer, F.: The Mathematics of Computerized Tomography. Nachdruck der Erstauflage von 1991, Philadelphia, 2001.
[Neidhart 2007] Neidhart, L.: Unendlichkeit im Schnittpunkt von Mathematik und Theologie. Göttingen, 2007.
[Nelson 1977] Nelson, E.: Internal set theory, a new approach to NSA. Bull. Amer. Math. Soc. 83, S. 1165–1198, 1977.
[Nette 1982] Nette, H.: Elisabeth I. Reinbek, 1982.
[Netz/Noel 2007] Netz, R.; Noel, W.: The Archimedes Codex. How a Medieval Prayer Book is Revealing the True Genius of Antiquity's Greatest Scientist. Philadelphia, 2007.
[Neugebauer 1969a] Neugebauer, O.: Vorlesungen über Geschichte der antiken mathematischen Wissenschaften. Erster Band: Vorgriechische Mathematik. Zweite, unveränderte Auflage, Berlin, Heidelberg, New York 1969.
[Neugebauer 1969b] Neugebauer, O.: The Exact Sciences in Antiquity. New York, Nachdruck der zweiten Auflage 1969.
[Neugebauer 1975] Neugebauer, O.: A History of Ancient Mathematical Astronomy, Part Two. New York, Heidelberg, Berlin 1975.
[Newton 1979] Newton, Sir I.: Opticks. Nachdruck der Ausgabe von 1931 mit überarbeitetem Vorwort, New York, 1979.
[Newton 1999] Newton, I.: The Principia – Mathematical Principles of Natural Philosophy. Translated by I. Bernard Cohen and Anne Whitman, Berkeley, Los Angeles, London, 1999.
[Nikolaus 1952] Nikolaus von Kues: Die mathematischen Schriften – übersetzt von Josepha Hofmann. Hamburg, 1952.
[Nikolaus 2002] Nikolaus von Kues: Philosophisch-Theologische Werke. Vier Bände, Lateinisch-Deutsch, Hamburg, 2002.
[Nikolaus 2002a] Nikolaus von Kues: De docta ignorantia/Die belehrte Unwissenheit. in: [Nikolaus 2002].
[Noether/Cavaillés 1937] Noether, E.; Cavaillés, J. (Hrsg.): Briefwechsel Cantor-Dedekind. Paris, 1937.

[Nowaki 2008] Nowaki, H.: Leonhard Euler und die Theorie des Schiffes. in: [Biegel/Klein/Sonar 2008].

[Odefey 2008] Odefey, A.: Eulers Musiktheorie. in: [Biegel/Klein/Sonar 2008].

[Oeser 1971] Oeser, E.: Kepler – die Entstehung der modernen Wissenschaft. Zürich, Göttingen, 1971.

[Oestmann 2004] Oestmann, G.: Heinrich Rantzau und die Astrologie. Braunschweig, 2004.

[Parmenides 2016] Parmenides. in: [Early Greek Philosophy 2016, Vol. V, p. 3–151].

[Pascal 1997] Pascal, B.: Gedanken. Ditzingen, 1997.

[Peckhaus 2005] Peckhaus, V. (Hrsg.): Oskar Becker und die Philosophie der Mathematik. München, 2005.

[Pepper 1968] Pepper, J. V.: Harriot's Calculation of Meridional Parts as Logarithmic Tangents. Archive for the History of Exact Sciences, Vol.4, No.5, 1968. Nachgedruckt in: [Shirley 1974].

[Pepper 1974] Pepper, J. V.: Harriot's Earlier Work on Mathematical Navigation. in: [Shirley 1974].

[Phillips 2000] Phillips, G. M.: Two Millennia of Mathematics: From Archimedes to Gauss. New York, Berlin, Heidelberg, 2000.

[Pichler/von Renteln 2009] Pichler, F., von Renteln, M. (Hrsg.): Kosmisches Wissen von Peurbach bis Laplace - Astronomie, Mathematik, Physik. Peurbach Symposium 2008, Linz, 2009.

[Pies 1996] Pies, E.: Der Mordfall Descartes – Dokumente, Indizien, Beweise. Solingen, 1996.

[Plato 1929] Plato: Timaeus. Critias. Cleitophon. Menexenus. Epistles. Loeb Classical Library LCL 234, Plato IX. With an English translation by R.G. Bury. Cambridge, Mass., London, 1929.

[Plato 1939] Plato: Cratylus. Parmenides. Greater Hippias. Lesser Hippias. Loeb Classical Library LCL 167, Plato IV. With an English translation by H.N. Fowler. Cambridge, Mass., London, First published 1926. Revised and reprinted 1939.

[Ploetz 2008] Der große Ploetz – Die Enzyklopädie der Weltgeschichte. Göttingen, 2008.

[Plofker 2007] Plofker, K.: Mathematics in India. in: [Katz 2007, S. 385–514].

[Plutarch 2004] Plutarch: Lives, Vol. V., Agesilaus and Pompey, Pelopidas and Marcellus. With an English Translation by Bernadotte Perrin. Loeb Classical Library LCL 87, First published 1917, Cambridge, Mass., London, 2004.

[Popper 2006] Popper, K. R.: Die Welt des Parmenides. 2. Auflage, München, Zürich, 2006.

[Pumfrey 2002] Pumfrey, S.: Latitude & The Magnetic Earth. Cambridge, 2002.

[Purkert/Ilgauds 1987] Purkert, W.; Ilgauds, H. J.: Georg Cantor 1845–1918. Basel, Boston, Stuttgart, 1987.

[Reich 1994] Reich, K.: Die Entwicklung des Tensorkalküls – Vom absoluten Differentialkalkül zur Relativitätstheorie. Basel, Boston, Berlin, 1994.

[Reich/Wiederkehr 2008] Reich, K.; Wiederkehr, K. H.: Der Achromat, die bedeutendste Erfindung des 18. Jahrhunderts in der Optik. in: [Biegel/Klein/Sonar 2008].

[Rice/González-Valesco/Corrigan 2017] Rice, B.; González-Valesco, E.; Corrigan, A.: The Life and Works of John Napier. Cham, 2017.

[Robert 2003] Robert, A. M.: Nonstandard Analysis. Reprint of the American edition 1988, 2003.

[Robinson 1996] Robinson, A.: Non-standard Analysis. Princeton, 1996.

[Rosenberger 1987] Rosenberger, F.: Isaac Newton und seine Physikalischen Prinzipien – Ein Hauptstück aus der Entwicklungsgeschichte der modernen Physik. Nachdruck der Erstauflage 1895, Darmstadt, 1987.

[Runciman 1951-4] Runciman, S.: A History of the Crusades. Three Vols., Cambridge, 1951-1954, numerous reprints and editions.

[Russell 1903] B. A. W. Russell: The Principles of Mathematics. Cambridge, 1903.

[Scharlau 1981] Scharlau, W. (Hrsg.): Richard Dedekind 1831–1981. Braunschweig, 1981.

[Schmieden/Laugwitz 1958] Schmieden, C.; Laugwitz, D.: Eine Erweiterung der Infinitesimalrechnung. Math. Zeitschrift 69, S. 1–39, 1958.

[Schofield 1980] Schofield, M.: An Essay on Anaxagoras. Cambridge, 1980.

[Schramm 1994] Schramm, M.: Kritische Tage in Mikro- und Makrokosmos. in: [von Gotstedter 1994], 563-673, 1994.

[Scott 1981] Scott, J.: The Mathematical Work of John Wallis, D. D., F. R. S (1616–1703). New York, Nachdruck der Erstauflage 1938, 1981.

[Scriba 1970] Scriba, Chr. J..: The Autobiography of John Wallis, F. R. S.. Notes and Records of the Royal Society of London 25, 17–46, 1970.

[Scriba/Schreiber 2000] Scriba, Chr. J.; Schreiber, P.: 5000 Jahre Geometrie. Hrsg. von: Alten, Djafari Naini, Wesemüller-Kock. Berlin, Heidelberg, New York 2000, Nachdrucke 2002 und 2003, 2. Auflage 2005, 3. Auflage 2010.

[Simonyi 2001] Simonyi, K.: Kulturgeschichte der Physik – Von den Anfängen bis heute. Dritte Auflage, Frankfurt am Main, 2001.

[Seek 1975] Seek, G. A.: Die Naturphilosophie des Aristoteles. Darmstadt, 1975.

[Shirley 1974] Shirley, J. W. (edt.): Thomas Harriot: Renaissance Scientist. Oxford 1974.

[Shirley 1981] Shirley, J. W.: A Source Book for the Study of Thomas Harriot. New York 1981.

[Shirley 1983] Shirley, J. W.: Thomas Harriot: A Biography. Oxford 1983.

[Siegmund-Schultze 1999] Siegmund-Schultze, R.: The Origins of Functional Analysis. in: [Jahnke 2003a], S. 385–407.

[Siegmund-Schultze 2009] Siegmund-Schultze, R.: Mathematicians Fleeing from Nazi Germany - Individual Fates and Global Impact. Princeton and Oxford, 2009.

[Singh 1997] Singh, S.: Fermat's Last Theorem. London, 1997.

[Skelton 1962] Skelton, R. A.: Mercator and English Geography in the 16th Century. in: [Stadtarchiv Duisburg 1962].

[Sloane 2011] Sloane, B.: The Black Death in London. Brimscombe Port, 2011.

[Sobel 1995] Sobel, D.: Longitude. The True Story of a Lone Genius Who Solved the Greatest Problem of His Time. London, New York, 1995.

[Sobel/Andrewes 1999] Sobel, D: Andrewes, W. J. H.: Längengrad – Die illustrierte Ausgabe. Berlin, 1999.

[Sonar 2001] Sonar, Th.: Der fromme Tafelmacher: Die frühen Arbeiten des Henry Briggs. Berlin, 2001.

[Sonar 2006] Sonar, Th.: Some Differentials on Gottfried Wilhelm Leibniz's Death, Burial, and Remains. in: Mathematical Intelligencer, Vol.28, No.2, S. 37–40, 2006.

[Sonar 2007] Sonar, Th.: Richard Dedekind und seine Beziehungen in der Gelehrtenrepublik. in: [Harborth/Heuer/Löwe/Löwen/Sonar 2007], S. 13–24.

[Sonar 2007a] Sonar, Th.: Die Bändigung des Unendlichen – Richard Dedekind und die Geburt der Mengenlehre. in: [Harborth/Heuer/Löwe/Löwen/Sonar 2007], S. 85–97.

[Sonar 2008] Sonar, Th.: Der Tod des Gottfried Wilhelm Leibniz. Wahrheit und Legende im Licht der Quellen. in: Abhandlungen der Braunschweigischen Wissenschaftlichen Gesellschaft, Band LIX, S. 161–201, 2008.

[Sonar 2008a] Sonar, Th.: Eulers Arbeiten zur Strömungsmechanik. In: [Biegel/Klein/Sonar 2008].

[Sonar 2008b] Sonar, Th.: Die Entwicklung der Ballistik von Aristoteles bis Euler. in: Abhandlungen der Braunschweigischen Wissenschaftlichen Gesellschaft, Band LIX, S. 203–230, 2008.

[Sonar 2008c] Sonar, Th.: Leonhard Euler – Analysis und Mechanik. Mitt. Math. Gesell. in Hamburg, Band 27, S. 5–22, 2008.

[Sonar 2009] Sonar, Th.: Turbulenzen um die Fluidmechanik. in: Spektrum der Wissenschaft 04/09, 78-87, 2009.

[Sonar 2018] Sonar, Th.: The History of the Priority Dispute between Newton and Leibniz. Basel, New York, 2018.

[Sonar/Loewe 2008] Sonar, Th.; Löwe, H.: Eulers Arbeiten zur Ballistik. in: [Biegel/Klein/Sonar 2008].

[Sorell 1999] Sorell, T.: Descartes. Freiburg, Basel, Wien, 1999.

[Spalt 1990] Spalt, D. D.: Die Unendlichkeiten bei Bernard Bolzano. in: [König 1990, S. 189–218].

[Spalt 1996] Spalt, D. D.: Die Vernunft im Cauchy-Mythos. Frankfurt am Main, 1996.

[Spalt 2001] Spalt, D. D.: Curt Schmieden's Non-Standard Analysis – A Method of Dissolving the Standard Paradoxes of Analysis. Centaurus No.43, S. 137–174, 2001.

[Spalt 2002] Spalt, D. D.: Cauchys Kontinuum – Eine historiographische Annäherung via Cauchys Summensatz. Arch. Hist. Exact Sci. 56, S. 285–338, 2002.

[Spalt 2015] Spalt, D. D.: Die Analysis im Wandel und im Widerstreit. Eine Formierungsgeschichte ihrer Grundbegriffe. Freiburg, München, 2015.

[Spargo/Pounds 1979] Spargo, P. E.; Pounds, C. A.: Newton's 'Derangement of the Intellect' – New Light on an Old Problem. Notes and Records of the Royal Society, Vol.34, 11–32, 1979.

[Specht 2001] Specht, R.: Descartes. 9. Auflage, Reinbek, 2001.

[Stadtarchiv Duisburg 1962] Stadtarchiv Duisburg (Hrsg.): Duisburger Forschungen. Schriftenreihe für Geschichte und Heimatkunde Duisburgs, Band 6., Duisburg-Ruhrort, 1962.

[Stäckel 1976] Stäckel, P. (Hrsg.): Variationsrechnung. Darmstadt, 1976.

[Stamm 1937] Stamm, E.: Tractatus de Continuo von Thomas Bradwardina. Isis 26, 13–32, 1937.

[Staudacher 2018] Staudacher, F.: Jost Bürgi, Kepler und der Kaiser. Uhrmacher, Astronom, Mathematiker, Instrumentenbauer, Erz-Metallurgist, 1552-1632. 4th revised and enlarged edt., Zürich 2018.

[Stedall 2002] Stedall, J. A.: A Discourse Concerning Algebra. English Algebra to 1685. Oxford, New York, 2002.

[Stedall 2003] Stedall, J. A.: The Greate Invention of Algebra: Thomas Harriot's Treatise on Equations. Oxford, New York, 2003.

[Stedall 2004] Stedall, J. A.: The Arithmetic of Infinitesimals: John Wallis 1656. New York, 2004.

[Stedall 2005] Stedall, J. A.: John Wallis, Arithmetica Infinitorum (1656). in: [Grattan-Guinness 2005].

[Stedall 2008] Stedall, J.: Mathematics Emerging – A Sourcebook 1540–1900. Oxford, New York, 2008.
[Stegmüller 1978] Stegmüller, W. (Hrsg.): Das Universalien-Problem. Darmstadt 1978.
[Stein 1999] Stein, S.: Archimedes – What Did He Do Besides Cry Eureka?. Washington, 1999.
[Stifel 2007] Stifel, M.: Vollständiger Lehrgang der Arithmetik. Deutsche Übersetzung von Eberhard Knobloch und Otto Schönberger. Würzburg, 2007.
[Stolz 1881] Stolz, O.: B. Bolzano's Bedeutung in der Geschichte der Infinitesimalrechnung. Math. Annalen Bd. 18, S. 255–279, 1881.
[Strohmaier 2006] Strohmaier, G.: Avicenna. Zweite Auflage, München, 2006.
[Struik 1969] Struik, D. J.: A Source Book in Mathematics, 1200–1800. Cambridge, Mass., 1969.
[Suerbaum 2003] Suerbaum, U.: Das elisabethanische Zeitalter. Stuttgart, 2003.
[Sutter 1984] Sutter, B.: Der Hexenprozeß gegen Katharina Kepler. Zweite Auflage, Weil der Stadt, 1984.
[Swade 2000] Swade, D.: The Cogwheel Brain - Charles Babbage and the Quest to Build the First Computer. London, 2000.
[Swift 1974] Swift, J.: Gullivers Reisen. Frankfurt/Main, 1974.
[Sylla 1973] Sylla, E.: Medieval concepts of the latitude of forms: The Oxford calculators. Archives d'Histoire Doctrinal et Littéraire du Moyen Âge 40, 223–283, 1973.
[Szabó 1996] Szabó, I.: Geschichte der mechanischen Prinzipien. Dritte Auflage, Basel, Boston, Berlin, 1996.
[Szkocka] http://banach.univ.gda.pl/pdf/ks-szkocka/ks-szkocka3ang.pdf
[Tapp 2005] Tapp, Chr.: Kardinalität und Kardinäle. Wiesbaden, 2005.
[Taschner 2005] Taschner, R.: The Continuum – A Constructive Approach to Basic Concepts of Real Analysis. Wiesbaden, 2005.
[Taschner 2006] Taschner, R.: Das Unendliche: Mathematiker ringen um einen Begriff. Zweite Auflage. Berlin, Heidelberg, New York, 2006.
[Thiele 1982] Thiele, R.: Leonhard Euler. Leipzig, 1982.
[Thiele 2000] Thiele, R. (Hrsg.): Mathesis. Festschrift zum siebzigsten Geburtstag von Matthias Schramm. Berlin, Diepholz, 2000.
[Thiele 2003] Thiele, R.: Antiquity. in: [Jahnke 2003a], p. 1–39.
[Thiele 2005] Thiele, R.: Georg Cantor (1845–1918). in: [Koetsier/Bergmans 2005, S. 523–547].
[Thiele 2007] Thiele, R.: Von der Bernoullischen Brachistochrone zum Kalibrator-Konzept. Turnhout, 2007.
[Thiele 2008] Thiele, R.: Buchbesprechung zu [Tapp 2005]. Sudhoffs Archiv 92, S. 126–128, 2008.
[Thiele 2008a] Thiele, R.: *"'... unsere Mathematiker können es mit denen aller Akademien aufnehmen"'* – Leonhard Eulers Wirken an der Berliner Akademie als Mathematiker und Mechaniker 1741–1766. in: [Biegel/Klein/Sonar 2008].
[Thoren 1990] Thoren, V. E.: The Lord of Uraniborg. A Biography of Tycho Brahe. Cambridge, New York, etc., 1990.
[Toeplitz 2007] Toeplitz, O.: The Calculus. A Genetic Approach. (English translation of: Die Entwicklung der Infinitesimalrechnung: Erster Band. Berlin, Göttingen, Heidelberg, 1949), Chicago and London, 2007.
[Toomer 1998] Toomer, G. J.: Ptolemy's Almagest. Princeton, 1998.

[Tropfke 1980] Tropfke, J.: Geschichte der Elementarmathematik – Band I: Arithmetik und Algebra. Vierte Auflage, Berlin, New York, 1980.

[Turnbull 1939] Turnbull, H. W. (edt.): James Gregory Tercentenary Memorial Volume. London, 1939.

[Turnbull 1940] Turnbull, H. W.: Early Scottish Relations with the Royal Society: I. James Gregory, F. R. S. (1638–1675). Notes and Records of the Royal Society of London 3, S. 22–38, 1940/41.

[Turnbull 1959–77] Turnbull, H. W. et al. (edts): The Correspondence of Isaac Newton. 7 Vols. Cambridge 1959-1977.

[Tweddle 2003] Tweddle, I.: James Stirling's Methodus Differentialis – An Annotated Translation of Stirling's Text. London, Berlin, Heidelberg, 2003.

[van der Waerden 1940] van der Waerden, B. L.: Zenon und die Grundlagenkrise der Mathematik. Math. Ann. 117, S. 141–161, 1940.

[van der Waerden 1971] van der Waerden, B. L.: Science Awakening. Vol. I. English translation by Arnold Dresden with additions of the author. 3rd edt., Groningen, 1971.

[van der Waerden 1979] van der Waerden, B. L.: Die Phytagoreer. Zürich, 1979.

[van Stigt 1990] van Stigt, W. P.: Brouwer's Intuitionism. Amsterdam, New York, Oxford, Tokyo, 1990.

[Vergil 2001] Vergil: Bucolica/Hirtengedichte. Stuttgart, 2001.

[Viertel 2014] Viertel, K.: Geschichte der gleichmäßigen Konvergenz. Ursprünge und Entwicklungen des Begriffs in der Analysis des 19. Jahrhunderts. Wiesbaden, 2014.

[Virgil 1999] Virgil I. Eclogues, Georgics, Aeneid Books 1-6. Loeb Classical Library LCL 63. Revised Edition, with new Introduction. Cambridge, Mass., London, 1999.

[Vitruvius 1914] Vitruvius: The Ten Books on Architecture. Translated by Morris Hicky Morgan. Cambridge, Mass., London, Oxford, 1914.

[Volkert 1988] Volkert, K.: Geschichte der Analysis. Mannheim, Wien, Zürich, 1988.

[von Fritz 1971] von Fritz, K.: Grundprobleme der Geschichte der antiken Wissenschaften. Berlin, New York, 1971.

[von Gotstedter 1994] von Gotstedter, A. (Hrsg.): Ad radices. Festband zum 50jährigen Bestehen des Institutes für Geschichte der Naturwissenschaften der Johann Wolfgang Goethe-Universität Frankfurt/Main. Stuttgart, 1994.

[Waldvogel 2016] Waldvogel, J.: Jost Bürgi's Artificium of 1586 in modern view, an ingenious algorithm for calculating tables of the sine function. (Elem. Math. 71, 89–99, 2016).

[Wenskus 2016] Wenskus, O.: Die angebliche Vorhersage einer Sonnenfinsternis durch Thales von Milet. Warum sich diese Legende so hartnäckig hält und warum es wichtig ist, ihr nicht zu glauben. (Hermes 144, 2016/1, 2–17, 2016).

[Westfall 2006] Westfall, R. S.: Never at Rest – A Biography of Isaac Newton. 18th printing, Cambridge, 2006.

[Weyl 1917] Weyl, H.: Das Kontinuum. in: Das Kontinuum und andere Monographien, New York, 1973.

[Weyl 1921] Weyl, H.: Über die neue Grundlagenkrise der Mathematik. Math. Zeitschr. 10, S. 39–79, 1921.

[Weyl 1994] Weyl, H.: The Continuum. A Critical Examination of the Foundation of Analysis. Translated by Stephen Pollard and Thomas Bole. Unabridged and corrected republication of the first edition 1987, New York, 1994.

[Weyl 2009] Weyl, H.: Philosophy of Mathematics and Natural Science. Translated by Olaf Helmer with a New Introduction by Frank Wilczek. Princeton and Oxford, 2009.

[Whiteside 1960–62] Whiteside, D. T.: Patterns of Mathematical Thought in the later Seventeenth Century. Archive for History of Exact Sciences, No.1, S. 179–388, 1960–1962.

[Whiteside 1967–1981] Whiteside, D. T. (edt.): The Mathematical Papers of Isaac Newton. 8 Vols., Cambridge, 1967–1981.

[Wickert 1995] Wickert, J.: Isaac Newton. Reinbek, 1995.

[Wieland 1965] Wieland, W.: Das Kontinuum in der Aristotelischen Physik. In: [Seek 1975].

[Williams 1997] Williams, M.R.: A History of Computing Technology. 2nd edt., Los Alamitos, 1997.

[Woolley 2001] Woolley, B.: The Queen's Conjuror: The Science and Magic of Dr Dee. London, 2001.

[Wußing/Arnold 1978] Wußing, H.; Arnold, W.: Biographien bedeutender Mathematiker. Köln, 1978.

[Wußing 1984] Wußing, H.: Isaac Newton. Leipzig, 1984.

[Wußing 2008] Wußing, H.: 6000 Jahre Mathematik – Eine kulturgeschichtliche Zeitreise. Band I: Von den Anfängen bis Leibniz und Newton. Berlin, Heidelberg, 2008.

[Wußing 2009] Wußing, H.: 6000 Jahre Mathematik – Eine kulturgeschichtliche Zeitreise. Band II: Von Euler bis zur Gegenwart. Berlin, Heidelberg, 2009.

[Young/Chisholm Young 1972] Young, W. H.; Chisholm Young, G.: The Theory of Sets of Points. Bearbeiteter Nachdruck von 1906, New York, 1972.

[Zhmud 1997] Zhmud, L.: Wissenschaft, Philosophie und Religion im frühen Pythagoreismus. Berlin, 1997.

List of Figures

The abbreviation PD in the source reference is used to mark pictures or figures being in the public domain on the internet or otherwise license-free at the time of the production of this book. The abbreviation HWK stands for the co-editor Heiko Wesemüller-Kock. The mathematical plots in many figures were done by the author; the figures 1.4.2 until 1.4.4, 2.2.3, 2.3.10, 2.3.16, 3.2.7, 5.3.13, 5.3.15 bis 5.3.17, 5.5.12 und 6.3.16 were designed by HWK.

T. Sonar, *3000 Years of Analysis*, https://doi.org/10.1007/978-3-030-58223-4

Index of persons

T. Sonar, *3000 Years of Analysis*, https://doi.org/10.1007/978-3-030-58223-4

Subject index

T. Sonar, *3000 Years of Analysis*, https://doi.org/10.1007/978-3-030-58223-4

The manufacturer's authorised representative in the EU is Springer Nature Customer Service Centre GmbH, Europaplatz 3, 69115 Heidelberg, Germany. If you have any concerns regarding our products, please contact ProductSafety@springernature.com

Printed and bound by CPI Group (UK) Ltd, Croydon, CR0 4YY
30/09/2026
02225240-0016